T0271203

Probability and Statistics
for
Economists

Probability and Statistics
for
Economists

Yongmiao Hong

Cornell

 **World Scientific**

NEW JERSEY · LONDON · SINGAPORE · BEIJING · SHANGHAI · HONG KONG · TAIPEI · CHENNAI · TOKYO

Published by

World Scientific Publishing Co. Pte. Ltd.

5 Toh Tuck Link, Singapore 596224

USA office: 27 Warren Street, Suite 401-402, Hackensack, NJ 07601

UK office: 57 Shelton Street, Covent Garden, London WC2H 9HE

Library of Congress Cataloging-in-Publication Data
Names: Hong, Yongmiao, author.
Title: Probability and statistics for economists / by Yongmiao Hong (Cornell).
Description: First Edition. | New Jersey : World Scientific Publishing Co., [2017] |
    Includes bibliographical references.
Identifiers: LCCN 2017031765 | ISBN 9789813228818 (hardcover : alk. paper)
Subjects: LCSH: Econometrics. | Probabilities. | Statistics.
Classification: LCC HB139 .H636 2017 | DDC 519.5--dc23
LC record available at https://lccn.loc.gov/2017031765

British Library Cataloguing-in-Publication Data
A catalogue record for this book is available from the British Library.

Printed in Singapore

Preface

This book is an introduction to probability and statistics for graduate students in economics, finance, management, statistics, applied mathematics, and other related fields.

Statistics is a science about data. It consists of descriptive statistics and inferential statistics. The former is on collection, summary, analysis and presentation of often a large amount of observed data in a simple and interpretable manner. The latter, on the other hand, makes use of fundamental laws of probability and statistical inference to draw conclusions about the underlying system that generates the observed data. Probability and statistics is often the first course in a graduate econometrics sequence in North American universities. Why do we need to teach probability and statistics to graduate students in economics? Put it simply, such a course provides necessary probability and statistics background for first-year graduate students for their courses in econometrics, microeconomics, macroeconomics, and finance. Statistics and calculus are two basic analytic tools in economics. Statistics is an essential tool to study situations involving uncertainties, in the same way as calculus is essential to characterize optimizing behaviors in economics. For example, probability theory is needed in the study of game theory. In macroeconomics, as Robert Lucas points out, the introduction of stochastic factors can provide much new insights into dynamic economic systems. In certain sense, a course on probability and statistics might not be called *Econometrics I*, because they are necessary analytic tools in every field of modern economics. Of course, the demand for probability and statistics varies from field to field in economics, with econometrics most heavily using it. Specifically, those who are attracted to theoretical econometrics would be expected to study probability and statistics in more depth by taking further graduate courses in mathematics and statistics. Those who are not attracted to should find this book an adequate preparation in the theory of probability and statistics for both their applied courses in econometrics and their courses in microeconomics, macroeconomics, and financial economics.

The aims of this book are two-folds: First, it provides essentials of probability theory and mathematical statistics needed for a graduate student in economics and finance. Second, it offers tuitions, explanations, and applications of important probability and statistical concepts and tools from an economic perspective. Indeed, there have been many textbooks on probability and statistics. It is the second aim that motivates the writing of this textbook. It is strongly believed that the second aim is an indispensable part of training for graduate students in economics, finance and management when they study probability and statistics.

The book is written as a one semester course. It consists of two parts: Part I is probability theory, and Part II is statistical theory. Probability theory is a natural mathematical tool to describe stochastic phenomena. A solid background in probability theory allows students to have a better understanding of statistical inference. Without some formalism of probability theory, students cannot appreciate valid interpretation from data analysis through modern statistical methods. Chapter 1 is an introduction to probability and statistics, arguing why probability and statistics are basic analytic tools for economics, finance and management. Chapters 2-5 are probability theory, and Chapters 6-10 are statistics theory. Chapter 2 lays down the foundation of probability theory, which is important for understanding subsequent materials. Chapter 3 introduces random variables and probability distributions in a univariate context. Chapter 4 discusses important examples of discrete and continuous probability distributions that are commonly used in economics, finance and management. Chapter 5 introduces random vectors and multivariate probability distributions. In most cases, we consider bivariate distributions, which offer much insight for multivariate distributions. Chapter 6 is an introduction to sampling theory, focusing on the classical statistical sampling theory under the normality distribution assumption. Chapter 7 introduces basic analytic tools for large sample or asymptotic theory suitable for analysis under the nonnormality distribution assumption. Chapter 8 discusses parameter estimation methods and methods to evaluate parameter estimators. Chapter 9 deals with hypothesis testing. Chapter 10 presents the classical linear regression model and related statistical inferences. Finally, Chapter 11 concludes the book. The book contains discussions ranging from basic concepts to advanced asymptotic analysis, with much intuition and explanation provided for important probability and statistical concepts and ideas from an economic perspective.

The purpose of this book is to develop a deep understanding of probability and statistics and a solid intuition for statistical concepts from an economic perspective. One year of calculus is a prerequisite for understanding the materials in this book, and an additional year of advanced calculus and some basic background in probability and statistics will be very helpful. The analysis is conducted in a relatively rigorous manner. Proofs will be given for some important theorems, because the proofs themselves can aid understanding and in some cases, the proof techniques or methods have practical value, particularly for the students who are later interested in econometric theory. On the other hand, graphical representation is also useful to understand the abstract and subtle concepts of probability and statistics.

Many students taking this course are experiencing the ideas of probability and statistics for the first time. It is important for students in economics, finance and management to build up stochastic thinking and statistical thinking. Essentially, this requires, among many other things, students to view observed economic data to be generated from a stochastic economic system or process and the observed data, which represents limited information of the system, can be used to make inferences on the stochastic system or process. It will be helpful for them to spend some time learning how the mathematical ideas of probability and statistics carry over into the world of applications in economics, finance and management. Thus, in addition to develop a fundamental understanding of probability and mathematical statistics that are most relevant to modern econometrics, this book also tries to develop a sound intuition for statistical concepts from an economic perspective. For example, why are statistical concepts (e.g., conditional mean, conditional variance) useful in economics? What are economic intuition and interpretation for probability and statistical relations? The book will provide many economic and financial examples to illustrate how probability tools and statistical methods can be used in economic analysis. This is in fact a most important feature that distinguishes this book from many other textbooks on probability and statistics.

The book is based on my lecture notes that I have been using in teaching probability and statistics for the first year doctoral students in Department of Economics at Cornell University. I thank the comments from the students who have taken this course. Moreover, I thank Weiping Bao, Daumantas Bloznelis, Biqing Cai, Yuhan Chi, Liyuan Cui, Ying Fang, Zhenghui Feng, Zhonghao Fu, Muyi Li, Ming Lin, Xia Wang, Yun Wang, Ke Xiao and Zengguang Zhong for their comments and suggestions. I'm also grateful for their comments and suggestions.

Yongmiao Hong
Ernest S. Liu Professor of Economics & International Studies
Department of Economics & Department of Statistical Science
Cornell University, Ithaca, U.S.A.

Contents

Chapter 1

Introduction to Probability and Statistics

Abstract: Probability perhaps has become the best analytic tool to describe any system involving uncertainties, and statistics provides a mathematical foundation to model situations involving uncertainty. As the beginning of this book, this chapter will introduce two fundamental axioms behind modern econometrics, emphasizes the important role of statistics in economics and also discusses the limitation of statistical analysis in economics.

Key words: Chaos, Data generating process, Econometrics, Quantitative analysis, Probability law, Uncertainty.

1.1 Quantitative Analysis in Economics

The most important feature of modern economics and finance is the wide use of quantitative analysis. Quantitative analysis consists of mathematical modeling of economic theory and empirical study of economic data. This is due to the cumulative effort of many generations of economists to make economics a science, something like or close to physics, chemistry and biology, which can make accurate predictions or forecasts. Economic theory, when formulated via mathematical tools, can achieve its logical consistency among assumptions, theories, and its implications. Indeed, as Karl Marx points out, the use of mathematics is an indication of the mature stage of a science. On the other hand, for any economic theory to be a science, it must be able to explain important empirical stylized facts and to predict future economic evolutions. This requires validating economic models using the observed economic phenomena, usually in form of data. Mathematical tools alone cannot achieve this objective. Instead, statistical tools have proven to be rather useful. The history of the development of economics is

1

a continuous process of refuting the existing economic theory that cannot explain new empirical stylized facts and developing new economic theories that can explain new observed empirical stylized facts. Empirical analytic tools play a vital role in such a process. In a sense, statistical methods and techniques are really the heart of the scientific research in economics.

As a matter of fact, the main empirical analytic tool in economics is econometrics. Econometrics is the statistical analysis of economic data in combination with economic theory. There is a lot of uncertainty in real economies and financial markets, and economic agents usually have to make decisions under uncertainty. Probability is a natural quantitative tool to describe uncertainty in economics. Historically, probability was motivated by interest in games of chance. Scholars then began to apply probability theory to actuarial problems and some aspects of social sciences. Later, probability and statistics were introduced into physics by L. Boltzmann, J. Gibbs, and J. Maxwell, and by last century, they had found applications in all phases of human endeavor that in some way involve an element of uncertainty or risk. Indeed, probability theory has become the best analytic tool to describe any system involving uncertainty.

Modern statistics has encompassed the science of basing inferences on observed data and the entire problem of making decisions in the face of uncertainty. It would be presumptuous to say that statistics, in its present state of development, can handle all situations involving uncertainty, but new techniques are constantly being developed and modern statistics can, at least, provides a framework for looking at the situations involving uncertainty in a logical and systematic fashion. It can be said that statistics provides mathematical models that are needed to study situations involving uncertainty in the same way as calculus provides mathematical models that are needed to describe, say the concepts of the Newtonian physics. Indeed, as Robert Lucas points out, the introduction of stochastic factors into a dynamic economic system can provide new insight into dynamic economic laws.

1.2 Fundamental Axioms of Statistical Analysis in Economics

There are two fundamental axioms behind modern econometrics:

- Axiom A: Any economy can be viewed as a random or stochastic system governed by some probability law;

- Axiom B: Economic phenomenon, often summarized in form of data, can be viewed as a realization of this stochastic data generating process.

Economics is about resource allocation in an uncertain environment. When an economic agent makes a decision, he or she usually does not know precisely the outcome of his or her action, which usually will arise in an unpredictable manner with time lags. As a consequence, uncertainty and time are two of the most important features of an economy. Therefore it seems reasonable to assume Axiom A. With Axiom A, it is natural to assume Axiom B under which one can call the economic system a "data generating process". It is impossible to prove these two axioms. They are the philosophic views of econometricians and economists about an economy. We note that not all economists may agree with these two axioms. For example, some economists view that an economic system is a chaotic process, which is deterministic but can generate seemingly random numbers.

To illustrate the different implications between a stochastic view and a chaotic view on an economy, we consider an example. As a well-known empirical stylized fact, high frequency stock returns are found to have little autocorrelation with their own lagged returns. Figure 1.1 plots the observations on the Standard & Poor 500 daily closing price and daily return over time respectively. To explain this empirical stylized fact, there are at least two possible hypotheses or conjectures. The first is to assume that the stock price follows a geometric random walk, that is,

$$\ln P_t = \ln P_{t-1} + X_t,$$

where its log-return series $\{X_t = \ln(P_t/P_{t-1})\}$ is a statistically independent sequence over time, which implies zero correlation between stock returns over time. Figure 1.2 plots the observations on the price level P_t and the return X_t generated from this geometric random walk model using a random number generator on a personal computer. Comparing Figures 1.1 and 1.2, we can observe some similarity between the real data series and the artificial series generated from the computer.

Alternatively, one may assume that the stock return follows a deterministic chaotic logistic map:

$$X_t = 4X_{t-1}(1 - X_{t-1}).$$

If we generate a large set of observations from this logistic map and calculate the sample autocorrelations between observations over time, we would

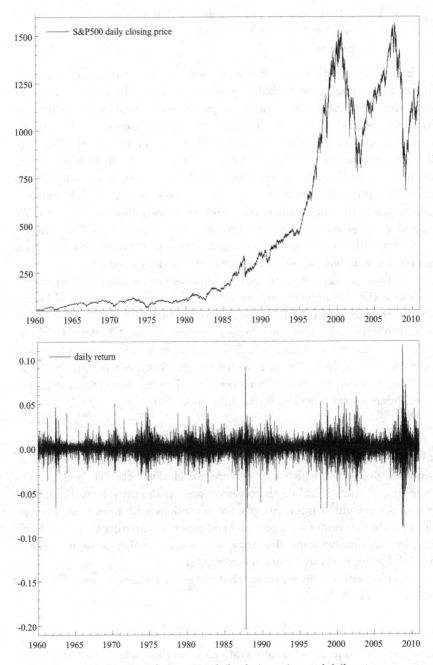

Figure 1.1: Standard & Poor 500 daily closing prices and daily returns

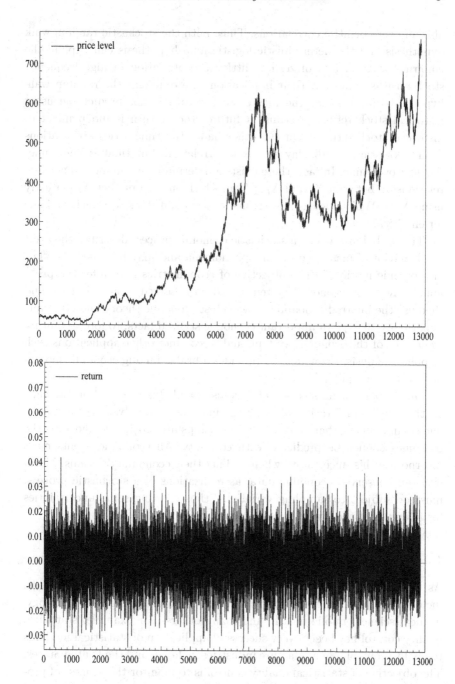

Figure 1.2: Price series P_t and return series X_t

also find zero or little correlations. Thus, both the stochastic random walk hypothesis and the deterministic logistic map hypothesis can explain the empirical stylized fact of zero or little autocorrelation in high-frequency stock returns. However, their implications are different: the random walk hypothesis implies that the future stock return is not predictable using historical stock returns, because a future stock return is independent of historical stock returns. On the other hand, the time series observations from a logistic map display zero autocorrelation, but they are not independent over time. In fact, there exists a deterministic nonlinear quadratic relationship between X_t and X_{t-1} from which one can predict X_t perfectly using X_{t-1}. Which view is more realistic to explain stock returns is an issue for empirical study.

The probability law of a stochastic economic process describes the average behavior of massive economic phenomena and may be called the "law of economic motions". The objective of econometrics is to infer the probability law of an economic system based on observed economic data, and then use the inferred probability law to test economic theory and economic hypotheses, to explain important economic stylized facts, to forecast future evolutions of the economic system, and to conduct other applications such as policy analysis. Econometrics provides a bridge linking economic models and economic reality.

One important implication of Axioms A and B is the need of "stochastic thinking" and "statistical thinking" in economic analysis. For example, one should expect that economic relationships are stochastic and thus the outcomes cannot be predicted with certainty. All economic agents must incorporate this uncertainty when making their economic decisions. Moreover, any observed economic data, as realizations of a stochastic process, must be subject to sampling variations, thus creating some uncertainties for inference of the law of economic motions.

1.3   Role of Statistics in Economics

As a science, statistics has been widely applied in many different fields, including physics, engineering, economics, finance, management, biology, medical science, public health, and many others. For example, statistical quality control has been a very successful application of statistical methods to quality control in manufacture industries in Japan and the United States. The objective of statistical quality control is to monitor the process of production and decides whether the production is in-control or out-of-control by setting some control limits. The idea of control limits is analogous to that of hypothesis testing in statistics.

Below, we briefly discuss what roles statistics can play in economics and related fields.

First of all, economic phenomena is usually rather complicated, and there is a vast amount of information in daily economic life. It is very important to summarize observed economic data in a simple, intuitive and interpretable way so as to convey information to economists, decision makers, and the public. Statistics, by its very nature, is an effective tool for doing so. Economic indicators such as the Consumer Price Index (CPI) and the unemployment rate are two examples of statistical description of the state of an macroeconomy which have been indispensable means for governments and central banks to decide their fiscal and monetary policies. Moreover, many important empirical stylized facts in economics are presented in form of a statistical relationship. One example is the well-known Phillips curve in macroeconomics, which documents that the inflation rate is negatively correlated with the unemployment rate. Figure 1.3 is a scatter plot of the U.S. inflation rate, the log-difference of the U.S. CPI series, and the U.S. unemployment rate, which indicates a negatively correlated relationship. Another example is volatility clustering in finance, which documents that a large asset volatility today tends to be followed by another large volatility tomorrow; a small volatility tends to be followed by another small volatility tomorrow, and the patterns alternate over time. Figure 1.4 plots the time series observations of the absolute and squared values of the Standard & Poor 500 daily return. They indicate the phenomena of volatility clustering. Concepts in probability and statistics can also provide simple characterizations of economic ideas and economic theory. Examples include the Lorenz curve of income inequality and the representation of stochastic dominance via the probability distribution function. In fact, most financial theory is represented with the concepts of probability. It would be difficult to imagine what will be left in modern finance if there is no use of probability concepts and tools.

Second, a key feature of economic data is the existence of variability in economics. Much of variability is due to inherent uncertainty in the economy, which results in risk given the risk-averse nature of economic agents. A central theme of modern financial risk management is to quantify and price financial risk via statistical methods and modeling. For such purposes, statistical models and measures of volatility are indispensable quantitative tools.

   Third, an ingredient methodology in statistics is sampling. The basic idea of sampling theory is to use a subset of information called "sample" to infer knowledge of the entire system or process called "population", and then use the knowledge of the system for various applications. This statistical method is consistent with the economic principle of cost minimization. An example is the aforementioned stochastic quality control, which uses a similar idea to hypothesis testing to ensure the quality of manufactured products while minimizing the cost of inspection and monitoring.

   Fourth, perhaps the most important objective of economic analysis is to discover or validate economic relationships, particularly causal economic relationships. Statistical inference can play an instrumental role in this regard. For example, statistics can estimate demand elasticities of output with respect to price for certain product, which may be useful for the marketing strategy of a company. Combined with well-designed experiments which can control important economic factors, statistical analysis can identify causal economic relationships. On the other hand, for observational or historical economic data for which economists cannot control economic factors, it is much more challenging if not impossible to identify causal economic relationships. Nevertheless, aided with economic theory, statistics can be still very helpful in identifying economic relationships.

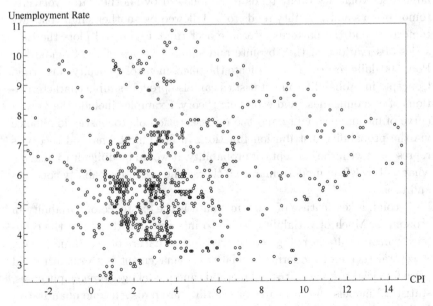

Figure 1.3: Scatter plots of the U.S. inflation rate and unemployment rate

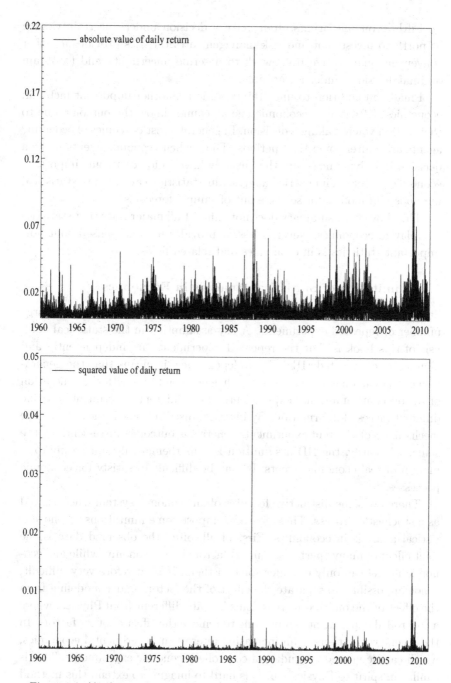

Figure 1.4: Absolute and squared values of the Standard & Poor 500 daily return

Fifth, economic agents often make a decision under uncertainty, such as portfolio investment and risk management. Statistics provides a rather convenient quantitative framework to describe uncertainty and to obtain optimal decisions under uncertainty.

Finally, in addition to uncertainty, time is another important factor in economics. Oftentimes, economic agents cannot know the outcome due to their action when making a decision. In general, most outcomes of economic agents arise after some time periods. Thus, when economic agents make a decision in a dynamic setup, they usually have to forecast some important economic factors. Time series analysis in statistics can provide statistical methods and models for sensible out-of-sample forecasts.

The above brief summary does not exhaust all major roles that statistics can play in economics. Nevertheless, it provides us some sense about how important statistics is in economics and related fields.

1.4   Limitation of Statistical Analysis in Economics

Statistics is an analysis of the "average behavior" of the outcomes of a large number of repeated experiments. A key assumption in the statistical analysis of this book is that the repeated experiments are independently and identically distributed (IID). By independence, it means that the generation of the outcome of an experiment has nothing to do with the generation of the outcome of another experiment. Thus, different experiments provide different pieces of information. By identical distributions, it means that the mechanisms of different experiments generating outcomes are essentially the same. Obviously, the IID assumption is a mathematical limiting approximation of real economic events. It may be difficult to satisfy for economic processes.

There are some distinctive features of an economic system when viewed as a stochastic process. These features impose some limitations on the statistical analysis in economics. First of all, often the observed data is the joint effect of many (perhaps infinite) factors in an economy, while any economic model can only consider some of them. It is therefore very difficult, if not impossible, to separate the effect of the factors under modeling from the effect of the omitted factors. This is quite different from Physics, where a controlled experiment can be done to remove the effect of other factors. In the recent years, there is an increasing interest in experimental economics, which can be viewed as studies of controlled economic experiments. This is similar in spirit to Physics, but it is hard to imagine to extend this method to the whole economy like China which has 1.3 billion of population!

The second feature of economic systems is that an economic process is irreversible. As a consequence, many important economic variables, such as Gross Domestic Product (GDP) in each year, only have one realization. For example, consider the observed time series data for the Chinese GDP growth rate during 1980-2010. Here, the GDP growth rates in different years are considered as different random variables, and as a result, each random variable only has one observation. Thus, it is impossible to conduct statistical analysis of economic data without making assumptions. In practice, it is often assumed that economic variables over different years or over different cross-sectional units follow the same population distribution or at least have the same attributes in certain aspects of their probability distributions. These are usually called "homogeneity" and "stationarity" assumptions. Thus, the data for variables over different time periods or different cross-sectional units can be viewed as generated from the same population or share the same attributes, so that statistical inference of observed data is possible. Obviously, the validity and accuracy of statistical inference depend on how reasonable these underlying assumptions are in describing economic systems.

The third feature of economic systems is their time-varying features. There are many economic events, such as government policy changes, changes in foreign exchange rate systems, oil shocks, financial crisis, etc, which may cause economic behaviors to change fundamentally. These are the so-called economic structural changes or regime shifts. Structural changes and regime shifts will render the existing economic models invalid in describing or predicting future economic evolutions.

Finally, most observed economic data may be subject to nontrivial measurement errors, due to the incentive problem of economic agents to report the true information (e.g., income), aggregation over time or space, the qualitative or subjective nature of some economic variables (e.g., happiness) that are difficult to measure, and etc.

Because of these distinctive features of economic systems, we should be aware of the practical relevance of underlying assumptions and their limitations when making statistical inference from observed economic data.

1.5 Conclusion

In this chapter, we provide a brief introduction to probability and statistics. As a natural quantitative tool to describe uncertainty in economics, probability theory has become the best analytic tool to describe any system involving uncertainty, and statistics provides a framework for modeling the situations involving uncertainty.

Statistics has wide applications in economics and related fields, such as summarizing and interpreting data efficiently, modeling variabilities, making inference of a system or process using a subset of sample information, identifying causal economic relationships, and making out-of-sample forecasts. However, the non-experimental nature of most economic phenomena, unobserved heterogeneity among economic agents, unstable economic relationships over time, and data quality problems impose certain limitations on validity and accuracy of statistical analysis in economics.

EXERCISE 1

1.1. What are the rationales that probability and statistics are useful in economics? What are your justifications for these rationales?

1.2. What major roles statistics can play in economics and related fields?

1.3. What are the limitations of statistical analysis in economics?

Chapter 2

Foundation of Probability Theory

Abstract: Probability theory is the foundation of statistical science, providing a mathematical means of modeling random experiments or uncertainty. Through these mathematical models, researchers are able to draw inferences about the random experiments using observed data. The aim of this chapter is to outline the basic ideas of probability theory that are fundamental to the study of statistics. The entire structure of probability, and therefore of statistics, can be built on the relatively straightforward foundation given in this chapter.

Key words: Bayes' theorem, Combination, Complement, Conditional probability, Event, Independence, Intersection, Multiplication rule, Permutation, Probability function, Probability space, Random experiment, Sample space, Sets, Sigma algebra, σ-field, Union.

2.1 Random Experiments

Many kinds of scientific research may be characterized in part by the fact that repeated experiments, under more or less the same conditions, are standard practice. For instance, an economist may be concerned with the prices of three specified assets (e.g., short-term, medium-term and long-term bonds) over various time periods. The only way in which a researcher can elicit information about such a phenomenon is to perform or observe repeated experiments (in the aforementioned example, observations of asset prices can be obtained in different time periods). Each experiment terminates with an outcome. But it is characteristic of these experiments that the outcome cannot be predicted with certainty prior to the performance of the experiment, although the experiment is of such a nature that a collection of every possible outcome can be described prior to its performance.

Definition 2.1. [**Random Experiment**]: A random experiment is a mechanism which has at least two possible outcomes. When a random experiment is performed, one and only one outcome will occur, but which outcome to occur is unknown in advance. In other words, a random experiment is a mechanism for which the outcome cannot be predicted with certainty.

The word "experiment" here means a process of observation or measurement in a broad sense. It is not necessarily a real experiment as encountered in (e.g.) physics. For example, it can be a process of observing and recording the price values of iPhone 6 over different months.

There are two essential elements of a random experiment:

- the set of all possible outcomes;
- the likelihood with which each outcome will occur.

As is well-known, modern economics is a study on scarce resource allocation in an uncertain environment. When an economic agent makes a decision, he or she usually does not know precisely the outcome of his or her action, which usually will arise in an uncertain manner. This, to some extent, is similar to a random experiment.

As mentioned in Chapter 1, modern econometrics is built upon the following two fundamental axioms:

- Axiom A: An economic system can be viewed as a random experiment governed by some probability law.
- Axiom B: Any economic phenomenon (often in form of data) can be viewed as an outcome of this random experiment. The random experiment is usually called a "data generating process".

In statistics, the purpose of mathematical statistics is to provide mathematical models for random experiments of interest. Once a model for such an experiment has been provided and the theory worked out in detail, the statistician may, within this framework, make inference (i.e., draw conclusions) about the probability law of the random experiment based on observed data. In the context of econometrics, the objective of econometric analysis is to infer the probability law of the stochastic economic system based on the observed data, and then use it to explain important empirical stylized facts, to test economic theory or hypotheses, to predict future behavior of the economy, and to other applications such as policy analysis.

2.2 Basic Concepts of Probability

Definition 2.2. [**Sample Space**]: The possible outcomes of a random experiment are called "basic outcomes", and **the set of all basic outcomes** constitutes "the sample space", which is denoted by S. When an experiment is performed, the realization of the experiment is one (and only one) outcome in the sample space. If the experiment is performed a number of times, different outcomes may occur each time or some outcomes may repeat.

A sample space S is sometimes called an outcome space. Each outcome in S is called an element of S, or simply a sample point. The concept of sample space was first introduced by Ludwig von Mises, an Austrian mathematician and engineer, in 1931. It is important to note that for a random experiment, one knows the set of all possible basic outcomes, but one does not know which outcome will arise before performing the random experiment.

Example 2.1. [**Throwing a Coin**]: In this experiment, there are two possible outcomes: "Head(H)", and "Tail(T)". The sample space is $S = \{H, T\}$.

Example 2.2. [**Direction of Changes**]: Let $Y = 1$ if the U.S. GDP growth rate is positive and $Y = 0$ if the U.S. GDP growth rate is negative. Then we can make inference of possibly asymmetric business cycles using information of the state variable Y.

Example 2.3. [**Rolling a Die**]: The basic outcomes are the numbers 1, 2, 3, 4, 5, 6. The sample space $S = \{1, 2, 3, 4, 5, 6\}$.

Example 2.4. [**Throwing Two Coins**]: We have a sample space

$$S = \{(H, H), (H, T), (T, H), (T, T)\}.$$

Example 2.5. The number of babies who are born at a given interval of time in a city will be $\{0, 1, 2, \cdots\}$.

Example 2.6. Suppose t_0 is the lowest temperature in an area, and t_1 is the highest temperature of the area. Let T denotes the possible temperature of the area. Then the sample space of T is

$$S = \{t \in \mathbb{R} : t_0 \leq t \leq t_1\},$$

where $\mathbb{R}$ denotes the real line.

A sample space S can be countable or uncountable. The sample space in Example 2.6 is uncountable, whereas the sample spaces in Examples 2.1 to 2.4 are finite and so countable, the sample space in Example 2.5 is countable but infinite. The distinction between a countable sample space and an uncountable sample space dictates the ways in which probabilities will be assigned.

We next define the concept of event that will allow us to investigate what we may be interested in.

Definition 2.3. [**Event**]: An event A is a collection of **basic outcomes** from the sample space S that share certain common features or equivalently obey certain restrictions. The event A is said to occur if the random experiment gives rise to one (and only one) of the constituent basic outcomes in A. That is, an event occurs if any of its basic outcomes has occurred (or equivalently if the outcome of the random experiment is an element of event A).

Mathematically speaking, an event is equivalent to a set. Thus, throughout this book, the words "set" and "event" are interchangeable.

Example 2.7. [**Rolling a Die**]: Event A is defined as "the number resulting is even". Event B is "the number resulting is at least 4". Then $A = \{2, 4, 6\}$ and $B = \{4, 5, 6\}$.

Obviously, we have the following relationships among the sample space, a basic outcome and an event:

$$\text{Basic outcome} \subseteq \text{Event} \subseteq \text{Sample space.}$$

Question: What are the properties of basic outcomes?

The basic outcomes are the basic building blocks for a sample space. They cannot be divided (partitioned or separated) into more primitive or more elementary kinds of outcomes. In other words, it is preferable that an element of a sample space does not represent two or more outcomes that are distinguishable in some way.

2.3 Review of Set Theory

Probability theory builds upon set theory, or the algebra of sets, which we begin now. Before going on, let us first get familiar with the definition of **Venn Diagram**. The Venn Diagram, originally introduced by Venn in his book, *Symbolic Logic*, published in 1881, can be used to depict a sample point, a sample space, an event, and related concepts. Specifically, the Venn Diagram is made up of two or more overlapping circles, where one circle denotes one set. It is often used in mathematics to show relationships between sets.

Let A and B be two events in the sample space S. Then we have the following definitions.

Definition 2.4. [**Intersection**]: Intersection of A and B, denoted $A \cap B$, is the set of basic outcomes in S that belong to both A and B. The intersection occurs if and only if both events A and B occur.

The intersection of A and B is also called the logical product. It can be represented by the Venn Diagram in Figure 2.1(a).

Definition 2.5. [**Exclusiveness**]: If A and B have no common basic outcomes, they are called mutually exclusive and their intersection is empty set $\varnothing$, i.e., $A \cap B = \varnothing$, where $\varnothing$ denotes an empty set that contains nothing. By convention, an empty set $\varnothing$ is a subset of any set.

Two exclusive events A and B can be represented by the Venn diagram in Figure 2.1(b).

Mutually exclusive events are also called *disjoint* because they do not overlap when represented in the Venn diagram. By definition, any mutually exclusive events cannot occur simultaneously. As an example, any pair of the basic outcomes in sample space S are mutually exclusive, because when a random experiment is performed, one and only one basic outcome will occur.

Definition 2.6. [**Union**]: The union of A and B, $A \cup B$, is the set of all basic outcomes in S that belong to either A or B. The union of A and B occurs if and only if either A or B (or both) occurs.

The union of A and B is also called the logical sum. It can be represented by the Venn diagram in Figure 2.1(c).

Definition 2.7. [**Collective Exhaustiveness**]: Suppose $A_1, A_2, \cdots, A_n$ are n events in the sample space S, where n is any positive integer. If $\cup_{i=1}^{n} A_i = S$, then these n events are said to be collectively exhaustive.

Definition 2.8. [**Complement**]: The complement of A is the set of basic outcomes of a random experiment belonging to S but not to A, denoted as A^c.

The complement of event A is also called the negation of A. It can be represented by the Venn diagram in Figure 2.1(d).

Obviously, any event A and its complement A^c are mutually exclusive and collectively exhaustive. That is, $A \cap A^c = \varnothing$ and $A \cup A^c = S$.

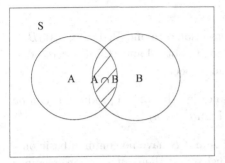

Figure 2.1(a): $A \cap B$ Figure 2.1(b): Two exclusive events

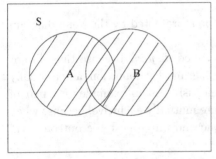
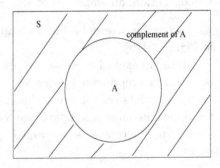

Figure 2.1(c): $A \cup B$ Figure 2.1(d): A^c

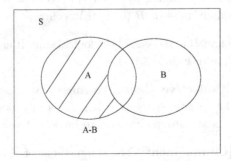

Figure 2.1(e): $A - B$

Definition 2.9. [Difference]: The difference of A and B, denoted as $A - B = A \cap B^c$, is the set of basic outcomes in S that belong to A but not to B.

The difference $A - B$ can be represented by the Venn diagram in Figure 2.1(e).

Example 2.8. [Rolling a Die]: The sample space $S = \{1, 2, 3, 4, 5, 6\}$. Define event A as "the resulting number is even", and define event B as "the resulting number is at least 4". Then it follows that

$$A = \{2, 4, 6\}, \ B = \{4, 5, 6\},$$
$$A^c = \{1, 3, 5\}, \ B^c = \{1, 2, 3\},$$
$$A - B = \{2\}, \ B - A = \{5\},$$
$$A \cap B = \{4, 6\}, \ A \cup B = \{2, 4, 5, 6\},$$
$$A \cap A^c = \varnothing, \ A \cup A^c = \{1, 2, 3, 4, 5, 6\} = S.$$

Theorem 2.1. *[Laws of Sets Operations]:* *For any three events* A, B, C *defined on a sample space* S,

(1) [*Complementation*]:

$$(A^c)^c = A,$$
$$\varnothing^c = S,$$
$$S^c = \varnothing.$$

(2) [*Commutativity* of union and intersection]:

$$A \cup B = B \cup A,$$
$$A \cap B = B \cap A.$$

(3) [*Associativity* of union and intersection]:

$$(A \cup B) \cup C = A \cup (B \cup C),$$
$$(A \cap B) \cap C = A \cap (B \cap C).$$

(4) [*Distributivity* laws]:

$$A \cap (B \cup C) = (A \cap B) \cup (A \cap C),$$
$$A \cup (B \cap C) = (A \cup B) \cap (A \cup C).$$

More generally, for any $n \geq 1$,

$$B \bigcap \left(\bigcup_{i=1}^{n} A_i \right) = \bigcup_{i=1}^{n} \left(B \bigcap A_i \right),$$

$$B \bigcup \left(\bigcap_{i=1}^{n} A_i \right) = \bigcap_{i=1}^{n} \left(B \bigcup A_i \right).$$

(5) [*De Morgan's laws*]:

$$(A \cup B)^c = A^c \cap B^c,$$
$$(A \cap B)^c = A^c \cup B^c.$$

More generally, for any $n \geq 1$,

$$\left(\bigcup_{i=1}^{n} A_i \right)^c = \bigcap_{i=1}^{n} A_i^c,$$

$$\left(\bigcap_{i=1}^{n} A_i \right)^c = \bigcup_{i=1}^{n} A_i^c.$$

Proof: For space, we shall only show Result (5) here; other results can be proved in a similar manner. If $a \in (A \cup B)^c$, then $a \in A^c$ and $a \in B^c$. It follows that $a \in A^c \cap B^c$.

Next, if $a \in A^c \cap B^c$, then $a \in A^c$ and $a \in B^c$. It follows that a does not belong to A and B. Therefore, $a \in (A \cup B)^c$ by the definition of complement.

Alternatively, these results could be intuitively understood by using Venn diagrams (please try it!).

Example 2.9. Suppose the events A and B are disjoint. Under what condition are A^c and B^c also disjoint?

Solution: A^c and B^c are disjoint if and only if $A \cup B = S$.

(1) We first show that if A^c and B^c are disjoint, then $A \cup B = S$. By De Morgan's law, $A^c \cap B^c = \emptyset \Rightarrow (A \cup B)^c = \emptyset$. It follows that $A \cup B = S$.

(2) Next, we show that if $A \cup B = S$, then A^c and B^c are disjoint. Note that $A \cup B = S \Rightarrow (A \cup B)^c = S^c = \emptyset$. It follows $A^c \cap B^c = \emptyset$ by De Morgan's law.

Example 2.10. Let A and B be two events in S. Answer the following questions:

(1) Are $A \cap B$ and $A^c \cap B$ mutually exclusive?

(2) Is $(A \cap B) \cup (A^c \cap B) = B$?

(3) Are A and $A^c \cap B$ mutually exclusive?

(4) Is $A \cup (A^c \cap B) = A \cap B$?

Example 2.11. Let the set of events $\{A_i, i = 1, \cdots, n\}$ be mutually exclusive and collectively exhaustive, and let A be an event in S.
(1) Are $A_1 \cap A, \cdots, A_n \cap A$ mutually exclusive?
(2) Is the union of $A_i \cap A, i = 1, \cdots, n$, equal to A? That is, do we have

$$\bigcup_{i=1}^{n} (A_i \cap A) = A?$$

Intuitively, a sequence of collectively exhaustive and mutually exclusive events forms a partition of sample space S. In certain sense, a set of collectively exhaustive and mutually exclusive events can be viewed as a set of orthogonal bases which can represent any event A in the sample space S, and $A_i \cap A$ could be viewed as the projection of event A onto the base A_i.

2.4 Fundamental Probability Laws

We will assign a probability to an event A in S. The probability function is a function or a mapping from an event to a real number. More precisely, we are interested in assigning probabilities to events, complements of events, unions and intersections of events. Hence, we want our collection of events to include these combinations of events. Such a collection of events is called a σ-field of subsets of the sample space S, which will constitute the domain of the probability function. Webster's *Third New International Dictionary* defines "probability" as "the quality or state of being probable". If the concept of probability is to be used in scientific applications, we need a more precise definition.

We will take an axiomatic approach, in which probabilities are defined as "mathematical objects" that behave according to certain well-defined rules.

Definition 2.10. [**Sigma Algebra**]: A *sigma* (σ) *algebra*, denoted by $\mathbb{B}$, is a collection of subsets (events) of S that satisfy the following properties:
(1) $\varnothing \in \mathbb{B}$ (i.e., the empty set is contained in $\mathbb{B}$);
(2) if $A \in \mathbb{B}$, then $A^c \in \mathbb{B}$ (i.e., $\mathbb{B}$ is closed under countable complement);
(3) if $A_1, A_2, \cdots \in \mathbb{B}$, then $\cup_{i=1}^{\infty} A_i \in \mathbb{B}$ (i.e., $\mathbb{B}$ is closed under countable unions).

A σ-algebra is also called a σ-field. It is a collection of events in S that satisfy certain properties and the domain of a probability function on which we will assign a probability to any event. It is important to note that a σ-field is a collection of subsets in S, but itself is not a subset of S. The sample space S is only an element of a σ-field. In probability theory, the event space is a σ-field. The pair, $(S, \mathbb{B})$, is called a measurable space.

Example 2.12. Show that for any sample space S, then set $\mathbb{B} = \{\varnothing, S\}$ is always a σ-field.

Solution: We verify the three properties of a σ-field:
  (1) $\varnothing \in \{\varnothing, S\}$. Thus $\varnothing \in \mathbb{B}$;
  (2) $\varnothing^c = S \in \mathbb{B}$ and $S^c = \varnothing \in \mathbb{B}$;
  (3) $\varnothing \cup S = S \in \mathbb{B}$.

Example 2.13. Suppose the sample space $S = \{1, 2, 3\}$. Show that a set containing the following eight subsets $\{1\}$, $\{2\}$, $\{3\}$, $\{1, 2\}$, $\{1, 3\}$, $\{2, 3\}$, $\{1, 2, 3\}$, and $\varnothing$ is a σ-field.

Example 2.14. Define $\mathbb{B}$ as the collection of all possible subsets (including an empty set $\varnothing$) in sample space S. Is $\mathbb{B}$ a σ-field?

Next, we define a probability measure using an axiomatic approach.

Definition 2.11. [**Probability Function**]: Suppose a random experiment has a sample space S and an associated σ-field $\mathbb{B}$. A probability function $P : \mathbb{B} \to [0, 1]$ is defined as a mapping that satisfies the following properties:
  (1) $0 \leq P(A) \leq 1$ for any event A in $\mathbb{B}$;
  (2) $P(S) = 1$;
  (3) if $A_1, A_2, \cdots \in \mathbb{B}$ are mutually exclusive, then $P(\cup_{i=1}^{\infty} A_i) = \sum_{i=1}^{\infty} P(A_i)$.

Here, Condition (1) means that "everything is possible" or "any event is possible to happen". For two extreme cases, $P(A) = 0$ means that "event A is unlikely to occur", while $P(A) = 1$ means that "event A is certain to occur". Condition (2) means that "something always occurs whenever a random experiment is performed", and Condition (3) means that the probability of the "sum (i.e., union)" of exclusive events is equal to the sum of their individual probabilities. Recall that mutually exclusive events cannot happen simultaneously.

A probability function tells how the probability of occurrence is distributed over the set of events, $\mathbb{B}$. In this sense we speak of a distribution of probabilities. Any function $P(\cdot)$ that satisfies the axioms of probability is called a probability function. For a given measurable space $(S, \mathbb{B})$, many different probability functions can be defined. Which one reflects what is likely to be observed in a particular experiment is still to be determined. The goal of econometrics and statistics is to find a probability function that most accurately describes the underlying economic process. This probability function is usually called the true probability function or the true probability distribution model.

2.4.1 *Interpretation of Probability*

How can one interpret the probability of an event?

There is a classical or *a priori* interpretation of probability. This is based on the notion of mutually exclusive and equally likely experimental outcomes. If sample space S consists of n mutually exclusive and equally likely outcomes, then the probability of any single outcome, or sample point, is $1/n$. The probability of an event in such a setting is simply the sum of the probabilities of the sample points that result in the occurrence of the event.

The classical interpretation is called *a priori* interpretation since the probabilities are derived from purely deductive reasoning, or simply by the structure of the event. One does not have to conduct the experiments since, logically, all basic outcomes are equally likely to occur.

However, the classical interpretation has some deficiencies. What happens if the outcomes of an experiment are not finite or are not equally likely? Below, we provide two interpretations of probability in a broader context.

Approach 1: Relative Frequency Interpretation

There are two basic interpretations for probability. First, the probability can be interpreted as the relative frequency of occurrences of events of the same kind in the long-run under the same set of conditions. When an experiment is performed, the realization of the experiment is an outcome in S. If the random experiment is performed a number of times under the same condition, different outcomes may occur each time, or some outcomes may repeat. This "frequency of occurrence" of an outcome may be thought of as a probability when the same experiments are repeated many times in such a way that the outcome of an experiment is not affected by the outcomes of other experiments. In other words, the probability of an outcome can be viewed as the limit of the "relative frequency" of occurrence of the outcome in a large number of repeated independent experiments under essentially the same condition.

For example, suppose we throw a coin. Each time, either "Head" or "Tail" occurs. Now let us repeat N times, where N is a very large number, say 100 million. Suppose that among the N trials, "Head" shows up N_h times. Then the proportion of occurrences of "Head" in the N trials is N_h/N. When $N \to \infty$, there will be little variation for the ratio N_h/N. If a coin is a fair coin, this ratio will approach 0.5 when a large number of trials is made. This relative frequency will coincide with the probability of the event that "Head" occurs. In other words, the frequency interpretation views that the probability of an event is the proportion of the times that events of the same kind will occur in the long run.

The above relative frequency interpretation is valid under the assumption of a large number of repeated experiments under the same condition. In statistics, such an assumption is often termed as "independence and identical distribution (IID)", where the same kind of experiments performed ensures the identical probability law for the outcomes of various experiments, and the outcome of an experiment is not affected by the outcomes of any other experiments ensures independence among different experiments.

Example 2.15. When the weather forecast bureau predicts that there is a 30% chance of raining, it means that under the same weather conditions it will rain 30% of the times. We cannot guarantee what will happen on any particular occasion, but if we keep records over a long period of time, we should find that the proportion of "raining" is very close to 0.30 for the days with the same weather condition.

Approach 2: Subjective Probability Interpretation

Application of the frequency interpretation requires a well-documented history of the outcomes of an event over a large number of experimental trials under the same conditions. It is also called an empirical interpretation of probability. A more recently employed method of calculating probabilities is called the **subjective method**, which, instead of thinking of probability as relative frequency, thinks of probability as a belief in the chance of an event occurring. Here, a personal or subjective assessment is made of the probability of an event which is difficult or impossible to estimate in any way. For example, the probability that the S&P500 price index will go up in a given future period of time cannot be estimated very well by using the frequency interpretation because economic conditions rarely replicate themselves closely.

Subjective probability is the foundation of Bayesian statistics, which is a rival to classical statistics and is to be introduced in Section 2.6 below.

Subjective probabilities should be used only when all other methods fail, and then only with a high level of skepticism.

There exist examples of subjective probability by economic agents in practice. These are nice quantitative tools to describe the subjective beliefs of economic agents about economic or financial events.

Example 2.16. [Professional Forecasts]: The U.S. central bank — Fed issues professional forecasts for important macroeconomic indicators such as GDP growth rate, inflation rate and unemployment rate. In each quarter, they send surveys to professional forecasters, asking their views on probability distributions of these important macroeconomic indicators. Specifically, each forecaster will be asked what is his/her forecast of the probability that the inflation rate lies in various intervals.

Example 2.17. [Risk Neutral Probability]: During the 1997-1998 Asian financial crisis, many investors were very concerned with the collapse of the Hong Kong peg exchange rate system with U.S. dollars and devaluations of Hong Kong dollars. In other words, their subjective probabilities of Hong Kong dollar devaluation were higher than the objective probabilities of the Hong Kong dollar downward movements. The former are called risk-neutral probability distributions and the latter are called objective or physical probability distributions in finance. The gap between these two distributions reflects the risk attitude of market investors. The risk-neutral probability distribution is a financial instrument in derivative pricing.

Example 2.18. [Allais' Paradox]: In experimental economics, suppose a set of prizes is X = {$0, $1,000,000, $5,000,000}. Which probability distribution do you prefer: $P_1 = (0.00, 1.00, 0.00)$ or $P_2 = (0.01, 0.89, 0.10)$? And which probability do you prefer: $P_3 = (0.90, 0.00, 0.10)$ or $P_4 = (0.89, 0.11, 0.00)$? Many subjects in the experiment report that they prefer P_1 over P_2, and P_3 over P_4. This is inconsistent with the well-known expected utility theory in microeconomics.

Obviously, individuals tend to overweight low-probability events and underweight high-probability events. Formally, suppose there is a prospect of payoff $\{(x_1, p_1), (x_2, p_2), \ldots, (x_n, p_n)\}$ with $x_1 > \cdots > x_n$, where x_i is the payoff in state i and p_i is the probability of state i. We define a rank-dependent weighting:

$$\pi_i = w\left(\sum_{j=1}^{i} p_j\right) - w\left(\sum_{j=1}^{i-1} p_j\right),$$

where $w: [0, 1]$ $[0, 1]$ is a strictly increasing and continuous weighting function, with $w(0) = 0$ and $w(1) = 1$. Then the value of the prospect is characterized as $\sum_{i=1}^{n} \pi_i x_i$. Here, the rank-dependent weightings $\{\pi_i\}_{i=1}^{n}$ can reasonably be interpreted as subjective probabilities. This is pretty much like the way we interpret the prices of the Arrow securities as subjective probabilities.

Geometrically, based on the Venn diagram, the probability of any event A in sample space S can be viewed as equal to the area of event A in S, with the normalization that the total area of S is equal to unity. More precisely, the properties of a probability function are akin to those of an area if all sets under consideration are confined to some bounded region that has total area 1.

2.4.2 Basic Probability Laws

So far we have defined and discussed the concepts of sample space S, sigma-field $\mathbb{B}$, and probability measure P, respectively. They together constitute a so-called probability space.

Definition 2.12. [Probability Space]: A *probability space* is a triple $(S, \mathbb{B}, P)$ where:

(1) S is the sample space corresponding to the outcomes of the underlying random experiment;

(2) $\mathbb{B}$ is the σ-field of subsets of S. These subsets are called events;

(3) $P : \mathbb{B} \to [0, 1]$ is a probability measure.

A probability space $(S, \mathbb{B}, P)$ completely describes a random experiment associated with sample space S.

Because the probability function $P(\cdot)$ is defined on $\mathbb{B}$, the collection of sets or events, it is also called a set function. We now discuss some of its properties.

Theorem 2.2. *If $\varnothing$ denotes the empty set, then $P(\varnothing) = 0$.*

Proof: Given $S = S \cup \varnothing$, and S and $\varnothing$ are mutually exclusive, we have $P(S) = P(S \cup \varnothing) = P(S) + P(\varnothing)$. It follows that $P(\varnothing) = 0$.

Intuitively, Theorem 2.2 means that it is unlikely that nothing occurs when a random experiment is implemented. In other words, something always occurs when a random experiment is implemented.

While $P(\varnothing) = 0$, it is important to note that it does not necessarily follow from $P(A) = 0$ that $A = \varnothing$. This can be easily understood from the relative frequency interpretation for probability. It will be also clearly seen when we introduce a so-called continuous random variable in Chapter 3 where a continuous random variable taking a single value has probability zero.

Theorem 2.3. $P(A) = 1 - P(A^c)$.

Proof: Observe $S = A \cup A^c$. Then

$$P(S) = P(A \cup A^c).$$

Because $P(S) = 1$ and A and A^c are mutually exclusive, we have

$$1 = P(A) + P(A^c).$$

The desired result follows immediately.

To appreciate how useful Theorem 2.3 is in practice, we consider a simple example.

Example 2.19. Suppose X denotes the outcome of some random experiment. The following is the probability distribution for X, namely, the probability that X takes various values:

$$P(X = i) = \frac{1}{2^i}, \qquad i = 1, 2, \cdots.$$

Find the probability that X is larger than 3.

Solution: The sample space $S = \{1, 2, \cdots\}$. Let A be the event that $X > 3$. Then $A = \{4, 5, \cdots\}$. It follows that

$$
\begin{aligned}
P(A) &= P(X > 3) \\
&= P(X = 4) + P(X = 5) + P(X = 6) + \cdots \\
&= \sum_{i=4}^{\infty} P(X = i) \\
&= \sum_{i=4}^{\infty} \frac{1}{2^i}.
\end{aligned}
$$

Direct calculation of this infinite sum may be a bit tedious. Instead, we can apply Theorem 2.3 and compute

$$P(A) = 1 - P(A^c)$$
$$= 1 - P(X \leq 3)$$
$$= 1 - [P(X = 1) + P(X = 2) + P(X = 3)]$$
$$= 1 - \left(\frac{1}{2} + \frac{1}{2^2} + \frac{1}{2^3} \right)$$
$$= \frac{1}{8}.$$

In probability, the ratio of the probability of an event A to the probability of its complement,

$$\frac{P(A)}{P(A^c)} = \frac{P(A)}{1 - P(A)}$$

is called the ratio of *odds*.

Theorem 2.4. *If A and B are two events in $\mathbb{B}$, and $A \subseteq B$, then $P(A) \leq P(B)$.*

Proof: Using the fact that $S = A \cup A^c$ and the distributive law, we have

$$B = S \cap B = (A \cup A^c) \cap B$$
$$= (A \cap B) \cup (A^c \cap B)$$
$$= A \cup (A^c \cap B),$$

where the last equality follows from $A \subseteq B$ so that $A \cap B = A$. Because A and $A^c \cap B$ are mutually exclusive, we have

$$P(B) = P(A) + P(A^c \cap B)$$
$$\geq P(A)$$

given that $P(A^c \cap B) \geq 0$. This completes the proof.

Corollary 2.1. *For any event $A \in \mathbb{B}$ such that $\varnothing \subseteq A \subseteq S$, $0 \leq P(A) \leq 1$.*

Theorem 2.5. *For any two events A and B in $\mathbb{B}$,*

$$P(A \cup B) = P(A) + P(B) - P(A \cap B).$$

Proof: Since $A \cup B = A \cup (A^c \cap B)$, and A and $A^c \cap B$ are mutually exclusive, we have

$$P(A \cup B) = P(A) + P(A^c \cap B). \tag{1}$$

On the other hand, because $B = S \cap B = (A \cap B) \cup (A^c \cap B)$, and both $A \cap B$ and $A^c \cap B$ are mutually exclusive, we have

$$P(B) = P(A \cap B) + P(A^c \cap B). \qquad (2)$$

Adding both Equations (1) and (2) yields

$$P(A \cup B) + P(A \cap B) + P(A^c \cap B)$$
$$= P(A) + P(A^c \cap B) + P(B).$$

This delivers the desired result.

In fact, Theorem 2.5 can be illustrated via a Venn diagram, keeping in mind that the probability of an event is equal to the area it occupies in the sample space S. This is illustrated in Figure 2.2.

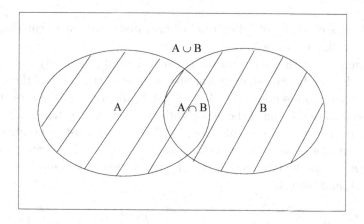

Figure 2.2: Venn diagram for the illustration of Theorem 2.6

Example 2.20. [Bonferroni's Inequality]: Show $P(A \cup B) \geq P(A) + P(B) - 1$.

Solution: Since $A \cap B \subseteq S$, we have $P(A \cap B) \leq P(S) = 1$ by Theorem 2.4. It follows from Theorem 2.5 that

$$P(A \cup B) = P(A) + P(B) - P(A \cap B)$$
$$\geq P(A) + P(B) - 1.$$

Example 2.21. Suppose there are two events A and B in S, with $P(A) = 0.20, P(B) = 0.30$ and $P(A \cap B) = 0.10$. Then

(1) Are A and B mutually exclusive?

(2) $P(A^c) = ?$ $P(B^c) = ?$
(3) $P(A \cup B) = ?$
(4) $P(A^c \cup B^c) = ?$
(5) $P(A^c \cap B^c) = ?$

Theorem 2.6. *[Rule of Total Probability]: If $A_1, A_2, \cdots \in \mathbb{B}$ are mutually exclusive and collectively exhaustive, and A is an event in S, then*

$$P(A) = \sum_{i=1}^{\infty} P(A \cap A_i).$$

Proof: Noting $S = \cup_{i=1}^{\infty} A_i$ and $A = A \cap S = A \cap (\cup_{i=1}^{\infty} A_i) = \cup_{i=1}^{\infty} (A \cap A_i)$, where the last equality follows by the distributive law. The result follows because $A \cap A_i$ and $A \cap A_j$ are disjoint for all $i \neq j$. This completes the proof.

The rule of total probability can be illustrated clearly in a Venn diagram (with $n = 3$, see Figure 2.3).

Intuitively, with a set of mutually exclusive and collectively exhaustive events $A_1, \cdots, A_n$, we can represent any event A as the union of the mutually exclusive intersections $A \cap A_1, \cdots, A \cap A_n$. As a result, the probability of A is equal to the sum of the probabilities of these intersections. In certain sense, a set of mutually exclusive and collectively exhaustive events could be viewed as a set of complete orthogonal bases on which we can span any event A, where the intersection $A \cap A_i$ can be viewed as the projection of event A onto base A_i.

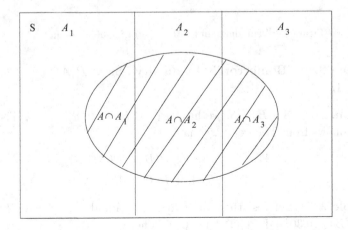

Figure 2.3: Venn diagram for the rule of total probability

Example 2.22. If $A = \{$students in a probability and statistics course whose scores > 90 points$\}$, $A_i = \{$students from country $i\}$, then $A \cap A_i = \{$students from country i whose scores are > 90 points$\}$.

Theorem 2.7. *[Subadditivity: Boole' Inequality]: For any sequence of events* $\{A_i \in \mathbb{B}, i = 1, 2, \cdots\}$,

$$P\left(\bigcup_{i=1}^{\infty} A_i\right) \le \sum_{i=1}^{\infty} P(A_i).$$

Proof: Put $B = \cup_{i=2}^{\infty} A_i$. Then $\cup_{i=1}^{\infty} A_i = A_1 \cup B$. It follows that

$$\begin{aligned}
P\left(\bigcup_{i=1}^{n} A_i\right) &= P(A_1 \cup B) \\
&= P(A_1) + P(B) - P(A_1 \cap B) \\
&\le P(A_1) + P(B),
\end{aligned}$$

where the inequality follows given $P(A_1 \cap B) \ge 0$. Again, put $C = \cup_{i=3}^{\infty} A_i$. Then

$$P(B) = P(A_2 \cup C) \le P(A_2) + P(C).$$

It follows that

$$P\left(\bigcup_{i=1}^{\infty} A_i\right) \le P(A_1) + P(A_2) + P(C).$$

Repeating this process, we have $P(\cup_{i=1}^{\infty} A_i) \le \sum_{i=1}^{\infty} P(A_i)$. This completes the proof.

Intuitively, the probability of the "sum (i.e., union)" of events is less than the sum of their individual probabilities. Equality occurs only when all events are mutually exclusive. Whenever there are overlapping events, the probability of total union will be strictly smaller than the sum of individual probabilities.

2.5 Methods of Counting

Question: How to calculate the probability of event A?

Suppose event A includes k basic outcomes $A_1, \cdots, A_k$ in sample space S. Then

$$P(A) = \sum_{i=1}^{k} P(A_i).$$

This is the basic formula to compute the probability of any event A.

If in addition S consists of n equally likely basic outcomes $A_1, \cdots, A_n$, then

$$P(A_i) = \frac{1}{n} \text{ for all } i = 1, \cdots, n.$$

In such a scenario, suppose event A consists of k basic outcomes. Then

$$P(A) = \frac{k}{n}.$$

Thus, in the cases where each basic outcome is equally likely to occur, the calculation of probability for event A boils down to the counting of the numbers of basic outcomes in event A and in sample space S respectively.

Motivated by problems associated with games of chance, the theory of probability first was developed under the assumption of equal likelihood. Under this assumption, one only has to count the number of basic outcomes of events. Furthermore, the classical probability concept applies also in a great variety of situations where gambling devices are used to make random selections — when office space is assigned to TAs by lots, when some of the families in a township are chosen in such a way that each one has the same chance of being included in a sample study, when machine parts are chosen for inspection so that each part produced has the same chance of being selected, and so forth.

Theorem 2.8. *[Fundamental Theorem of Counting]: If a random experiment consists of k separate tasks, the i-th of which can be done in n_i ways, $i = 1, 2, \cdots, k$, then the entire job can be done in $n_1 \times n_2 \times \cdots \times n_k = \Pi_{i=1}^{k} n_i$ ways.*

Proof: We shall first prove it for $k = 2$; then by induction. The first task can be done in n_1 ways, and for each of these ways, there are n_2 ways to do the second task. Therefore, the total ways for doing the first and second jobs is $n_1 \times n_2$. This completes the proof.

We will consider two important counting methods — permutation and combination — to calculate probability of events from a random experiment where each basic outcome is equally likely to occur.

2.5.1 Permutations

Example 2.23. Suppose we will choose two letters from four letters $\{A, B, C, D\}$ in different orders, with each letter being used at most once each time. How many possible orders could we obtain?

Solution: There are 12 ways: AB, BA, AC, CA, AD, DA, BC, CB, BD, DB, CD, DC.

We emphasize that the word *"ordered"* means that *"AB"* and *"BA"* are distinct outcomes.

Example 2.24. How many different ways to choose 20 letters from the 26 letters?

We now consider **a general problem:** Suppose there are x **boxes** arranged in row and there are n **objects**, where $x \leq n$. We are going to choose x from the n objects to place them in the x boxes. Each object can be used at most once in each arrangement (i.e., no replacement is allowed). How many possible different **sequences** could we obtain? That is, how many different ways can we fill the x boxes?

- First, how many ways can we fill Box 1? There are n objects available, so there are n ways.
- Second, suppose we have filled Box 1. How many different ways can we fill Box 2?
  Because there has been one object used to fill box 1, $n - 1$ objects remain and each of these $n - 1$ objects can be used to box 2. Therefore, there are $(n - 1)$ ways to fill box 2.
  Thus, there are $n(n - 1)$ ways to fill the first two boxes.
- Third, suppose we have filled the first two boxes. Then there are $(n - 2)$ ways to fill the Box 3.
  Thus, there are $n(n - 1)(n - 2)$ ways to fill the first three boxes.
- $\cdots$
- For the last box (i.e. Box x), given that $x - 1$ objects have been used to fill the first $x - 1$ boxes, there are $[n - (x - 1)]$ objects left, so there are $[n - (x - 1)]$ ways to fill the last box.

To sum up, the total number of possible orderings of choosing x out of n objects is:

$$P_n^x = \frac{n!}{(n - x)!} = n(n - 1)(n - 2) \cdots [n - (x - 1)].$$

The notation $k!$ is called "k factorial" and computed as

$$k! = k \times (k - 1) \times \cdots \times 2 \times 1.$$

Note that, as a convention, $0! = 1$.

Now, we can verify the answer in Example 2.23: with $n = 4, x = 2$, we have

$$P_n^x = P_4^2 = \frac{4!}{(4-2)!} = 12.$$

We can also obtain the answer for Example 2.24: with $n = 26, x = 20$, we have

$$P_n^x = \frac{26!}{(26-20)!} = \frac{26!}{6!}.$$

Example 2.25. A company has six sales representatives and has the following incentive scheme. It decides that the most successful representative during the previous year will be awarded a January vacation in Hawaii, while the second most successful representative will win a vacation in Las Vegas. The other representatives will be required to take a course on probability and statistics. How many different outcomes are possible?

Solution: Ordering matters here because who goes to Hawaii and who goes to Las Vegas will be considered as different outcomes. Thus, we use permutations: with $n = 6$, $x = 2$, we have

$$P_n^x = \frac{6!}{(6-2)!} = 30.$$

Example 2.26. [Birthday Problem]: Suppose there are k students in a class, where $2 \le k \le 365$. What is the probability that at least two students have the same birthday? Here, by the same birthday, we mean the same day of the same month, but not necessarily of the same year. Moreover, we make the following assumptions:

(1) no twins in the class;

(2) each of the 365 days is equally likely to be the birthday of anyone in the class;

(3) anyone born on Feb. 29 will be considered as on March 1.

Solution: First, how many possible ways in which the whole class could be born? This is a problem of ordering with replacement:

$$365^k = 365 \times 365 \times \cdots \times 365,$$

where each student has 365 days to be born. This is the total number of basic outcomes in the sample space S.

Second, the event A that at least 2 students have the same birthday is complement to the event A^c that all k students have different birthdays.

How many ways that k students can have different birthdays? This is a problem of choosing k different days out of 365 days to k students: By permutations, this number is

$$\frac{365!}{(365-k)!}.$$

Therefore,

$$P(A) = 1 - P(A^c)$$
$$= 1 - \frac{365!/(365-k)!}{365^k}.$$

To illustrate, we consider some concrete probabilities for various class sizes, as shown in the Table below:

k	20	30	40	50
$P(A)$	0.411	0.706	0.891	0.970

For a class with more than 50 students, the probability that at least two students have the same birthday is quite high!

2.5.2 *Combinations*

Example 2.27. Suppose now we will choose two letters from four letters $\{A, B, C, D\}$. Each letter is used at most once in each arrangement but now we are not concerned with their ordering. In other words, we are choosing a set that contains two different letters. How many possible such sets could we obtain?

Solution: There are six sets that contain two different letters:

$$\{A, B\}, \{A, C\}, \{A, D\}, \{B, C\}, \{B, D\}, \{C, D\}.$$

More generally, suppose we are interested in the number of different ways of choosing x objects from n objects but are not concerned about the order of the selected x objects. Here, each object can be used at most once in each arrangement. How many sets of x objects can we obtain? That is, how many sets each of which contains x different objects can we obtain?

We consider the following basic formula:

The total number of choosing x from n objects with ordering
= The total number of choosing x from n objects without ordering $\times$ the number of ordering x objects.

First, to choose x from n objects and place x objects in order, there are $n!/(n-x)!$ possible ordered sequences. Next, for a set of x objects, there are $x!/(x-x)! = x!$ ways to place them in different orders. It follows that the number of combinations of choosing x from n without ordering is

$$C_n^x = \binom{n}{x} = \frac{P_n^x}{x!} = \frac{n!}{x!(n-x)!}.$$

Lemma 2.1. *[Properties of Combinations]:*

(1) $\binom{n}{k} = \binom{n}{n-k}$;

(2) $\binom{n}{1} = n$;

(3) $\binom{n}{k} = \frac{P_n^k}{k!}$.

We can now use the combination formula to verify the answer to Example 2.27: with $n = 4, x = 2$, we have $C_n^x = n!/[(n-x)!x!] = 6$.

Example 2.28. A personnel officer has 8 candidates to fill 4 positions. Five candidates are men, and three are women. If in fact, every combination of candidates is equally likely to be chosen, what is the probability that no women will be hired?

Solution: Since we do not care who will fill a specific position, ordering does not matter.

(1) How many ways to select 4 out of the total of 8 candidates? The total number of possible combinations of choosing 4 out of 8 candidates is

$$C_8^4 = \frac{8!}{4!4!} = 70.$$

(2) When no women are hired, the four successful candidates must come from the available five men. The number of choosing 4 out of 5 male candidates is

$$C_5^4 C_3^0 = \frac{5!}{4!1!} = 5.$$

Therefore, the probability that no women will be hired is

$$P(A) = \frac{C_5^4 C_3^0}{C_8^4}$$
$$= \frac{5}{70}$$
$$= \frac{1}{14}.$$

Question: In this example, could we use the permutation method as an alternative? The answer is yes, but the same result will be obtained. Please verify it!

Example 2.29. Suppose a class contains 15 boys and 30 girls, and 10 students will be selected randomly to form a team. Here, by "randomly" we mean that in each case, all possible selections are equally likely. What is the probability that exactly 3 boys will be selected?

Solution: (1) How many ways to form a team with 10 members? It is given by

$$C_{45}^{10} = \frac{45!}{10!35!}.$$

(2) How many ways that exactly 3 boys (and therefore 7 girls) will be selected? Here, we choosing 3 out of 15 boys and 7 out of 30 girls. The number is $C_{15}^3 C_{30}^7$.

(3) It follows that the probability that exactly 3 boys will be selected is

$$P(A) = \frac{C_{15}^3 C_{30}^7}{C_{45}^{10}} = 0.2904.$$

Example 2.30. A manager has four assistants — John, George, Mary and Jean to assign to four tasks. Each one will be assigned to one task.

(1) How many different arrangements of assignments will be possible?

(2) What is the probability that Mary will be assigned to a *specific* task?

Solution: We shall use the permutation method.

(1) There are $P_4^4 = 4! = 24$ different arrangements in total.

(2) If Mary is assigned to a specific task, the manager has to arrange the other three candidates to the remaining 3 tasks. There are $P_3^3 = 3! = 6$ different ways for the managers to make such arrangements. It follows that the probability that Mary will be assigned to a specific task is given by

$$P(A) = \frac{P_3^3}{P_4^4}$$

$$= \frac{6}{24}$$

$$= \frac{1}{4}.$$

In fact, by common sense, given that everyone will be assigned a job, each one (including Mary) has equal chance to be assigned to each of the four tasks. Thus, Mary has a probability of $1/4$ to be assigned to any specific job.

Example 2.31. Suppose a team of 12 people is selected in a random manner from a group of 100 people. Determine the probability that two particular persons A and B will be selected.

Solution: (1) How many ways to form the team? There are C_{100}^{12}.

(2) Suppose two persons, say John and Tom, are included in the team, how many ways to select the other 10 members? There are C_{98}^{10}.

(3) Therefore, the probability that two particular persons, John and Tom, will be selected is given by

$$P(A) = \frac{C_{98}^{10}}{C_{100}^{12}}.$$

Example 2.32. The U.S. senate has 2 senators from each of the 50 states.

(1) If a committee of 8 senators is selected at random, what is the probability that it will contain at least one of the two senators from the New York state?

(2) What is the probability that a group of 50 senators selected at random will contain one senator from each state?

Solution: (1) (a) How many ways to form a 8-member committee? There are C_{100}^{8} different ways.

(b) Suppose A denotes the event that at least one of the 2 senators from the New York state will be selected. Then A^c is the event that no senator from the New York state will be selected. How many ways to select a 8-member committee if no senator from the New York state is considered? There are C_{98}^{8} different ways.

It follows that

$$P(A) = 1 - P(A^c)$$

$$= 1 - \frac{C_{98}^{8}}{C_{100}^{8}}.$$

(2) (a) How many ways to select a group of 50 senators? There are C_{100}^{50} different ways.

(b) If the group contains one senator from each state, then there are 2 possible choices for each state. Thus, the total number of choosing one senator from each state is

$$2^{50} = 2 \times 2 \times \cdots \times 2.$$

It follows that the probability that a group of 50 senators will contain one senator from each state is given by

$$P(A) = \frac{2^{50}}{C_{100}^{50}}.$$

Example 2.33. Choose an integer randomly from 1 to 2000. How many possible ways to choose an integer that can be divided exactly neither by 6 nor by 8?

Solution: Define $A = \{$the integer that can be divided exactly by 6$\}$, and $B = \{$the integer can be divided exactly by 8$\}$. Then by De Morgan's law,

$$
\begin{aligned}
P(A^c \cap B^c) &= P[(A \cup B)^c] \\
&= 1 - P(A \cup B) \\
&= 1 - [P(A) + P(B) - P(A \cap B)].
\end{aligned}
$$

Because

$$333 < \frac{2000}{6} < 334,$$

then

$$P(A) = \frac{333}{2000}.$$

Similarly,

$$P(B) = \frac{250}{2000}.$$

Moreover, an integer that can be divided exactly both by 6 and 8 is an integer that can be divided exactly by 24. Because

$$83 < \frac{2000}{24} < 84,$$

we have

$$P(A \cap B) = \frac{83}{2000}.$$

It follows that

$$P(A^c \cap B^c) = 1 - \left(\frac{333}{2000} + \frac{250}{2000} - \frac{83}{2000} \right) = \frac{3}{4}.$$

Example 2.34. Suppose we throw a fair coin 10 times independently. (1) What is the probability of obtaining exactly three heads? (2) What is the probability of obtaining three or fewer heads?

Solution: (1) How many possible outcomes in the experiment of throwing a coin 10 times? There are 2 possible outcomes each time, so there are 2^{10} possible outcomes.

Now, how many possible ways to obtain exactly three heads in the experiment? Because the heads are **indistinguishable**, we have to use the combination method. Thus, we have $C_{10}^3 = 120$ different ways to obtain three heads in the total of 10 trials. It follows that

$$P(3 \text{ heads obtained}) = \frac{120}{2^{10}} = 0.1172.$$

(2)
$$P(3 \text{ or fewer heads}) = P(0 \ head) + P(1 \ head) + P(2 \ heads) + P(3 \ heads)$$
$$= \frac{176}{2^{10}} = 0.1719.$$

Example 2.35. Suppose we throw a fair coin n times independently. What is the probability that exactly x heads will show up?

Solution: (1) How many possible outcomes in total? There are $2 \times 2 \times \cdots \times 2 = 2^n$ outcomes.

(2) How many possible ways to obtain exactly x heads in the experiment? Since heads are indistinguishable, we have to use the combination method. There are C_n^x different ways to obtain x heads.

It follows that

$$P(\text{exactly } x \text{ heads}) = \frac{C_n^x}{2^n}$$
$$= C_n^x \left(\frac{1}{2}\right)^x \left(1 - \frac{1}{2}\right)^{n-x}.$$

This is a special case of the so-called binomial distribution $B(n, p)$ with $p = \frac{1}{2}$. See Chapter 4 for more discussion on the latter.

Example 2.36. We would like to choose r elements out of n elements. For the following cases, how many ways do we have? (1) ordered, without replacement; (2) unordered, without replacement; (3) ordered, with replacement; (4) unordered, with replacement.

Solution: (1) $P_n^r = \frac{n!}{(n-r)!}$; (2) $C_n^r = \frac{P_n^r}{r!} = \frac{n!}{r!(n-r)!}$; (3) n^r; (4) C_{r+n-1}^r.

There is a key difference between permutations and combinations. Permutations are order specific, whereas combinations are not. So, although "ABC" and "CBA" are the same combinations, they are different permutations.

2.6 Conditional Probability

Economic events are generally related to each other. For example, there may exist causal relationships between economic events. Because of the connection, the occurrence of event B may affect or contain the information about the probability that event A will occur. Thus, if we have information about event B, then we can know better about the occurrence of event A. This can be described by the concept of conditional probability.

Example 2.37. [Volatility Spillover]: Shocks that originate from the financial market can have impact on output fluctuations in the real economy, and vice versa.

Example 2.38. [Financial Contagion]: A large drop of the price in one market can cause a large drop of the price in another market, given the speculations and reactions of market participants. This can occur regardless of market fundamentals.

Intuitively, a sample space is a description of uncertainty we are faced with. The original sample space S of a random experiment describes the largest degree of uncertainty with which we view the experiment. When new information arrives, some uncertainty is eliminated. Specifically, when event B has occurred, the uncertainty is then reduced from S to B. In this case, we are in a position to update the sample space based on new information. As a result, we want to be able to update probability calculations as well. These updated probabilities are called conditional probabilities.

Definition 2.13. [Conditional Probability]: Let A and B be two events in probability space $(S, \mathbb{B}, P)$. Then the conditional probability of event A given event B, denoted as $P(A|B)$, is defined as

$$P(A|B) = \frac{P(A \cap B)}{P(B)}$$

provided that $P(B) > 0$. Similarly, the conditional probability of event B given A is defined as

$$P(B|A) = \frac{P(A \cap B)}{P(A)}$$

provided $P(A) > 0$.

In defining the conditional probability $P(A|B)$, we assume $P(B) > 0$. This is because $P(B) = 0$ implies that B is unlikely to happen, and conditioning on an unlikely event is meaningless from a practical point of view.

The conditional probability $P(A|B)$ can be represented in the Venn diagram. It is the area occupied by event A within the area occupied by B relative to the area occupied by B. Specifically, when event B has occurred, the complement B^c will never occur. The uncertainty has been reduced from S to B. Thus, we will treat B as a new sample space when we consider $P(A|B)$; the intuition is that our original sample space S has been updated to B. All further occurrences are then calibrated with respect to their relationships to B, as is depicted in Figure 2.4.

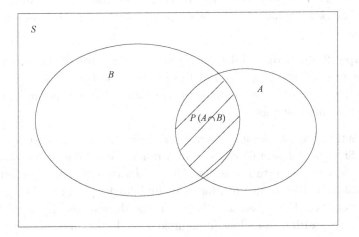

Figure 2.4: Venn diagram for conditional probability

Now, the triple $(S \cap B, \mathbb{B} \cap B, P(\cdot|B))$ is a probability space associated with the conditional probability function $P(A|B)$. In particular, $P(A|B)$ satisfies all probability laws defined on the sample space B. Indeed it can be shown that all conditional counterparts of Theorems 2.2, 2.3, 2.4 and 2.5 hold. For example, we have

$$P(A^c|B) = 1 - P(A|B).$$

To show this, first, given the identity $(A^c \cap B) \cup (A \cap B) = B$, and the fact that the intersections $A^c \cap B$ and $A \cap B$ are mutually exclusive, we have

$$P(A^c \cap B) + P(A \cap B) = P(B),$$
$$P(A^c \cap B) = P(B) - P(A \cap B).$$

It follows that

$$P(A^c|B) = \frac{P(A^c \cap B)}{P(B)}$$

$$= 1 - \frac{P(A \cap B)}{P(B)}$$

$$= 1 - P(A|B).$$

Example 2.39. Suppose the sample space S contains 25 sample points, which are chosen equally. Moreover, event A contains 15 points, event B contains 7 points, while $A \cap B$ contains 5 points, i.e.,

$$P(A) = \frac{15}{25}, \quad P(B) = \frac{7}{25}, \quad P(A \cap B) = \frac{5}{25}.$$

Then we have

$$P(A|B) = \frac{P(A \cap B)}{P(B)} = \frac{5}{7},$$

$$P(B|A) = \frac{P(B \cap A)}{P(A)} = \frac{1}{3}.$$

Question: Is $P(A) = P(A|S)$?

The answer is yes. By definition, we have

$$P(A|S) = \frac{P(A \cap S)}{P(S)} = P(A)$$

given $A \cap S = A$ and $P(S) = 1$. Intuitively, S represents the "least knowledge" among all possible events, or equivalently the highest degree of uncertainty in a random experiment. Therefore, it is the same as not conditioning on any useful specific information.

Example 2.40. Let A and B be disjoint and $P(B) > 0$. What is $P(A|B)$?

Solution: Given $P(A \cap B) = 0$, we have

$$P(A|B) = \frac{P(A \cap B)}{P(B)} = 0.$$

Intuitively, mutually exclusive events cannot occur simultaneously. If event B has occurred, then event A will never occur.

The conditional probability $P(A|B)$ describes how to use the information on event B to predict the probability of event A. This is a predictive relationship between A and B. It is not necessarily a causal relationship from B to A, even if the information of event B can be used to predict event A. Obviously, if there exists a causal relationship from B to A, then one can use B to predict A. On the other hand, it is also possible that there exists a causal relationship from C to A, but not from B to A. Nevertheless, both B and C are highly related to each other, and as a result, one can also use B to predict A. In order to characterize a causal relationship, we have to use economic theory outside probability and statistics.

With the definition of conditional probability, we can state the following multiplication rules.

Lemma 2.2. *[Multiplication Rules]:*

(1) If $P(B) > 0$, then $P(A \cap B) = P(A|B)P(B)$.

(2) If $P(A) > 0$, then $P(A \cap B) = P(B|A)P(A)$.

These formula can be used to compute the joint probability of events A and B, that is, $P(A \cap B)$.

Example 2.41. [Selecting Two Balls]: Suppose two balls are to be selected, without replacement, from a box containing r red balls and b blue balls. What is the probability that the first is red and the second is blue?

Solution: Define $A = \{$the first ball is red$\}$, $B = \{$the second ball is blue$\}$. Then

$$P(A) = \frac{r}{r+b},$$

$$P(B|A) = \frac{b}{r+b-1}.$$

$$P(A \cap B) = P(B|A)P(A)$$

$$= \frac{rb}{(r+b)(r+b-1)}.$$

The multiplication rule can be repeatedly used to obtain the joint probability of multiple events.

Theorem 2.9. *Suppose $\{A_i \in \mathbb{B}, i = 1, \ldots, n\}$ is a sequence of n events. Then the joint probability of these n events*

$$P\left(\bigcap_{i=1}^{n} A_i\right) = \prod_{i=1}^{n} P\left(A_i \,\middle|\, \bigcap_{j=1}^{i-1} A_j\right),$$

with the convention that $P(A_1 | \cap_{j=1}^{0} A_j) = P(A_1)$.

Example 2.42. [Computation of Joint Probabilities]: For $n = 3$, we have

$$P(A_1 \cap A_2 \cap A_3) = P(A_3|A_2 \cap A_1)P(A_2|A_1)P(A_1).$$

In fact, in expressing the joint probability $P(\cap_{i=1}^n A_i)$, there are $n!$ different ways of conditioning sequences, and the above theorem is just one of them. In the so-called time series analysis, where i is an index for time, the partition that the event A_i is conditional on $\cap_{j=1}^{i-1} A_j$ has a nice interpretation: A_i is conditional on the past information available at time $i - 1$.

Joint probability calculations are important for the so-called maximum likelihood estimation (MLE), to be introduced in Chapter 8. In econometrics, a probability model is often specified, and the MLE will choose a parameter estimate to maximize the joint probability that the n random variables take the values of the observed data. Here, it is essential to calculate the joint probability of n events, and the multiplication rule is very useful for such a purpose. See Chapter 8 for more discussion on MLE.

Theorem 2.10. [Rule of Total Probability]: Let $\{A_i\}_{i=1}^\infty$ be a partition (i.e., mutually exclusive and collectively exhaustive) of sample space S, with $P(A_i) > 0$ for $i \geq 1$. Then for any event A in $\mathbb{B}$,

$$P(A) = \sum_{i=1}^\infty P(A|A_i)P(A_i).$$

Proof: We have shown in Theorem 2.6 that

$$P(A) = \sum_{i=1}^\infty P(A \cap A_i).$$

The desired result follows immediately from the multiplication rule that $P(A \cap A_i) = P(A|A_i)P(A_i)$.

Intuitively, this theorem is called the rule of total probability because it says that if event A can be partitioned as a set of mutually exclusive subevents, then the probability of event A is equal to the sum of probabilities of this set of mutually exclusive subevents contained in A. It is also called the rule of elimination.

Example 2.43. Let $B_1, \cdots, B_k$ be mutually exclusive, and let $B = \cup_{i=1}^k B_i$. Suppose $P(B_i) > 0$ and $P(A|B_i) = p$ for $i = 1, \cdots, k$. Find $P(A|B)$.

Solution: By the definition of conditional probability, we have

$$P(A|B) = \frac{P(A \cap B)}{P(B)}$$

$$= \frac{P\left[\bigcup_{i=1}^{k} (A \cap B_i)\right]}{\sum_{i=1}^{k} P(B_i)}$$

$$= \frac{\sum_{i=1}^{k} P(A \cap B_i)}{\sum_{i=1}^{k} P(B_i)}$$

$$= \frac{\sum_{i=1}^{k} P(A|B_i)P(B_i)}{\sum_{i=1}^{k} P(B_i)}$$

$$= \frac{p \sum_{i=1}^{k} P(B_i)}{\sum_{i=1}^{k} P(B_i)}$$

$$= p.$$

Example 2.44. Suppose $B_1, B_2,$ and B_3 are mutually exclusive. If $P(B_i) = \frac{1}{3}$ and $P(A|B_i) = \frac{i}{6}$ for $i = 1, 2, 3$, what is $P(A)$?

Solution: Noting that B_1, B_2, B_3 are collectively exhaustive (why?), we have

$$A = S \cap A$$
$$= (B_1 \cup B_2 \cup B_3) \cap A$$
$$= \bigcup_{i=1}^{3} (A \cap B_i).$$

It follows that

$$P(A) = P\left[\bigcup_{i=1}^{3} (A \cap B_i)\right]$$

$$= \sum_{i=1}^{3} P(A \cap B_i)$$

$$= \sum_{i=1}^{3} P(A|B_i)P(B_i)$$

$$= \frac{1}{3} \sum_{i=1}^{3} P(A|B_i)$$

$$= \frac{1}{3} \left(\frac{1}{6} + \frac{2}{6} + \frac{3}{6}\right)$$

$$= \frac{1}{3}.$$

2.7 Bayes' Theorem

An interesting application of conditional probability is the famous Bayes' Theorem, which was originally propounded by Thomas Bayes (1701-1761), a nonconformist minister in Tunbridge Wells, England. This theorem, published after Bayes' death, leads to the so-called Bayesian school of statistics and Bayesian school of econometrics, a rival to the school of classical statistics and classical econometrics, which still predominates statistics and econometrics teaching in most universities in the world. The knowledge that an event B has occurred can be used to revise or update the prior probability that an event A will occur is the essence of Bayes' theorem.

The simplest version of Bayes' theorem is given below.

Theorem 2.11. *[Bayes' Theorem]: Suppose A and B are two events with $P(A) > 0$ and $P(B) > 0$. Then*

$$P(A|B) = \frac{P(B|A)P(A)}{P(B|A)P(A) + P(B|A^c)P(A^c)}.$$

The interest here is the update of the probability of event A when another event B has occurred. Bayes' theorem shows how conditional probabilities of the form $P(B|A)$ may be combined with initial probability of A (i.e., $P(A)$) to obtain the final probability $P(A|B)$. Here, $P(A)$ is called a "prior" probability (i.e., before the fact or evidence) about event A since it is the probability of A before new information B arrives. The conditional probability $P(A|B)$ is called a "posterior" probability (i.e., after the fact or evidence) since it represents the revised assignment of probability of A after the new information that B has occurred is obtained. Essentially, Bayes' theorem can be verbally stated as the posterior probability of event A is proportional to the probability of the sample evidence B after A has occurred times the prior probability of A.

As an example, let A be the event that "the IBM stock price will increase by 30% this year". An investor is interested in the IBM stock, and, based on the historical data, he has some prior probability judgement about the return on the IBM stock $(P(A))$. Now suppose he attends a workshop by a stock analyst, and knows that the analyst highly recommends the IBM stock. Given that the analyst has recommended the IBM stock (event B), the investor may become more confident about the IBM stock, and therefore may modify his original judgement about the return on the IBM stock $(P(A|B))$.

Bayes' theorem has been the subject of extensive controversy. There can be no question about the validity of Bayes' theorem, but considerable arguments have been raised about the assignment of the prior probabilities. Also, a good deal of mysticism surrounds Bayes' theorem because it entails a "backward" or "inverse" sort of reasoning, that is, reasoning "from effect to cause". Interestingly, this is a rather useful approach in economics and finance given irreversibility of an economic process.

Theorem 2.12. *[Alternative Statement of Bayes' Theorem]: Suppose $A_1, \cdots, A_n$ are n mutually exclusive and collectively exhaustive events in the sample space S, and A is an event with $P(A) > 0$. Then the conditional probability of A_i given A is*

$$P(A_i|A) = \frac{P(A|A_i)P(A_i)}{\sum_{j=1}^{n} P(A|A_j)P(A_j)}, \qquad i = 1, \cdots, n.$$

Proof: By the conditional probability definition and multiplication rule, we have

$$P(A_i|A) = \frac{P(A_i \cap A)}{P(A)}$$
$$= \frac{P(A|A_i)P(A_i)}{P(A)}.$$

Because $\{A_i\}_{i=1}^{n}$ are collectively exhaustive and mutually exclusive, from the rule of total probability in Theorem 2.10, we have

$$P(A) = \sum_{j=1}^{n} P(A \cap A_j)$$
$$= \sum_{j=1}^{n} P(A|A_j)P(A_j).$$

The desired result then follows immediately.

Our interest here is to update the probability about A_i given that event A has occurred. When event A has occurred, we may have better knowledge about the occurrence of A_i. Event A provides useful information for our updating knowledge on A_i.

Example 2.45. [How to Determine Auto-insurance Premium?]: Suppose an insurance company has three types of customers — high risk, medium risk and low risk. From the company's historical consumer database, it is known that 25% of its customers are high risk, 25% are medium risk, and 50% are low risk. Also, the database shows that the probability that a customer has at least one speeding ticket in one year is 0.25 for high risk, 0.16 for medium risk, and 0.10 for low risk.

Now suppose a new customer wants to be insured and reports that he has had one speeding ticket this year. What is the probability that he is a high risk customer, given that he has had one speeding ticket this year?

Solution: It is important for the auto-insurance company to determine whether the new customer belongs to the category of high risk customers, because it will affect the insurance premium to be charged. We denote events $H = \{$the customer is of high risk$\}$, $M = \{$the customer is of medium risk$\}$, $L = \{$the customer is of low risk$\}$, and $A = \{$the customer has received a speeding ticket this year$\}$.

Then

$$P(H) = 0.25, \quad P(M) = 0.25, \quad P(L) = 0.50.$$
$$P(A|H) = 0.25, \; P(A|M) = 0.16, \; P(A|L) = 0.10.$$

It follows that

$$P(H|A) = \frac{P(A|H)P(H)}{P(A)}$$

$$= \frac{P(A|H)P(H)}{P(A|H)P(H) + P(A|M)P(M) + P(A|L)P(L)}$$

$$= 0.410.$$

Without the speeding ticket information reported by the new customer, the auto-insurance company, based on its customer database, only has a prior probability $P(H) = 0.25$ for the new customer. With the new information (A), the auto-insurance company has an updated probability $P(H|A) = 0.41$ for the new customer.

Example 2.46. [Is It Useful for Publishers to Send Free Sample Textbooks to Professors?]: A publisher sends a sample statistics textbook to 80% of all statistics professors in the U.S. schools. 30% of the professors who receive this sample textbook adopt the book, as do 10% of the professors who do not receive the sample book. What is the probability that a professor who adopts the book has received a sample book?

Solution: Define event $A = \{$A professor has received a sample copy$\}$. Then

$$P(A) = 0.80, P(A^c) = 1 - 0.8 = 0.2.$$

Also define $B = \{$the professor adopts the textbook$\}$. Then

$$P(B|A) = 0.3, P(B|A^c) = 0.1.$$

It follows from Bayes' theorem that

$$
\begin{aligned}
P(A|B) &= \frac{P(B|A)P(A)}{P(B|A)P(A) + P(B|A^c)P(A^c)} \\
&= \frac{0.3 \cdot 0.8}{0.3 \cdot 0.8 + 0.1 \cdot 0.2} \\
&= 0.923
\end{aligned}
$$

Example 2.47. [Are Stock Analysts Helpful?]: Data evidence shows that last year 25% of the stocks in a stock exchange performed well, 25% poorly, and the remaining 50% performed on average. Moreover, 40% of those that performed well had been rated "good buy" by a stock analyst at the beginning of last year, as had been 20% of those that performed on average, and 10% of those that performed poorly. What is the probability that a stock rated a "good buy" by the stock analyst will perform well this year?

Solution: Define events $A = \{$the stock is rated as "good buy" by the stock analyst$\}$, $A_1 = \{$the stock performs better than the market average$\}$, $A_2 = \{$the stock performs as the market average$\}$, $A_3 = \{$the stock performs worse than the market average$\}$.
     Then

$$P(A_1) = 0.25, \quad P(A_2) = 0.5, \quad P(A_3) = 0.25.$$
$$P(A|A_1) = 0.4, \ P(A|A_2) = 0.2, \ P(A|A_3) = 0.1.$$

By Bayes' theorem, we have

$$
\begin{aligned}
P(A_1|A) &= \frac{P(A|A_1)P(A_1)}{\sum_{i=1}^{3} P(A|A_i)P(A_i)} \\
&= \frac{0.4 \cdot 0.25}{0.4 \cdot 0.25 + 0.2 \cdot 0.50 + 0.1 \cdot 0.25} \\
&= 0.444.
\end{aligned}
$$

Without the recommendation by the stock analyst, an investor, based on the historical data of the stock market, will only have the prior probability $P(A_1) = 0.25$. With the recommendation by the stock analyst (A), the investor will update his belief to $P(A_1|A) = 0.444$.

2.8 Independence

Independence is perhaps one of the most important concepts in statistics and econometrics. Suppose we believe that the probability that event A occurs is $P(A)$. Now we have an additional information that event B has occurred. Because events A and B are "related", we can use the new information B to update our probability about A, namely, we can obtain $P(A|B)$ via Bayes' Theorem. In general, $P(A|B) \neq P(A)$.

Now suppose that event B is "unrelated" to event A. That is, there exists no relationship between A and B. Then we expect that the information of B is irrelevant to predicting $P(A)$. In other words, we expect that $P(A|B) = P(A)$. If this is indeed the case, we say that events A and B are "independent". For example, the stock change in NYSE and raining in Ithaca are likely to be independent events, because they are irrelevant to each other. Intuitively, two events A and B are independent if the occurrence or non-occurrence of either one does not affect the probability of the occurrence of the other.

Definition 2.14. [**Independence**]: Events A and B are said to be statistically independent if $P(A \cap B) = P(A)P(B)$.

Events that are independent are called statistically independent, stochastically independent, or independent in a probability sense. In most instances, we use the word "independent" without a modifier if there is no possibility of misunderstanding. Independence is a probability notion to describe nonexistence of any kind of relationship between two events. It plays a fundamental role in probability theory and statistics.

Question: What is the implication of independence?

Suppose $P(B) > 0$. Then by definition of independence,

$$
\begin{aligned}
P(A|B) &= \frac{P(A \cap B)}{P(B)} \\
&= \frac{P(A)P(B)}{P(B)} \\
&= P(A).
\end{aligned}
$$

Therefore, **the knowledge of B does not help in predicting** A, because the occurrence of B does not affect the probability of the occurrence of A. Similarly, we have $P(B|A) = P(B)$, i.e. the occurrence of A has no effect on the occurrence or probability of B. Intuitively, independence implies that

A and B are "irrelevant", or there exists no relationship between them. In particular, there is no causal relationship between them.

There are different ways to define independence. The advantage of the definition of $P(A \cap B) = P(A)P(B)$ is that it treats the events symmetrically and will be easier to generalize to more than two events.

Example 2.48. Let $A = \{$Raining in Ithaca$\}$, $B = \{$Standard & Poor 500 price index going up$\}$. These two events are likely to be independent.

Example 2.49. Let $A = \{$Oil price goes up$\}$, $B = \{$Output growth slows down$\}$. These two events are likely to be dependent of each other.

Example 2.50. [Phillips Curve]: Let $A = \{$Inflation rate increases$\}$, $B = \{$Unemployment decreases$\}$. These two events are most likely to be dependent of each other.

Question: Why is the concept of independence useful in economics and finance?

We now provide an example to illustrate how useful the concept of independence is in economics.

Example 2.51. [Random Walk Hypothesis (Fama 1970)]: A stock price P_t will follow a random walk if $P_t = P_{t-1} + X_t$, where $\{X_t\}$ is independent across different time periods. Note that here $X_t = P_t - P_{t-1}$ is the stock price change from time $t-1$ to time t.

A closely related concept is the geometric random walk hypothesis. The stock price $\{P_t\}$ is called a geometric random walk if

$$\ln P_t = \ln P_{t-1} + X_t,$$

where X_t is independent across different time periods. Note that

$$\begin{aligned}
X_t &= \ln(P_t/P_{t-1}) \\
&= \ln\left(1 + \frac{P_t - P_{t-1}}{P_{t-1}}\right) \\
&\simeq \frac{P_t - P_{t-1}}{P_{t-1}}.
\end{aligned}$$

Thus, X_t can be interpreted as the relative stock price change. Here, we have made use of the first order Taylor series expansion.

The most important implication of the random walk hypothesis is: if $\{X_t\}$ is serially independent across different time periods, then a future stock price change X_t is not predictable using the historical stock price

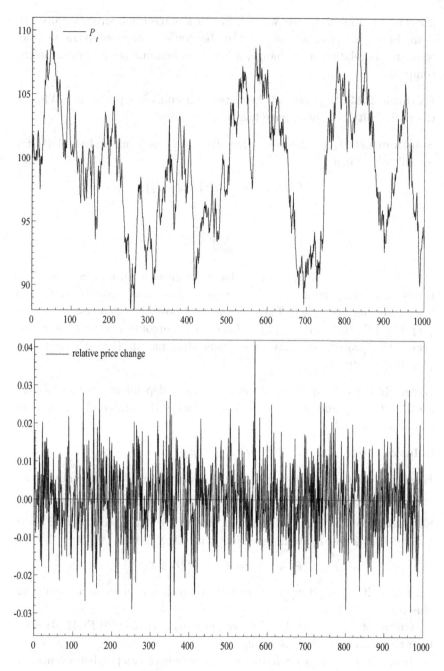

Figure 2.5: Observations for a geometric random walk and its relative price change

information. In such a case, we call the stock market is informationally efficient. Figure 2.5 plots a time series of observations for a geometric random
walk and its relative price change, using a random number generator on the
computer.

Example 2.52. Suppose we throw two fair coins independently. What is
the probability that both have heads showing up?

Solution: Let H_1 be the event that first coin has head, and similarly for
second coin. Then

$$P(H_1 \cap H_2) = P(H_1)P(H_2)$$
$$= \frac{1}{2} \cdot \frac{1}{2}$$
$$= \frac{1}{4}.$$

There is an alternative method based counting with replacement: Two
heads showing up is one possible outcome. How many possible outcomes
from throwing two coins? There are $2^2 = 4$ possible outcomes: $\{H_1, H_2\}$,
$\{H_1, T_2\}, \{T_1, H_2\}$ and $\{T_1, T_2\}$. Since each outcome is equally likely to
occur, the probability that two heads showing up (i.e., the outcome
$\{H_1, H_2\}$ occurs) is $\frac{1}{4}$.

Example 2.53. Suppose we throw two coins independently with probabilities $P(H_1) = p$ and $P(H_2) = q$. Then what is the probability that both
coins have heads?

Solution: $P(H_1 \cap H_2) = P(H_1)P(H_2) = pq$.

Example 2.54. Can two independent events A and B be mutually exclusive? Can two mutually exclusive events A and B be independent?

Solution: We first consider a regular case where $P(A) > 0$ and $P(B) > 0$:
Case (1): If A and B are independent with $P(A) > 0$ and $P(B) > 0$, then

$$P(A \cap B) = P(A)P(B) > 0.$$

Therefore, if A and B are independent, then they cannot be mutually exclusive.

On the other hand, if A and B are mutually exclusive (so $P(A \cap B) = 0$),
then they cannot be independent.

However, there exists a pathological case where independent events can
be mutually exclusive as well. This happens when $P(A) = 0$ or $P(B) = 0$.

Case (2): Suppose $P(A) = 0$, or $P(B) = 0$. If A and B are independent, then

$$P(A \cap B) = P(A)P(B) = 0.$$

This implies that A and B could be mutually exclusive. On the other hand, if A and B are mutually exclusive, they are independent.

When $P(A) > 0$ and $P(B) > 0$, independent events cannot be mutually exclusive. This implies that independent events contain common basic outcomes and so can occur simultaneously. For example, Standard & Poor 500 price index can increase when there is raining in Ithaca. Intuitively, two independent events can occur simultaneously, so they are not mutually exclusive. On the other hand, two mutually exclusive events cannot occur simultaneously, so they are not independent.

Below we present another example.

Example 2.55. There are four cards, numbered 1, 2, 3, 4. The experiment is to select one card randomly. Define events $A_1 = \{1 \text{ or } 2\}$, $A_2 = \{1 \text{ or } 3\}$. Then

$$P(A_1) = 1/2 = P(A_2).$$

Since $A_1 \cap A_2 = \{1\}$, we have $P(A_1 \cap A_2) = 1/4$. Therefore, events A_1 and A_2 are independent, although they have a common element.

Theorem 2.13. *Let A and B are two independent events. Then*
(1) A and B^c are independent;
(2) A^c and B are independent;
(3) A^c and B^c are independent.

Proof: (1) If $P(A \cap B^c) = P(A)P(B^c)$, then A and B^c are independent. Because $(A \cap B) \cup (A \cap B^c) = A$, we have

$$P(A \cap B) + P(A \cap B^c) = P(A).$$

It follows from the multiplication rule that

$$\begin{aligned}
P(A \cap B^c) &= P(A) - P(A \cap B) \\
&= P(A) - P(A)P(B) \\
&= P(A)[1 - P(B)] \\
&= P(A)P(B^c).
\end{aligned}$$

(2) By symmetry.

(3) Because $(A \cap B^c) \cup (A^c \cap B^c) = B^c$, we have

$$P(A \cap B^c) + P(A^c \cap B^c) = P(B^c).$$

It follows that

$$
\begin{aligned}
P(A^c \cap B^c) &= P(B^c) - P(A \cap B^c) \\
&= P(B^c) - P(A)P(B^c) \\
&= P(A^c)P(B^c).
\end{aligned}
$$

This completes the proof.

In fact, the above theorem could be understood intuitively: Suppose A and B are independent. Then A and B^c should be independent as well because if not, one would be able to predict the probability of B^c using A, and thus predict the probability of B using A via the complement probability formula $P(B|A) = 1 - P(B^c|A)$.

We now provide a definition of independence for more than two events.

Definition 2.15. [Independence Among Several Events]: k events $A_1, A_2, \cdots, A_k$ are mutually independent if, for every possible subset $A_{i_1}, \cdots, A_{i_j}$ of j of those events $(j = 2, 3, \cdots, k)$,

$$P(A_{i_1} \cap \cdots \cap A_{i_j}) = P(A_{i_1}) \times \cdots \times P(A_{i_j}).$$

For three or more events, independence is called mutual independence or joint independence. If there is no possibility of misunderstanding, independence is often used without the modifier "mutual" or "joint" when considering several events.

A collection of events are mutually independent if the joint probability of any subcollection of the events is equal to the product of the individual probabilities. We need $(2^k - 1 - k)$ conditions to characterize independence among k events (because $\sum_{j=0}^{k} \binom{k}{j} = 2^k, \binom{k}{0} = 1, \binom{k}{1} = k$).

For example, three events A, B, and C are independent, if the following $2^3 - (1 + 3) = 4$ conditions are satisfied:

$$P(A \cap B) = P(A)P(B),$$

$$P(A \cap C) = P(A)P(C),$$

$$P(B \cap C) = P(B)P(C),$$

$$P(A \cap B \cap C) = P(A)P(B)P(C).$$

Example 2.56. [Mooncake Betting]: In the city of Xiamen (also called Amoy), Southeast China, there is a traditional activity called *Mooncake*

Betting to celebrate Mid-Autumn Festival. This is essentially a game of rolling 5 dices. The final winner, called "Zhuang-Yuan", will result if she rolls 5 dices with at least 4 dices showing number 4 simultaneously. If at least two candidates have 4 dices with number 4, then the one who gets a larger number on the fifth dice will win. Suppose 2 friends play such a game. On average, how many rounds of rolling that both of them have to complete in order to produce a winner?

Example 2.57. [Reliability]: A project, such as launching a satellite, consists of k independent subprojects, denoted as $A_1, A_2, \cdots, A_k$. Suppose subproject i has a failure rate f_i, where $i = 1, \cdots, k$. What is the probability that the project will be successfully implemented?

Solution: The success of the project requires that all subprojects be successful. Thus, the probability of a successful project is given by

$$P\left(\bigcap_{i=1}^{k} A_i\right) = \prod_{i=1}^{n} P(A_i)$$

$$= \prod_{i=1}^{n} [1 - P(A_i^c)]$$

$$= \prod_{i=1}^{n} (1 - f_i).$$

As a numerical example, suppose there are $n = 10$ subprojects each of which has an equal failure rate f_i. Then to ensure the probability of a successful launching to be at least 0.99, each subproject should have a failure rate of less than 0.0001.

Obviously, by definition, joint independence implies pairwise independence. However, the converse is not true. It is possible to find that three events are pairwise independent but not jointly independent. Below is an example.

Example 2.58. Suppose

$$S = \{aaa, bbb, ccc, abc, bca, cba, acb, bac, cab\}$$

and each basic outcome is equally likely to occur. For $i = 1, 2, 3$, define $A_i = \{i\text{-th place in the triple is occupied by letter } a\}$. For example,

$$A_1 = \{aaa, abc, acb\}.$$

It is then easy to see that

$$P(A_1) = P(A_2) = P(A_3) = \frac{3}{9} = \frac{1}{3},$$

and

$$P(A_1 \cap A_2) = P(A_1 \cap A_3) = P(A_2 \cap A_3) = \frac{1}{9},$$

so that A_1, A_2 and A_3 are pairwise independent. However,

$$P(A_1 \cap A_2 \cap A_3) = \frac{1}{9} > P(A_1)P(A_2)P(A_3) = \frac{1}{27}.$$

Therefore, A_1, A_2, A_3 are not mutually independent.

The implication of Example 2.56 is that if one uses A_2 or A_3 to predict A_1, then A_2 or A_3 is not helpful. However, if one uses both A_2 and A_3 jointly to predict A_1, then A_1 is predictable. Generally speaking, the difference between pairwise independent events and joint independent events is that, in the former, knowledge about the joint occurrence of any two of them may be useful in predicting the chance of the occurrence of the remaining one, but in the latter it would not.

Example 2.59. [Complementarity Between Economic Reforms]: In the fields of economic growth and development, many studies find that one economic policy usually necessities another policy to stimulate economic growth, which is called policy complementarities. In transitional economics, individual reforms or sequential reforms may not be effective or fully effective, or even back-firing. Reforms must be packed together in order to be effective. For example, in order to improve firm productivity (A_1), changing a manager (A_2) should be together with granting autonomy to the firm (A_3).

There are many other examples of economic complementarities. Harrison (1996), Rodriguez and Rodrik (2000), Loayza *et al.* (2005), Chang *et al.* (2005) document that international trade openness, only when combined with other policies that improve a country's educational investment, financial depth, inflation stabilization, public infrastructure, governance, labor market flexibility, and ease of firm entry and exit, can promote economic growth.

We have explained that joint independence implies pairwise independence, but not vice versa. However, in some special cases, it is possible that pairwise independence implies joint dependence.

Question: When can we have pairwise independence imply joint independence? If so, please give an example.

In checking multiple joint independence, it is important to check the joint probability of every possible subset of events should be equal to the product of the probabilities of all individual events in the subset. For example, when checking independence among three events A, B, C, it is not sufficient to only check the condition that $P(A \cap B \cap C) = P(A)P(B)P(C)$. Conditions on all possible pairs of events should be considered as well.

Question: Suppose A, B, C are three events. Does $P(A \cap B \cap C) = P(A)P(B)P(C)$ imply independence among the three events A, B, C? If yes, prove it. If not, give an example.

2.9 Conclusion

In this chapter, we have laid down a foundation of probability theory. Based on the fundamental axiom that any economy can be viewed as a random experiment, we first characterize a random experiment by a probability space $(S, \mathbb{B}, P)$, where S is the sample space, $\mathbb{B}$ is a σ-field, and $P : \mathbb{B} \to [0, 1]$ is a probability measure. Interpretations for probabilities are provided. Given each pair of $(S, \mathbb{B})$, a measurable space, one can define many probability functions. The main objective of econometrics is to use the observed economic data to infer a suitable probability function which truly or most closely represents the true probability distribution for the underlying data generating process. We discuss the properties of a probability function, and the methods of counting which are very useful in calculating probabilities of interested events. We also introduce the concept of conditional probability function, as a way to characterize predictive relationships between or among economic events, and discuss Bayes' theorem. Finally, we introduce the concept of independence and its implications in economics and finance.

EXERCISE 2

2.1. For events A and B, find formula for the probabilities of the following events in terms of the quantities $P(A)$, $P(B)$, and $P(A \cap B)$:

    (1) either A or B or both;

    (2) either A or B but not both;

    (3) at least one of A or B;

    (4) at most one of A or B.

2.2. Formulate and prove a version of De Morgan's Laws that applies to a finite collection of sets $A_1, \cdots, A_n$.

2.3. Let S be a sample space.

    (1) Show that the collection $\mathbb{B} = \{\varnothing, S\}$ is a sigma algebra.

    (2) Let $\mathbb{B} = \{$all subsets of S, including S itself$\}$. Show that $\mathbb{B}$ is a sigma algebra.

    (3) Show that the intersection of two sigma algebra is a sigma algebra.

2.4. Consider two events A and B such that $P(A) = \frac{1}{3}$ and $P(B) = \frac{1}{2}$. Determine the value of $P(B \cap A^c)$ for each of the following conditions:

    (1) A and B are disjoint;

    (2) $A \subset B$;

    (3) $P(A \cap B) = \frac{1}{8}$.

2.5. Let A and B be two events. Check if the following relations hold:

    (1) $A \cup B = A \cup (A^c \cap B)$;

    (2) $B = (A \cap B) \cup (A^c \cap B)$.

    Please give your reasoning clearly.

2.6. Suppose events A and B are mutually exclusive.

    (1) Can we say that A^c and B^c mutually exclusive? Give your reasoning.

    (2) Please find some example such that A^c and B^c are mutually exclusive as well.

2.7. Suppose $P(A) = \frac{1}{3}$ and $P(B^c) = \frac{1}{4}$. Is it possible that A and B are mutually exclusive?

2.8. Which of the following statements is true? If it is true, prove it; otherwise, give a counter example.

    (1) If $P(A) + P(B) + P(C) = 1$, then the events A, B, C are mutually exclusive.

    (2) If $P(A \cup B \cup C) = 1$, then A, B, C are mutually exclusive.

2.9. Check if the following statements are true? If a statement is true, prove it; otherwise, give a counter example.

    (1) If A is an event with probability 1, then A is the sample space.

(2) If B is an event with probability 0, then B is an empty set.

(3) If $P(A) = 1$ and $P(B) = 1$, then $P(A \cap B) = 1$.

2.10. Suppose A, B, C are three events such that A and B are disjoint, A and C are independent, and B and C are independent. Suppose also that $4P(A) = 2P(B) = P(C)$, $P(A \cup B \cup C) = 5P(A)$, and $P(A) > 0$. Determine the value of $P(A)$.

2.11. Two players, A and B, alternately and independently flip a coin and the first player to obtain a head wins. Assume player A flips first.

(1) If the coin is fair, what is the probability that A wins?

(2) Suppose that $P(\text{Head}) = p$, not necessarily $\frac{1}{2}$. What is the probability that A wins?

(3) Show that for all $p, 0 < p < 1, P(A \text{ wins}) > \frac{1}{2}$. (Hint: Try to write $P(A \text{ wins})$ in terms of the events $E_1, E_2, \ldots,$ where $E_i = \{\text{Head first appears on the } i\text{th toss}\}$.

2.12. A couple have two children. At least one of them is a boy. What is the probability that both children are boys?

2.13. Suppose any conditioning event has positive probability. Prove each of the following statements.

(1) If $P(B) = 1$, then $P(A \cap B) = P(A)$ for any A.

(2) If $A \subset B$, then $P(B|A) = 1$ and $P(A|B) = P(A)/P(B)$.

(3) If A and B are mutually exclusive, then

$$P(A|A \cup B) = \frac{P(A)}{P(A) + P(B)}.$$

2.14. Show that $P(A|B) > P(A)$ if and only if $P(B|A) > P(B)$. In probability, A and B are said to be positively correlated if $P(A|B) > P(A)$, and are said to be negatively correlated if $P(A|B) < P(A)$.

2.15. Find an example for which $P(A \cap B) < P(A)P(B)$.

2.16. Show that if $P(A_1|A_3) \geq P(A_2|A_3)$ and $P(A_1|A_3^c) \geq P(A_2|A_3^c)$, then $P(A_1) \geq P(A_2)$.

2.17. Let A, B, C be any three events defined on sample space S. Express $P(A \cup B \cup C)$ in terms of $P(A), P(B), P(C), P(A \cap B), P(B \cap C), P(C \cap A)$ and $P(A \cap B \cap C)$. Give your reasoning.

2.18. Let A_1, A_2, A_3, A_4 be four events defined on sample space S. Express $P(A_1 \cup A_2 \cup A_3 \cup A_4)$ in terms of $P(A_i), P(A_i \cap A_j), P(A_i \cap A_j \cap A_k)$ and $P(A_1 \cap A_2 \cap A_3 \cap A_4)$.

2.19. Let A_i, $i = 1, 2, \ldots, n$, be a sequence of events of an experiment, where n is an integer. Show that $P\left(\cup_{i=1}^{n} A_i\right) \leq \Sigma_{i=1}^{n} P(A_i)$.

2.20. Let A_i, $i = 1, 2, \ldots, n$, be a sequence of events of an experiment, where n is an integer. Show that $P\left(\cap_{i=1}^{n} A_i\right) \geq 1 - \Sigma_{i=1}^{n} P(A_i)$.

2.21. A secretary types four letters to four people and addresses the four envelopes. If he inserts the letters at random, one into each envelope, what is the probability that exactly two letters will go into the correct envelopes? Exactly three?

2.22. An elevator in a building starts with five passengers and stops at seven floors. If each passenger is equally likely to get off at any floor and all passengers leave independently of each other, what is the probability that no two passengers will get off at the same floor?

2.23. A mutual fund company has 6 funds that invest in the U.S. market, and 4 funds that invest in foreign markets. A customer wants to invest in two U.S. funds and two foreign funds.

(1) How many different sets of funds from this company could the investor choose?

(2) Unknown to this investor, one of the U.S. funds and one of the foreign funds will seriously under-perform next year. If the investor selects funds for purchase at random, what is the probability that at least one of the chosen funds will seriously under-perform next year?

2.24. Suppose that a box contains r red balls and w white balls. Suppose also that balls are drawn from the box one at a time at random, without replacement.

(1) What is the probability that all r red balls will be obtained before any white balls are obtained?

(2) What is the probability that all r red balls will be obtained before two white balls are obtained?

2.25. Suppose that 5% of men and 0.25% of women are color-blind. A person is chosen at random and that person is color-blind. What is the probability that the person is a male? (Assume that males and females are in equal numbers.)

2.26. If 50% of the families in a city subscribe to the morning newspaper, 65% of the families subscribe to the afternoon newspaper, and 85% of the families subscribe to at least one of the two newspapers, what is the proportion of the families subscribe to both newspapers?

2.27. If the probability of hinting a target is $\frac{1}{5}$, and ten shots are fired independently, what is the probability of the target being hit at least twice? What is the conditional probability that the target is hit at least twice, given that it is hit at least once?

2.28. Standardized tests provide an interesting application of probability theory. Suppose first that a test consists of 20 multiple-choice questions, each with 4 possible answers. If the student guesses on each question, then the taking of the exam can be modeled as a sequence of 20 independent events. Find the probability that the student gets at least 10 questions correct, given that he is guessing.

2.29. Two events A and B are independent, and $B \subset A$. Find $P(A)$.

2.30. Suppose $0 < P(B) < 1$. Show that event A and B are independent if and only if $P(A|B) = P(A|B^c)$.

2.31. Give an example to show that $P(A \cap B \cap C) = P(A)P(B)P(C)$ for three given events A, B, C does not necessarily guarantee $P(A \cap B) = P(A)P(B)$.

2.32. Person A and Person B shoot to the same target independently. Person A hits the target with a probability of 0.8, while Person B hits the target with a probability of 0.9. What is the probability that the target can be hit?

2.33. Police plan to enforce speed limits by using radius traps at 4 different locations within the city limits. The radar traps at each of the locations L_1, L_2, L_3, and L_4 are operated 40%, 30%, 20%, and 30% of the time, and if a person who is speeding on his way to work has probabilities of 0.2, 0.1, 0.5 and 0.2, respectively, of passing through these locations, what is the probability that he will receive a speeding ticket? Give your reasoning.

2.34. In a large city, it is established that 0.5% of the population has contracted AIDS. The available tests give the correct diagnosis for 80% of healthy persons and for 98% of sick persons. Suppose a person is tested and found sick. Find the probability that the diagnosis is wrong, that is, that the person is actually healthy.

2.35. A bank gives a test to screen prospective employees. Among those who perform their jobs satisfactorily, 65 percent passed the test. Among those who do not perform satisfactorily (and are fired), 25 percent passed the test. According to the bank's records, 90 percent of its employees perform their jobs satisfactorily. What is the probability that a prospective employee who passed the test will *not* perform satisfactorily?

2.36. A market research group specializes in providing assessments of the prospects of sites for new clothing stores in shopping centers. The group assesses prospects as either good, fair or poor. The records of requests for assessments made to this group were examined, and it was found that for all stores that turned out to be successful, the assessment was good for 60%, fair for 30%, and poor for 10%. For all stores that turned out to be unsuccessful, the assessment was good for 10%, fair for 30%, and poor for 60%. It is also known that 70% of new clothing stores are successful and 30% are unsuccessful.

(1) For a randomly chosen store, what is the probability that prospects will be assessed as good?

(2) If prospects for a store are assessed as good, what is the probability that it will be successful?

Chapter 3

Random Variables and Univariate Probability Distributions

Abstract: In this and next two chapters, we will use advanced calculus to formalize and extend the probability theory introduced in Chapter 2. The use of calculus enables us to investigate probability more deeply. A number of quantitative-oriented probability concepts will be introduced. In this chapter, we first introduce the concept of a random variable and characterize the probability distributions of a random variable and functions of a random variable by the cumulative distribution function, the probability mass function or probability density function, the moment generating function and the characteristic function, respectively. We also introduce a class of moments and discuss their relationships with a probability distribution. This chapter focuses on univariate distributions.

Key words: Characteristic function, Continuous random variable, Cumulative distribution function, Discrete random variable, Kurtosis, Mean, Moment generating function, Moments, Probability density function, Probability mass function, Quantile, Random variable, Skewness, Variance.

3.1 Random Variables

Recall that the probability space triple $(S, \mathbb{B}, P)$ completely characterizes a random experiment. In general, the space S and the associated σ-algebra $\mathbb{B}$ differ according to the natures of random experiments. For example, when one throws a coin, the sample space $S = \{H, T\}$, where H denotes head, and T denotes tail; for the election of some candidate, $S = \{\text{Win, Fail}\}$; if one throws three coins, then $S = \{HHH, HHT, HTH, HTT, THH, THT, TTH, TTT\}$.

It is inconvenient to work with different sample spaces. In particular, a sample space S may be tedious to describe if the elements of S are not

67

real numbers. In many experiments, it is easier to deal with a summary variable than with the original probability structure. To develop a unified probability theory, we need to unify different sample spaces. For this purpose, we need to formulate a rule, or a set of rules, by which elements of S may be represented by numbers. This can be achieved by assigning a real number to each possible outcome in S. In other words, we construct a mapping from the original sample space S to a new sample space Ω, a set of real numbers. This transformation is called a random variable. A random variable is a function defined on a sample space S. Its purpose is to facilitate the solution of a problem by transferring considerations to a new probability space with a simpler or more convenient structure.

On the other hand, in many applications, we are interested only in a particular aspect of the outcomes of experiments, rather than the outcomes themselves. For example, when we roll a number of dice, we are usually interested in the total number obtained, and not in the outcome of each die. In such applications, a suitably defined random variable will better serve for our purpose.

Definition 3.1. [Random Variable]: A random variable, $X(\cdot)$, is a $\mathbb{B}$-measurable mapping (or point function) from the sample space S to the real line $\mathbb{R}$ such that to each outcome $s \in S$, there exists a corresponding unique real number, $X(s)$. The collection of all possible values that the random variable X can take, also called the range of $X(\cdot)$, constitutes a new sample space, denoted as Ω.

By the definition of a function, the function $X : S \to \Omega$ need not be a one-to-one mapping. It is possible that two basic outcomes $s_1, s_2 \in S$ will deliver the same value for random variable X; that is, $X(s_1) = X(s_2)$. See Figure 3.1.

Example 3.1. When we throw a coin, the sample space $S = \{H, T\}$. Define a random variable $X(\cdot)$ by $X(H) = 1$ and $X(T) = 0$. Then we obtain a new sample space $\Omega = \{1, 0\}$.

Example 3.2. For the election of a candidate, the sample space $S = \{$Win, Fail$\}$. Define a random variable $X(\cdot)$ by $X($Win$) = 1$ and $X($Fail$) = 0$. Then $\Omega = \{1, 0\}$.

It is not necessary to have the same number of basic outcomes for both S and Ω. In some cases, it may be more convenient to work with a suitably defined new sample space Ω.

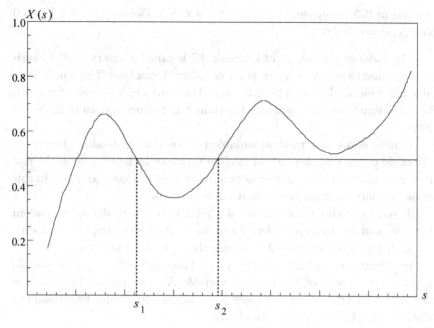

Figure 3.1: $X(s_1) = X(s_2)$

Example 3.3. Suppose we throw three fair coins. Then the space sample

$$S = \{TTT, TTH, THT, HTT, HHT, HTH, THH, HHH\}.$$

Let $X(\cdot)$ be the number of heads shown up. Then $X(TTT) = 0$, $X(TTH) = 1$, $X(THT) = 1$, $X(HTT) = 1$, $X(HHT) = 2$, $X(HTH) = 2$, $X(THH) = 2$, $X(HHH) = 3$. We have $\Omega = \{0, 1, 2, 3\}$.

In this example, $X(s)$ is the number of heads, where s is a basic outcome in S. Therefore, $P(X = 3) = P(A)$, where $A = \{s \in S : X(s) = 3\} = \{HHH\}$, denotes the probability that exactly three heads occur in the experiment.

Example 3.4. When a die is rolled, the sample space $S = \{1, 2, 3, 4, 5, 6\}$. Define $X(s) = s$. Then $\Omega = S$. This is an identity transformation.

Suppose the number of basic outcomes in S is countable. Then (1) is it possible that the number of basic outcomes in Ω larger than that of S? (2) Is it possible that the number of basic outcomes in Ω smaller than that of S?

Example 3.5. Suppose $S = \{s : -\infty < s < \infty\}$. Define $X(s) = 1$ if $s > 0$ and $X(s) = 0$ if $s \leq 0$.

The random variable X in Example 3.5 is called a binary random variable because there are only two possible values X can take. The binary variable has wide applications in economics, For example, it is useful for directional forecasts or investigation of asymmetric business cycles (e.g., Neftci 1984).

Definition 3.1 of a random variable is limited to real-valued functions. We could define complex-valued random variables by looking upon the real and imaginary parts separately as two real-valued random variables. In this book, we only consider real-valued random variables.

Throughout this book, we use a capital letter X to denote a random variable, and use a lowercase letter x to denote its realization, i.e., a possible value that random variable X can take (i.e., $x = X(s)$ for some $s \in S$).

In defining a random variable X, we have also defined a new sample space Ω, the range of the random variable X. The probability function defined on the original sample space S can be used to obtain the probability distribution of the random variable X.

First, suppose we have a sample space with a finite number of basic outcomes

$$S = \{s_1, \cdots, s_n\}$$

and a probability function $P : \mathbb{B} \to [0, 1]$, where $\mathbb{B}$ is a σ-field associated with S. Also, we define a random variable $X : S \to \mathbb{R}$ with the range

$$\Omega = \{x_1, \cdots, x_m\},$$

where m may not be the same as n. Then we can define the probability function $P_X : \Omega \to \mathbb{R}$ for the random variable X in the following way:

$$\begin{aligned} P_X(x_i) &\equiv P(X = x_i) \\ &= P(C_i), \end{aligned}$$

where C_i is an event in S such that

$$C_i = \{s \in S : X(s) = x_i\}.$$

Note that the left hand side, $P_X(\cdot)$, an induced probability function on the new sample space Ω, is defined in terms of the original probability function $P(\cdot)$.

More formally, for any set $A \in \mathbb{B}_\Omega$, where $\mathbb{B}_\Omega$ is a σ-field generated from Ω, we can define a probability function $P_X : \mathbb{B}_\Omega \to \mathbb{R}$

$$P_X(A) = P(C_A)$$
$$= P\left[s \in S : X(s) \in A\right],$$

where $C_A = \{s \in S : X(s) \in A\}$ is the set of basic outcomes in the original sample space S for which the values of random variable X fall in the set $A \subset \Omega$. Thus, a random variable X is a function that carries the probability from the original sample space S to a new space Ω of real numbers. In this sense, with $A \subset \Omega$, the probability $P_X(A)$ is often called an induced probability function. It can be shown that when S is a countable sample space, the induced probability function $P_X(\cdot)$ satisfies the three axioms of the probability function. (**Question**: How can one prove this?)

When S is continuous and so is uncountable (e.g., $S = \mathbb{R}$), we cannot use the above expressions immediately unless we can be sure that the set $C_A = \{s \in S : X(s) \in A\}$ belongs to the σ-algebra $\mathbb{B}$, which is associated with the original sample space S. Whether or not $C_A \in \mathbb{B}$ depends on, of course, the form of mapping $X : S \to \Omega$.

Question: What functional form $X(\cdot)$ will ensure $C_A \in \mathbb{B}$? Or what condition on $X(\cdot)$ will ensure that the set C_A belongs to $\mathbb{B}$?

The following measurability condition ensures that for any set $A \in \mathbb{B}_\Omega$, the set C_A always belongs to $\mathbb{B}$.

Definition 3.2. [Measurable Function]: A function $X : S \to \mathbb{R}$ is $\mathbb{B}$-measurable (or measurable with respect to the σ-field $\mathbb{B}$ generated from S) if for every real number a, the set $\{s \in S : X(s) \leq a\} \in \mathbb{B}$.

As a convention, a $\mathbb{B}$-measurable function is simply called a measurable function if it does not cause any confusion. When a function X is $\mathbb{B}$-measurable, we can express the probability of an event $A \subset \Omega$, say "$X \leq a$", in terms of the probability of an event C in $\mathbb{B}$, where $C = \{s \in S : X(s) \leq a\}$. In other words, the measurable function ensures that $P(X \in A)$ is always well-defined for all subsets A in $\mathbb{B}_\Omega$. If $X(\cdot)$ is not a measurable function, then there exist subsets in the σ-field in $\mathbb{R}$ for which probabilities are not defined. However, constructing such sets is very complicated and beyond the scope of this book.

Throughout this book, we assume that the random variable X is always measurable with respect to some σ-algebra of S. In fact, in the advanced probability theory, the term "random variable" is restricted to being a $\mathbb{B}$-measurable function from S to $\mathbb{R}$.

Theorem 3.1. *Let* $\mathbb{B}$ *be a σ-algebra associated with sample space S. Let $f(\cdot)$ and $g(\cdot)$ be $\mathbb{B}$-measurable real valued functions, and c be a real number. Then the functions $c \cdot f(\cdot), f(\cdot) + g(\cdot), f(\cdot) \cdot g(\cdot)$ and $|f(\cdot)|$ are also $\mathbb{B}$-measurable.*

Proof: See White (1984, Proposition 3.23), or Bartle (1966, Lemma 2.6).

If one starts with functions $X(s)$ and $Y(s)$ that are measurable mappings from S to Ω, then new functions $Z(s)$ can be constructed by ordinary algebraic operations, such as $Z(s) = aX(s), Z(s) = X(s) + Y(s), Z(s) = X(s)Y(s)$, and $Z(s) = X(s)/Y(s)$. These random variables are all measurable. If one has a sequence $X_1(s), X_2(s), \cdots$ of measurable functions and constructs $Z(s)$ through limiting operations such as $Z(s) = \lim_{i \to \infty} Z_i(s)$ or $Z(s) = \lim_{n \to \infty} \sup_{1 \le i \le n} |Z_i(s)|$, then $Z(s)$ is also measurable.

A proof of these claims is not particularly difficult, but is beyond this course. All that is important for our purposes is to know that standard functions are measurable and that any standard sequence of countable operations on such functions will not destroy measurability.

It can be shown that the induced probability $P_X(\cdot)$ of the random variable X satisfies the definition of a probability function. First, Condition (1) of a probability function in Definition 2.11 can be easily verified by observing that for $C_A = \{s \in S : X(s) \in A\}$,

$$1 \ge P_X(A) = P(C_A) \ge 0$$

given $0 \le P(C_A) \le 1$ for any $C_A \in \mathbb{B}$, where $\mathbb{B}$ is a σ-field generated from the original sample space S. Next, Condition (2) of a probability function also holds for $P_X(A)$ because $S = \{s \in S : X(s) = \Omega\}$ implies

$$P_X(\Omega) = P(S) = 1.$$

Finally, in discussing Condition (3) of a probability function, let us restrict our attention to two mutually exclusive events A_1 and A_2 in $\mathbb{B}_\Omega$. Here, the induced probability of $A_1 \cup A_2$ is given by

$$P_X(A_1 \cup A_2) = P(C),$$

where $C = \{s \in S : X(s) \in A_1 \cup A_2\}$. However, we can also write

$$C = \{s \in S : X(s) \in A_1\} \cup \{s \in S : X(s) \in A_2\}$$
$$= C_1 \cup C_2, \text{ say.}$$

We now claim that events C_1 and C_2 are disjoint sets in S. This must be so, because if some basic outcome $s_0 \in S$ were common to both C_1 and C_2, then $X(s_0) \in A_1$ and $X(s_0) \in A_2$. That is, the same real number $X(s_0)$

would belong to both A_1 and A_2. This is a contradiction because A_1 and A_2 are disjoint sets in Ω. Accordingly,

$$P(C) = P(C_1) + P(C_2).$$

By definition of induced probability, $P(C_1) = P_X(A_1)$ and $P(C_2) = P_X(A_2)$, and thus we have

$$P_X(A_1 \cup A_2) = P_X(A_1) + P_X(A_2).$$

This is Condition (3) for two disjoint sets A_1 and A_2 when defining a probability function.

The reason that we need the random variable X to be a $\mathbb{B}$-measurable function is to assure that we can find the well-behaved induced probabilities on the sigma field $\mathbb{B}_\Omega$ generated from the subsets of Ω. We need this requirement throughout this course for every function that is a random variable. In the rest of this book, we will abuse the notations for the original probability function $P(\cdot)$ and the induced probability function $P_X(\cdot)$; we will denote both probability functions as $P(\cdot)$.

Example 3.6. Suppose we throw three coins. Then the sample space

$$S = \{HHH, HTH, HHT, THH, THT, TTH, HTT, TTT\}.$$

Define a random variable $X(\cdot)$ to be the number of heads obtained from the experiment. Then the new sample space, or the range of X, is given by

$$\Omega = \{0, 1, 2, 3\}.$$

Now, suppose we are interested in calculating the probability that $P(0 \leq X \leq 1)$. Denote

$$C = \{s \in S : 0 \leq X(s) \leq 1\}$$
$$= \{TTT, TTH, THT, HTT\}.$$

It follows that

$$P(0 \leq X \leq 1) = P(C)$$
$$= P(TTT) + P(TTH) + P(THT) + (HTT)$$
$$= \frac{1}{2}.$$

3.2 Cumulative Distribution Function

Question: How to characterize a random variable X?

We can always use the induced probability function $P_X(A)$ for any $A \in \Omega_{\mathbb{B}}$ but this is not most convenient in most cases. Below, we introduce an alternative function to characterize the probability distribution of X. This is the so-called cumulative distribution function.

Definition 3.3. [**Cumulative Distribution Function (CDF)**]: The CDF of a random variable X is defined as:

$$F_X(x) = P(X \leq x) \text{ for all } x \in \mathbb{R}.$$

The subscript X of the F function indicates that it is the CDF of the random variable X.

We first examine the properties of the CDF $F_X(x)$.

Theorem 3.2. *[**Properties of** $F_X(\cdot)$]: Suppose $F_X(\cdot)$ is the CDF of some random variable X. Then*

(1) $\lim_{x \to -\infty} F_X(x) = 0, \lim_{x \to +\infty} F_X(x) = 1;$

(2) $F_X(x)$ *is non-decreasing, i.e., for any* $x_1 < x_2$, $F_X(x_1) \leq F_X(x_2);$

(3) $F_X(x)$ *is right-continuous, i.e., for all* x *and* $\delta > 0$,

$$\lim_{\delta \to 0^+} [F_X(x + \delta) - F_X(x)] = 0.$$

Theorem 3.3. *Let* $a < b$. *Then*

$$P(a < X \leq b) = F_X(b) - F_X(a).$$

Proof: Observing that the event $\{X \leq b\} = \{X \leq a\} \cup \{a < X \leq b\}$ and the events $\{X \leq a\}$ and $\{a < X \leq b\}$ are disjoint, we have

$$P(X \leq b) = P(X \leq a) + P(a < X \leq b).$$

The desired result follows immediately from the definition of the CDF $F_X(x)$.

Theorem 3.4. $P(X > b) = 1 - F_X(b).$

Proof: The result follows immediately from the formula $P(A^c) = 1 - P(A)$ and the definition of the CDF $F_X(x)$, with $A = \{X \leq a\}$.

Example 3.7. Suppose $F(x)$ is a CDF. Define $G(x) = 1 - F(-x)$. Is $G(x)$ a CDF too?

Solution: We verify whether the three properties of a CDF holds for function $G(\cdot)$:

(1) $G(-\infty) = 1 - F[-(-\infty)] = 1 - F(+\infty) = 1 - 1 = 0$,
  $G(+\infty) = 1 - F(-\infty) = 1 - 0 = 1$.
(2) For $x_1 < x_2$, we have

$$G(x_1) = 1 - F(-x_1),$$
$$G(x_2) = 1 - F(-x_2).$$
$$G(x_2) - G(x_1) = [1 - F(-x_2)] - [1 - F(-x_1)]$$
$$= F(-x_1) - F(-x_2)$$
$$\geq 0,$$

where the inequality follows because $-x_1 > -x_2$ and $F(\cdot)$ is non-decreasing. Thus, $G(\cdot)$ is non-decreasing.

(3) However, $G(x)$ is left-continuous but not necessarily right-continuous (why?). Thus, $G(\cdot)$ is not necessarily a CDF.

Example 3.8. Is $P(X \geq b) = 1 - F_X(b)$?

Solution: Given $\{X \geq b\} = \{X > b\} \cup \{X = b\}$ and the events $\{X > b\}$ and $\{X = b\}$ are disjoint, we have

$$P(X \geq b) = P(X > b) + P(X = b)$$
$$= 1 - F_X(b) + P(X = b).$$

Therefore, $P(X \geq b) = 1 - F_X(b)$ if and only if $P(X = b) = 0$.

Example 3.9. [Mixture of Distributions]: Suppose $F_1(x)$ and $F_2(x)$ are two CDF's. Is the linear combination

$$F(x) = pF_1(x) + (1 - p)F_2(x)$$

a CDF? Here p is a constant.

Solution: The answer is yes, if $0 \leq p \leq 1$. Here, the distribution $F(x)$ is usually called a mixture of distributions $F_1(x)$ and $F_2(x)$. A mixture of two distributions can provide a great deal of flexibility such as capturing skewness and heavy tails.

One possibility for a mixed distribution to arise is that in an observed data, some observations are generated from one distribution, and the remaining observations are generated from another distribution. Another possibility for a mixed distribution is that there exist two mutually exclusive states, state 1 and state 2, which arise with probabilities p and $1 - p$ respectively. The random variable X will follow the distribution $F_1(x)$ when state 1 occurs and will follow distribution $F_2(x)$ when state 2 occurs. Then the distribution $F(x)$ of X is a mixture of distributions $F_1(x)$ and $F_2(x)$.

A well-known example in econometrics is the so-called Markov regime-switching model in which p depends on a state variable characterizing the business cycles (e.g., Hamilton 1994).

In practice, p may depend on some economic variables. An example of $p = p(Z)$ is

$$p(Z) = \frac{1}{1 + \exp(-\alpha'Z)},$$

where Z is a state variable or vector of the economy.

When we repeat the same kind of a random experiment, the outcome of each trial will follow the same probability law or the same probability distribution. We now formally define the concept of identical distribution.

Definition 3.4. [Identical Distributions]: Two random variables X and Y are identically distributed if for every set A in $\mathbb{B}_\Omega$, where $\mathbb{B}_\Omega$ is the smallest σ-field containing all intervals of real numbers of the form $(a, b), [a, b), (a, b],$ and $[a, b]$, one has

$$P(X \in A) = P(Y \in A).$$

Question: Does the identical distribution imply $X = Y$?
The answer is no. This can be seen from the example below.

Example 3.10. Suppose a penny and a nickel are each tossed n times, and consider the following two definitions of X and Y respectively:

(1) X is the number of heads obtained with the penny, and Y is the number of heads obtained with the nickel.

(2) Both X and Y are the number of heads obtained with the penny.

In both cases, X and Y have the identical distribution. However, X and Y are independent in Case (1) while $X = Y$ in Case (2).

It is important to note that identical distribution does not imply $X = Y$, although $X = Y$ of course implies that X and Y have the same distribution.

Question: Why can the CDF $F_X(x)$ characterize the probability distribution of random variable X?

The definition of the CDF makes it clear that the probability function $P(\cdot)$ of the random variable X determines the CDF $F_X(\cdot)$. It is also true, although not so obvious, that a probability function $P(\cdot)$ of the random variable X can be found from a CDF $F_X(\cdot)$. That is, $P(\cdot)$ and $F_X(\cdot)$ contain the same information about the distribution of probability, and which function is used is a matter of convenience.

Theorem 3.5. *Two random variables X and Y are identically distributed if and only if*

$$F_X(x) = F_Y(x) \text{ for all } -\infty < x < \infty.$$

The concept of CDF has been widely used in economic analysis. Below we provide some examples which use the CDF to characterize income distributions and stochastic dominance.

Example 3.11. [Income Distribution and the Lorenz Curve]: In economics, the Lorenz curve and the Gini coefficient are two popular measures of income inequality. The Lorenz curve is a graphical representation of the CDF of the (empirical) probability distribution of income. It graphically shows that for the bottom $x\%$ of households or population, what percentage $y\%$ of the total income they have. Every point on the Lorenz curve represents a statement like "the bottom 20% of all households have 10% of the total income". A perfectly equal income distribution would be one in which every household has the same income. In this case, the bottom $x\%$ of the society would always have $x\%$ of the income for all $x \in [0, 100]$. This can be depicted by the 45-degree straight line $y = x$, called the "line of perfect equality". It is shown in Figure 3.2.

Example 3.12. [First Order Stochastic Dominance]: If two distributions $F(\cdot)$ and $G(\cdot)$ satisfy $F(x) \leq G(x)$ for all x on the real line, then we say that the distribution $F(\cdot)$ has first order stochastic dominance over distribution $G(\cdot)$.

Consider two probability distributions depicted in Figure 3.3:

$$F(x) = \begin{cases} 1 - e^{-x}, & \text{if } x \geq 0, \\ 0, & \text{if } x < 0, \end{cases}$$

and

$$G(x) = \begin{cases} 1 - e^{-2x}, & \text{if } x \geq 0, \\ 0, & \text{if } x < 0. \end{cases}$$

Then $F(x) \leq G(x)$ for all $x \in (-\infty, \infty)$ and so $F(\cdot)$ dominates $G(\cdot)$ in first order.

The first order stochastic dominance is widely used in decision analysis, welfare economics, finance, and so on. For example, a very common application of stochastic dominance is to the analysis of income distributions. If

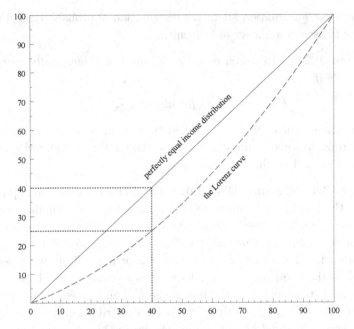

Figure 3.2: The Lorenz curve and income inequality

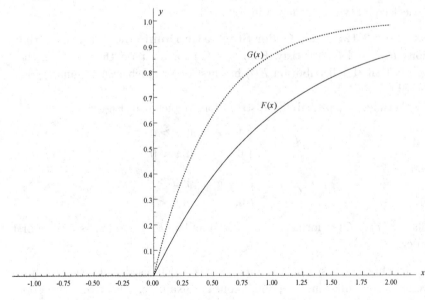

Figure 3.3: $F(\cdot)$ stochastically dominates $G(\cdot)$ in first order

x denotes an income level, then the inequality in the definition means that the proportion of individuals in distribution $F(\cdot)$ with income no greater than x is no larger than the proportion of such individuals in $G(\cdot)$. In other words, there is at least as high a proportion of poor people in $G(\cdot)$ as in $F(\cdot)$. Thus, $F(\cdot)$ stochastic dominates $G(\cdot)$ in first order. This implies that whatever poverty line we may choose, there is always more poverty in $G(\cdot)$ than in $F(\cdot)$. For another example, if portfolio $F(\cdot)$ has first order stochastic dominate over portfolio $G(\cdot)$, then $P(X_F > x) \geq P(X_G > x)$ for all x; that is, the probability that the return of portfolio $F(\cdot)$ exceeds any level x is always larger than the probability that the return of portfolio $G(\cdot)$ exceeds x. As a result, an expected utility maximizing agent will prefer $F(\cdot)$ over $G(\cdot)$.

Example 3.13. [Second Order Stochastic Dominance]: A probability distribution $F(\cdot)$ stochastically dominates another probability distribution $G(\cdot)$ in second order if, for all $x \in (-\infty, \infty)$,

$$\int_{-\infty}^{x} F(y)dy \leq \int_{-\infty}^{x} G(y)dy.$$

This definition requires the dominant distribution $F(\cdot)$ to have a smaller area beneath the distribution function for all x. Consider two probability distributions depicted in Figure 3.4:

$$F(x) = \begin{cases} 0, & \text{if } x < 1, \\ x - 1, & \text{if } 1 \leq x < 2, \\ 1, & \text{if } x \geq 2, \end{cases}$$

and

$$G(x) = \begin{cases} 0, & \text{if } x < 0, \\ \frac{x}{3}, & \text{if } 0 \leq x < 3, \\ 1, & \text{if } x \geq 3. \end{cases}$$

Then distribution $F(\cdot)$ stochastically dominates distribution $G(\cdot)$ in second order. Risk-averse economic agents will always prefer distribution $F(\cdot)$ because for any increasing and concave utility function $u(\cdot)$, we have $\int_{-\infty}^{\infty} u(x)dF(x) \geq \int_{-\infty}^{\infty} u(x)dG(x)$ if and only if $F(\cdot)$ has second order stochastic dominance over $G(\cdot)$.

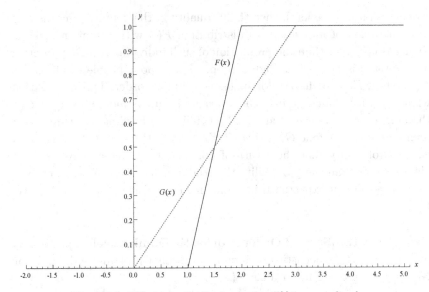

Figure 3.4: $F(\cdot)$ stochastically dominates $G(\cdot)$ in second order

3.3 Discrete Random Variables

We now consider two basic categories of random variables. The first is called discrete random variables.

Definition 3.5. [**Discrete Random Variables (DRV)**]: If a random variable X can only take a countable number of values, then X is called a DRV.

Example 3.14. For a discrete random variable X, the associated sample space Ω must contain only a countable number of basic outcomes. For example, $\Omega = \{1, 2, 3, 4, 5, 6\}$ or $\Omega = \{0, 1, 2, \cdots\}$.

The probability distribution of a discrete random variable X can also be described by the probability mass function defined below.

Definition 3.6. [**Probability Mass Function (PMF)**]: The PMF of a DRV X is defined as

$$f_X(x) = P(X = x) \text{ for all } x \in \mathbb{R}.$$

The PMF $f_X(x)$ is also called a probability function. We first examine the properties of the PMF $f_X(x)$.

Theorem 3.6. *[Properties of PMF]:*

(1) $0 \leq f_X(x) \leq 1$ *for all* $x \in \mathbb{R}$;

(2) $\sum_{x \in \Omega} f_X(x) = 1$.

Definition 3.7. [**Support**]: The collection of the points on the real line $\mathbb{R}$ at which a DRV X has a positive probability is called the support of X, denoted as

$$\text{Support}(X) = \{x \in \mathbb{R} : f_X(x) > 0\}.$$

Therefore, we have

$$\text{Support}(X) = \Omega.$$

Intuitively, the support of X is the set of all possible values that X can take with strictly positive probability. Although $f_X(x)$ is defined on the entire real line $\mathbb{R}$, it suffices to know the support of a DRV X and the probabilities of all points in the support.

The PMF can be represented graphically via a so-called probability histogram.

Definition 3.8. [**Probability Histogram**]: A probability histogram is a plot to represent a discrete probability distribution where rectangles are constructed so that their bases of equal width are centered at each value x and their heights are equal to the corresponding probabilities given by the PMF. The bases are constructed so as to have no space between the rectangles.

Among other things, histogram can show whether there are more than one mode, or "high points". A mode is a bar in a histogram that is surrounded by bars of lower frequency. A histogram exhibiting two modes is said to be bimodal, and one having more than two modes is said to be multimodes.

Question: Given the PMF $f_X(x)$ of a DRV X, what information can we extract from it?

Theorem 3.7. *Suppose* $f_X(x)$ *is the PMF of a DRV X. Then CDF of a DRV X is*

$$F_X(x) = P(X \leq x)$$
$$= \sum_{y \leq x} f_X(y) \text{ for any } x \in \mathbb{R},$$

where the summation is over all values y *in* Ω *that are less than or equal to* x.

This theorem provides a method to compute the CDF $F_X(\cdot)$ from the PMF $f_X(\cdot)$. That is, given the PMF $f_X(\cdot)$, we can obtain the CDF $F_X(x)$ by summing up $f_X(y)$ over all possible y in Ω such that $y \leq x$. Note that the argument of $F_X(\cdot)$ runs continuously from $-\infty$ to $+\infty$, i.e., $F_X(x)$ is well-defined for all $x \in (-\infty, \infty)$ in the entire real line. The CDF $F_X(\cdot)$ is not merely defined over the support of X, which is only a set of countable numbers on the real line.

Example 3.15. Suppose a random variable X follows the following probability distribution:

x	1	2	3	4
$f_X(x)$	0.1	0.3	0.4	0.2

The probability histogram is shown in Figure 3.5.

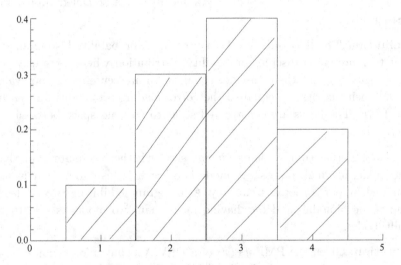

Figure 3.5: Probability distribution

Then its CDF is:

$$F_X(x) = \begin{cases} 0, & x < 1, \\ 0.1, & 1 \leq x < 2, \\ 0.4, & 2 \leq x < 3, \\ 0.8, & 3 \leq x < 4, \\ 1, & x \geq 4. \end{cases}$$

Example 3.16. Suppose a random variable X has the following PMF

$$f_X(x) = \begin{cases} \frac{1}{N}, & x = 1, \cdots, N, \\ 0, & \text{otherwise.} \end{cases}$$

Find its CDF $F_X(x)$.

Solution: This is called a discrete uniform distribution because each of the N integers $\{1, \cdots, N\}$ is equally likely to occur. To compute the CDF $F_X(x)$, where x ranges from $-\infty$ to ∞, we divide the real line $\mathbb{R}$ into $N+1$ intervals:

Case 1: $x < 1$. Then the event $\{X \leq x\}$ is an empty set $\varnothing$, and

$$F_X(x) = \sum_{x_i \leq x} f_X(x_i) = 0.$$

Case 2: $1 \leq x < 2$. Then the event $\{X \leq x\} = \{1\}$, and so

$$F_X(x) = \sum_{x_i \leq x} f_X(x_i) = f_X(1) = \frac{1}{N}.$$

Case 3: $2 \leq x < 3$. Then the event $\{X \leq x\} = \{1, 2\}$, and

$$F_X(x) = \frac{2}{N}.$$

Case j: $j-1 \leq x < j$, $2 \leq j \leq N$. Then the event $\{X \leq x\} = \{1, \cdots, j-1\}$, and

$$F_X(x) = \frac{j-1}{N}.$$

Case $N+1$: $x \geq N$. Then the event $\{X \leq x\} = \{1, \cdots, N\}$, and so

$$F_X(x) = 1.$$

To sum up, we have

$$F_X(x) = \begin{cases} 0, & x < 1, \\ j/N, \, j \leq x < j+1, 1 \leq j < N, \\ 1, & x \geq N. \end{cases}$$

This function is a step function, where jumps occur at the points with strictly positive probabilities, that is, when jumps occur at the points contained in the support of X. See Figure 3.6 (for the case of $N = 6$).

Since CDF $F_X(x)$ is a function defined on the entire real line, it is incomplete if one just computes $F_X(x)$ on the support of X, which contains $\{1, \cdots, N\}$ in the above example.

Question: Suppose X is a DRV with CDF $F_X(x)$, is it possible to obtain $f_X(x)$ from $F_X(x)$?

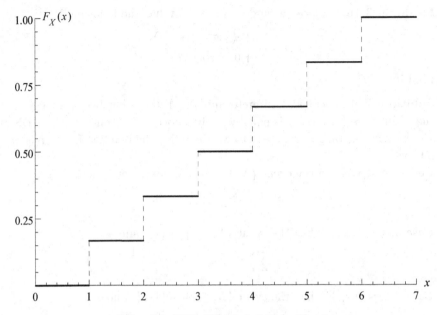

Figure 3.6: CDF of Example 3.16 for $N = 6$

For a DRV X, $F_X(x)$ is a step function. The points $\{x_i\}$ where $f_X(x_i) > 0$ will be the jump points for $F_X(x)$. Without loss of generality, we can arrange these points in an increasing sequential order, that is, $x_1 < x_2 < \cdots$. Then observing that the event $\{x_{i-1} < X \leq x_i\}$ only contains the point x_i, we can obtain the probability

$$P(X = x_i) = f_X(x_i)$$
$$= P(x_{i-1} < X \leq x_i)$$
$$= F_X(x_i) - F_X(x_{i-1}), \ i = 2, 3, \ldots .$$

For $i = 1$, we have $P(X = x_1) = f_X(x_1) = F_X(x_1)$ because $f_X(x_1) = P(X = x_1) = P(X \leq x_1) = F_X(x_1)$ when x_1 is the smallest value that X can take.

We state the above result in the following theorem:

Theorem 3.8. *Suppose X is a DRV with CDF $F_X(x)$, and its support contains a sequence of points $\{x_1 < x_2 < \cdots\}$. Then its PMF*

$$f_X(x_i) = \begin{cases} F_X(x_i), & i = 1, \\ F_X(x_i) - F_X(x_{i-1}), & i > 1. \end{cases}$$

This theorem shows that one can obtain the PMF $f_X(x)$ by differencing the CDF $F_X(x)$. Therefore, $f_X(x)$ and $F_X(x)$ are equivalent ways to describe the DRV X. Given $f_X(x)$ or $F_X(x)$, we know the entire probability law of X.

The above theorem implies that for $i > 1$,

$$F_X(x_i) = F_X(x_{i-1}) + f_X(x_i).$$

This suggests that for the DRV X, the CDF $F_X(x)$ always have jumps at points with strictly positive probabilities. Therefore, for the DRV X, $F_X(x)$ is always a step function with jumps occurring at points with positive probabilities.

3.4 Continuous Random Variables

Next, we consider the second basic category of random variables called continuous random variables.

Definition 3.9. [Continuous Random Variables (CRV)]: A random variable X is called continuous if its distribution function $F_X(x)$ is continuous for all x in the real line. In contrast, a random variable X is discrete if $F_X(x)$ is a step function of x.

A CRV X provides a proper description of income, temperature, stock return, and so on.

Question: Can we define a PMF $f_X(x)$ for a CRV X?

Obviously, for any constant $\epsilon > 0$, we have $\{X = x\} \subset \{x - \frac{\epsilon}{2} < X \leq x + \frac{\epsilon}{2}\}$, we have

$$
\begin{aligned}
0 &\leq P(X = x) \\
&\leq P\left(x - \frac{\epsilon}{2} < X \leq x + \frac{\epsilon}{2}\right) \\
&= F_X\left(x + \frac{\epsilon}{2}\right) - F_X\left(x - \frac{\epsilon}{2}\right) \\
&\to 0 \text{ as } \epsilon \to 0
\end{aligned}
$$

by continuity of $F_X(\cdot)$. Thus,

$$P(X = x) = 0 \text{ for all } x.$$

That is, if X is a continuous random variable, the probability that X takes a single point is zero. To understand this intuitively, let us consider an analogous example of a satellite flying over the U.S. continental territories. Suppose it takes one hour for the satellite to fly over the U.S. continental territories, takes 10 minutes to fly over the New York state, and takes 0.1 second to fly over Ithaca. It is conceivable that it takes almost zero second for the satellite to fly over Uris Hall at Cornell University.

The result that $P(X = x) = 0$ for all x for a CRV X has important implications. For example, the PMF $f_X(\cdot)$ is not a useful concept for a CRV because it is a singular function. For another example, for a CRV X, we have

$$P(a < X \leq b) = P(a \leq X < b)$$
$$= P(a \leq X \leq b).$$

Question: Because the PMF $f_X(\cdot)$ is not a useful concept to describe a CRV X, we have to look for an alternative tool. We now ask the following question: under what conditions, can we write $F_X(x) = \int_{-\infty}^{x} f_X(y)dy$ for some function $f_X(x)$? And if so, what is the interpretation for such a function $f_X(x)$?

Definition 3.10. [Absolute Continuity (AC)]: A function $F : \mathbb{R} \to \mathbb{R}$ is called absolutely continuous with respect to the Lebesgue measure if $F(x)$ is continuous on $\mathbb{R}$ and is differentiable almost everywhere (i.e. for almost all x).

What is meant by "almost everywhere"? Intuitively, in any finite interval of $\mathbb{R}$, there are a finite number of points or an infinite but countable number of points where $F_X(x)$ is not differentiable. Note that a continuously differentiable function (i.e., a function is differentiable everywhere and the derivative is continuous) is absolutely continuous.

Question: What is an intuitive explanation for the Lebesgue measure?

Definition 3.11. [Probability Density Function (PDF)]: Suppose the distribution function $F_X(x)$ of a CRV X is absolutely continuous. Then there exists a function $f_X(x)$ such that

$$F_X(x) = \int_{-\infty}^{x} f_X(y)dy \text{ for all } x \in (-\infty, \infty).$$

The function $f_X(x) : \mathbb{R} \to \mathbb{R}$ is called a PDF of X.

The relationship in the above definition is a fundamental result in calculus. When the absolute continuity condition for $F_X(x)$ does not hold, X may not have the above relationship. Throughout this course, we will assume that $F_X(x)$ is absolutely continuous.

For those x's where the derivative $F'_X(x)$ exists, we have from the above definition that

$$f_X(x) = \frac{dF_X(x)}{dx} = F'_X(x).$$

Given that $F_X(x)$ is absolutely continuous, the PDF $f_X(x)$ exists almost everywhere. Because $f_X(x)$ is a slope of the CDF $F_X(x)$, it can take values greater than 1.

Question: What is the interpretation of the PDF $f_X(x)$?

For any small constant $\epsilon > 0$, we have, by the mean value theorem,

$$P\left(x - \frac{\epsilon}{2} < X \leq x + \frac{\epsilon}{2}\right) = F_X\left(x + \frac{\epsilon}{2}\right) - F_X\left(x - \frac{\epsilon}{2}\right)$$
$$= \int_{x-\frac{\epsilon}{2}}^{x+\frac{\epsilon}{2}} f_X(y)dy$$
$$= f_X(\bar{x})\epsilon$$

for some $\bar{x} \in (x - \frac{\epsilon}{2}, x + \frac{\epsilon}{2}]$. Although $f_X(x)$ itself is not a probability measure, it is proportional to the probability that X takes values in a small interval centered at point x. Thus, $f_X(x)$ characterizes the relative magnitude of the probability that X takes values in a small interval centered at x. In particular, the plot of $f_X(x)$ can be used to describe the shape of the probability distribution of a CRV X. It is indicative of mode, skewness or symmetry, heavy tails, and etc. Geometrically, the probability that a CRV X takes values in an interval, say $(a, b]$, is the area under the curve of PDF $f_X(x)$ from $x = a$ to $x = b$, as shown in Figure 3.7.

For a CRV X with well-defined PDF $f_X(x)$, $f_X(x)$ and $F_X(x)$ are equivalent to each other in the following sense: given one function, we can always recover the other. Specifically, we can obtain the CDF $F_X(x)$ by integrating the PDF $f_X(y)$ for y from $-\infty$ to x. Given the CDF $F_X(x)$, we can obtain the PDF $f_X(x)$ by differentiating $F_X(x)$. Mathematically speaking, these operations of integration and differentiation are similar to those of summation and difference in the case of discrete random variables.

Question: Is $f_X(x)$ a unique function for a given $F_X(x)$?

Given a CDF $F_X(x)$, we can obtain the PDF $f_X(x) = F'_X(x)$ at the points where $F_X(x)$ is differentiable. When $F_X(x)$ differentiable on the entire real line, $f_X(x)$ is uniquely determined. However, when $F_X(x)$ is not differentiable at some points, $f_X(x)$ is not defined at those points.

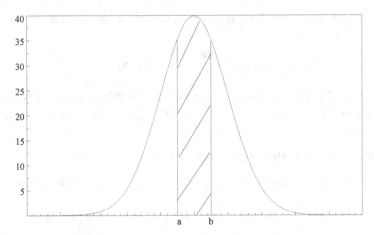

Figure 3.7: Geometric interpretation for probability that a CRV X takes values in $(a, b]$

Question: How to define the values of PDF $f_X(x)$ at the points where $F'_X(x)$ does not exist?

We can define $f_X(x)$ arbitrarily at those points. The fact that $P(X = x) = 0$ for a CRV X allows us to change the value of the PDF $f_X(x)$ of a continuous random variable X at a single point or a sequence of infinite but countable points without altering the probability distribution of X. For instance, consider the PDF

$$f_X(x) = \begin{cases} e^{-x} & \text{for } 0 < x < \infty, \\ 0 & \text{elsewhere,} \end{cases}$$

and the PDF

$$f_X(x) = \begin{cases} e^{-x} & \text{for } 0 \le x < \infty, \\ 0 & \text{elsewhere.} \end{cases}$$

These two PDF's have the same CDF $F_X(x)$ or equivalently the same probability measure $P_X(A)$ for any subset A on the real line. We observe that these two PDF's differ only at point $x = 0$ and $P(X = 0) = 0$. More generally, if two PDF's differ only at a finite number of points or an infinite but countable sequence of points, their corresponding CDF's or probability set functions are exactly the same. (In contrast, the PMF of a DRV cannot be changed at any point in the support, since such a change will alter the probability distribution of a DRV.)

To the extent just described, the PDF $f_X(x)$ is not unique. In many problems, however, there will be one version of the PDF $f_X(x)$ that is more natural than any other PDF. For this reason, the PDF $f_X(x)$ will, wherever possible, be made continuous on the real line $\mathbb{R}$.

We now examine the properties of the PDF $f_X(x)$.

Theorem 3.9. [Properties of PDF]: *A function $f_X(x)$ is a PDF of a CRV X if and only if*

(1) $f_X(x) \geq 0$ *for all $x \in \mathbb{R}$;*

(2) $\int_{-\infty}^{\infty} f_X(x)dx = 1$.

Proof: [*Necessity*] Suppose $f_X(\cdot)$ is a PDF. Then $F_X(x) = \int_{-\infty}^{x} f_X(y)dy$ is a CDF. By the mean value theorem, $F_X(x + \delta) - F_X(x) = f_X(\bar{x})\delta$, where $\bar{x}$ lies between x and $x + \delta$. It follows that $f_X(\bar{x}) \geq 0$ since $F_X(x)$ is non-decreasing. Moreover, Property (2) follows because $F_X(\infty) = 1$.

[*Sufficiency*] The converse can be proved similarly by constructing a function $F_X(x) = \int_{-\infty}^{x} f_X(y)dy$ and showing that $F_X(x)$ is a CDF of X. First, $F_X(x)$ is non-decreasing by the mean valued theorem and $f_X(x) \geq 0$; second, $F_X(-\infty) = 0$ and $F_X(\infty) = 1$ by $\int_{-\infty}^{\infty} f_X(x)dx = 1$. Also, $F_X(x) = \int_{-\infty}^{x} f_X(y)dy$ implies that $F_X(x)$ is continuous, because

$$F_X(x + \delta) - F_X(x) = \int_{x}^{x+\delta} f_X(y)dy \to 0 \text{ as } \delta \to 0,$$

and so it is right-continuous. Therefore, $F_X(x)$ is a CDF.

Example 3.17. Is it possible to construct a PDF $f(x)$ from any nonnegative function $g(x)$ with a finite integral (i.e., $0 < \int_{-\infty}^{\infty} g(x)dx < \infty$)?

Solution: The answer is yes, by defining

$$f(x) = \frac{g(x)}{\int_{-\infty}^{\infty} g(y)dy}.$$

Example 3.18. Is it possible to construct a symmetric PDF $f(x)$ about zero from any nonnegative function $g(x)$ with a finite integral? Recall that function $f(x)$ is symmetric about zero if $f(x) = f(-x)$ for all $x \in \mathbb{R}$.

Solution: Yes. We can construct a symmetric PDF $f(x)$ from any nonnegative function $g(x)$ by defining

$$f(x) = \frac{g(x) + g(-x)}{2\int_{-\infty}^{\infty} g(y)dy}.$$

Definition 3.12. [Support]: The support of a CRV X is defined as

$$\text{Support}(X) = \{x \in \mathbb{R} : f_X(x) > 0\},$$

where $f_X(x)$ is the PDF of X.

The support of a CRV X is the set of all possible points on $\mathbb{R}$ with strictly positive PDF $f_X(x)$. This implies that the probability that a CRV X takes values in a small neighborhood of any point in its support is always positive. In contrast, the probability that X takes values in some small neighborhood of any point outside the support will be zero. Thus, it suffices to focus on the support of X when calculating the probabilities of a CRV.

We now state a lemma and discuss its implications.

Example 3.19. [Location-Scale Family]: Let $f(x)$ be a PDF, and let μ and $\sigma > 0$ be any given constants. Then the function

$$g(x) = \frac{1}{\sigma} f\left(\frac{x - \mu}{\sigma}\right)$$

is a PDF.

Proof: By change of variable.

Each pair of (μ, σ) delivers a valid PDF. The family of $f(x - \mu)$, indexed by the parameter μ, is called the location family with standard PDF $f(x)$, where the parameter μ is called the location parameter for the family. On the other hand, the family of $\frac{1}{\sigma} f(\frac{x}{\sigma})$, indexed by the parameter σ, is called the scale family with standard PDF $f(x)$. where the parameter σ is called the scale parameter for the family. Finally, the family of $\frac{1}{\sigma} f(\frac{x-\mu}{\sigma})$, indexed by parameters (μ, σ), is called the location-scale family with standard PDF $f(x)$, where μ and σ are called the location and scale parameters respectively. The effect of introducing the location parameter μ is to shift the probability distribution so that the point that was above 0 is now above μ, and the effect of introducing the scale parameter is to stretch or contract the probability distribution with the scale parameter σ.

If X is a CRV, the PDF $f_X(x)$ has at most a finite number of discontinuities or an infinite but countable number of discontinuities in every finite interval. This means that (1) the CDF $F_X(x)$ is everywhere continuous, and (2) the derivative of $F_X(x)$ exists and is equal to $f_X(x)$ at each continuity point of $f_X(x)$. That is, $F'_X(x) = f_X(x)$ at each continuity point of $f_X(x)$.

If X is a DRV, the PMF $f_X(x)$ is not the derivative of the CDF $F_X(x)$ with respect to x. However, the PMF $f_X(x)$ is the Radon-Nikodym derivative of $F_X(x)$ with respect to some counting measure (more precisely, it is the Radon-Nikodym derivative of $F_X(x)$ with respect to the Lebesgue measure). Because a derivative is often called a density, the PMF $f_X(x)$ can also be called a PDF in a broader sense.

We have restricted ourselves to random variables of either the discrete or continuous type, so that we can use the PMF and PDF respectively. This affords an enormous simplification, but it should be recognized that this simplification is obtained at a considerable cost from a mathematical point of view. Not only shall we exclude from consideration many random variables that do not have these types of distributions, but we shall also exclude many interesting subsets of the space. One example of random variables which we have ruled out here is a random variable with a mixture of both discrete and continuous components.

Definition 3.13. [Mixed Distribution of Discrete and Continuous Components]: A random variable X is said to follow a mixed distribution if its CDF is discontinuous at each point having a nonzero probability and continuous elsewhere. As in a DRV case, the height of the step at a discontinuity point gives the probability that X will take on that particular value and otherwise X behaves like a CRV elsewhere. For a mixed random variable X of discrete and continuous components, the CDF is the weighted sum of two CDF's, one of which is the CDF of a DRV, and the other is the CDF of a CRV, and the weights have to be nonnegative and sum to unity. Figure 3.8 plots the CDF of a mixed distribution of discrete and continuous components.

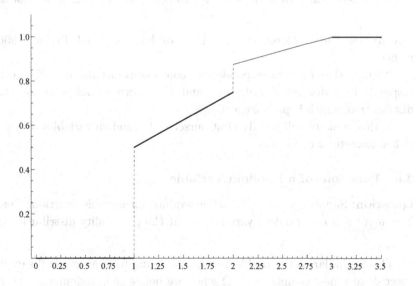

Figure 3.8: CDF of a mixed distribution of discrete and continuous components

Distributions that are mixtures of the discrete and continuous types do occur frequently in practice. For example, an economic random variable X^* may follow a continuous distribution, but we observe X^* when $X^* > c$ and observe value c if $X^* \leq c$. In this case, the observed variable

$$X = \begin{cases} X^* \text{ if } X^* > c, \\ c \ \text{ if } X^* \leq c, \end{cases}$$

follows a mixture distribution of discrete and continuous components because $P(X = c) > 0$ and $P(X = x) = 0$ for all $x > c$. This is called a left-censored distribution. Such distributions arise when X^* is the unemployment spell of a layoff worker, the reservation wage of an employee, and etc.

We now state a general result.

Theorem 3.10. *[Lebesgue's Decomposition]: Any CDF $F_X(x)$ may be written in the form*

$$F_X(x) = \sum_{i=1}^{3} a_i F_i(x),$$

where the weights $a_i \geq 0$, $i = 1, 2, 3$, $a_1 + a_2 + a_3 = 1$, $F_1(x)$ is "absolutely continuous", $F_2(x)$ is a "step function" with a finite or countably infinite number of "jumps", and $F_3(x)$ is a "singular" function, that is, it is continuous with zero derivative "almost everywhere" (i.e. at almost all points).

Proof: See Parzen (1960, pp. 170, 174); or Kingman and Taylor (1966, p. 294).

We note that $F_1(\cdot)$ corresponds to a continuous distribution, $F_2(\cdot)$ corresponds to a discrete distribution, and $F_3(\cdot)$ corresponds to a singular distribution, which is pathological.

In this book we will mainly limit ourselves to random variables that are either discrete or continuous.

3.5   Functions of a Random Variable

Question: Suppose $g : \mathbb{R} \to \mathbb{R}$ is a real-valued measurable function. Then $Y = g(X)$ is a new random variable. What the probability distribution of Y?

This is similar to the problem of transforming from an original sample space S to a new sample space Ω when we define the random variable in Section 3.1.

We now provide a few examples to illustrate why it is important to study the probability distribution of $Y = g(X)$.

Example 3.20. [**Consumption Function**]: $Y = g(X)$, where X is income and Y is consumption. Often it is assumed that consumption is a linear function of income, i.e.,

$$Y = \alpha + \beta X.$$

Example 3.21. [**Options Prices**]: The European call option price Y is a nonlinear function of volatility X of an underlying stock price according to the well-known Black-Scholes (1973) formula:

$$Y = S_0 \Phi(d_1) - K e^{-rT} \Phi(d_2),$$

where

$$d_1 = \frac{\ln(S_0/K) + (r + X^2/2)T}{X\sqrt{T}},$$

$$d_2 = \frac{\ln(S_0/K) + (r - X^2/2)T}{X\sqrt{T}}.$$

The function $\Phi(x) = \int_{-\infty}^{x} \frac{1}{\sqrt{2\pi}} e^{-u^2/2} du$ is the CDF for a standard normal distribution (see Chapter 4), S_0 is the stock price at time zero, K is the strike price, r is the continuously compounded risk-free interest rate, and T is the time to maturity of the option.

Example 3.22. [**Asset Returns**]: Suppose P_t is an asset price in period t. Then the return

$$X_t = \ln(P_t) - \ln(P_{t-1})$$

$$\approx \frac{P_t - P_{t-1}}{P_{t-1}}$$

is approximately the relative price change from period $t - 1$ to period t.

To find the probability distribution of $Y = g(X)$, we consider the cases where X is a DRV and a CRV respectively.

3.5.1   The Discrete Case

Question: How to find the PMF $f_Y(y)$ given the PMF $f_X(x)$ of a DRV X?

For a DRV X, a general method is to use the formula:

$$f_Y(y) = \sum_{x \in \Omega_X(y)} f_X(x),$$

where for any given y, $\Omega_X(y) = \{x \in \Omega_X : g(x) = y\}$ is the set of all possible values of x in the support Ω_X of X that obey the restriction $g(x) = y$.

Intuitively, $Y = g(X)$ is a transformation from the sample space Ω_X to a new sample space Ω_Y (say). The PMF $f_Y(y)$ of Y can be defined as an induced probability function using the PMF $f_X(x)$ of X, and this is exactly the formula

$$f_Y(y) = P(Y = y) = P[X \in \Omega_X(y)].$$

Example 3.23. Suppose X follows the probability distribution:

X	-2	-1	0	1	2
$f_X(x)$	0.2	0.1	0.1	0.3	0.3

Find the PMF of $Y = X^2 + X$.

Solution: For $X = -2, -1, 0, 1, 2$, $Y = X^2 + X = 2, 0, 0, 2, 6$ respectively. Thus, the support of Y is $\Omega_Y = \{0, 2, 6\}$. Also,

$$P(Y = 0) = P_X(X = -1) + P_X(X = 0) = 0.2,$$
$$P(Y = 2) = P_X(X = -2) + P_X(X = 1) = 0.5,$$
$$P(Y = 6) = P_X(X = 2) = 0.3.$$

It follows that the probability distribution of Y is:

Y	0	2	6
$f_Y(y)$	0.2	0.5	0.3

3.5.2 The Continuous Case

We now assume that $g(\cdot)$ is a continuous function so that Y is a CRV when X is a CRV. Then what is the PDF $f_Y(y)$ of Y given the PDF $f_X(x)$ of X?

Below we discuss two important methods.

(1) The CDF Approach

The basic idea is first to find the CDF $F_Y(y)$ of Y and then its PDF $f_Y(y) = F_Y'(y)$ by differentiation.

Step 1: Find the expression of $F_Y(y)$ in terms of $F_X(x)$:

$$F_Y(y) = P(Y \leq y)$$
$$= P[g(X) \leq y]$$
$$= P[X \in \Omega_{g^{-1}}(y)],$$

where

$$\Omega_{g^{-1}}(y) = \{x \in \Omega_X : g(x) \le y\}$$

is a subset in Ω_X that contains all x's satisfying the inequality $g(x) \le y$.

The trick here is to transform the probability statement about Y into a probability statement about X, using $Y = g(X)$.

Step 2: Differentiating the CDF $F_Y(y)$ with respect to y, we obtain

$$f_Y(y) = F_Y'(y).$$

Step 3: Check if $f_Y(y)$ is a PDF (i.e., check $f_Y(y) \ge 0$ for all $y \in (-\infty, \infty)$, and $\int_{-\infty}^{\infty} f_Y(y)dy = 1$).

Below, we illustrate this general method via a variety of numerical examples.

Example 3.24. Suppose a CRV X has a PDF

$$f_X(x) = \begin{cases} 1, & -\frac{1}{2} < x < \frac{1}{2}, \\ 0, & \text{otherwise.} \end{cases}$$

Find the PDF $f_Y(y)$ for each of the following random variables:
(1) $Y = a + bX, \quad b \ne 0$;
(2) $Y = X^2$;
(3) $Y = |X|$.

Solution: It is important to identify first the possible values of the curve Y (i.e., the support of Y). For this purpose, it is extremely useful to draw a plot of $Y = g(X)$.

(1) Given $Y = a + bX$, we have

$$\begin{aligned} F_Y(y) &= P(Y \le y) \\ &= P(a + bX \le y) \\ &= P(bX \le y - a). \end{aligned}$$

Since the division by b will change the direction of inequality when $b < 0$, we consider two cases: $b > 0$ and $b < 0$ respectively.

Case (i) $b > 0$: We have

$$\begin{aligned} F_Y(y) &= P(Y \le y) \\ &= P(bX \le y - a) \\ &= P[X \le (y - a)/b] \\ &= F_X\left(\frac{y - a}{b}\right). \end{aligned}$$

It follows that

$$F_Y(y) = F_X\left(\frac{y-a}{b}\right)$$
$$= F_X(z),$$

where $z = (y-a)/b$. Then using the chain rule of differentiation, we obtain

$$f_Y(y) = F_Y'(y)$$
$$= F_X'(z)\frac{dz}{dy}$$
$$= f_X(z)\frac{1}{b}$$
$$= f_X\left(\frac{y-a}{b}\right)\frac{1}{b}$$
$$= 1 \cdot \frac{1}{b} \quad \text{if} \quad -\frac{1}{2} < \frac{y-a}{b} < \frac{1}{2}$$

It follows that

$$f_Y(y) = \begin{cases} \frac{1}{b}, & a - \frac{b}{2} < y < a + \frac{b}{2} \\ 0, & \text{otherwise} \end{cases}$$

The distribution of X is called a uniform probability distribution on $(-\frac{1}{2}, \frac{1}{2})$ because its PDF is a constant over $(-\frac{1}{2}, \frac{1}{2})$. Because Y is a linear transformation of X, it also follows a uniform distribution on $(a - \frac{b}{2}, a + \frac{b}{2})$. The parameters a and b shift the mean and change the scale of the uniform distribution respectively, where a is called a location parameter, and b is a scale parameter.

Case (ii) $b < 0$: We have

$$P(Y \leq y)$$
$$= P(bX \leq y - a)$$
$$= P\left(X \geq \frac{y-a}{b}\right)$$
$$= 1 - F_X\left(\frac{y-a}{b}\right).$$

It follows that $F_Y(y) = 1 - F_X(z)$, where $z = (y-a)/b$. By differentiation, we obtain

$$f_Y(y) = 0 - F_X'(z)\frac{dz}{dy}$$

$$= -f_X(z)\frac{1}{b}$$

$$= -f_X\left(\frac{y-a}{b}\right)\frac{1}{b}$$

$$= -\frac{1}{b} \text{ if } 0 < \frac{y-a}{b} < 1.$$

It follows that the PDF of Y is given by

$$f_Y(y) = \begin{cases} -\frac{1}{b}, & a + \frac{b}{2} < y < a - \frac{b}{2}, \\ 0, & \text{otherwise.} \end{cases}$$

(2) Given $Y = X^2$ and the support$(X) = (-\frac{1}{2}, \frac{1}{2})$, Y always takes nonnegative values between $[0, \frac{1}{4})$. Let $y \geq 0$. Then we have

$$\begin{aligned} F_Y(y) &= P(Y \leq y) \\ &= P(X^2 \leq y) \\ &= P(-\sqrt{y} \leq X \leq \sqrt{y}) \\ &= F_X(\sqrt{y}) - F_X(-\sqrt{y}), \end{aligned}$$

where we have made use of the fact that the event $\{X^2 \leq y\}$ is equivalent to the event $\{-\sqrt{y} \leq X \leq \sqrt{y}\}$. By differentiation,

$$\begin{aligned} f_Y(y) &= \frac{d}{dy}[F_X(\sqrt{y}) - F_X(-\sqrt{y})] \\ &= F_X'(\sqrt{y})\left(\frac{1}{2\sqrt{y}}\right) - F_X'(-\sqrt{y})\left(-\frac{1}{2\sqrt{y}}\right) \\ &= f_X(\sqrt{y})\frac{1}{2\sqrt{y}} + f_X(-\sqrt{y})\frac{1}{2\sqrt{y}}. \end{aligned}$$

Thus, we have

$$f_Y(y) = \begin{cases} \frac{1}{\sqrt{y}}, & 0 < y < \frac{1}{4}, \\ 0, & \text{otherwise.} \end{cases}$$

(3) Given $Y = |X|$, Y is nonnegative. Letting $y \geq 0$, we have

$$\begin{aligned} F_Y(y) &= P(Y \leq y) \\ &= P(|X| \leq y) \\ &= P(-y \leq X \leq y) \\ &= F_X(y) - F_X(-y). \end{aligned}$$

By differentiation,

$$f_Y(y) = F'_X(y) - F'_X(-y)(-1)$$
$$= f_X(y) + f_X(-y).$$

It follows that

$$f_Y(y) = \begin{cases} 2, 0 \leq y < \frac{1}{2}, \\ 0, \text{ otherwise.} \end{cases}$$

This is a uniform density on $[0, \frac{1}{2})$. That is, the absolute value of a uniform random variable is still uniform (but on a smaller interval).

Example 3.25. Suppose a random variable X has a PDF

$$f_X(x) = \frac{1}{2}\alpha e^{-\alpha|x|}, \quad -\infty < x < \infty,$$

where $\alpha > 0$. This is called the double exponential (or Laplace) distribution. Find the PDF for each of the following transformations:
(1) $Y = |X|$;
(2) $Y = X^2$.

Solution: (1) Since $Y = |X|$ is always nonnegative, letting $y \in [0, \infty)$, we have

$$F_Y(y) = P(Y \leq y)$$
$$= P(|X| \leq y)$$
$$= P(-y \leq X \leq y)$$
$$= F_X(y) - F_X(-y).$$

By differentiation,

$$f_Y(y) = f_X(y) + f_X(-y)$$
$$= \frac{1}{2}\alpha e^{-\alpha|y|} + \frac{1}{2}\alpha e^{-\alpha|-y|}$$
$$= \alpha e^{-\alpha y}, \text{ for } y > 0.$$

It follows that

$$f_Y(y) = \begin{cases} \alpha e^{-\alpha y}, \text{ if } y > 0, \\ 0, \quad \text{ if } y \leq 0. \end{cases}$$

This is called an exponential distribution, which is widely used in labor economics (e.g., modeling unemployment duration) and finance (e.g., modeling transaction time). In other words, the absolute value of a double exponential random variable follows an exponential distribution.

(2) For $Y = X^2$, again letting $y \geq 0$, we have

$$
\begin{aligned}
P(Y \leq y) &= P(X^2 \leq y) \\
&= P(-\sqrt{y} \leq X \leq \sqrt{y}) \\
&= F_X(\sqrt{y}) - F_X(-\sqrt{y}).
\end{aligned}
$$

It follows that

$$
\begin{aligned}
f_Y(y) &= f_X(\sqrt{y}) \frac{1}{2\sqrt{y}} + f_X(-\sqrt{y}) \frac{1}{2\sqrt{y}} \\
&= \frac{\alpha}{2} \frac{1}{\sqrt{y}} e^{-\alpha\sqrt{y}}, \quad \text{for } y > 0.
\end{aligned}
$$

Hence, the PDF of Y is given by

$$
Y = \begin{cases} \frac{\alpha}{2} \frac{1}{\sqrt{y}} e^{-\alpha\sqrt{y}}, & \text{if } y > 0, \\ 0, & \text{if } y \leq 0. \end{cases}
$$

This is a special case of the so-called Weibull distribution for a nonnegative continuous random variable. A Weibull distribution has a PDF

$$
f_X(x) = \begin{cases} \frac{\beta}{\delta}(\frac{x-\gamma}{\delta})^{\beta-1} e^{-\left(\frac{x-\gamma}{\delta}\right)^{\beta}}, & x > \gamma, \\ 0, & \text{otherwise.} \end{cases}
$$

The Weibull distribution is widely used in survival analysis or duration analysis. Obviously, the distribution obtained in this example is a special case of Weibull (β, δ, γ) distribution with $\beta = 1/2, \delta = \alpha^{-2}, \gamma = 0$.

Example 3.26. A random variable X has a PDF

$$
f_X(x) = \frac{1}{\sqrt{2\pi}} e^{-x^2/2}, \qquad -\infty < x < \infty.
$$

Find the PDF $f_Y(y)$ of $Y = X^2$.

Solution: Given $Y = X^2$ always takes nonnegative values, we can let $y \geq 0$ and obtain

$$
\begin{aligned}
P(Y \leq y) &= P(X^2 \leq y) \\
&= P(-\sqrt{y} \leq X \leq \sqrt{y}) \\
&= F_X(\sqrt{y}) - F_X(-\sqrt{y}).
\end{aligned}
$$

Therefore, by the chain rule of differentiation, we obtain

$$f_Y(y) = F_X'\left(\sqrt{y}\right)\frac{1}{2\sqrt{y}} + F_X'\left(-\sqrt{y}\right)\frac{1}{2\sqrt{y}}$$

$$= \frac{1}{\sqrt{2\pi}}\frac{1}{\sqrt{y}}e^{-y/2}, \quad y \geq 0.$$

It follows that

$$f_Y(y) = \begin{cases} \frac{1}{\sqrt{2\pi}}\frac{1}{\sqrt{y}}e^{-y/2}, & y \geq 0, \\ 0, & y < 0. \end{cases}$$

The random variable X is called a standard normal variable, denoted as $N(0,1)$, and $Y = X^2$ is called, a chi-square random variable with degree of freedom 1, denoted as χ_1^2.

Example 3.27. Suppose a CRV X has a PDF

$$f_X(x) = \frac{1}{\sqrt{2\pi}\sigma}e^{-\frac{(x-\mu)^2}{2\sigma^2}}, \quad -\infty < x < \infty.$$

Find the PDF $f_Y(y)$ of $Y = e^X$.

Solution: Given $Y = e^X$ is always positive, we let $y > 0$. Then

$$F_Y(y) = P(Y \leq y)$$
$$= P(e^X \leq y)$$
$$= P(X \leq \ln y)$$
$$= F_X(\ln y).$$

By the chain rule of differentiation, we obtain

$$f_Y(y) = F_X'(\ln y)\frac{1}{y}$$

$$= f_X(\ln y)\frac{1}{y}$$

$$= \frac{1}{\sqrt{2\pi}\sigma}\frac{1}{y}e^{-(\ln y-\mu)^2/2\sigma^2}, \quad y > 0.$$

It follows that

$$f_Y(y) = \begin{cases} \frac{1}{\sqrt{2\pi}\sigma}\frac{1}{y}e^{-(\ln y-\mu)^2/2\sigma^2}, & \text{if } y > 0, \\ 0, & \text{if } y \leq 0. \end{cases}$$

The random variable X is called a normal random variable with mean μ and variance σ^2, denoted $N(\mu, \sigma^2)$, which includes $N(0,1)$ as a special case. The random variable $Y = e^X$, which is always nonnegative, is called the lognormal variable. That is, the logarithm of Y follows a normal distribution. The lognormal distribution has been widely used to model the probability distributions of asset prices, incomes, and other nonnegative economic variables. One example is the Black and Scholes (1973) formula for the European call option price, where it is assumed that the stock price follows a lognormal distribution.

Example 3.28. Suppose a CRV X has a PDF $f_X(x) = \frac{3}{8}(x+1)^2, -1 < x < 1$, and $f_X(x) = 0$ otherwise. Define

$$Y = \begin{cases} 1 - X^2, & \text{if } X \leq 0, \\ 1 - X, & \text{if } X > 0. \end{cases}$$

Find the PDF $f_Y(y)$.

Solution: Step 1: The support of Y is the interval $0 < y \leq 1$, as can be seen from Figure 3.9.

Step 2: For $0 < y \leq 1$, we have

$$\begin{aligned} F_Y(y) &= P(Y \leq y) \\ &= P(-1 < X \leq -\sqrt{1-y} \text{ or } 1 - y \leq X < 1) \\ &= P(-1 < X \leq -\sqrt{1-y}) + P(1 - y \leq X < 1) \\ &= F_X(-\sqrt{1-y}) - F_X(-1) + F_X(1) - F_X(1-y). \end{aligned}$$

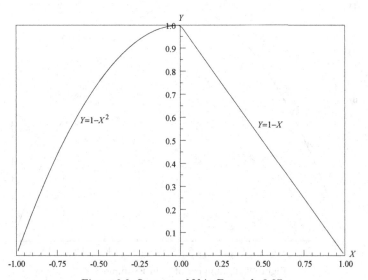

Figure 3.9: Support of Y in Example 3.27

It follows that

$$f_Y(y) = F_Y'(y)$$

$$= F_X'(-\sqrt{1-y})\frac{1}{2\sqrt{1-y}} + F_X'(1-y)$$

$$= f_X(-\sqrt{1-y})\frac{1}{2\sqrt{1-y}} + f_X(1-y)$$

$$= \frac{3}{8}(1-\sqrt{1-y})^2\frac{1}{2\sqrt{1-y}} + \frac{3}{8}(2-y)^2, \ 0 < y \le 1.$$

Therefore, the PDF of Y is given by

$$f_Y(y) = \begin{cases} \frac{3}{8}(1-\sqrt{1-y})^2\frac{1}{2\sqrt{1-y}} + \frac{3}{8}(2-y)^2, & 0 < y \le 1, \\ 0, & \text{otherwise.} \end{cases}$$

Example 3.29. Suppose a CRV X has a PDF $f_X(x) = \frac{1}{2}$ for $0 < x < 2$, and $f_X(x) = 0$ otherwise. Find the PDF of $Y = X(2 - X)$.

Solution: Observe that the support of Y is the interval $0 < y \le 1$, as can be seen from Figure 3.10.

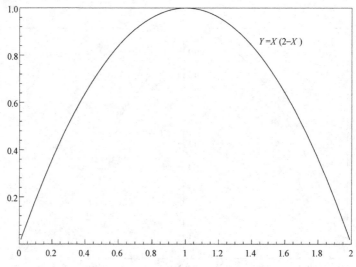

Figure 3.10: Support of Y in Example 3.28

Then

$$
\begin{aligned}
F_Y(y) &= P(Y \le y) \\
&= P[X(2 - X) \le y] \\
&= P[(X - 1)^2 \ge 1 - y] \\
&= P(X - 1 \ge \sqrt{1 - y} \text{ or } X - 1 \le -\sqrt{1 - y}) \\
&= P(X \ge 1 + \sqrt{1 - y}) + P(X \le 1 - \sqrt{1 - y}) \\
&= 1 - F_X(1 + \sqrt{1 - y}) + F_X(1 - \sqrt{1 - y}).
\end{aligned}
$$

Thus,

$$
\begin{aligned}
f_Y(y) &= F_Y'(y) \\
&= f_X(1 + \sqrt{1 - y})\frac{1}{2\sqrt{1 - y}} + f_X(1 - \sqrt{1 - y})\frac{1}{2\sqrt{1 - y}} \\
&= \frac{1}{2\sqrt{1 - y}} \text{ for } 0 < y \le 1.
\end{aligned}
$$

It follows that the PDF of Y is given by

$$
f_Y(y) = \begin{cases} \frac{1}{2\sqrt{1-y}}, & 0 < y \le 1, \\ 0, & \text{otherwise.} \end{cases}
$$

We now state an important result.

Theorem 3.11. [Probability Integral Transform]: *Suppose X has a continuous CDF $F_X(x)$ which is strictly monotonically increasing. Define $Y = F_X(X)$, that is,*

$$
Y = \int_{-\infty}^{X} f_X(x)dx.
$$

Then Y follows a uniform distribution on $[0, 1]$. That is, the PDF $f_Y(y) = 1$ for $0 \le y \le 1$ and 0 otherwise.

Proof: The support of $Y = F_X(X)$ is the unit interval $[0, 1]$. Letting $y \in [0, 1]$, we have

$$
\begin{aligned}
F_Y(y) &= P(Y \le y) \\
&= P[F_X(X) \le y].
\end{aligned}
$$

Because $F_X(x)$ is strictly increasing, its inverse function, denoted as $F_X^{-1}(y)$, exists and is also strictly increasing. For any real-value x, we have

$$
F_X^{-1}[F_X(x)] = x.
$$

By using $Y = F_X(X)$ and applying the inverse function operation, we obtain

$$
\begin{aligned}
F_Y(y) &= P(Y \leq y) \\
&= P[F_X(X) \leq y] \\
&= P\{F_X^{-1}[F_X(X)] \leq F_X^{-1}(y)\} \\
&= P[X \leq F_X^{-1}(y)] \\
&= F_X[F_X^{-1}(y)] \\
&= y, \text{ for } y \in [0, 1].
\end{aligned}
$$

It follows that the PDF of Y is given by

$$
f_Y(y) = \begin{cases} 1, \text{ for } 0 \leq y \leq 1, \\ 0, \text{ otherwise.} \end{cases}
$$

This is a uniform distribution on $[0, 1]$, which is called the standard uniform distribution, denoted $U[0, 1]$.

The CDF $F_X(x) = P(X \leq x)$ is not a random variable when x is an arbitrary but fixed number. However, $F_X(X)$ is a random variable because it is a function of the random variable X. The transformation $Y = F_X(X)$ is called the probability integral transformation. The result in Theorem 3.11 is not only of theoretical importance, but also facilitates the simulation of observed values of any CRV X.

The problem of generating uniform random numbers from computers has been worked on, with great success, by computer scientists. There exist many algorithms for generating pseudo-random numbers that will pass almost all tests for uniformity. Moreover, most good statistical packages have a reliable uniform random number generator. See Devroye (1985) or Ripley (1987) for more discussion on generating pseudo-random numbers.

From the uniform random numbers, we can generate random numbers with any given continuous distribution via the probability integral transform. Specifically, to generate an observation on X from a specified distribution $F_X(\cdot)$, we can first generate a realization y from the uniform distribution $U[0, 1]$ on the computer, and then solve for x from the equation $F_X(x) = y$. The value $x = F_X^{-1}(y)$ is a realization of X with the specified distribution $F_X(\cdot)$. For example, we consider how to generate an exponentially distributed random variable.

The random variable X follows a standard exponential distribution if its PDF $f_X(x) = e^{-x}$ for $x \geq 0$, and 0 otherwise. The CDF of X is

$$F_X(x) = \begin{cases} 1 - e^{-x} & \text{for } x \geq 0, \\ 0, & \text{otherwise.} \end{cases}$$

Define $Y = F_X(X) = 1 - e^{-X}$. Then Y follows a $U[0,1]$ distribution.

Now we generate a number y from the $U[0,1]$ distribution on a computer. Then

$$x = -\ln(1-y)$$

is a realization from the standard exponential distribution.

Figure 3.11 is the histogram of the probability based on 10,000 realizations generated using the aforementioned method. It is pretty close to the PDF $f_X(x) = e^{-x}$ for $x > 0$ and 0 otherwise.

The result that $F_X(X) \sim U[0,1]$ provides a basis for goodness-of-fit tests of distributional models. To check whether a prespecified probability model $F_0(\cdot)$ is correctly specified (i.e., whether X follows the hypothesized distribution $F_0(\cdot)$), one can first compute the probability integral transform $Y = F_0(X)$ and then check if Y follows the $U[0,1]$ distribution using a sample $\mathbf{Y}^n = (Y_1, \cdots, Y_n)$, where $Y_i = F_0(X_i)$ and n is called the sample size. If the random sample $\{X_i\}_{i=1}^n$ indeed follows the hypothesized distribution $F_0(\cdot)$, then Y_i will follow a $U[0,1]$ distribution. If X_i does not follow

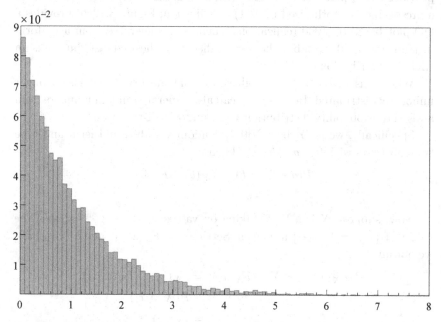

Figure 3.11: Probability histogram based on 1000 realizations generated from EXP(1)

the hypothesized distribution $F_0(\cdot)$, then Y_i will not follow a $U[0,1]$ distribution. This is the basic idea behind the popular Kolmogorov-Smirnov test for a hypothesized distribution model $F_0(\cdot)$. This result can be extended to test more sophisticated distributional models in economics and finance; see for example Hong and Li (2005) for more discussion.

Alternatively, there exists a simple graphical method called QQ plot that can be used to check whether $X_1, \cdots, X_n$ follows the prespecified hypothesized distribution $F_0(\cdot)$. Specifically, define the following function

$$\hat{F}_Y(y) = \frac{1}{n} \sum_{i=1}^{n} \mathbf{1}(Y_i \leq y),$$

where $\mathbf{1}(Y_i \leq y)$ is the indicator function that takes value 1 if $Y_i \leq y$ and value 0 otherwise. This is called the empirical CDF of $Y_1, \cdots, Y_n$; it is the proportion of those from $Y_1, \cdots, Y_n$ which take values less than or equal to y. When X_i follows the prespecified distribution $F_0(\cdot)$, Y_i will follow a $U[0,1]$ distribution, and $\hat{F}_Y(y)$ will be approximately a 45-degree line for $0 \leq y \leq 1$. If X_i does not follow the prespecified distribution $F_0(\cdot)$, then $\hat{F}_Y(y)$ will not be approximately a 45-degree line for $0 \leq y \leq 1$. The plot of $\hat{F}_Y(y)$ for $0 \leq y \leq 1$ is called a QQ-plot. Figure 3.12(a) provides a QQ-plot based on 1000 realizations when X_i is generated from a prespecified hypothesized EXP(1) distribution; Figure 3.12(b) provides a QQ-plot based on 1000 realizations when X_i is generated from a uniform distribution on $[0, 2\sqrt{3}]$ but the prespecified hypothesized distribution is an EXP(1) distribution.

When X is a DRV, the probability integral transform $F_X(X)$ is no longer uniformly distributed. However, we can also generate random numbers from a discrete probability distribution via a uniform distribution.

Specifically, we let Y be a $U[0,1]$ random variable, and let a and b be two constants with $0 \leq a < b \leq 1$. Then

$$P(a < Y \leq b) = F_Y(b) - F_Y(a)$$
$$= b - a.$$

Now, suppose X is a DRV taking on values $x_1 < x_2 < \cdots < x_k$. The values of $\{x_i, i = 1, 2, \cdots\}$ are given. Setting $a = F_X(x_{i-1})$ and $b = F_X(x_i)$, we obtain

$$P\left[F_X(x_{i-1}) < Y \leq F_X(x_i)\right] = F_X(x_i) - F_X(x_{i-1})$$
$$= P(X = x_i)$$
$$= f_X(x_i)$$

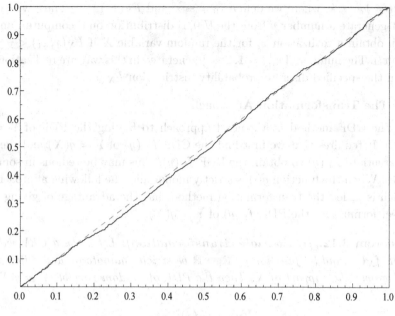

Figure 3.12(a): QQ-plot for $X_i \sim \text{EXP}(1)$

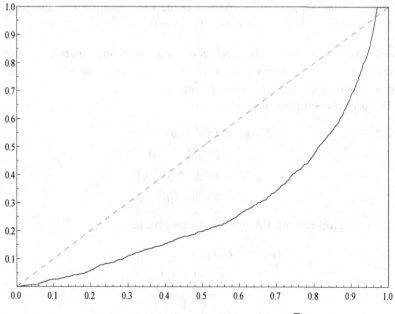

Figure 3.12(b): QQ-plot for $X_i \sim U[0, 2\sqrt{3}]$

for $i = 1, \cdots, k$, where we define $x_0 = -\infty$ and $F_X(x_0) = 0$. Thus, we can first generate a number y from the $U[0,1]$ distribution on a computer, and then obtain a realization x_i for the random variable X if $F_X(x_{i-1}) < y \leq F_X(x_i)$. The numbers $\{x_i, i = 1, 2, \cdots\}$ generated in this way are realizations from the specified discrete probability distribution $F_X(x)$.

(2) The Transformation Approach

The CDF method is a general approach to finding the PDF of $Y = g(X)$. It requires that we first find the CDF $F_Y(y)$ of $Y = g(X)$ and then differentiate $F_Y(y)$ to obtain the PDF $f_Y(y)$. This may be tedious in some cases. When the function $g(\cdot)$ is strictly monotonic, the following approach, which is called the transformation method, has the advantage of giving a direct formula for the PDF $f_Y(y)$ of $Y = g(X)$.

Theorem 3.12. *[Univariate Transformation]: Let X be a CRV with PDF $f_X(x)$ and let function $g : \mathbb{R} \to \mathbb{R}$ be strictly monotonic and differentiable over the support of X. Then the PDF of random variable $Y = g(X)$ is*

$$f_Y(y) = f_X(x) \frac{1}{|g'(x)|}$$

for any y in the support of Y, where x is the unique number in the support of X such that $g(x) = y$; otherwise, $f_Y(y) = 0$.

Proof: We first consider the case where $g(x)$ is strictly increasing. For a strictly increasing function $g(x)$, there exists a unique strictly increasing inverse function $g^{-1}(y)$ such that $g^{-1}[g(x)] = x$.

For y in the support of Y, we have

$$\begin{aligned}
F_Y(y) &= P(Y \leq y) \\
&= P[g(X) \leq y] \\
&= P[X \leq g^{-1}(y)] \\
&= F_X[g^{-1}(y)].
\end{aligned}$$

By the chain rule of differentiation, we obtain

$$\begin{aligned}
f_Y(y) &= F_Y'(y) \\
&= F_X'[g^{-1}(y)] \frac{d}{dy} g^{-1}(y) \\
&= f_X(x) \frac{1}{g'(x)},
\end{aligned}$$

where $x = g^{-1}(y)$, and we have made used of the fact that

$$\frac{d}{dy}g^{-1}(y) = \frac{1}{g'(x)},$$

which follows from differentiating the identity that

$$g^{-1}[g(x)] = x,$$

where $y = g(x)$.

Next, we consider the case when $g(X)$ is monotonically decreasing. Using a similar reasoning, we can obtain

$$f_Y(y) = -f_X(x)\frac{1}{g'(x)},$$

where $x = g^{-1}(y)$. Therefore, combining the cases of monotonic increasing and decreasing, we have

$$f_Y(y) = f_X(x)\frac{1}{|g'(x)|},$$

where $x = g^{-1}(y)$ for any y in the support of Y. This completes the proof.

Before applying this univariate transformation theorem, it is very important to check whether the function $g : \mathbb{R} \to \mathbb{R}$ is strictly monotonic over the support Ω_X of X. It cannot be directly applied to non-monotonic functions. For example, it cannot be applied to the case where $Y = X^2$ for X over the range of $(-\infty, \infty)$, because $g(x) = x^2$ is not strictly monotonic over $(-\infty, \infty)$. Of course, it can be applied to the function $Y = X^2$ if the support of X is $[0, \infty)$ or $(-\infty, 0]$.

However, if $Y = g(X)$ is strictly monotonic over several regions of the real line, we can extend the above univariate transformation theorem to cover such more general cases.

Theorem 3.13. *Suppose $g(x) = g_i(x)$ for all $x \in A_i$, where $i = 1, \cdots, k$, where for each i, $g_i(x)$ is strictly monotonic (strictly increasing or decreasing) and differentiable on region A_i. Further suppose the regions $\{A_i\}$ are disjoint and $\cup_{i=1}^{k} A_i = \mathbb{R}$. Then the PDF of $Y = g(X)$ is given by*

$$f_Y(y) = \sum_{i=1}^{k} f_X[g_i^{-1}(y)]\frac{1}{|g_i'[g_i^{-1}(y)]|}$$

for all y in the support Ω_Y of Y.

Proof: This is left as an exercise.

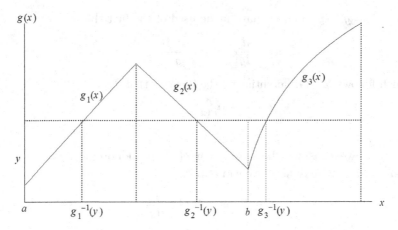

Figure 3.13: A simple example with $k = 3$ subintervals to illustrate Theorem 3.13

To illustrate Theorem 3.13, we consider a simple example with $k = 3$ subintervals, as shown in Figure 3.13.

For the y value as given, we have

$$
\begin{aligned}
P(Y \le y) &= P\left[g(X) \le y\right] \\
&= P\left[a \le X \le g_1^{-1}(y)\right] + P\left[g_2^{-1}(y) \le X \le b\right] \\
&\quad + P\left[b \le X \le g_3^{-1}(y)\right] \\
&= F_X\left[g_1^{-1}(y)\right] - F_X(a) + F_X(b) - F_X\left[g_2^{-1}(y)\right] \\
&\quad + F_X\left[g_3^{-1}(y)\right] - F_X(b),
\end{aligned}
$$

where constants a and b are as given in Figure 3.13. By differentiation, we obtain

$$
\begin{aligned}
F_Y(y) &= f_X\left[g_1^{-1}(y)\right] \frac{1}{g_1'(y)} + f_X\left[g_2^{-1}(y)\right] \frac{1}{-g_2'(y)} \\
&\quad + f_X\left[g_3^{-1}(y)\right] \frac{1}{g_3'(y)} \\
&= \sum_{i=1}^{3} f_X\left[g_i^{-1}(y)\right] \frac{1}{|g_i'(y)|}.
\end{aligned}
$$

3.6   Mathematical Expectations

We now introduce an important concept called mathematical expectation or simply expectation. Intuitively, expectation is the long-run average of the observed values for a random variable in repeated independent experiments with a very large number of repetitions. It is the key to understand what happens in the long run when we repeatedly observe the values of a random variable.

We first provide the definition of expectation for a measurable function $g(X)$.

Definition 3.14. [Expectation]: Suppose X is a random variable with PMF or PDF $f_X(x)$. Then the expectation of a measurable function $g(X)$ is defined as

$$E[g(X)] = \int_{-\infty}^{\infty} g(x) dF_X(x)$$

$$= \begin{cases} \sum_{x \in \Omega_X} g(x) f_X(x), & \text{if } X \text{ is a DRV,} \\ \int_{-\infty}^{\infty} g(x) f_X(x) dx, & \text{if } X \text{ is a CRV,} \end{cases}$$

provided that the sum or integral exists. Here, for the DRV case, the summation is over all possible values in the support Ω_X of X.

If $E|g(X)| = \infty$, we say that $E[g(X)]$ does not exist. In other words, to avoid convergence problems, we require that $\sum_{x \in \Omega_X} |g(x)| f_X(x) < \infty$ for a DRV X, and $\int_{-\infty}^{\infty} |g(x)| f_X(x) dx < \infty$ for a CRV X.

There exists another method to compute the expectation $E[g(X)]$. Since $Y = g(X)$ is a random variable and has a PMF or PDF $f_Y(y)$. It follows that

$$E[g(X)] = E(Y)$$

$$= \begin{cases} \sum_{y \in \Omega_Y} y f_Y(y), & \text{if } Y \text{ is a DRV,} \\ \int_{-\infty}^{\infty} y f_Y(y) dy, & \text{if } Y \text{ is a CRV,} \end{cases}$$

where for the discrete case, the summation is over all possible values of y in the support Ω_Y of Y. This method will give the same result as the above definition using the probability distribution of X.

What is the interpretation for the expectation $E[g(X)]$? For simplicity, we consider the case of a discrete random variable X. Suppose we generate a large number of realizations for the random variable X via a large number of repeated independent experiments, and consider the average of all realizations for $g(X)$ from these repeated experiments. Among all possible realizations generated, each specific value x (say), may repeat several times. It follows that there is an alternative way to compute the average value of realizations for $g(X)$ from a large number of repeated independent experiments, that is, it can be computed as a weighted average of various values of $g(X)$ with weights being the relative frequency of occurrence of each specific value x. This weighted average is approximately equal to the expectation $E[g(X)]$, where the weighting function $f_X(x)$ measures the relative frequency for each $x \in \Omega_X$.

Because the expectation $E(\cdot)$ is an integration or summation, it is a linear operator, namely

$$E\left[ag_1(X) + bg_2(X)\right] = aE\left[g_1(X)\right] + bE\left[g_2(X)\right]$$

for any two measurable functions $g_1(\cdot)$ and $g_2(\cdot)$, where a, b are constants. Linearity of the expectation operator can simplify calculation and derivation in many cases.

Suppose $g(\cdot)$ is a concave function (**Question:** what is the definition of a concave function?). An example of the concave function is $g(X) = \ln X$. Then for any concave function and any probability distribution with the existence of $E[g(X)]$, we have

$$E\left[g(X)\right] \leq g\left[E(X)\right].$$

That is, the average of the values of a concave function is less than the value of the concave function evaluated at the average. This is called Jensen's inequality, which is widely used in economics.

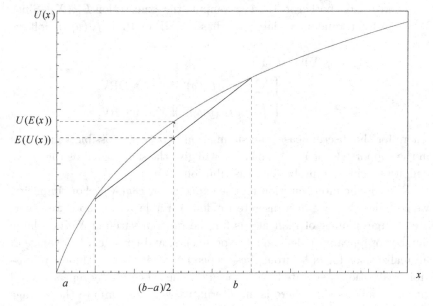

Figure 3.14: An economic interpretation for Jensen's inequality

Suppose $g(\cdot)$ is a utility function of an economic agent. Then the risk-averseness of the economic agent requires that the utility function $g(\cdot)$ be concave. In this case, Jensen's inequality has a nice economic interpretation: The expected utility under an uncertain environment is less than the utility of a sure pay equal to $E(X)$. The economic agent prefers a sure pay of $E(X)$ in an uncertain environment. This can be illustrated in Figure 3.14.

3.7 Moments

We now consider some important special cases of expectation.

Definition 3.15. [Mean]: The mean of a random variable X is defined as

$$\mu_X = E(X)$$
$$= \begin{cases} \sum_{x \in \Omega_X} x f_X(x), & \text{if } X \text{ is a DRV,} \\ \int_{-\infty}^{\infty} x f_X(x) dx, & \text{if } X \text{ is a CRV,} \end{cases}$$

where the summation is over all possible values in the support Ω_X of X.

The mean μ_X is also the expectation of X; it is also called the first moment of X. It can be viewed as a "location" parameter. It is a measure of the central tendency for the distribution of X in the following sense: $\mu_X = E(X)$ is the limit of the average of the realizations of the random variable X from a large number of repeated independent experiments. When X is an asset return, μ_X will represent the long-run average return for the asset, under the assumption that the distribution of the asset return does not change over time.

The terminology of expectation has its origin in games of chance. This can be illustrated as follows. Four small similar chips, numbered 1, 1, 1, and 2, respectively, are placed in a bowl and are mixed. A player is blindfolded and is to draw a chip from the bowl. If she draws one of the three chips numbered 1, she will receive one dollar. If she draws the chip numbered 2, she will receive two dollars. It seems reasonable to assume that the player has a "$\frac{3}{4}$ claim" on the \$1 and a "$\frac{1}{4}$ claim" on the \$2. Her "total claim" is $1 \times \frac{3}{4} + 2 \times \frac{1}{4} = \frac{5}{4} = \$1.25$. Thus the expectation of X is precisely the player's claim in this game.

Recall that the expectation μ_X exists for a continuous distribution if and only if

$$\int_{-\infty}^{\infty} |x| f_X(x) dx < \infty.$$

Whenever X is a bounded random variable, that is, whenever there exist constants a and b with $-\infty < a < b < \infty$ such that $\Pr\left(a \leq X \leq b\right) = 1$, then μ_X must exist. In contrast, suppose X has a PDF

$$f_X(x) = \frac{1}{\pi}\frac{1}{1+x^2}, \text{ for } -\infty < x < \infty.$$

This is called a Cauchy(0,1) distribution. Then

$$\int_{-\infty}^{\infty} |x|\, f_X(x)dx = \frac{2}{\pi}\int_0^{\infty}\frac{x}{1+x^2}dx = \infty.$$

Therefore, the expectation $E(X)$ does not exist for the Cauchy(0,1) distribution.

The next theorem shows that μ_X is the optimal solution to the least squares problem.

Theorem 3.14. *Suppose $E(X^2)$ exists. Then*

$$\mu_X = \arg\min_{a} E\left(X - a\right)^2.$$

Proof: The first order condition of the above minimization problem is

$$\left.\frac{dE(X-a)^2}{da}\right|_{a=a^*} = \left.\frac{d}{da}\int_{-\infty}^{\infty}(x-a)^2 dF_X(x)\right|_{a=a^*} = 0.$$

Exchanging differentiation and integration, we obtain

$$\left.-2\int_{-\infty}^{\infty}(x-a)dF_X(x)\right|_{a=a^*} = -2\int_{-\infty}^{\infty}(x-a^*)dF_X(x) = 0.$$

It follows that

$$a^* = \frac{\int_{-\infty}^{\infty}x dF_X(x)}{\int_{-\infty}^{\infty}dF_X(x)} = \mu_X.$$

This completes the proof.

Question: Does $X = \mu_X$ have the largest probability to occur? That is, is the probability $P(X = \mu_X)$ the largest?

The answer is no. For example, suppose X takes values 1 and 0 with equal probability, that is, $P(X = 1) = P(X = 0) = \frac{1}{2}$. Then $\mu_X = E(X) = \frac{1}{2}$. As a result, $P(X = \mu_X) = 0$ has the minimum probability.

Definition 3.16. [Variance and Standard Deviation]: The variance of random variable X is defined as

$$\sigma_X^2 = E(X - \mu_X)^2$$

$$= \begin{cases} \sum_{x \in \Omega_X} (x - \mu_X)^2 f_X(x), & \text{if } X \text{ is a DRV}, \\ \int_{-\infty}^{\infty} (x - \mu_X)^2 f_X(x) dx, & \text{if } X \text{ is a CRV}, \end{cases}$$

where Ω_X is the support of X. The quantity

$$\sigma_X = \sqrt{\sigma_X^2}$$

is called the standard deviation of X.

The variance σ_X^2 is a measure of the degree of spread of a distribution around its mean. It is a scale parameter for the distribution of X.

In economics, σ_X^2 is often interpreted as a measure of uncertainty. More specifically, it is usually called a measure of "volatility" of X. A larger value of σ_X^2 means that X is more variable. In contrast, at the extreme, if $\sigma_X^2 = 0$, then $X = \mu_X$ with probability 1 and there is no variation in X. To see this, consider the case that X is a DRV. Then

$$\sigma_X^2 = \sum_{x \in \Omega_X} (x - \mu_X)^2 f_X(x)$$

$$= 0$$

if and only if

$$(x - \mu_X)^2 f_X(x) = 0 \text{ for all } x \in \Omega_X,$$

which implies

$$x = \mu_X \text{ with } f_X(\mu_X) = 1.$$

That is, there is only one possible value μ_X for X when $\sigma_X^2 = 0$. This is an example of the so-called degenerate distribution.

The standard deviation $\sigma_X = \sqrt{\sigma_X^2}$ is easier to interpret in that the measurement unit of σ_X is the same as that for the variable X.

Since variance (or standard deviation) depends on the units of measurement, it may be difficult to compare two or more standard deviations if they are expressed in different units of measurement. To overcome this difficulty, a measure of relative variation is used. This is called the coefficient of variation, defined as

$$V = \frac{\sigma_X}{\mu_X}.$$

This ratio is measurement unit-free.

We should emphasize that σ_X^2 is not the only measure for uncertainty of a distribution. In fact, it is possible that two different distributions have the

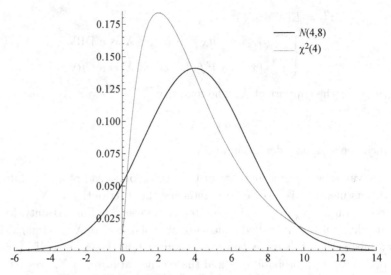

Figure 3.15: Probability distributions with same variance

same variance, as illustrated in Figure 3.15. In this case, σ_X^2 cannot distinguish the two distributions, alternative measures for uncertainty should be used.

The following theorem provides a convenient way to compute the variance of X.

Theorem 3.15. $\sigma_X^2 = E(X^2) - \mu_X^2$.

Proof: By the formula that $(a - b)^2 = a^2 - 2ab + b^2$, we have

$$
\begin{aligned}
\sigma_X^2 &= E\left[(X - \mu_X)^2\right] \\
&= E(X^2) - E(2\mu_X X) + E(\mu_X^2) \\
&= E(X^2) - 2\mu_X E(X) + \mu_X^2 \\
&= E(X^2) - \mu_X^2.
\end{aligned}
$$

This completes the proof.

We note that σ_X^2 is called the second central moment, while $E(X^2)$ is called the second moment of X.

We now consider the mean and variance of a linear transformation $Y = a + bX$.

Theorem 3.16. *If* $Y = a + bX$. *Then (1)* $\mu_Y = a + b\mu_X$; *(2)* $\sigma_Y^2 = b^2\sigma_X^2$.

The parameters a and b can be interpreted as the location and scale parameters respectively. Both location and scale parameters affect the mean of Y, but the variance of Y is only affected by the scale parameter b but not by the location parameter a.

There are many economic examples why we may be interested in the mean and variance of the linear transformation $Y = a + bX$. For example, suppose there are two assets: a risk-free asset with gross return equal to unity, and a risky asset with random return X. Suppose a and b are the proportions of investments on risk-free and risky assets respectively (they are called portfolio weights). Then Y is the return on such a portfolio. It is important to know the expected return μ_Y and risk (measured by σ_Y^2) of such a portfolio. Note that the expected return μ_Y consists of the expected return from the risky asset and the return from the risk-free asset, but as expected, the risk of the portfolio Y, as measured by $\sigma_Y^2 = b^2 \sigma_X^2$, only comes from the risky asset X and has nothing to do with the risk-free asset.

We now consider moments of a general order.

Definition 3.17. [k-th Moment and k-th Central Moment]: The k-th moment of a random variable X is defined as

$$E(X^k) = \begin{cases} \sum_{x \in \Omega_X} x^k f_X(x), & \text{if } X \text{ is a DRV}, \\ \int_{-\infty}^{\infty} x^k f_X(x) dx, & \text{if } X \text{ is a CRV}, \end{cases}$$

where Ω_X is the support of X.

Similarly, the k-th central moment of a random variable X is defined as

$$E(X - \mu_X)^k = \begin{cases} \sum_{x \in \Omega_X} (x - \mu_X)^k f_X(x), & \text{if } X \text{ is a DRV}, \\ \int_{-\infty}^{\infty} (x - \mu_X)^k f_X(x) dx, & \text{if } X \text{ is a CRV}. \end{cases}$$

It is of interest to note that the term "moment" comes from the field of physics: if the quantities $f_X(x)$ in the discrete case were point masses acting perpendicularly to the x-axis at distances x from the origin, then $\mu_X = E(X)$ would be the x-coordinate of the center of gravity, that is, the first moment divided by $\sum_{x \in \Omega_X} f_X(x)$, and $E(X^2)$ would be the moment of inertia. This also explains why the k-th moment $E(X^k)$ is called the k-th moment about the origin: in the analogy to physics, the length of the lever arm is in each case the distance from the origin. The analogy applies also in the continuous case, where $E(X)$ and $E(X^2)$ might be the x-coordinate of the center of gravity and the moment of inertia of a rod with variable mass density.

Question: What is the relationship between uncentered moments and centered moments?

Using the binomial formula that $(u + v)^k = \sum_{i=0}^{k} \binom{k}{i} u^i v^{k-i}$, we obtain

$$E(X - \mu_X)^k = E\left[\sum_{i=0}^{k} \binom{k}{i} X^i (-\mu_X)^{k-i}\right]$$

$$= \sum_{i=0}^{k} \binom{k}{i} E(X^i)(-\mu_X)^{k-i}.$$

Thus, the k-th central moment $E(X - \mu_X)^k$ is a linear combination of the first k uncentered moments $\{E(X^m), m = 1, \ldots, k\}$.

Similarly, we have

$$E(X^k) = E(X - \mu_X + \mu_X)^k$$

$$= \sum_{i=0}^{k} \binom{k}{i} E(X - \mu_X)^i \mu_X^{k-i}.$$

That is, the k-th moment is a linear combination of the first k-th central moments.

In econometrics, primary interest has been in the first moment and the second moment of X. We now provide some examples to highlight why the first two moments are important in economics.

Example 3.30. [St. Petersburg Paradox]: A player is to receive 2^x dollars when, in a series of flips of a balanced coin, the first head appears on the x-th flip. Then the probability that the first head appears on the x-th flip is $f_X(x) = (\frac{1}{2})^x$, for $x = 1, 2, \cdots$. The expected payoff from this game

$$E(2^X) = \sum_{x=1}^{\infty} 2^x \left(\frac{1}{2}\right)^x$$

$$= \infty.$$

However, a risk-averse player usually would not like to play such a game if he is required to pay \$1000.

Why?

This is because a risk-averse player will not only consider the expected return but also take into account the risk of this game.

Risk is often measured by variance. To illustrate this idea, we now consider an example of optimal portfolio decision in a classical setting.

Example 3.31. [Portfolio Selection]: How will an investor make an optimal portfolio decision?

Suppose μ and σ^2 are the mean and variance of a portfolio return over a holding period. Assume that an investor is risk-averse in the sense that he likes a higher return but a lower risk. Specifically, his utility function $U(\mu, \sigma^2)$ is a function of μ and σ^2 such that $\frac{\partial}{\partial \mu} U(\mu, \sigma^2) > 0$ and $\frac{\partial}{\partial \sigma^2} U(\mu, \sigma^2) < 0$. Here, $\frac{\partial}{\partial \mu} U(\mu, \sigma^2) > 0$ implies that the higher the expected return, the higher the utility, and $\frac{\partial}{\partial \sigma^2} U(\mu, \sigma^2) < 0$ implies that the higher the risk, the lower the utility. An example of $U(\mu, \sigma^2)$ is

$$U(\mu, \sigma^2) = a\mu - \frac{b}{2}\sigma^2,$$

where $a > 0$ and $b > 0$ are preference parameters. The investor will maximize the utility function $U(\mu, \sigma^2)$ subject to the budget constraint.

Now suppose that in the market there is a risky asset with random return X and a risk-free asset with constant return r. Further suppose the investor has totally I dollars to be split between the risky asset (z) and the risk-free asset $(I - z)$. The total return of the investment is

$$Y = (I - z)r + zX.$$

The expected return of the portfolio is $\mu_Y = (I-z)r + z\mu_X$ and the variance $\sigma_Y^2 = z^2\sigma_X^2$. The associated utility function

$$U(\mu_Y, \sigma_Y^2) = a[(I - z)r + z\mu_X] - \frac{b}{2}z^2\sigma_X^2.$$

The optimal weight z^* can be determined by maximizing the utility of the investor. That is, the investor will choose z to maximize $U(\mu_Y, \sigma_Y^2)$. The first order condition is

$$a(\mu_X - r) - bz^*\sigma_X^2 = 0.$$

The optimal investment on the risky asset is

$$z^* = \frac{a(\mu_X - r)}{b\sigma_X^2}.$$

We now investigate two special cases:

Case (1): $b = 0$. In this case, the investor is risk-neutral. Then he will only invest on the risky asset (i.e., $z^* = I$ or $z^* = \infty$ if borrowing is allowed) if $\mu_X > r$.

Case (2): $b = \infty$. In this case, the investor is extremely risk-averse. He will choose $z^* = 0$. That is, he will only invest on the risk-free asset even if $\mu_X > r$.

Under a regular case, z^* is an interior solution, which implies that the investor splits the investment between the risky asset and risk-free asset. The relative proportion of investments between the risky and risk-free assets depends on the preference parameters a, b, as well as the gap between the expected return of the risky asset and the return of the risk-free asset.

In fact, a criterion that has been rather popular in decision-making for portfolio investment is the so-called Sharpe ratio, defined as

$$\text{Sharpe ratio} = \frac{\mu_Y}{\sigma_Y},$$

where μ_Y is the expected excess return of a portfolio, and σ_Y is the standard deviation of the excess return. Intuitively, the Sharpe ratio can be interpreted as the expected return per unit of risk. An investor will prefer a portfolio with a higher Sharpe ratio. This is consistent with the optimal portfolio selection considered in the above example.

In addition to the first two moments, higher moments, particularly the third and fourth order moments, are also of interest in economics.

In economics and finance, the systematic risk of a security is measured by the contribution to the variance of a well-diversified portfolio. However, there is considerable evidence that the unconditional return distribution cannot be adequately characterized by mean and variance alone. This leads economists to the next moment — skewness. Everything else being equal, investors should prefer portfolios that are right-skewed to portfolios that are left-skewed. This is consistent with the Arrow-Pratt notion of risk aversion. Hence, assets that decrease a portfolio's right skewness are less desirable and should command higher expected returns.

For a random variable X, the third central moment $E(X - \mu_X)^3$ is a measure of "skewness" or "lack of symmetry" of the probability distribution of X. Positive skewness indicates a relatively long right tail compared to the left tail of the distribution; negative skewness indicates the opposite. In economics, skewness has been used to measure financial crashes because when more negative large values than positive large values occur, $E(X - \mu_X)^3$ will be negative and large in absolute value.

Skewness is defined as a standardized third central moment:

$$S_X = \frac{E(X - \mu_X)^3}{\sigma_X^3}.$$

The standardization by σ_X^3 makes the skewness invariant to the scale of the distribution. A positive S_X indicates a relatively long right tail compared to the left, and a negative S_X indicates the opposite. Assuming that $\mu_X = 0$, where X is an asset return, then a positive (negative) skewness indicates a higher (lower) probability of experiencing large gains than large losses of the same magnitude.

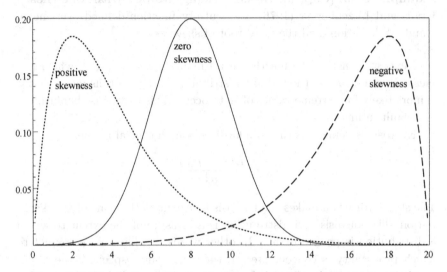

Figure 3.16: Distributions with zero, positive and negative skewness respectively

Figure 3.16 provides plots of three probability distributions which have zero skewness, positive skewness and negative skewness respectively.

Skewness in the returns of financial assets can arise from many sources. Brennan (1993) points out that managers have option-like features in their compensation. The impact of financial distress on firms and the choice of projects can also induce skewness in the returns. More fundamentally, skewness can be induced through asymmetric risk preferences among investors. Singleton and Wingender (1986) consider information contained in conditional skewness measures. They find that the pack of persistence in skewness hinders any chance of successfully constructing portfolios to take advantage of *ex-post* skewness found in returns. Bekaert *et al.* (1998) try to derive a trading strategy based on a conditional measure of skewness in financial markets. Bond (2000) uses models incorporating time-varying conditional skewness to calculate a measure of downside risk and then from this calculate portfolio weights for an investor with loss aversion. It is found that the loss-averse investor derives a higher expected utility for using some models incorporating conditional skewness.

Skewness in returns may have important implications for asset pricing (Harvey and Siddique 2000), portfolio construction (Kraus and Litzenkerger 1976, Markowitz 1991), and risk management.

Example 3.32. [**Capital Asset Pricing Model with Skewness**]: Kraus and Litzenberger (1976) use a utility function defined over mean, standard deviation, and the third root of skewness.

Next, we consider the fourth central moment $E(X - \mu_X)^4$, which is a measure of how heavy the tail of a distribution is. Heavy tails mean that it is more likely for extreme values of X to occur. This may cause bankruptcy or default in finance.

Kurtosis is defined as the standardized fourth central moment:

$$K_X = \frac{E(X - \mu_X)^4}{\sigma_X^4}.$$

The standardization makes the kurtosis invariant to the scale of the distribution. The kurtosis of a distribution is a measure of the extent to which a distribution is peaked or flat. In other words, it indicates the extent to which probability is concentrated in the center and especially the thickness of the tails of the distribution. A distribution with $K_X < 3$ is called *playtykurtic* (fat or short-tailed), and a distribution with $K_X > 3$ is called *leptokurtic* (slim or long-tailed). For a distribution which has $K_X = 3$, it is called *mesokurtic*.

Why is $K_X = 3$ used as a benchmark? This is mainly because the most well-known normal distribution has $K_X = 3$.

Example 3.33. Suppose X is a normal random variable with mean μ and variance σ^2. That is, its PDF is

$$f_X(x) = \frac{1}{\sqrt{2\pi\sigma^2}} e^{-\frac{(x-\mu)^2}{2\sigma^2}}, \quad -\infty < x < \infty.$$

Then $\mu_X = \mu$, $\sigma_X^2 = \sigma^2$, $S_X = 0$ and $K_X = 3$.

The excess kurtosis of a random variable X is defined as $K_X - 3$. Thus, an excess kurtosis implies that extraordinary gains or losses are more common than a normal distribution predicts.

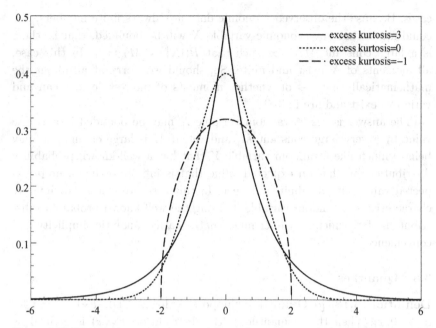

Figure 3.17: Distributions with zero, positive and negative excess kurtosis respectively

Figure 3.17 provides the plots of three distributions with zero excess kurtosis, positive excess kurtosis and negative excess kurtosis.

Both skewness and kurtosis are closely related to the tails of the distribution of X. They measure the shape of a probability distribution. Recent studies find evidence that skewness and kurtosis of financial time series are time-varying. See (e.g.) Hansen (1994), Harvey and Siddique (2000). These findings are important for modeling the distribution of financial time series, which is important in derivative pricing and risk management.

Various moments of a probability distribution constitute an important class of expectations. The class of moments, when they exist, is a summary characteristic of a probability distribution. They can intuitively describe the features of a probability distribution. In particular, the first four moments, $\mu_X, \sigma_X^2, S_X, K_X$, when they exist, describe the center, scale, asymmetry, and thickness of the tail of the distribution.

Economic and particularly high-frequency financial data usually have heavy-tails, namely they appear to follow distributions with larger tails than the normal distributions. These distributions may not have finite higher order moments (e.g., skewness and kurtosis), even the variance in some cases. Because the observed economic data are always finite numbers, one could argue that any economic variable X will be bounded, that is, there is a large constant $M < \infty$ such that $P(|X| \leq M) = 1$. In this case, all moments of X exist and finite. So, should we worry at all about the mathematically niceties of whether moments of interest (e.g., mean and variance) exist and are finite?

The answer is yes. A random variable X may be bounded in absolute value by a very large constant M, and yet, if M is large enough, X may behave much like a random variable Y that has a well-known probability distribution but has an expected value that is infinite or has a finite expected value but an infinite variance. In this case, we can approximately characterize the random variable X using the well-known probability distribution of Y, which may offer much mathematical analytic simplicity and convenience.

3.8   Quantiles

Definition 3.18. [α-Quantile]: Suppose X has a CDF $F_X(x)$, and let $\alpha \in (0,1)$. Then the α-quantile of the distribution $F_X(x)$ is defined as $Q(\alpha)$, which satisfies the following equation:

$$P[X \leq Q(\alpha)] = \alpha,$$

or equivalently

$$F_X[Q(\alpha)] = \alpha.$$

When $F_X(x)$ is strictly increasing, we have

$$Q(\alpha) = F_X^{-1}(\alpha),$$

where $F^{-1}(\alpha)$ is the inverse function of $F_X(x)$.

If $F_X(x)$ is not strictly increasing, we can define its α-quantile as

$$Q(\alpha) = \inf\{x \in \mathbb{R} : \alpha \leq F_X(x)\}.$$

Figure 3.18 shows that the α-quantile is a cutoff point that divides the whole population into two proportions of α and $1 - \alpha$ respectively.

Suppose $f_X(x) = F_X'(x)$ exists almost everywhere. Then

$$\int_{-\infty}^{Q(\alpha)} f_X(x)dx = \alpha.$$

Figure 3.19 is an alternative graphical representation of the α-quantile in the probability density plot.

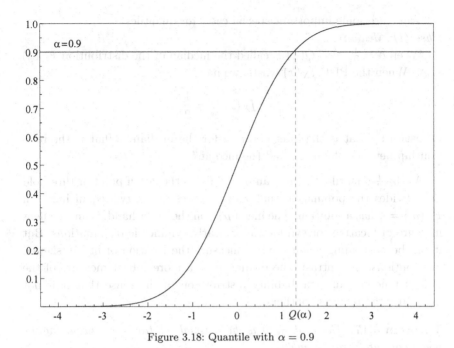

Figure 3.18: Quantile with $\alpha = 0.9$

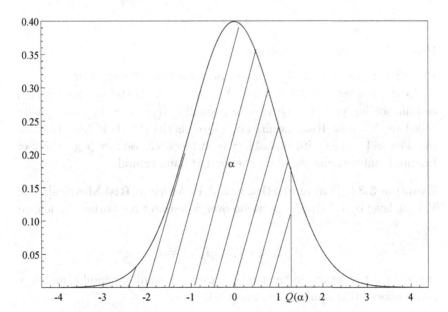

Figure 3.19: An alternative graphical representation of the α-quantile

There are some important special cases for quantiles:

Case (1): Median

When $\alpha = \frac{1}{2}$, $m = Q(\frac{1}{2})$ is called the median of the distribution $F_X(x)$ of X. When the PDF $f_X(x)$ exists, we have

$$\int_{-\infty}^{m} f_X(x)dx = \frac{1}{2}.$$

Question: What is the interpretation for the median? What is the relationship between the mean and the median?

As the $\frac{1}{2}$-quantile, the median $m = Q(\frac{1}{2})$ is the cutoff point or threshold that divides the population that $F_X(x)$ represents in two equal halves if $F_X(m) = \frac{1}{2}$ has a solution. The mean μ_X, on the other hand, is an excellent measure of location for symmetric or nearly symmetric distributions. But it can be misleading when used to measure the location of highly skewed distributions. In contrast, the median m is a more robust measure of the central tendency of a probability distribution in the sense that it is not much affected by a few outliers.

Theorem 3.17. *The median m is the optimal solution for minimizing the mean absolute error, namely,*

$$m = \arg\min_{a} E|X - a|.$$

Proof: This is left as an exercise.

Case (2): Value at Risk (VaR) and Financial Risk Management

In the finance literature, when X is a return on the portfolio over a certain holding period, $-Q(\alpha)$ with a small α (e.g., $\alpha = 0.01$) is usually called the Value-at-Risk. Intuitively, $-Q(\alpha)$ is the threshold level that actual loss will exceed with probability α. It has been used by (e.g.) Bank of International Settlements to set the level of bank capital.

Example 3.34. [Value at Risk and J.P. Morgan RiskMetrics]: The VaR at level α, $V_t(\alpha)$, of a portfolio over a certain time horizon is defined as

$$P[X_t < -V_t(\alpha)|I_{t-1}] = \alpha,$$

where X_t is the return on the portfolio over the holding period t, and I_{t-1} is the information available at time $t - 1$.

Figure 3.20 provides a graphical representation for $V_t(\alpha)$.

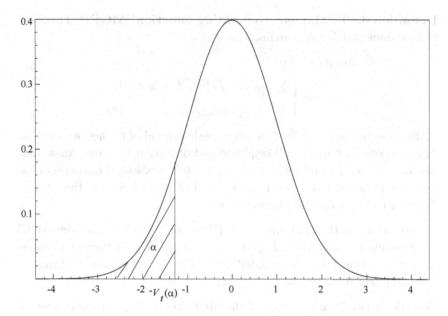

Figure 3.20: Graphical representation for $V_t(\alpha)$

The VaR, $V_t(\alpha)$, is the threshold which actual loss will exceed with probability α. This forms a basis for determining the suitable level of risk capital to prevent extreme adverse market events which would otherwise cause bankruptcy or default with probability higher than α. Obviously, V_t is the negative conditional quantile at level α.

In addition to VaR, quantiles have been widely used in economics. For example, suppose X is the income of the population. Then the difference $Q(0.75) - Q(0.25)$ can be used as a robust measure for the income gap or income inequality between the bottom 25% and top 25% of the population. In statistics, the quantiles at the levels of 25%, 50%, 75% are called quartitles, and the difference $Q(0.75) - Q(0.25)$ is called the interquartile range.

3.9 Moment Generating Function

When we are interested in various moments of X, the mathematical manipulations may become quite involved as we calculate higher order moments. Fortunately, it is sometimes more convenient to define the moment generating function and use it to calculate moments.

128   *Probability and Statistics for Economists*

Definition 3.19. [Moment Generating Function (MGF)]: The MGF of a random variable X is defined as

$$M_X(t) = E\left(e^{tX}\right)$$

$$= \begin{cases} \sum_{x \in \Omega_X} e^{tx} f_X(x), & \text{if } X \text{ is a DRV,} \\ \int_{-\infty}^{\infty} e^{tx} f_X(x)dx, & \text{if } X \text{ is a CRV.} \end{cases}$$

If the expectation exists for t in some neighborhood of 0, then we say that $M_X(t)$ exists for t in a small neighborhood of 0 (that is, there exists some constant $\epsilon > 0$ such that for all $t \in (-\epsilon, \epsilon)$, $E(e^{tX})$ exists). If the expectation does not exist for any small neighborhood of 0, then we say that $M_X(t)$ does not exist for the distribution of X.

Note that for the existence of $M_X(t)$, we only need to find some small neighborhood of 0 on which $M_X(t)$ is well defined. This is because all information about the probability distribution of X is contained in the behavior of $M_X(t)$ near 0, as will be seen below.

Question: Does the existence of the MGF $M_X(t)$ imply the existence of an infinite set of moments?

The answer is yes: the existence of $M_X(t)$ for $t \in (-\epsilon, \epsilon)$ implies that derivatives of $M_X(t)$ of all orders exist at $t = 0$.
Why?
By MacLaurin's series expansion, we have for all real numbers $t \in (-\epsilon, \epsilon)$,

$$e^{tx} = \sum_{k=0}^{\infty} \frac{(tx)^k}{k!}.$$

It follows that MacLaurin's series expansion for $M_X(t)$ is

$$M_X(t) = \sum_{k=0}^{\infty} \frac{t^k}{k!} E(X^k).$$

If, for some k such that $E(X^k) = \infty$, then $M_X(t)$ does not exist for all $t \neq 0$. Therefore, if $M_X(t)$ exists for all $t \in (-\epsilon, \epsilon)$, then all moments of X must exist. The expansion above thus reveals that $M_X(t)$ contains all moment information in the neighborhood of 0, and provides a way to characterize all moments from the MGF uniquely.

We now investigate the properties of the MGF $M_X(t)$.

Theorem 3.18. *Suppose the MGF $M_X(t)$ exists for all t in some small neighborhood of 0. Then $M_X(0) = 1$.*

Proof: By definition, $M_X(t) = \int_{-\infty}^{\infty} e^{tx} dF_X(x)$, it follows that $M_X(0) = \int_{-\infty}^{\infty} dF_X(x) = 1$. This completes the proof.

Theorem 3.19. *If the MGF $M_X(t)$ exists for all t in some neighborhood of 0, then for all positive integers $k = 1, 2, \cdots$,*

$$M_X^{(k)}(0) = E(X^k).$$

Proof: For any given integer $k > 0$ and all $t \in (-\epsilon, \epsilon)$, we have

$$M_X^{(k)}(t) = \frac{d^k}{dt^k} \int_{-\infty}^{\infty} e^{tx} dF_X(x)$$
$$= \int_{-\infty}^{\infty} \frac{d^k}{dt^k} \left(e^{tx} \right) dF_X(x)$$
$$= \int_{-\infty}^{\infty} x^k e^{tx} dF_X(x).$$

Setting $t = 0$, we obtain

$$M_X^{(k)}(0) = \int_{-\infty}^{\infty} x^k dF_X(x) = E(X^k).$$

This theorem shows that every moment of X can be computed by differentiating $M_X(t)$ at the origin, provided $M_X(t)$ exists for $t \in (-\epsilon, \epsilon)$. This is exactly the reason why we call $M_X(t)$ a MGF, that is, the function can be used to generate moments.

For example, for $k = 1, 2$, we have

$$M_X^{(1)}(0) = \mu_X,$$
$$M_X^{(2)}(0) = E(X^2) = \sigma_X^2 + \mu_X^2.$$

Note that the second derivative $M_X^{(2)}(0)$ is equal to the second moment of X, not its variance σ_X^2 (unless $\mu_X = 0$).

Theorem 3.20. *Suppose $Y = a + bX$, where a and b are two constants, and the MGF $M_X(t)$ of X exists for all t in a small neighborhood of 0. Then the MGF*

$$M_Y(t) = e^{at} M_X(bt)$$

for all t in a small neighborhood of 0.

Proof: The proof is left as an exercise.

Example 3.35. Suppose a random variable X has mean μ, variance σ^2, and the MGF $M_X(t)$ for all t in a small neighborhood of 0. Define a standardized random variable

$$Y = \frac{X - \mu}{\sigma},$$

which corresponds to the case of $Y = a + bX$ with $a = -\frac{\mu}{\sigma}, b = \frac{1}{\sigma}$. Then the MGF

$$M_Y(t) = e^{-\frac{\mu}{\sigma}t} M_X\left(\frac{t}{\sigma}\right).$$

In addition to using the MGF to compute moments, we can also use the MGF to characterize the probability distribution.

Question: How to use the MGF to characterize the probability distribution of X?

Theorem 3.21. *[Uniqueness of MGF]: Suppose two random variables X and Y have their MGF's $M_X(t)$ and $M_Y(t)$ existing for all t in some neighborhood of 0 denoted as $N_\epsilon(0) = \{t : -\epsilon < t < \epsilon\}$. Then X and Y have the same $M_X(t)$ and $M_Y(t)$ for all t in $N_\epsilon(0)$, if and only if $F_X(z) = F_Y(z)$ for all $z \in \mathbb{R}$ (i.e., they have an identical distribution).*

Proof: The proof of this theorem relies on the theory of Laplace transforms (see, e.g., Feller 1971). The defining equation of MGF $M_X(t)$, when PMF or PDF $f_X(x)$ exists, is given by

$$M_X(t) = \int_{-\infty}^{\infty} e^{tx} f_X(x) dx,$$

which defines a Laplace transform. A key fact about Laplace transforms is their uniqueness. If $M_X(t) = \int_{-\infty}^{\infty} e^{tx} f_X(x) dx$ is valid for all t with $|t| < \epsilon$, where ϵ is some positive number, then given $M_X(t)$, there is only one function $f_X(x)$ that satisfies the Laplace transform. Given this fact, the theorem is reasonable. However, the formal proof is rather technical and provides no additional insight.

An important implication of the above uniqueness theorem is as follows: given some MGF $M_X(t)$, suppose we can find some CDF $F_X(x)$ that corresponds to $M_X(t)$. Then $F_X(x)$ must be the only distribution that generates $M_X(t)$.

We now consider a few examples to show how to recover the probability distribution from the MGF $M_X(t)$.

Example 3.36. A DRV X has

$$M_X(t) = \frac{1}{2} + \frac{1}{4}e^{-t} + \frac{1}{4}e^t, \qquad -\infty < t < \infty.$$

Find the probability distribution of X and justify your answer.

Solution: By definition,

$$M_X(t) = E(e^{tX})$$
$$= \sum_x e^{tx} f_X(x)$$
$$= \frac{1}{2}e^{0 \cdot t} + \frac{1}{4}e^{(-1) \cdot t} + \frac{1}{4}e^{1 \cdot t}.$$

We guess the PMF

$$f_X(x) = \begin{cases} \frac{1}{4}, & \text{if } x = -1, \\ \frac{1}{2}, & \text{if } x = 0, \\ \frac{1}{4}, & \text{if } x = 1. \end{cases}$$

It can be checked easily that this guessed PMF has the MGF $M_X(t) = \frac{1}{2} + \frac{1}{4}e^{-t} + \frac{1}{2}e^t$. We now argue that the guessed PMF is the unique distribution with the given $M_X(t)$. Suppose there exists an alternative distribution $h_X(x)$ that yields the same MGF $M_X(t)$. Then by the uniqueness theorem of the MGF, this alternative distribution $h_X(x)$ must coincide with the guessed PMF $f_X(x)$. This justifies that the guessed PMF $f_X(x)$ is the unique distribution associated with the MGF $M_X(t)$.

Example 3.37. Suppose a random variable X has mean 0, variance 2, and MGF

$$M_X(t) = a(1 + be^{-2t} + e^{-t} + e^t + ce^{2t}), \qquad -\infty < t < \infty.$$

(1) Find the values of a, b, and c.
(2) What is the PMF of X? Justify why your answer is indeed the PMF of X.

Solution: (1) From the above theorems, we have

$$M(0) = 1,$$
$$M'(0) = \mu = 0,$$
$$M''(0) = E(X^2)$$
$$= \sigma^2 + \mu^2$$
$$= 2 + 0^2$$
$$= 2.$$

It follows that

$$M_X(0) = 1 = a(3 + b + c),$$

$$M_X'(0) = a(-2b - 1 + 1 + 2c) = 0,$$

$$M_X''(0) = a(4b + 1 + 1 + 4c) = 2.$$

Solving these equations, we can obtain $a = \frac{1}{5}, b = c = 1$.

(2) Using the reasoning analogous to that of Example 3.19, we can find a guessed PMF $f_X(x) = \frac{1}{5}$ for $x = -2, -1, 0, 1, 2$, and $f_X(x) = 0$ otherwise. This is the unique distribution by the uniqueness theorem of the MGF.

Example 3.38. Suppose a DRV X has a MGF $M_X(t) = \frac{1}{n}\left(\frac{1-e^{nt}}{1-e^t}\right)$ for $|t| < 1$. What is the PMF $f_X(x)$?

Solution: By the formula of the partial sum of geometric series,

$$\sum_{i=1}^{n} z^{i-1} = \frac{1 - z^n}{1 - z}, \text{ for } |z| < 1,$$

we have

$$\frac{1 - e^{nt}}{1 - e^t} = \sum_{i=1}^{n} e^{t(i-1)}$$

by putting $z = e^t$. It follows that

$$M_X(t) = \frac{1}{n}\left(\frac{1 - e^{nt}}{1 - e^t}\right)$$

$$= \sum_{i=1}^{n} \frac{1}{n} e^{t(i-1)}.$$

On the other hand, recall that if X is a DRV, then

$$M_X(t) = \sum_{i=1}^{n} f_X(x_i) e^{tx_i},$$

where $x_i = i - 1$ for $i = 1, \cdots, n$. We thus guess that the PMF is $f_X(x) = \frac{1}{n}$ for $x = 0, 1, \cdots, n - 1$ and $f_X(x) = 0$ otherwise. This is a discrete uniform distribution. It can generate the MGF $M_X(t)$ given above. By the uniqueness theorem of the MGF, this guessed PMF is the unique distribution associated with the MGF $M_X(t) = \frac{1}{n}\frac{1-e^{nt}}{1-e^t}$ for $|t| < 1$.

Example 3.39. If a DRV X has a MGF $M_X(t) = \frac{1-r}{1-re^t}$ for $|t| < \frac{1}{r}$, where $r > 0$, what is the probability distribution of X?

Solution: Using the formula that for $|a| < 1$,

$$\frac{1}{1-a} = \sum_{x=0}^{\infty} a^x,$$

we have

$$M_X(t) = (1-r) \sum_{x=0}^{\infty} (re^t)^x$$

$$= \sum_{x=0}^{\infty} (1-r) r^x e^{tx}$$

$$= \sum_{x=0}^{\infty} f_X(x) e^{tx},$$

where $f_X(x) = (1-r) r^x$, for $x = 0, 1, \cdots$. By the uniqueness theorem of the MGF, this guessed PMF is the unique probability distribution associated with the MGF $M_X(t) = \frac{1-r}{1-re^t}$ for $|t| < \frac{1}{r}$. This distribution is actually called the geometric distribution; see discussion in Chapter 4.

Next, we consider a convergence theorem that is very useful in investigating the asymptotic or limiting behavior of a sequence of distributions.

Theorem 3.22. *[Convergence of MGF]: Suppose $\{X_n, n = 1, 2, \cdots\}$ is a sequence of random variables, each with MGF $M_n(t)$ and CDF $F_n(x)$. Furthermore, suppose that*

$$\lim_{n \to \infty} M_n(t) = M_X(t)$$

for all t in a neighborhood of 0, where $M_X(t)$ is the MGF of a random variable X with CDF $F_X(x)$. Then

$$\lim_{n \to \infty} F_n(x) = F_X(x)$$

for all continuity points x (i.e., all points where $F_X(x)$ is continuous).

Suppose $\{X_n, n = 1, 2, \cdots\}$ is a sequence of random variables with distribution functions $\{F_n(x)\}$, and X is a random variable with CDF $F_X(x)$. If $F_n(x) \to F_X(x)$ as $n \to \infty$ for all continuity points (i.e., all $x's$ where $F_X(x)$ is continuous), then we say X_n converges in distribution to the random variable X, denoted $X_n \xrightarrow{d} X$. See formal definition and related discussion in Chapter 7.

An important implication of the theorem is that convergence in MGF implies convergence in distribution. In practice, both $F_n(x)$ and $M_n(t)$ are generally unknown, but if we can show that they converge to some well-known limit functions $F_X(x)$ and $M_X(t)$, then we can use these well-known limit functions as the approximations to the unknown functions $F_n(x)$ and $M_n(t)$ respectively.

Question: How to check the convergence in MGF?

The following two theorems on the expectations of limits provide nice tools to check convergence in MGF.

Theorem 3.23. *[Monotone Convergence]: If* $0 \leq g_n(x) \leq g_{n+1}(x)$ *for* $n \geq 1$, *then*

$$\lim_{n \to \infty} \int_{-\infty}^{\infty} g_n(x) dF(x) = \int_{-\infty}^{\infty} \lim_{n \to \infty} g_n(x) dF(x).$$

Theorem 3.24. *[Dominated Convergence]: If* $|g_n(x)| \leq \bar{g}(x)$ *for all* $n \geq 1$, $\int_{-\infty}^{\infty} \bar{g}(x) dF(x) < \infty$, *and* $\lim_{n \to \infty} g_n(x) = g(x)$ *except for* $x \in N$, *where* N *is a set with probability zero, then*

$$\lim_{n \to \infty} \int_{-\infty}^{\infty} g_n(x) dF(x) = \int_{-\infty}^{\infty} g(x) dF(x).$$

We have seen that if the MGF $M_X(t)$ exists in a neighborhood of 0, it characterizes an infinite set of moments. Thus, a natural question is whether we can use the infinite set of moments to uniquely characterize a distribution function? In other words, if $E(X^k) = E(Y^k)$ for all positive integers $k > 0$, are random variables X and Y identically distributed? Unfortunately, the answer is "no" in general. In other words, characterizing the set of moments is not enough to determine a distribution uniquely because there may be two distinct random variables having the same moments. This is illustrated in the example below.

Example 3.40. Consider two PDF's:

$$f_X(x) = \begin{cases} \frac{1}{\sqrt{2\pi}x} \exp[-(\ln x)^2/2], & x > 0, \\ 0, & x \leq 0, \end{cases}$$

$$f_Y(x) = \begin{cases} f_X(x)[1 + \sin(2\pi \ln x)], & x > 0, \\ 0, & x \leq 0. \end{cases}$$

Show $E(X^k) = E(Y^k)$ for all positive integers k, where X, Y have PDF's $f_X(x)$ and $f_Y(x)$ respectively.

Note that $f_X(x)$ and $f_Y(x)$ are two different distributions, as is obvious in Figure 3.21.

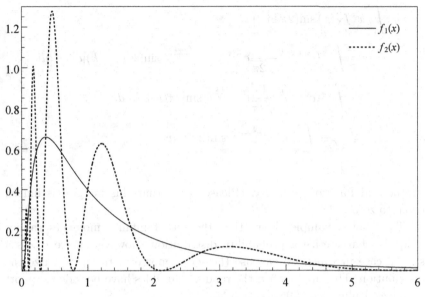

Figure 3.21: $f_X(x)$ and $f_Y(x)$ of Example 3.40

Solution: (1) First of all, from Example 3.27, observe that X follows a lognormal distribution$(0,1)$, and $Z = \ln X \sim N(0,1)$, the standard normal distribution. It follows that

$$\begin{aligned}
E(X^k) &= E[(e^Z)^k] \\
&= E(e^{kZ}) \\
&= M_Z(k) \\
&= e^{\frac{1}{2}k^2},
\end{aligned}$$

where we have made use of the fact that $M_Z(t) = e^{\frac{1}{2}t^2}$ is the MGF of the standard normal random variable $Z \sim N(0,1)$.

(2) Next, observe that

$$E(Y^k) = E(X^k) + \int_0^\infty x^k f_X(x) \sin(2\pi \ln x) dx.$$

To show $E(Y^k) = E(X_1^k)$, it suffices to show

$$\int_0^\infty x^k f_X(x) \sin(2\pi \ln x) dx = 0.$$

Define the transformation $v = \ln x - k$. Then the integral

$$\int_0^\infty x^k f_X(x) \sin(2\pi \ln x)\,dx$$

$$= \int_{-\infty}^\infty (e^{v+k})^k \frac{1}{\sqrt{2\pi}} e^{-(v+k)} e^{-\frac{(v+k)^2}{2}} \sin[2\pi(v+k)] e^{(v+k)}\,dv$$

$$= \int_{-\infty}^\infty (e^{v+k})^k \frac{1}{\sqrt{2\pi}} e^{-\frac{(v+k)^2}{2}} \sin[2\pi(v+k)]\,dv$$

$$= \frac{1}{\sqrt{2\pi}} \int_{-\infty}^\infty e^{-\frac{1}{2}(v^2-k^2)} \sin(2\pi v)\,dv$$

$$= 0$$

by the odd function property. (Please verify that this integral is indeed equal to zero.)

The above example shows that the set of infinite moments cannot uniquely characterize a probability distribution. However, there exists a special case in which we can use the set of moments to characterize the distribution. This arises when the random variables have bounded support (see, e.g., Billingsley 1995, Section 30).

Theorem 3.25. *Let $F_X(x)$ and $F_Y(y)$ be two CDF's both of which have bounded supports. Then $F_X(z) = F_Y(z)$ for all $z \in (-\infty, \infty)$ if and only if $E(X^k) = E(Y^k)$ for all integers $k = 1, 2, \cdots$.*

Proof: Since X and Y have bounded supports, there exists a sufficiently large constant $M > 0$ such that $P(|X| \le M) = 1$ and $P(|Y| \le M) = 1$. Because $E|X|^k \le M^k$ and $E|Y|^k \le M^k$, we have

$$M_X(t) = \sum_{k=0}^\infty \frac{t^k}{k!} E(X^k)$$

$$\le \sum_{k=0}^\infty \frac{t^k}{k!} E|X|^k$$

$$\le \sum_{k=0}^\infty \frac{(tM)^k}{k!}$$

$$= e^{tM} < \infty \text{ for any } t < \infty.$$

Similarly, we have

$$M_Y(t) \le e^{tM} < \infty.$$

Now, if $E(X^k) = E(Y^k)$ for all integers $k > 0$, we have $M_X(t) = M_Y(t)$ for all t in a neighborhood of 0, given $M_X(t) = \sum_{k=0}^\infty \frac{t^k}{k!} E(X^k)$. This in turn implies $F_X(z) = F_Y(z)$ for all $z \in (-\infty, \infty)$ by the uniqueness theorem of the MGF.

On the other hand, if $F_X(z) = F_Y(z)$ for all $z \in (-\infty, \infty)$, and $P(|X| \leq M) = 1, P(|Y| \leq M) = 1$, then the moments $E(X^k)$ and $E(Y^k)$ exist, and $E(X^k) = E(Y^k)$ for all $k > 0$. It follows that for two random variables X and Y with bounded supports, $F_X(z) = F_Y(z)$ for all $z \in (-\infty, \infty)$ if and only if $E(X^k) = E(Y^k)$ for all integers $k > 0$. This completes the proof.

We have shown above that if $M_X(t)$ exists for all t in some small neighborhood of 0, then all moments $E(X^k)$ exist. Now, we ask whether the converse holds. That is, does the existence of all moments of a probability distribution imply the existence of the MGF $M_X(t)$ for t in some neighborhood of zero?

In general, the existence of all moments does not ensure the existence of the MGF. Intuitively, the MGF $M_X(t)$ is a weighted sum of all moments. If $M_X(t)$ is finite for all t in some small neighborhood of 0, the all moments must be finite. If some moment does not exist, then the MGF does not exist. Furthermore, it is possible that all moments are finite but their weighted sum is infinite. In this case, the MGF does not exist. For example, assume the k-th moment of X is given by $E(X^k) = (k+1)!$, but the MGF $M_X(t) = \sum_{k=0}^{\infty} \frac{t^k}{k!} E(X^k)$ goes to infinity, and so $M_X(t)$ does not exist.

Question: Can you give an example of a probability distribution for which all moments exist but the MGF does not exist?

We now provide a solution to this difficulty, by introducing a so-called characteristic function, which maintains the nice features of the MGF $M_X(t)$ and exists for all distributions (no matter whether moments exist).

3.10 Characteristic Function

For some distributions (e.g., the Cauchy and lognormal distributions), the MGF does not exist. We now introduce a function which exists for any probability distribution.

Definition 3.20. [Characteristic Function]: The characteristic function of a random variable X with CDF $F_X(x)$ is defined as

$$\varphi_X(t) = E(e^{itX})$$
$$= \int_{-\infty}^{\infty} e^{itx} dF_X(x),$$

where $\mathbf{i} = \sqrt{-1}$, and

$$e^{\mathbf{i}tx} = \cos(tx) + \mathbf{i}\sin(tx).$$

By definition, $\varphi_X(t)$ is the Fourier transform of the distribution function $F_X(x)$. As a fundamental property of the Fourier transformation, both $\varphi_X(t)$ and $F_X(x)$ contain the same information about the probability distribution of X. Given one, we can always recover the other. In particular, if X is a random variable with PDF/PMF $f_X(x)$, then the inverse Fourier transform will recover the PDF/PMF, that is,

$$f_X(x) = \frac{1}{2\pi} \int_{-\infty}^{\infty} e^{-\mathbf{i}tx} \varphi_X(t)dt.$$

This inversion formula is rather useful in calculating probabilities for the distributions whose PDF's have no closed form.

Theorem 3.26. *[Properties of Characteristic Function]:*

(1) For any probability distribution, the characteristic function $\varphi_X(t)$ always exists and is bounded, that is,

$$|\varphi_X(t)| \leq 1, \text{ for } -\infty < t < \infty.$$

(2) $\varphi_X(0) = 1$.

(3) $\varphi_X(t)$ is continuous over $(-\infty, \infty)$.

(4) $\varphi_X(-t) = \varphi_X(t)^*$, where $\varphi_X(t)^*$ denotes the complex conjugate of $\varphi_X(t)$.

(5) Suppose $Y = a + bX$, where a and b are any real constants. Then

$$\varphi_Y(t) = e^{\mathbf{i}at}\varphi_X(bt).$$

(6) If the MGF $M_X(t)$ exists for all t in some neighborhood of 0, then $\varphi_X(t) = M_X(\mathbf{i}t)$ for all $t \in (-\infty, \infty)$.

The characteristic function $\varphi_X(t)$ is more general than the MGF $M_X(t)$ in the sense that $\varphi_X(t)$ always exists for all distributions and all t on the real line while $M_X(t)$ may not exist for some distributions. We note that Result (1) follows because

$$|\varphi_X(t)| \leq \int_{-\infty}^{\infty} |e^{\mathbf{i}tx}|dF_X(x) = \int_{-\infty}^{\infty} dF_X(x) = 1 < \infty.$$

Example 3.41. Suppose a random variable X follows a Cauchy$(0,1)$ distribution with PDF

$$f_X(x) = \frac{1}{\pi}\frac{1}{1+x^2}, \text{ for } -\infty < x < \infty.$$

Then its characteristic function exists and is given by

$$\varphi_X(t) = e^{-|t|}, \text{ for } -\infty < t < \infty.$$

Note that $\varphi_X(t)$ has a kink point at $t = 0$ so it is not differentiable at the origin, which is consistent with the fact that none of the moments $E(X^k)$ with $k \geq 1$ of X exists (as a consequence, the MGF do not exist for the Cauchy distribution).

Like the MGF $M_X(t)$, the characteristic function $\varphi_X(t)$ can be used to generate moments when the moments exist.

Theorem 3.27. *Suppose the k-th moment of X exists. Then $\varphi_X(t)$ is differentiable with respect to $t \in (-\infty, \infty)$ up to order k, and*

$$\varphi_X^{(k)}(0) = \mathbf{i}^k E(X^k).$$

Proof: Given $E|X|^k < \infty$, the k-th derivative $\varphi_X^{(k)}(t)$ exists because

$$\left| \varphi_X^{(k)}(t) \right| = \left| \frac{d^k}{dt^k} \int_{-\infty}^{\infty} e^{\mathbf{i}tx} dF_X(x) \right|$$

$$= \left| \int_{-\infty}^{\infty} (\mathbf{i}x)^k e^{\mathbf{i}tx} dF_X(x) \right|$$

$$\leq \int_{-\infty}^{\infty} |x|^k dF_X(x)$$

$$= E|X|^k < \infty.$$

Because $\varphi_X^{(k)}(t) = \int_{-\infty}^{\infty} (\mathbf{i}x)^k e^{\mathbf{i}tx} dF_X(x)$, by setting $t = 0$, we obtain

$$\varphi_X^{(k)}(0) = \int_{-\infty}^{\infty} (\mathbf{i}x)^k dF_X(x)$$

$$= \mathbf{i}^k E(X^k).$$

We also have a uniqueness theorem for the characteristic function $\varphi_X(t)$.

Theorem 3.28. *[Uniqueness of Characteristic Function]: Suppose two random variables X and Y have the characteristic functions $\varphi_X(t)$ and $\varphi_Y(t)$ respectively. Then X and Y are identically distributed if and only if $\varphi_X(t) = \varphi_Y(t)$ for all $t \in (-\infty, \infty)$.*

Proof: By definition, the characteristic function $\varphi_X(t)$ is the Fourier transform of the CDF $F_X(x)$, i.e.,

$$\varphi_X(t) = \int_{-\infty}^{\infty} e^{\mathbf{i}tx} dF_X(x).$$

By the one-to-one correspondence between a function and its Fourier transform, we have a unique characteristic function $\varphi_X(t)$ for each distribution $F_X(x)$. This completes the proof.

It is important to check all t on the entire real line for the characteristic functions $\varphi_X(t)$ and $\varphi_Y(t)$. This is quite different from the case of MGF's, where it is only necessary to check all t in some neighborhood of 0.

The following theorem shows that the convergence in characteristic function is equivalent to convergence in distribution.

Theorem 3.29. *[Convergence in Characteristic Function]: Let $\{X_n\}$ be a sequence of random variables with CDF's $F_n(x)$ and characteristic functions $\varphi_n(t)$. Let X be a random variable with CDF $F_X(x)$ and characteristic function $\varphi_X(t)$. Let $n \to \infty$.*

(1) If $F_n(x) \to F_X(x)$ for all points x where $F_X(x)$ is continuous, then for every $t \in (-\infty, \infty)$, $\varphi_n(t) \to \varphi_X(t)$.

(2) Further, if $\varphi_n(t) \to \varphi_X(t)$ for all $t \in (-\infty, \infty)$, then $F_n(x) \to F_X(x)$ for all continuity points $x \in \mathbb{R}$.

Proof: See Lukacs (1970, pp. 49-50).

We are going to use this theorem to prove a central limit theorem in Chapter 7.

The characteristic function is useful because sometimes it is more convenient to study the behaviors of characteristic functions $\varphi_X(t)$ than distribution functions. For example, a class of so-called stable distributions is defined in terms of their characteristic functions (see Section 4.3.4, Chapter 4). In finance, a general class of popular continuous-time models called affine jump diffusion models (see, e.g., Duffie, Pan and Singleton 2000) is well known for not having a closed form for its (conditional) probability density function, which renders maximum likelihood estimation unfeasible. However, this class of models has a closed form expression for its (conditional) characteristic function. Thus, inference and estimation can be based on the characteristic function.

Another important example of a stochastic process whose characteristic function has a closed form is the class of Levy processes, which serve as a building block for modern financial modeling. The most famous Levy process is the Brownian motion, which is the basic element in many famous financial models including those of Black and Scholes (1973) and Cox, Ingersoll and Ross (1985).

There has been increasing interest in using the characteristic function in econometrics and finance. See Hong (1999) and the references there.

3.11   Conclusion

In this chapter, we introduce the key concept of random variables as a quantitative measure of the outcome of a random experiment. To describe the probability distribution of a random variable, we introduce the CDF. Depending on whether a random variable is discrete or continuous, we further introduce the PMF and the PDF to characterize the probability distribution of the random variable. Methods to derive the probability distribution of a measurable transformation of a random variable are also developed.

With the CDF (or equivalently the PMF or the PDF), we define a class of moments, which is a summary characteristic of a probability distribution. Economic intuitions and applications for first few moments are provided. We also discuss the MGF and the characteristic function. Both of them can be differentiated to generate moments (when moments exist), and more importantly, both of them can be used to uniquely characterize the probability distribution of a random variable.

EXERCISE 3

3.1. Seven balls are distributed randomly into seven cells. Let X_i be the number of cells containing exactly i balls. What is the probability distribution of X_3? (That is, find $P(X = x)$ for every possible x.)

3.2. Prove that the following functions are cumulative distribution functions (CDF's):

(1) $\frac{1}{2} + \frac{1}{\pi} \tan^{-1}(x), x \in (-\infty, +\infty)$;

(2) $(1 + e^{-x})^{-1}, x \in (-\infty, +\infty)$;

(3) $e^{-e^{-x}}, x \in (-\infty, +\infty)$;

(4) $1 - e^{-x}, x \in (0, +\infty)$, and $0, x \leq 0$.

3.3. A CDF F_X is stochastically greater than a CDF F_Y if $F_X(t) \leq F_Y(t)$ for all t and $F_X(t) < F_Y(t)$ for some t. Prove that if $X \sim F_X$ and $Y \sim F_Y$, then $P(X > t) \geq P(Y > t)$ for every t and $P(X > t) > P(Y > t)$ for some t, that is, X tends to be bigger than Y.

3.4. An appliance store receives a shipment of 30 microwave ovens, 5 of which are (unknown to the manager) defective. The store manager selects 4 ovens at random, without replacement, and tests to see if they are defective. Let X be the number of defectives found. Calculate the PMF and CDF of X and plot the CDF.

3.5. Suppose $X = X_1$ with probability p and $X = X_2$ with probability $1 - p$, where $p \in (0, 1)$, X_1 and X_2 are random variables with CDF's $F_1(x)$ and $F_2(x)$ respectively. Find the CDF of X.

3.6. Suppose $F_1(x)$ and $F_2(x)$ are two CDF's. Define $F(x) = F_1(x)F_2(x)$. Is $F(x)$ a CDF? Give your reasoning.

3.7. Let $f(x) = \frac{c}{x}$ for $x = 1, 2, \cdots$ and c is a constant. Can you find a finite value for constant c so that $f(x)$ is a valid PMF? If yes, give the value of c. Otherwise, explain why not.

3.8. An investment firm offers its customers municipal bonds that mature after varying numbers of years. Given that the cumulative distribution of T, the number of years to maturity for a randomly selected bond is

$$F(t) = \begin{cases} 0, & t < 1, \\ \frac{1}{4}, & 1 \leq t < 3, \\ \frac{1}{2}, & 3 \leq t < 5, \\ \frac{3}{4}, & 5 \leq t < 7, \\ 1, & t \geq 7. \end{cases}$$

Find: (1) $P(T = 5)$; (2) $P(T > 3)$; (3) $P(1.4 < T < 6)$. Give your reasoning.

3.9. A grocery store sells X hundred kilograms of rice every day, where the distribution of X is of the following form:

$$F(x) = \begin{cases} 0, & x < 0, \\ kx^2, & 0 \leq x < 3, \\ k(-x^2 + 12x - 3), & 3 \leq x < 6, \\ 1, & x \geq 6. \end{cases}$$

Suppose this grocery store's total sales of rice do not reach 600 kilograms on any given day.

(1) Find the value of k.

(2) What is the probability that the store sells between 200 and 400 kilograms of rice next Thursday?

(3) What is the probability that the store sells over 300 kilograms of rice next Thursday?

(4) We are given that the store sold at least 300 kilograms of rice last Friday. What is the probability that it did not sell more than 400 kilograms on that day?

3.10. A random variable X is called symmetric if for all x on the real line, $P(X \geq x) = P(X \leq -x)$. Show that if X is symmetric, then for all $x > 0$, its CDF $F(\cdot)$ satisfies the following relations:

(1) $P(|X| \leq x) = 2F(x) - 1$;

(2) $P(|X| > t) = 2[1 - F(x)]$;

(3) $P(X = x) = F(x) + F(-x) - 1$.

3.11. For each of the following functions, determine the value of c that makes $f(x)$ a PDF:

(1) $f(x) = c\sin x, 0 < x < \frac{\pi}{2}$;

(2) $f(x) = ce^{-|x|}, -\infty < x < \infty$.

3.12. Suppose X has the geometric PMF $f_X(x) = \frac{1}{3}(\frac{2}{3})^x, x = 0, 1, 2, ...$ Determine the probability distribution of $Y = X/(X + 1)$.

3.13. In each of the following distributions of X, find the PDF of Y and show that the PDF integrates to 1:

(1) $f_X(x) = \frac{1}{2}e^{-|x|}, -\infty < x < \infty; Y = |X|^3$;

(2) $f_X(x) = \frac{3}{8}(x + 1)^2, -1 < x < 1; Y = 1 - X^2$.

3.14. Let X have PDF $f_X(x) = \frac{2}{9}(x + 1), -1 \leq x \leq 2$. Find the PDF of $Y = X^2$.

3.15. If the random variable X has PDF

$$f(x) = \begin{cases} \frac{x-1}{2}, \text{if } 1 < x < 3, \\ 0, \text{otherwise.} \end{cases}$$

Find a monotone function $g(x)$ such that the random variable $Y = g(X)$ has a uniform (0,1) distribution.

3.16. Let $g(\cdot)$ be a real-valued function that satisfies the relation $\int_{-\infty}^{\infty} g(t)dt = 1$. Show that if, for a random variable X, the random variable $Y = \int_{-\infty}^{X} g(t)dt$ is uniform, then $g(\cdot)$ is the PDF of X.

3.17. Let $Y = a + bX$, where the random variable X has a PDF $f_X(x)$. Find the PDF $f_Y(y)$ of Y.

3.18. Let X have the PDF

$$f_X(x) = \frac{4}{\beta^3 \sqrt{\pi}} x^2 e^{-x^2/\beta^2}, \qquad 0 < x < \infty, \beta > 0.$$

Verify that $f_X(x)$ is indeed a PDF. [Hint: you may use the property that the integral of the PDF of a normal random variable is 1.]

3.19. Suppose $f(x) = c[1+2\sin(x)]$ for $-\pi < x < \pi$ and $f(x) = 0$ otherwise. Can you find a value of c so that $f(x)$ is a valid PDF. If yes, find the value of c. If not, give your reasoning.

3.20. Check for what value(s) of k that the following function can be a PDF:

$$f(x) = \begin{cases} \frac{1}{2} + kx, & -1 \le x \le 1, \\ 0, & \text{otherwise.} \end{cases}$$

Give your reasoning.

3.21. Suppose $f_X(x)$ and $f_Y(y)$ are two PDF's. Define $g(z) = \int_{-\infty}^{\infty} f_X(z - y)f_Y(y)dy$. Is $f(z)$ a PDF? Explain.

3.22. Suppose $f_X(x)$ is a PDF and $f_Y(y)$ is a PMF. Define $f(z) = \sum_{i=1}^{k} y_i^{-1} f_X(z/y_i)f_Y(y_i)$, where $y_i > 0$ for all $i = 1, \cdots, k$. Is $f(z)$ a PDF? Explain.

3.23. Suppose X has a PDF $f_X(x)$. Define $Y = X \cdot 1(a \le X \le b)$, where a, b are constants, and $1(\cdot)$ is the indicator function, taking value 1 if $a \le X \le b$, and taking value 0 otherwise. Show the PDF of Y is

$$f_Y(y) = \frac{f_X(x)}{F_X(b) - F_X(a)}, \qquad a \le y \le b.$$

3.24. $f_X(x) = \frac{1}{2}$ for $0 < x < 2$. Find the PDF of $Y = X(2 - X)$.

3.25. Suppose a continuous random variable X has PDF $f(x)$ and CDF $F(x)$ under the null hypothesis H_0, and has PDF $g(x)$ and CDF $G(x)$ under the alternative hypothesis H_1, where both $F(x)$ and $G(x)$ are strictly increasing. Define

$$Y = \int_{-\infty}^{X} f(x)dx = F(X).$$

(1) Show that under the alternative hypothesis H_1, the CDF of Y is given by

$$H(y) = P(Y \le y) = G[F^{-1}(y)],$$

where $F^{-1}(y)$ is the inverse function of $F(x)$.

(2) Show that under the alternative hypothesis H_1, the PDF of Y is

$$h(y) = \frac{g[F^{-1}(y)]}{f[F^{-1}(y)]}, \text{ for } 0 < y < 1.$$

3.26. Suppose X is a random variable with PDF $f_X(\cdot)$ and $Y = g(X)$ where $g(\cdot)$ is a strictly increasing function. Show that the α-th quantile of Y is $Q_\alpha(Y) = g[Q_\alpha(X)]$, where $Q_\alpha(X)$ is the α-th quantile of X.

3.27. (1) Let X be a continuous, nonnegative random variable, i.e. $f(x) = 0$ for $x < 0$. Show that $E(X) = \int_0^\infty [1 - F_X(x)] \, dx$, where $F_X(x)$ is the CDF of X.

(2) Let X be a discrete random variable whose range is the set of nonnegative integers. Show $E(X) = \sum_{k=0}^\infty [1 - F_X(k)]$, where $F_X(k) = P(X \le k)$. Compare this with Part (1).

3.28. Let X be a nonnegative random variable with CDF $F(\cdot)$. Define the indicator function $\mathbf{1}(t) = 1$ if $X > t$ and 0 otherwise.

(1) Show $\int_0^\infty \mathbf{1}(t)dt = X$.

(2) Taking expectation of both sides of the equation in Part (1), show $E(X) = \int_0^\infty [1 - F(t)]dt$.

(3) For $k > 0$, use Part (2) to show $E(X^k) = k \int_0^\infty t^{k-1}[1 - F(t)]dt$.

3.29. Show that for any continuous random variable X with CDF $F(\cdot)$ and PDF $f(\cdot)$, $E(X) = \int_0^\infty [1 - F(t)]dt - \int_0^\infty F(-t)dt$.

3.30. The median of a distribution is a value m such that $P(X \le m) \ge \frac{1}{2}$ and $P(X \ge m) \ge \frac{1}{2}$. (If X is continuous, m satisfies $\int_{-\infty}^m f(x)dx = \int_m^\infty f(x)dx = \frac{1}{2}$.) Find the median of the following distributions:

(1) $f(x) = 3x^2, 0 < x < 1$;

(2) $f(x) = \frac{1}{\pi(1+x^2)}, -\infty < x < +\infty$.

3.31. Show that if X is a continuous random variable, then

$$\min_{a} E\,|X - a| = E\,|X - m|,$$

where m is the median of X.

3.32. Let X have the PDF

$$f(x) = \frac{4}{\beta^3 \sqrt{\pi}} x^2 e^{-x^2/\beta^2}, \quad 0 < x < \infty, \beta > 0$$

Find $E(X)$ and $\mathrm{Var}(X)$.

3.33. Suppose that $X \sim N(0,1)$, namely, X has a PDF

$$f_X(x) = \frac{1}{\sqrt{2\pi}} \exp(-\frac{1}{2}x^2), \quad -\infty < x < \infty.$$

Define

$$Y = \frac{1}{\sqrt{2\pi}} \exp(-\frac{1}{2}X^2).$$

(1) Find the mean μ_Y of Y.
(2) Find the probability density function $f_Y(y)$ of Y.

3.34. Let X have the PDF

$$f(x) = \begin{cases} \frac{2}{\sqrt{2\pi}} e^{-\frac{x^2}{2}}, & \text{if } 0 < x < \infty \\ 0, & \text{otherwise.} \end{cases}$$

(1) Find the mean and variance of X.
(2) Find the PDF of $Y = X^2$.

3.35. Suppose the PDF $f_X(x)$ is an *even* function. ($f_X(x)$ is an *even* function if $f_X(x) = f_X(-x)$ for every x.) Show:
(1) X and $-X$ are identically distributed;
(2) $M_X(t)$ is symmetric about 0.

3.36. Let $f(x)$ be a PDF, and let a be a number such that, for all $\epsilon > 0$, $f(a+\epsilon) = f(a-\epsilon)$. Such a PDF is said to be symmetric about the point a.
(1) Give three examples of symmetric PDF's.
(2) Show that if $X \sim f(x)$, then the median of X is the number a.
(3) Show that if $X \sim f(x)$, and $E(X)$ exists, then $E(X) = a$.

3.37. (1) Suppose a PDF $f(x)$ is symmetric about constant a, namely $f(x - a) = f(-(x - a))$ for all x. Show that the mean $E(X) = a$ and the skewness is zero. Give your reasoning.

(2) Suppose a probability distribution has zero skewness. Can we conclude that the distribution is symmetric about the mean? If yes, show it. If not, give a counter example.

3.38. Suppose a continuous random variable X has the following PDF

$$f(x) = \begin{cases} \frac{1}{\sqrt{2\pi}}e^{-\frac{x^2}{2}}, & x < 0, \\ \frac{1}{\sqrt{8\pi}}e^{-\frac{x^2}{8}}, & x \geq 0. \end{cases}$$

Find: (1) $E(X)$; (2) var(X). Give your reasoning. [Hint: The following normal density formula may be useful: $\int_{-\infty}^{\infty} \frac{1}{\sqrt{2\pi\sigma^2}}e^{-\frac{(x-\mu)^2}{2\sigma^2}} dx = 1$ for all μ and σ^2.]

3.39. A random variable X is said to have a two-piece normal distribution with parameters $\alpha, \sigma_1, \sigma_2$ if its PDF

$$f_X(x) = \begin{cases} Ae^{-(x-\alpha)^2/(2\sigma_1^2)}, & x \leq \alpha, \\ Ae^{-(x-\alpha)^2/(2\sigma_2^2)}, & x > \alpha. \end{cases}$$

Find: (1) the constant A; (2) the mean of X; (3) the variance of X.

3.40. The Student t_ν-distribution normalized to have unit variance has the following probability density function

$$f_Z(z) = \frac{\Gamma\left(\frac{\nu+1}{2}\right)}{\Gamma\left(\frac{\nu}{2}\right)\sqrt{\pi(\nu-2)}}\left(1+\frac{z^2}{\nu-2}\right)^{-\frac{\nu+1}{2}}, \quad -\infty < z < \infty,$$

where the degree of freedom $2 < \nu < \infty$.

Now we define a skewed generalization of the Student t_ν density function:

$$f_X(x) = \begin{cases} BC\left[1+\frac{1}{\nu-2}\left(\frac{Bx+A}{1-\lambda}\right)^2\right]^{-\frac{\nu+1}{2}}, & x < -\frac{A}{B}, \\ BC\left[1+\frac{1}{\nu-2}\left(\frac{Bx+A}{1+\lambda}\right)^2\right]^{-\frac{\nu+1}{2}}, & x \geq -\frac{A}{B}, \end{cases}$$

where $2 < \nu < \infty$ and $-1 < \lambda < 1$, and the constants A, B, C are given by

$$A = 4\lambda C\frac{\nu-2}{\nu-1},$$
$$B^2 = 1+3\lambda^2 - A^2,$$
$$C = \frac{\Gamma\left(\frac{\nu+1}{2}\right)}{\Gamma\left(\frac{\nu}{2}\right)\sqrt{\pi(\nu-2)}}.$$

(1) Show $f_X(x)$ is a valid probability density function.

(2) Compute the mean of $f_X(x)$.

(3) Compute the variance of $f_X(x)$.

(4) Compute the skewness of $f_X(x)$.

For each part above, give your reasoning.

3.41. Let X and Y are two random variables, and a be a given constant. If for all $t > 0$, $P(|Y - a| \leq t) \leq P(|X - a| \leq t)$, then X is said to be more concentrated about a than Y. Suppose $E(X) = E(Y) = \mu$, and X is more concentrated about μ than Y. Show $\mathrm{var}(X) \leq \mathrm{var}(Y)$.

3.42. Let X and Y be two discrete random variables with the identical set of possible values $\Omega = \{a_1, a_2, \ldots, a_n\}$, where the a_i's are n different real numbers. Show that if $E(X^k) = E(Y^k)$ for $k = 1, 2, \ldots, n-1$, then X and Y are identically distributed; that is, $P(X = t) = P(Y = t)$ for $t \in \{a_1, \ldots, a_n\}$.

3.43. Suppose a discrete random variable X has the following PMF

$$f_X(x) = (1 - \gamma)\gamma^x, x = 0, 1, \cdots,$$

where γ is a fixed parameter and $0 < \gamma < 1$. Find: (1) the MGF of X; (2) the mean and the variance. [Hint: use the formula $\sum_{x=0}^{\infty} a^x = \frac{1}{1-a}$ for $|a| < 1$.]

3.44. Suppose a discrete random variable X has variance $\sigma_X^2 = \frac{1}{2}$ and moment generating function

$$M_X(t) = a + b(e^{-t} + e^t), \quad -\infty < t < \infty.$$

Find the PMF $f_X(x)$ of X and justify your answer.

3.45. Suppose a discrete random variable X has mean $\mu_X = 1$ and the moment generating function

$$M_X(t) = a + \frac{1}{5}e^{-3t} + \frac{2}{5}e^t + \frac{1}{5}e^{bt}, \text{ for all } -\infty < t < \infty,$$

where a and b are unknown constants. Find:

(1) the values of a and b;

(2) the PMF $f_X(x)$. Justify your solution.

3.46. Assume that the moment generating functions $M_X(t)$ and $M_Y(t)$ of two random variables X and Y exist for all t in some small neighborhood of 0. This assumption covers both part (1) and part (2) below.

(1) If X and Y are identically distributed, does this imply $E(X^k) = E(Y^k)$ for all positive integers k? Give your reasoning.

(2) If $E(X^k) = E(Y^k)$ for all positive integers k, does this imply that X and Y are identically distributed? Give your reasoning.

3.47. The cumulant generating function $K_X(t)$ of a random variable X is defined as the logarithm of its MGF $M_X(t)$, i.e., $K_X(t) = \ln M_X(t)$. The coefficient of $t^k/k!$ in the Taylor expansion of $K_X(t)$ is called the k-th cumulant of X and is denoted as κ_k. Show:

(1) $\kappa_1 = E(X)$;
(2) $\kappa_2 = E(X^2)$;
(3) $\kappa_3 = E(X^3)$;
(4) $\kappa_4 = E(X^4) - 3[E(X^2)]^2$;
(5) $\kappa_5 = E(X^5) - 10E(X^3)E(X^2)$.

Chapter 4

Important Probability Distributions

Abstract: In this chapter, we introduce a variety of discrete probability distributions and continuous probability distributions that are commonly used in economics and finance. Examples of discrete probability distributions include Bernoulli, Binomial, Negative Binomial, Geometric and Poisson distributions. Examples of continuous probability distributions include Beta, Cauchy, Chi-square, Exponential, Gamma, generalized Gamma, normal, lognormal, Weibull, and uniform distributions. The properties of these distributions as well as their applications in economics and finance are discussed. We also show some important techniques of obtaining moments and MGF's for various probability distributions.

Key words: Beta distribution, Bernoulli distribution, Binomial distribution, Double exponential distribution, Exponential distribution, Gamma distribution, Generalized Gamma distribution, Geometric distribution, Negative Binomial distribution, Normal distribution, Lognormal distribution, Poisson distribution, Poisson process, Weibull distribution.

4.1 Introduction

Recall that the probability space $(S, \mathbb{B}, P)$ completely characterizes a random experiment. In practice, the true probability distribution for the random experiment is usually unknown. One usually considers a class of probability measures, say PMF/PDF $f(x, \theta)$, indexed by parameter θ, where the functional form $f(\cdot, \cdot)$ is known. Each parameter value of θ yields a probability distribution, and different values of θ result in different distributions. The collection of these distributions constitute a family of probability distributions. This family of probability distributions is called a class of parametric probability distribution models. One main objective of statistics and econometrics is to use the observed economic data to estimate the true parameter value, say θ_0, under the assumption that there exists some parameter value θ_0 such that $f(x, \theta_0)$ coincides with the true probability distribution $f_X(x)$ of the random experiment. Before we come to the stage of estimating the true model parameter value θ_0 (we will do so in Chapters 6 and 8), we first introduce a number of important parametric distribution models and discuss their properties and applications in economics and finance. We emphasize that it is important to understand the meanings and roles that parameters play in each parametric distribution.

4.2 Discrete Probability Distributions

We start with discrete probability distributions.

4.2.1 *Bernoulli Distribution*

A DRV X follows a Bernoulli(p) distribution if its PMF

$$f_X(x) = \begin{cases} p & \text{if } x = 1, \\ 1 - p & \text{if } x = 0, \end{cases}$$
$$= p^x (1 - p)^{1-x} \text{ for } x = 0, 1,$$

where $0 < p < 1$. This is a binary random variable, taking value 1 with probability p, and taking value 0 with probability $1 - p$.

For a Bernoulli(p) random variable, we have

$$E(X) = p,$$
$$\text{var}(X) = p(1 - p).$$

All moments can be derived from the MGF of the Bernoulli(p) distribution

$$M_X(t) = pe^t + 1 - p \text{ for } -\infty < t < \infty.$$

We note that all moments of X are a function of p, the only parameter for a Bernoulli distribution. Because $E(X) = P(X = 1)$ and X only takes two possible values, the mean parameter $E(X) = p$ can fully characterize the probability distribution of a Bernoulli random variable.

This binary distribution can arise when one tosses a coin whose head shows up with probability p. It has wide applications in economics and finance. For example, one can define a random variable X to take value 1 if the IBM stock price goes up, and to take value 0 if the IBM stock price goes down. Then X is the directional indicator of the IBM stock price and follows a Bernoulli distribution. Das (2002), in a study of jumps in the Fed Funds rates to capture surprise effects, approximates the interest rate jumps by a Bernoulli distribution.

In many economic applications, the probability p will vary across individuals as a function of economic variables, say Z. For example, one may assume

$$P(X = 1|Z) = \frac{1}{1 + \exp(-\beta'Z)}.$$

This is called a logit model, which has been a popular econometric model for binary choice problems where the outcome only has two possibilities, and the interest is to explain the choice of economic agents using some economic characteristics.

4.2.2 Binomial Distribution

A DRV X follows a Binomial distribution, denoted as $B(n, p)$, if its PMF

$$f_X(x) = \binom{n}{x} p^x (1-p)^{n-x}, \quad x = 0, 1, \cdots, n,$$

where $n \geq 1$ and $0 < p < 1$.

A binomial random variable can take $n + 1$ possible integer values $\{0, 1, \cdots, n\}$. Thus, it is a distribution with finite support. Figure 4.1 shows the probability histogram of the Binomial(n, p) distribution with different choices of n.

When can this distribution arise? Suppose one throws a coin n times independently. Each time the head has probability p to occur. How many heads can one get from these n trials? Let X_i denote the number of heads for the i-th trial, and let X denote the total number of heads in the n trials. Then

$$X = \sum_{i=1}^{n} X_i,$$

where $\{X_i\}$ is a sequence of independently identically distributed (IID) Bernoulli(p) random variables. It can be shown that $X = \Sigma_{i=1}^{n} X_i$ follows a $B(n, p)$ distribution (how to show this?).

Question: How to verify

$$\sum_{x=0}^{n} f_X(x) = \sum_{x=0}^{n} \binom{n}{x} p^x (1-p)^{n-x} = 1 \text{ for all } n \geq 1 \text{ and all } p \in (0, 1)?$$

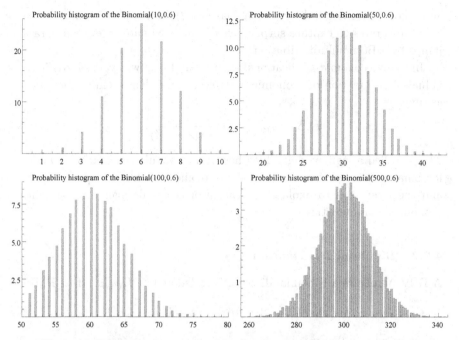

Figure 4.1: Probability histograms of Binomial(n, p) with different choices of n

By the binomial theorem that for any real numbers x and y and integer $n \geq 0$,

$$(x + y)^n = \sum_{i=0}^{n} \binom{n}{i} x^i y^{n-i},$$

we can obtain the identity immediately by setting $x = p$ and $y = 1 - p$. This binomial expansion is the reason why the distribution is called the binomial distribution.

We now calculate some moments for the Binomial(n, p) distribution. For the mean of $B(n, p)$, we have

$$
\begin{aligned}
E(X) &= \sum_{x=0}^{n} x f_X(x) \\
&= \sum_{x=0}^{n} x \binom{n}{x} p^x (1 - p)^{n-x} \\
&= \sum_{x=1}^{n} x \binom{n}{x} p^x (1 - p)^{n-x} + 0 \cdot \binom{n}{0} p^0 (1 - p)^{n-0}
\end{aligned}
$$

$$= \sum_{x=1}^{n} n\binom{n-1}{x-1}p^x(1-p)^{n-x} \quad \text{(by setting } y = x - 1)$$

$$= np \sum_{y=0}^{n-1} \binom{n-1}{y}p^y(1-p)^{(n-1)-y}$$

$$= np \sum_{y=0}^{n-1} f_Y(y) = np,$$

where the last sum can be viewed as the sum of the PMF $f_Y(\cdot)$ values of a Binomial $(n-1, p)$ random variable Y so that $\sum_{y=0}^{n-1} f_Y(y) = 1$.

Next, we compute the variance of $B(n, p)$. The second moment

$$E(X^2) = \sum_{x=0}^{n} x^2 f_X(x)$$

$$= \sum_{x=0}^{n} x^2 \binom{n}{x}p^x(1-p)^{n-x}$$

$$= \sum_{x=1}^{n} x \cdot x\binom{n}{x}p^x(1-p)^{n-x}$$

$$= \sum_{x=1}^{n} xn\binom{n-1}{x-1}p^x(1-p)^{n-x} \quad \text{(setting } y = x - 1)$$

$$= n \sum_{y=0}^{n-1} (y+1)\binom{n-1}{y}p^{y+1}(1-p)^{(n-1)-y}$$

$$= np \sum_{y=0}^{n-1} y\binom{n-1}{y}p^y(1-p)^{(n-1)-y} + np \sum_{y=0}^{n-1} \binom{n-1}{y}p^y(1-p)^{(n-1)-y}$$

$$= npE(Y) + np \sum_{y=0}^{n-1} f_Y(y)$$

$$= np(n-1)p + np$$

$$= np[(n-1)p + 1],$$

where the first sum can be viewed as the mean of a Binomial$(n-1, p)$ random variable Y, and the second sum is the sum of the PMF $f_Y(\cdot)$ of Y over its support. Therefore, we have the variance

$$\sigma_X^2 = E(X^2) - \mu_X^2$$

$$= \{np \cdot [(n-1)p] + np\} - (np)^2$$

$$= np(1-p).$$

Finally, we compute the MGF

$$M_X(t) = E(e^{tX})$$
$$= \sum_{x=0}^{n} e^{tx} \tbinom{n}{x} p^x (1-p)^{n-x}$$
$$= \sum_{x=0}^{n} \tbinom{n}{x} (pe^t)^x (1-p)^{n-x}$$
$$= (pe^t + 1 - p)^n, \qquad -\infty < t < \infty,$$

where the last equality follows from the binomial formula

$$\sum_{x=0}^{n} \tbinom{n}{x} u^x v^{n-x} = (u+v)^n.$$

The binomial distribution has been one of the oldest to have been the subject of study in statistics. It arises whenever underlying events have two possible outcomes, the chances of which remain constant. The importance of the binomial distribution has been extended from the original application of games to many other areas. The binomial distribution has wide applications in economics. Many experiments can be modeled as a sequence of Bernoulli trials, and the sum of Bernoulli trials follows the binomial distribution. For example, it can be used to approximate the distribution of the numbers of defective products in a total of n products each of which has probability p to be defective. It can also be used to model the cumulative number of jumps that occur in financial price movements during a given period of time (e.g., Das 2002).

4.2.3 *Negative Binomial Distribution*

A $B(n,p)$ random variable describes the probability distribution for the number of successes in a fixed number of n trials. Now we are interested in the probability distribution of the number of trials required to obtain a given number of successes. For this reason, we call this distribution as the negative binomial distribution, denoted as $NB(n,p)$.

Specifically, in a sequence of independent Bernoulli(p) trials, let the random variable X denote the number of trials such that at the X-th trial the r-th success occurs, where r is a fixed integer. In other words, $X-1$ is the number of trials right before the r-th success is obtained. Because there are $r-1$ successes in the first $X-1$ trials and the X-th trial is a success, the PMF of X

$$f_X(x) = \left[\binom{x-1}{r-1} p^{r-1}(1-p)^{(x-1)-(r-1)} \right] \cdot p$$

$$= \binom{x-1}{r-1} p^r (1-p)^{x-r}, \qquad x = r, r+1, \cdots,$$

where $\binom{x-1}{r-1} p^{r-1}(1-p)^{(x-1)-(r-1)}$ is the probability of getting $r-1$ successes with $x-1$ trials, and p is the probability of success in the x-th trial.

The negative binomial distribution is sometimes defined in terms of the number of failures when obtaining the r-th success, $Y = X - r$. As an example, this distribution can be used to model family size when a family prefers to have a given number of boys or a given number of girls (Rao *et al.* 1973).

The support of Y is the set of all nonnegative integers $\{0, 1, \cdots\}$. The PMF of Y

$$f_Y(y) = P(Y = y)$$

$$= P(X = y + r)$$

$$= \binom{y+r-1}{r-1} p^r (1-p)^y, \text{ for } y = 0, 1, \cdots.$$

4.2.4 *Geometric Distribution*

The geometric distribution is the probability distribution of the number of Bernoulli trials required to obtain the first success. This is a special case of the negative binomial distribution with $r = 1$. When $r = 1$, the negative binomial distribution becomes

$$f_X(x) = p(1-p)^{x-1}, \qquad x = 1, 2, \cdots.$$

In some developing countries, many rural families stop to give birth of a baby until they have the first boy. Thus, the geometric distribution is applicable to model rural populations in developing countries.

The geometric distribution is the simplest of the waiting time distributions. The random variable X can be interpreted as the number of trials required to obtain the first success, so we are "waiting for a success".

The geometric distribution has the so-called "memoryless" or "non-aging" property in the sense that for integers $s > t$, we have

$$P(X > s | X > t) = P(X > s - t).$$

That is, the probability of getting additional $s - t$ failures, having already observed t failures, is the same as the probability of observing $s - t$ failures at the start of the sequence. In other words, the probability of getting a run of failures depends only on the length of the run, not on its position.

Question: How to show the "memoryless" property?

Using the conditional probability formula $P(A|B) = P(A \cap B)/P(B)$ and the fact that when $s > t$, the event $\{X > t\}$ contains the event $\{X > s\}$ as a subset, we have

$$
\begin{aligned}
P(X > s | X > t) &= \frac{P(X > s, X > t)}{P(X > t)} \\
&= \frac{P(X > s)}{P(X > t)} \\
&= \frac{1 - P(X \le s)}{1 - P(X \le t)} \\
&= \frac{1 - \sum_{x=1}^{s} p(1-p)^{x-1}}{1 - \sum_{x=1}^{t} p(1-p)^{x-1}} \\
&= \frac{(1-p)^s}{(1-p)^t} \\
&= (1-p)^{s-t} \\
&= P(X > s - t).
\end{aligned}
$$

The geometric distribution is usually viewed as a discrete analog of the exponential distribution to be introduced below. It can be used to model births and populations.

4.2.5 *Poisson Distribution*

A DRV X follows a Poisson(λ) distribution if its PMF

$$
f_X(x) = e^{-\lambda} \frac{\lambda^x}{x!}, \qquad x = 0, 1, \cdots,
$$

where $\lambda > 0$. The parameter λ is called an intensity parameter.

The support of a Poisson(λ) random variable is the set of all nonnegative integers, and thus is infinite but countable. Its PMF is depicted in Figure 4.2, for different values of parameter λ.

We first verify $\sum_{x=0}^{\infty} f_X(x) = 1$ for any given $\lambda > 0$. Using MacLaurin's series expansion

$$
e^{\lambda} = \sum_{x=0}^{\infty} \frac{\lambda^x}{x!},
$$

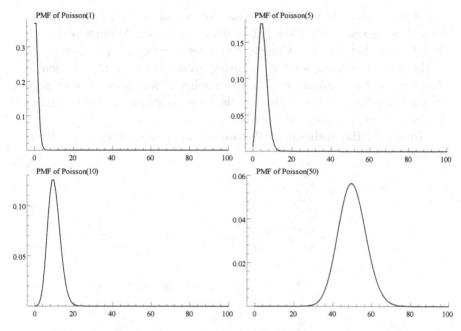

Figure 4.2: PMF's of Poisson distribution for different parameter values λ

we have

$$\sum_{x=0}^{\infty} f_X(x) = e^{-\lambda} \sum_{x=0}^{\infty} \frac{\lambda^x}{x!}$$
$$= e^{-\lambda} e^{\lambda}$$
$$= 1.$$

Next, we compute the mean of $X \sim \text{Poisson}(\lambda)$:

$$E(X) = \sum_{x=0}^{\infty} x e^{-\lambda} \frac{\lambda^x}{x!}$$
$$= \sum_{x=1}^{\infty} x e^{-\lambda} \frac{\lambda^x}{x!} \quad [\text{using } x! = x \cdot (x-1)!]$$
$$= \lambda \sum_{x=1}^{\infty} e^{-\lambda} \frac{\lambda^{x-1}}{(x-1)!}$$
$$= \lambda \sum_{y=0}^{\infty} e^{-\lambda} \frac{\lambda^y}{y!}$$
$$= \lambda,$$

where the last sum is obtained by change of variable $y = x - 1$ and it can be viewed as the sum of the PMF $f_Y(\cdot)$ values of a Poisson(λ) random variable Y. The fact that $E(X) = \lambda$ provides an interpretation for parameter λ: it is the average number for the occurring events following the Poisson(λ) distribution. For example, if X is the number of jumps in an asset price that follow a Poisson(λ) distribution, then λ is the average number of jumps in asset prices in a given time period.

To obtain the variance of Poisson(λ), we now compute the second moment:

$$
\begin{aligned}
E(X^2) &= \sum_{x=0}^{\infty} x^2 e^{-\lambda} \frac{\lambda^x}{x!} \\
&= \sum_{x=1}^{\infty} x e^{-\lambda} \frac{\lambda^x}{(x-1)!} \\
&= \lambda \sum_{x=1}^{\infty} x e^{-\lambda} \frac{\lambda^{x-1}}{(x-1)!} \quad \text{(setting } y = x - 1) \\
&= \lambda \sum_{y=0}^{\infty} (y+1) e^{-\lambda} \frac{\lambda^y}{y!} \\
&= \lambda \sum_{y=0}^{\infty} y e^{-\lambda} \frac{\lambda^y}{y!} + \lambda \sum_{y=0}^{\infty} e^{-\lambda} \frac{\lambda^y}{y!} \\
&= \lambda E(Y) + \lambda \sum_{y=0}^{\infty} f_Y(y) \\
&= \lambda^2 + \lambda,
\end{aligned}
$$

where the first sum can be viewed as the mean of a Poisson(λ) random variable Y, and the second sum can be viewed as the sum of the PMF $f_Y(\cdot)$ values of Y over its support.

Therefore, the variance

$$
\begin{aligned}
\sigma_X^2 &= E(X^2) - \mu_X^2 \\
&= (\lambda^2 + \lambda) - \lambda^2 \\
&= \lambda.
\end{aligned}
$$

It is interesting to note that both the mean and variance are equal to λ for a Poisson(λ) distribution.

Finally, the MGF of the Poisson(λ) distribution

$$M_X(t) = E(e^{tX})$$

$$= \sum_{x=0}^{\infty} e^{tx} f_X(x)$$

$$= \sum_{x=0}^{\infty} e^{tx} e^{-\lambda} \frac{\lambda^x}{x!}$$

$$= e^{-\lambda} \sum_{x=0}^{\infty} \frac{(\lambda e^t)^x}{x!} \qquad \left(\text{using } e^a = \sum_{x=0}^{\infty} \frac{a^x}{x!} \right)$$

$$= e^{-\lambda} e^{(\lambda e^t)}$$

$$= e^{\lambda(e^t - 1)}, \, -\infty < t < \infty.$$

The support of a Poisson(λ) random variable is the set of all nonnegative integers while the support of a Binomial $B(n, p)$ random variable is the set of nonnegative integers $\{0, 1, \cdots, n\}$. When $n \to \infty$, however, there exists a close link between these two distributions.

Recall the MGF of the binomial distribution $B(n, p)$ is

$$M_B(t) = \left(pe^t + 1 - p \right)^n, \qquad -\infty < t < \infty.$$

When $n \to \infty$ but $np \to \lambda$, i.e., when the occurrence of an event is small or rare but there are many trials, we can use a Poisson(λ) distribution to approximate a binomial distribution $B(n, p)$ because

$$M_B(t) = \left(pe^t + 1 - p \right)^n$$

$$= \left[1 + \frac{np(e^t - 1)}{n} \right]^n$$

$$\to e^{\lambda(e^t - 1)} = M_P(t),$$

where we have made use of the formula that $(1 + \frac{a}{n})^n \to e^a$ as $n \to \infty$. Therefore, the Binomial $B(n, p)$ distribution is approximately equivalent to a Poisson(λ) distribution when $n \to \infty, np \to \lambda$, i.e., when n is large but p is small. Among other things, this avoids the tedious calculation of the binomial probability formula. Indeed, Poisson (1837) derived the Poisson distribution by considering the limit of a sequence of binomial distributions. This is called the law of small numbers in Bortkiewicz (1898).

Figure 4.3 plots the Poisson(λ) approximation to the binomial distribution $B(n, p)$, for different values of n, where $np = \lambda = 5$. One can see that even if n is not very large, this approximation works well.

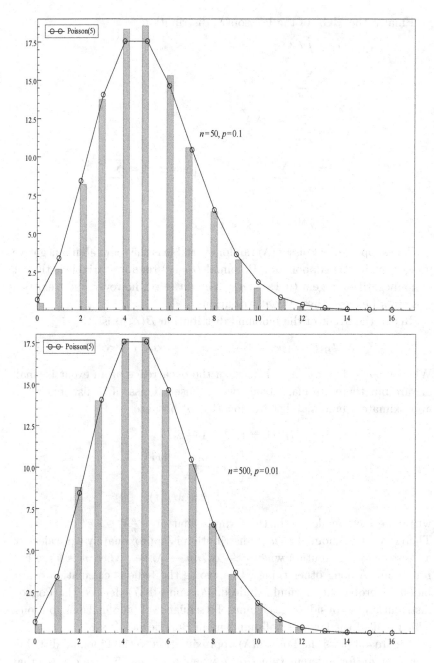

Figure 4.3: Poisson(λ) approximation to Binomial distribution $B(n,p)$

In fact, for a negative binomial distribution $NB(r,p)$, when $r \to \infty$ but $r(1-p) \to \lambda \in (0,\infty)$, $NB(r,p)$ will also become a Poisson(λ) distribution, because

$$f_Y(y) = \binom{y+r-1}{r-1} p^r (1-p)^y$$

$$\to \frac{e^{-\lambda} \lambda^y}{y!}, \qquad y = 0, 1, \cdots.$$

We can prove this by showing the convergence of the MGF of the negative binomial distribution $NB(r,p)$ to that of the Poisson(λ) distribution as $r \to \infty, r(1-p) \to \lambda$.

The Poisson distribution often appears in connection with the study of sequences of random events occurring over time or space. Suppose, starting from a time point $t = 0$, we start counting the number of events. Then for each value of t, we obtain an integer denoted by $N(t)$, which is the number of events that have occurred during the time period $[0,t]$. Clearly, for each value of t, $N(t)$ is a discrete random variable with the set of possible values $\{0, 1, 2, \ldots\}$. To study the distribution of $N(t)$, the number of events occurring in $[0,t]$, we make the following assumptions:

- *Stationarity*: For all $m \geq 0$, and for any two equal time intervals Δ_1 and Δ_2, the probability of m events in Δ_1 is equal to the probability of m events in Δ_2;
- *Independent Increments*: For all $m \geq 0$, and for all any time interval $(t, t+s)$, the probability of m events in $(t, t+s)$ is independent of how many events have occurred earlier or how they have occurred. In particular, suppose the times $0 \leq t_1 < t_2 < \cdots < t_k$ are given. For $1 \leq i \leq k-1$, let A_i be the event that m_i events of the process occur in $[t_i, t_{i+1})$. The independent increments mean that $\{A_1, A_2, \ldots, A_{k-1}\}$ is an independent set of events;
- *Sequencing*: The occurring of two or more events in a very small time interval is practically impossible. This condition is mathematically expressed by $\lim_{\delta \to \infty} P[N(\delta) > 1]/\delta = 0$. This implies that as $\delta \to 0$, the probability of two or more events, $P[N(\delta) > 1]$, approaches 0 faster than δ does. That is, if δ is negligible, then $P[N(\delta) > 1)$ is even more negligible.

A stochastic process $\{N(t)\}$ that satisfies these three assumptions is called a stationary Poisson process. Note that, by stationarity, the random variables $N(t_2) - N(t_1)$ and $N(t_2 + s) - N(t_1 + s)$ have the same probability distribution. The number of events in $[t_i, t_{i+1})$, $N(t_{i+1}) - N(t_i)$, is called the increment in the process $\{N(t)\}$ between t_i and t_{i+1}. It is worthwhile to note that stationarity and sequencing mean that the simultaneous occurrence of two or more events is impossible. Suppose $N(0) = 0$. Then there exists a positive number λ such that

$$P\left[N(t) = m\right] = \frac{(\lambda t)^m e^{-\lambda t}}{m!}, \qquad m = 0, 1, 2, \ldots.$$

That is, for all $t > 0, N(t)$ is a Poisson random variable with parameter λt. Here, $E[N(t)] = \lambda t$ and therefore $\lambda = E[N(1)]$. The parameter λ is the average number of events during a unit time period. Any process with this property is called a Poisson process with rate λ, and is often denoted by $\{N(t), t > 0\}$.

It has been claimed in Douglas (1980) that the Poisson distribution plays a similar role with respect to discrete distributions to that of the normal distribution for absolutely continuous distributions. The Poisson approximation for the binomial distribution makes the Poisson distribution rather useful in quality control for the number of defective items. More generally, the Poisson distribution has been used to model the distribution of the number of events in a specific period of time or a specific unit of space (e.g., the number of customers passing through a cashier counter, the number of telephone calls, the number of accidents, the number of earthquakes in Japan, the number of jumps in an asset price). One of the basic assumptions on which the Poisson distribution is built is that, for small time intervals, the probability of an arrival is proportional to the length of waiting time. This makes it a reasonable model for situations like those just mentioned above. Bortkiewicz (1898) considered circumstances in which the Poisson distribution might arise. In particular, he used the Poisson distribution to characterize the number of deaths from kicks by horses per annum in the Prussian Army Corps where the probability of death from this cause was small while the number of soldiers exposed to the risk was large. In finance the Poisson distribution has been widely used to model the probability distribution of the number of jumps of an asset price over a given period of time. Merton (1976), for example, proposes a model where in addition to a Brownian motion component, the price process of the underlying asset is assumed to have a Poisson jump component. The Poisson regression, namely the analysis of the relationship between an observed count and a set of explanatory variables (Z) by assuming $\lambda = \alpha + \beta Z$, is also popularly used in econometrics (e.g., Hausman *et al.* 1984).

4.3 Continuous Probability Distributions

We now introduce a variety of popular continuous probability distributions.

4.3.1 *Uniform Distribution*

A CRV X follows a uniform probability distribution on the interval $[a, b]$, denoted as $X \sim U[a, b]$, if its PDF

$$f_X(x) = \begin{cases} \frac{1}{b-a}, & a \leq x \leq b, \\ 0, & \text{otherwise.} \end{cases}$$

Figure 4.4 shows the PDF of the $U[0, 1]$ distribution. Due to the shape of its PDF, the uniform distribution is also called a rectangular distribution.

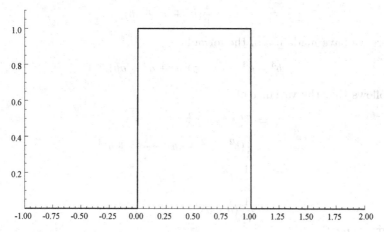

Figure 4.4: PDF of the $U[0, 1]$ distribution

Because X is a bounded random variable, all its moments exist. The k-th moment

$$\begin{aligned} E(X^k) &= \int_{-\infty}^{\infty} x^k f_X(x) dx \\ &= \frac{1}{b-a} \int_a^b x^k dx \\ &= \frac{1}{b-a} \frac{x^{k+1}}{k+1} \Big|_a^b \\ &= \frac{1}{b-a} \frac{b^{k+1} - a^{k+1}}{k+1}. \end{aligned}$$

When $k = 1$, we obtain the mean of X,

$$
\begin{aligned}
\mu_X &= \frac{1}{b-a} \frac{b^2 - a^2}{2} \\
&= \frac{1}{2} \frac{b^2 - a^2}{b - a} \\
&= \frac{1}{2}(a + b).
\end{aligned}
$$

When $k = 2$, we obtain the second moment

$$
\begin{aligned}
E(X^2) &= \frac{1}{b-a} \frac{b^3 - a^3}{3} \\
&= \frac{1}{3} \frac{b^3 - a^3}{b - a} \\
&= \frac{1}{3}(b^2 + a^2 + ab),
\end{aligned}
$$

where we have made use of the formula

$$
b^3 - a^3 = (b - a)(b^2 + a^2 + ab).
$$

It follows that the variance of X,

$$
\begin{aligned}
\sigma_X^2 &= E(X^2) - \mu_X^2 \\
&= \frac{1}{3}(b^2 + a^2 + ab) - \frac{1}{4}(a + b)^2 \\
&= \frac{1}{12}(b^2 + a^2 - 2ab) \\
&= \frac{1}{12}(b - a)^2.
\end{aligned}
$$

The MGF

$$
\begin{aligned}
M_X(t) &= \int_{-\infty}^{+\infty} e^{tx} f_X(x) dx \\
&= \int_a^b e^{tx} \frac{1}{b-a} dx \\
&= \frac{1}{t(b-a)} e^{tx} \Big|_a^b \\
&= \frac{1}{t(b-a)} (e^{tb} - e^{ta}), \qquad -\infty < t < \infty.
\end{aligned}
$$

When $a = 0, b = 1$, the distribution is called the standard uniform $U[0, 1]$ distribution, which has mean $\frac{1}{2}$ and variance $\frac{1}{12}$. In Chapter 3, we have shown that the probability integral transform $Y = F_X(X)$ follows a $U[0, 1]$

distribution. The uniform distribution plays a very important role in statistics and econometrics.

4.3.2 Beta Distribution

A CRV X follows a Beta(α, β) distribution if it has a PDF

$$f_X(x) = \frac{1}{B(\alpha, \beta)} x^{\alpha-1}(1-x)^{\beta-1}, \qquad 0 \le x \le 1,$$

where $\alpha > 0, \beta > 0$, and $B(\alpha, \beta)$ is called the Beta function defined as

$$\begin{aligned} B(\alpha, \beta) &= \int_0^1 x^{\alpha-1}(1-x)^{\beta-1} dx \\ &= \frac{\Gamma(\alpha)\Gamma(\beta)}{\Gamma(\alpha+\beta)}. \end{aligned}$$

The function $\Gamma(\alpha)$ is called the Gamma function, defined as

$$\Gamma(\alpha) = \int_0^\infty t^{\alpha-1} e^{-t} dt.$$

The following lemma states some useful properties of the Gamma function $\Gamma(\alpha)$.

Lemma 4.1. *[Properties of $\Gamma(\alpha)$]:*
(1) $\Gamma(\alpha + 1) = \alpha\Gamma(\alpha)$;
(2) $\Gamma(k) = (k-1)!$ if k is a positive integer;
(3) $\Gamma(\frac{1}{2}) = \sqrt{\pi}$.

Proof: The proof is left as an exercise.

Question: How to show the identity that

$$B(\alpha, \beta) = \frac{\Gamma(\alpha)\Gamma(\beta)}{\Gamma(\alpha+\beta)}?$$

For a Beta(α, β) distribution, the mean

$$\begin{aligned} E(X) &= \int_0^1 \frac{\Gamma(\alpha+\beta)}{\Gamma(\alpha)\Gamma(\beta)} x^{(\alpha+1)-1}(1-x)^{\beta-1} dx \\ &= \frac{\alpha}{\alpha+\beta} \int_0^1 \frac{(\alpha+\beta)\Gamma(\alpha+\beta)}{\alpha\Gamma(\alpha)\Gamma(\beta)} x^{(\alpha+1)-1}(1-x)^{\beta-1} dx \\ &= \frac{\alpha}{\alpha+\beta} \int_0^1 \frac{\Gamma(\alpha+1+\beta)}{\Gamma(\alpha+1)\Gamma(\beta)} x^{(\alpha+1)-1}(1-x)^{\beta-1} dx \\ &= \frac{\alpha}{\alpha+\beta}, \end{aligned}$$

where the last integral can be viewed as the integral of the PDF of a Beta($\alpha + 1, \beta$) random variable.

To compute the variance, we first compute the second moment:

$$
\begin{aligned}
E(X^2) &= \int_0^1 \frac{\Gamma(\alpha + \beta)}{\Gamma(\alpha)\Gamma(\beta)} x^{(\alpha+2)-1}(1-x)^{\beta-1} dx \\
&= \frac{\alpha}{\alpha + \beta} \int_0^1 \frac{\Gamma(\alpha + 1 + \beta)}{\Gamma(\alpha + 1)\Gamma(\beta)} x^{(\alpha+2)-1}(1-x)^{\beta-1} dx \\
&= \frac{\alpha(\alpha+1)}{(\alpha+\beta)(\alpha+\beta+1)} \int_0^1 \frac{\Gamma(\alpha + 2 + \beta)}{\Gamma(\alpha + 2)\Gamma(\beta)} x^{(\alpha+2)-1}(1-x)^{\beta-1} dx \\
&= \frac{\alpha(\alpha+1)}{(\alpha+\beta)(\alpha+\beta+1)},
\end{aligned}
$$

where the last integral can be viewed as the integral of the PDF of a Beta($\alpha + 2, \beta$) random variable.

Therefore, we have the variance

$$
\begin{aligned}
\mathrm{var}(X) &= E(X^2) - E^2(X) \\
&= \frac{\alpha(\alpha+1)}{(\alpha+\beta)(\alpha+\beta+1)} - \left(\frac{\alpha}{\alpha+\beta}\right)^2 \\
&= \frac{\alpha\beta}{(\alpha+\beta)^2(\alpha+\beta+1)}.
\end{aligned}
$$

The MGF

$$
\begin{aligned}
M_X(t) &= E(e^{tX}) \\
&= \int_0^1 \frac{\Gamma(\alpha + \beta)}{\Gamma(\alpha)\Gamma(\beta)} e^{tx} x^{\alpha-1}(1-x)^{\beta-1} dx \\
&= 1 + \sum_{j=1}^{\infty} \left(\prod_{i=0}^{j-1} \frac{\alpha+i}{\alpha+\beta+i} \right) \frac{t^j}{j!},
\end{aligned}
$$

where we have made use of MacLaurin's series expansion

$$
e^{tx} = \sum_{j=0}^{\infty} \frac{t^j x^j}{j!}.
$$

Like the uniform distribution, Beta(α, β) is one of the few well-known distributions that have bounded support. In fact, the standard uniform distribution on the unit interval $[0, 1]$, $U[0, 1]$, is a special case of Beta(α, β) with $\alpha = \beta = 1$. The shape of the Beta(α, β) distribution depends on the values of parameters (α, β), as can be seen in Figure 4.5. For this reason, α and β are called shape parameters.

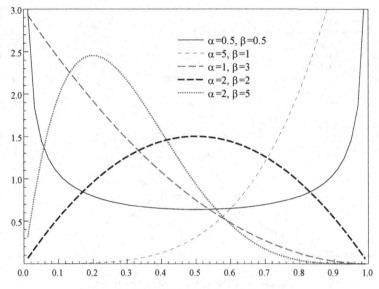

Figure 4.5: PDF of the Beta(α, β) distribution

Because the support of a Beta(α, β) is [0,1], the Beta distribution can be used to model the probability distribution of proportions or quantities whose values fall into the interval [0, 1]. For example, Granger (1980) uses the Beta distribution for the marginal propensities to consume for individual consumers and show that the sum of "short memory" time series processes displays a "long memory" property that the remote past consumption is still persistently correlated with the current consumption.

4.3.3 *Normal Distribution*

A CRV X is called to follow a normal distribution, denoted as $X \sim N(\mu, \sigma^2)$, if it has the PDF

$$f_X(x) = \frac{1}{\sqrt{2\pi\sigma^2}} e^{-(x-\mu)^2/2\sigma^2}, \quad -\infty < x < \infty,$$

where $-\infty < \mu < \infty$ and $\sigma^2 > 0$.

The parameters μ and σ^2 are location and scale parameters respectively. When $\mu = 0, \sigma^2 = 1$, $X \sim N(0, 1)$ is called a standard normal or unit normal distribution. Tables relating to the standard normal distribution are a necessary ingredient of any standard statistical textbook in statistical theory or its applications. Figure 4.6 shows the PDF of the standard normal and other normal distributions.

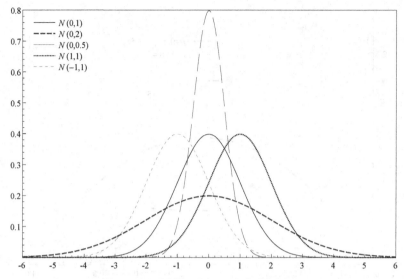

Figure 4.6: PDF's of normal distributions

The normal distribution was discovered in 1733 by Abraham De Moivre (1667-1754) in his investigation of approximating coin tossing probabilities. He named the PDF of his discovery the exponentially bell-shaped curve. In 1809, Carl Friedrich Gauss (1777-1855) firmly established the importance of the normal distribution by using it to predict the location of astronomical bodies. As a result, the normal distribution then became commonly known as the Gaussian distribution.

The normal distribution is the most important distribution in probability theory. For many decades the normal distribution have been holding a central position in statistics. Most theoretical arguments for the use of the normal distribution are based on the forms of the Central Limit Theorem (CLT), which says that under suitable conditions, a sample average of n independent and identically distributed (IID) random variables $\{X_1, \cdots, X_n\}$, with suitable centering and standardization, will converge in distribution to a standard normal distribution as the sample size n increases, that is,

$$\sqrt{n}\frac{\bar{X}_n - \mu_X}{\sigma_X} \xrightarrow{d} N(0,1) \text{ as } n \to \infty,$$

where $\bar{X}_n = n^{-1}\sum_{i=1}^{n} X_i$, and $\xrightarrow{d}$ denotes convergence in distribution, that is, the distribution of $\sqrt{n}(\bar{X}_n - \mu_X)/\sigma_X$ converges to the $N(0,1)$ distribution as $n \to \infty$. This follows no matter whether X_i is discrete or continuous, and whether or not X_i is compactly supported. See more discussion on CLT's in Chapter 7.

Question: How to verify

$$\int_{-\infty}^{\infty} \frac{1}{\sqrt{2\pi}\sigma} e^{-\frac{(x-\mu)^2}{2\sigma^2}} dx = 1 \text{ for all } (\mu, \sigma^2)?$$

Put $y = (x - \mu)/\sigma$. Then it becomes to check whether

$$\int_{-\infty}^{\infty} \frac{1}{\sqrt{2\pi}} e^{-\frac{y^2}{2}} dy = 1.$$

Because

$$\left(\int_{-\infty}^{\infty} e^{-\frac{x^2}{2}} dx\right)^2 = \int_{-\infty}^{\infty} e^{-\frac{x^2}{2}} dx \int_{-\infty}^{\infty} e^{-\frac{y^2}{2}} dy$$

$$= \int_{-\infty}^{\infty}\int_{-\infty}^{\infty} e^{-\frac{x^2+y^2}{2}} dxdy \text{ (set } x = r\cos(\theta), y = r\sin(\theta))$$

$$= \int_{0}^{\infty}\int_{0}^{2\pi} e^{-\frac{r^2}{2}} r\,dr\,d\theta$$

$$= 2\pi \int_{0}^{\infty} e^{-\frac{r^2}{2}} r\,dr$$

$$= 2\pi,$$

it follows that

$$\frac{1}{\sqrt{2\pi}} \int_{-\infty}^{\infty} e^{-\frac{x^2}{2}} dx = 1.$$

We now compute the moments of the normal distribution. In particular, we show that the mean and variance of X are equal to μ and σ^2 respectively.

First, the mean

$$E(X) = \int_{-\infty}^{\infty} x f_X(x) dx \qquad (\text{setting } x = (x - \mu) + \mu)$$

$$= \int_{-\infty}^{\infty} (x - \mu) f_X(x) dx + \mu \int_{-\infty}^{\infty} f_X(x) dx$$

$$= \int_{-\infty}^{\infty} (x - \mu) \frac{1}{\sqrt{2\pi\sigma^2}} e^{-\frac{(x-\mu)^2}{2\sigma^2}} dx + \mu \qquad (\text{setting } y = x - \mu)$$

$$= \int_{-\infty}^{\infty} y \frac{1}{\sqrt{2\pi\sigma^2}} e^{-\frac{y^2}{2\sigma^2}} dy + \mu$$

$$= \mu,$$

where the integral in the last second equality is identically zero because the integrand $g(y) = y\frac{1}{\sqrt{2\pi}\sigma}e^{-y^2/2\sigma^2}$ is an odd function (i.e. $g(-y) = -g(y)$ for all y).

Next, using integration by part, the variance

$$\sigma_X^2 = \int_{-\infty}^{\infty} (x - \mu)^2 f_X(x) dx$$

$$= \int_{-\infty}^{\infty} (x - \mu)^2 \frac{1}{\sqrt{2\pi\sigma^2}} e^{-\frac{(x-\mu)^2}{2\sigma^2}} dx \qquad \text{(setting } y = x - \mu)$$

$$= \int_{-\infty}^{\infty} y^2 \frac{1}{\sqrt{2\pi}\sigma} e^{-\frac{y^2}{2\sigma^2}} dy$$

$$= -\sigma^2 \int_{-\infty}^{\infty} y d\left(\frac{1}{\sqrt{2\pi}\sigma} e^{-\frac{y^2}{2\sigma^2}} \right) \qquad \left(\text{setting } u = y, v = \frac{1}{\sqrt{2\pi\sigma^2}} e^{-y^2/2\sigma^2} \right)$$

$$= -\sigma^2 \left(y \frac{1}{\sqrt{2\pi}\sigma} e^{-\frac{y^2}{2\sigma^2}} \Big|_{-\infty}^{+\infty} - \int_{-\infty}^{\infty} \frac{1}{\sqrt{2\pi}\sigma} e^{-\frac{y^2}{2\sigma^2}} dy \right)$$

$$= \sigma^2,$$

where the last integral can be viewed as the integral of the PDF of a $N(0, \sigma^2)$ random variable Y.

Finally, we derive the MGF of $X \sim N(0, \sigma^2)$.

Theorem 4.1. *Suppose $X \sim N(\mu, \sigma^2)$. Then*

$$M_X(t) = e^{\mu t + \frac{\sigma^2}{2} t^2}, \qquad -\infty < t < \infty.$$

Proof: There are at least two methods to calculate the MGF of X.

(1) Method 1: For the first method, we have

$$M_X(t) = \int_{-\infty}^{\infty} e^{tx} \frac{1}{\sqrt{2\pi}\sigma} e^{-\frac{(x-\mu)^2}{2\sigma^2}} dx$$

$$= \frac{1}{\sqrt{2\pi}\sigma} \int_{-\infty}^{\infty} e^{tx} e^{-\frac{1}{2\sigma^2}[x^2 - 2\mu x + \mu^2]} dx$$

$$= e^{-\frac{\mu^2}{2\sigma^2}} \frac{1}{\sqrt{2\pi}\sigma} \int_{-\infty}^{\infty} e^{-\frac{1}{2\sigma^2}[x^2 - 2(\mu+\sigma^2 t)x + (\mu+\sigma^2 t)^2 - (\mu+\sigma^2 t)^2]} dx$$

$$= e^{-\frac{\mu^2}{2\sigma^2}} e^{\frac{(\mu+\sigma^2 t)^2}{2\sigma^2}} \left\{ \frac{1}{\sqrt{2\pi}\sigma} \int_{-\infty}^{\infty} e^{-\frac{[x - (\mu+\sigma^2 t)]^2}{2\sigma^2}} dx \right\}$$

$$= e^{\frac{2\mu\sigma^2 t + \sigma^4 t^2}{2\sigma^2}} \times 1$$

$$= e^{\mu t + \frac{1}{2}\sigma^2 t^2}, \quad \text{for } t \in (-\infty, \infty),$$

where the second to last equality follows from the fact that

$$\frac{1}{\sqrt{2\pi}\sigma} \int_{-\infty}^{\infty} e^{-\frac{[x - (\mu+\sigma^2 t)]^2}{2\sigma^2}} dx = 1$$

for all μ, σ^2 and t.

(2) Method 2: For the second method, we note that $X = \mu + \sigma Y$, where $Y \sim N(0, 1)$. By Theorem 3.20, we have

$$
\begin{aligned}
M_X(t) &= E\left(e^{tX}\right) \\
&= E\left[e^{t(\mu + \sigma Y)}\right] \\
&= e^{\mu t} E(e^{\sigma t Y}) \\
&= e^{\mu t} M_Y(\sigma t).
\end{aligned}
$$

It suffices to find $M_Y(t)$:

$$
\begin{aligned}
M_Y(t) &= E\left(e^{tY}\right) \\
&= \frac{1}{\sqrt{2\pi}} \int_{-\infty}^{\infty} e^{ty} e^{-\frac{1}{2}y^2} \, dy \\
&= e^{\frac{1}{2}t^2} \frac{1}{\sqrt{2\pi}} \int_{-\infty}^{\infty} e^{-\frac{1}{2}(y-t)^2} \, dy \\
&= e^{\frac{1}{2}t^2}.
\end{aligned}
$$

It follows from Theorem 3.20 that

$$
M_X(t) = e^{\mu t} M_Y(\sigma t) = e^{\mu t + \frac{1}{2}\sigma^2 t^2}.
$$

Therefore, we have

$$
\begin{aligned}
M_X'(t)\big|_{t=0} &= e^{\mu t + \frac{1}{2}\sigma^2 t^2}(\mu + \sigma^2 t)\big|_{t=0} \\
&= \mu, \\
M_X''(t)\big|_{t=0} &= \left[e^{\mu t + \frac{1}{2}\sigma^2 t^2}\sigma^2 + e^{\mu t + \frac{1}{2}\sigma^2 t^2}(\mu + \sigma^2 t)^2\right]\Big|_{t=0} \\
&= \sigma^2 + \mu^2.
\end{aligned}
$$

All centered odd moments $E(X - \mu)^{2k+1} = 0$ for all integers $k \geq 0$ because the normal distribution is symmetric about μ. Suppose we are interested in computing the moments $E(X - \mu)^{2k}$ for an arbitrary positive integer $k > 0$. One can of course differentiate $M_X(t)$ up to $2k$ times, but this is rather tedious when k is large. We now consider some techniques to calculate the moments $E(X - \mu)^{2k}$ by exploiting the duality between integration and differentiation to obtain higher order moments of a normal random variable X.

For the first method, put $\beta = \frac{1}{2\sigma^2}$ or equivalently $\sigma = \frac{1}{\sqrt{2\beta}}$, we obtain

$$
\begin{aligned}
E(X-\mu)^{2k} &= \int_{-\infty}^{\infty} (x-\mu)^{2k} \frac{1}{\sqrt{2\pi\sigma^2}} e^{-\frac{1}{2\sigma^2}(x-\mu)^2} dx \\
&= \int_{-\infty}^{\infty} y^{2k} \frac{1}{\sqrt{2\pi\sigma^2}} e^{-\frac{y^2}{2\sigma^2}} dy \\
&= \int_{-\infty}^{\infty} \frac{1}{\sqrt{2\pi}\sigma} y^{2k} e^{-\beta y^2} dy \\
&= \frac{1}{\sqrt{2\pi}\sigma} \int_{-\infty}^{\infty} (-1)^k \frac{d^k}{d\beta^k} e^{-\beta y^2} dy \\
&= \frac{1}{\sqrt{2\pi}\sigma} (-1)^k \frac{d^k}{d\beta^k} \int_{-\infty}^{\infty} \sqrt{2\pi\sigma^2} \frac{1}{\sqrt{2\pi\sigma^2}} e^{-\frac{1}{2\sigma^2}y^2} dy \\
&= \frac{1}{\sqrt{2\pi}\sigma} (-1)^k \frac{d^k}{d\beta^k} \left(\sqrt{2\pi}\sigma \right) \\
&= \frac{1}{\sqrt{2\pi}\sigma} (-1)^k \sqrt{2\pi} \frac{d^k}{d\beta^k} \left(\frac{1}{\sqrt{2\beta}} \right) \quad (\text{noting } \sigma = \frac{1}{\sqrt{2\beta}}) \\
&= \frac{1}{\sqrt{2}\sigma} (-1)^k \frac{d^k}{d\beta^k} \left(\beta^{-\frac{1}{2}} \right) \\
&= \frac{1}{\sqrt{2}\sigma} (-1)^k \left(-\frac{1}{2} \right) \left(-\frac{3}{2} \right) \cdots \left(\frac{1}{2} - k \right) \beta^{-\frac{1}{2}-k} \\
&= \frac{1}{\sqrt{\pi}} \Gamma \left(k + \frac{1}{2} \right) 2^k \sigma^{2k},
\end{aligned}
$$

where the last equality is obtained by using $\beta = \frac{1}{2\sigma^2}$ and $\Gamma\left(\frac{1}{2}\right) = \sqrt{\pi}$ from Lemma 4.1(3).

For the special case of $k = 2$, we have

$$
E(X-\mu)^4 = (-1)^2 \left(-\frac{1}{2} \right) \left(-\frac{3}{2} \right) 4\sigma^4 = 3\sigma^4.
$$

It follows that the kurtosis of $N(\mu, \sigma^2)$ is

$$
K = \frac{E(X-\mu)^4}{\sigma^4} = 3.
$$

This method of exploiting the duality between integration and differentiation is applicable to many probability distributions, including discrete ones.

In fact, for the normal distribution, there is an alternative simpler method to compute various moments.

Lemma 4.2. *[Stein's Lemma]: Suppose* $X \sim N(\mu, \sigma^2)$, *and* $g(\cdot)$ *is a differentiable function satisfying* $E|g'(X)| < \infty$. *Then*

$$E[g(X)(X - \mu)] = \sigma^2 E[g'(X)].$$

Proof: Using integration by part, we have

$$
\begin{aligned}
E[g(X)(X - \mu)] &= \int_{-\infty}^{\infty} g(x)(x - \mu) \frac{1}{\sqrt{2\pi\sigma^2}} e^{-\frac{(x-\mu)^2}{2\sigma^2}} dx \\
&= -\int_{-\infty}^{\infty} g(x) d\left[\frac{\sigma^2}{\sqrt{2\pi\sigma^2}} e^{-\frac{(x-\mu)^2}{2\sigma^2}}\right] \\
&= -g(x) \frac{\sigma^2}{\sqrt{2\pi\sigma^2}} e^{-\frac{(x-\mu)^2}{2\sigma^2}} \Big|_{-\infty}^{\infty} + \int_{-\infty}^{\infty} \frac{\sigma^2}{\sqrt{2\pi\sigma^2}} e^{-\frac{(x-\mu)^2}{2\sigma^2}} dg(x) \\
&= \sigma^2 \int_{-\infty}^{\infty} g'(x) \frac{1}{\sqrt{2\pi\sigma^2}} e^{-\frac{(x-\mu)^2}{2\sigma^2}} dx \\
&= \sigma^2 E[g'(X)].
\end{aligned}
$$

This completes the proof.

As an example, let us apply this lemma to calculate $E(X - \mu)^4$. We write

$$
\begin{aligned}
E(X - \mu)^4 &= E\left[(X - \mu)^3 (X - \mu)\right] \\
&= E[g(X)(X - \mu)],
\end{aligned}
$$

where $g(X) = (X - \mu)^3$. By Stein's lemma, we have

$$
\begin{aligned}
E(X - \mu)^4 &= \sigma^2 E\left[3(X - \mu)^2\right] \\
&= 3\sigma^4.
\end{aligned}
$$

The differentiation of $g(X)$ reduces the order of moments.

4.3.4 Cauchy and Stable Distributions

A CRV X follows a Cauchy(μ, σ) distribution if its PDF

$$f_X(x) = \frac{1}{\pi\sigma} \frac{1}{1 + \left(\frac{x-\mu}{\sigma}\right)^2} \quad \text{for } -\infty < x < \infty,$$

where $\sigma > 0$.

The parameters μ and σ are location and scale parameters respectively. This distribution is symmetric about μ, with unbounded support. When $\mu = 0$ and $\sigma = 1$, the distribution is called a standard Cauchy distribution, denoted as Cauchy$(0, 1)$.

There has been little use of the Cauchy distribution in practice. However, it is of special theoretical importance. In particular, the Cauchy distribution has some peculiar properties and could provide counter examples to some generally accepted results and concepts in statistics.

Compared to the normal distribution, the Cauchy distribution has very long and heavy tails. The most notable difference between the normal and Cauchy distributions is in the longer and flatter tails of the latter. For a Cauchy(μ, σ) distribution, the tail of the PDF decays to zero at a very slow hyperbolic rate: $f_X(x) \sim x^{-2}$ as $|x| \to \infty$. As a consequence, all moments of order greater than or equal to 1 do not exist, and so its MGF does not exist either. For example, for the Cauchy$(0, 1)$ distribution, we have

$$
\begin{aligned}
E|X| &= \int_{-\infty}^{\infty} |x| f_X(x) dx \\
&= \int_{-\infty}^{\infty} |x| \frac{1}{\pi} \frac{1}{1+x^2} dx \\
&= \frac{2}{\pi} \int_{0}^{\infty} \frac{x}{1+x^2} dx \\
&= \frac{1}{\pi} \ln(1+x^2)|_0^{\infty} \\
&= \infty.
\end{aligned}
$$

This implies the mean and all higher order moments do not exist. Thus, the location parameter μ cannot be interpreted as the mean, and the scale parameter σ cannot be interpreted as the standard deviation.

The characteristic function of the Cauchy(μ, σ) distribution is

$$
\begin{aligned}
\varphi_X(t) &= E\left(e^{\mathrm{i}tX}\right) \\
&= e^{\mathrm{i}\mu t - \sigma|t|}.
\end{aligned}
$$

This characteristic function is not differentiable with respect to t at the origin, which is consistent with the fact that all moments of order greater than or equal to 1 do not exist.

Question: When can a Cauchy distribution arise?

A Cauchy distribution can arise when a ratio of two independent normal random variables is considered.

In fact, both the Cauchy distribution and the normal distribution belong to the so-called stable distribution, which has very considerable importance in probability theory.

Question: What is a stable distribution?

For a stable distribution, its PDF is usually unknown in closed form. However, its characteristic function has a closed form

$$\varphi_X(t) = e^{\mathbf{i}\mu t - \sigma|t|^c[1+\mathbf{i}\lambda\,\mathrm{sgn}(t)\omega(|t|,c)]},$$

where $\mathbf{i} = \sqrt{-1}$, $0 < c \le 2$, $-1 \le \lambda \le 1$, $\sigma > 0$,

$$\omega(|t|,c) = \begin{cases} \tan\left(\frac{1}{2}\pi c\right), & c \ne 1, \\ -\frac{2}{\pi}\ln(|t|), & c = 1, \end{cases}$$

and

$$\mathrm{sgn}(t) = \begin{cases} 1, & \text{if } t > 0, \\ 0, & \text{if } t = 0, \\ -1, & \text{if } t < 0. \end{cases}$$

Intuitively, μ is a location parameter, σ is a scale parameter, c is a tail parameter, and λ is a skew parameter. The shape of the PDF is determined by both c and λ. If $\lambda = 0$, the distribution is symmetric. When $\lambda = 0$, $c = 2$ gives a normal distribution, and $c = 1$ gives a Cauchy distribution. Figure 4.7 plots the PDF's for the stable distribution with various choices of parameters $(\mu, \sigma, c, \lambda)$.

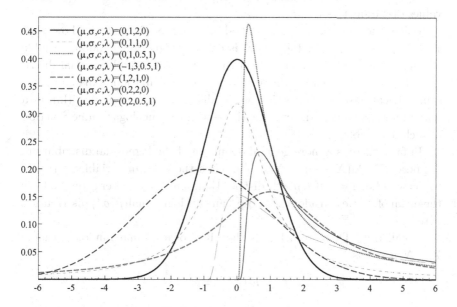

Figure 4.7: PDF's for the stable distribution with various choices of parameters $(\mu, \sigma, c, \lambda)$.

The moments of a stable distribution exist only when $c > 1$. When the sum of independent identically distributed stable random variables has a limiting distribution, it must be a stable distribution rather than the normal distribution. Thus nonnormal stable distributions generalize CLT to the cases where the second moments of the summed variables are infinite.

The stable distributions are closely related to the Levy process which has recently become a popular topic in financial econometrics. They are more appropriate to model heavy tails which are often observed in financial data. Mandelbrot (1963) and Fama (1965) have applied the stable distributions to model stock returns.

4.3.5 *Lognormal Distribution*

A CRV X follows a Lognormal(μ, σ^2) distribution if its PDF

$$f_X(x) = \begin{cases} \frac{1}{\sqrt{2\pi}\sigma} \frac{1}{x} e^{\frac{-1}{2\sigma^2}(\ln x - \mu)^2}, & x > 0, \\ 0, & x \leq 0. \end{cases}$$

Figure 4.8 shows the shape of the lognormal distribution with different values of parameters (μ, σ^2).

Using the transformation method in Chapter 3, we can show that $Y = \ln(X) \sim N(\mu, \sigma^2)$. Indeed, X is called a lognormal random variable because its logarithm follows a normal distribution. The lognormal distribution is sometimes called the anti-lognormal distribution. This name has some logical basis in that it is not the distribution of the logarithm of a normal variable but of the exponential — i.e., the anti-logarithmic function of such a variable.

In fact, there is a more general definition of the lognormal distribution. Suppose $Y = \ln(X - \alpha) \sim N(\mu, \sigma^2)$. Then the random variable X is said to follow a Lognormal(α, μ, σ^2) distribution. Since parameter α only affects the mean of X, we consider the two-parameter lognormal(μ, σ^2) distribution here.

Recall from Theorem 4.1 that the MGF of a normal random variable $Y \sim N(\mu, \sigma^2)$ is

$$M_Y(t) = E\left(e^{tY}\right) = e^{\mu t + \frac{\sigma^2}{2}t^2}.$$

It follows that all moments of the lognormal(μ, σ^2) random variable X exist and are given by

$$E(X^k) = E\left(e^{kY}\right)$$
$$= M_Y(k)$$
$$= e^{k\mu + \frac{\sigma^2}{2}k^2}, \qquad k = 1, 2, \cdots.$$

In particular, we have the mean

$$\mu_X = e^{\mu + \frac{\sigma^2}{2}},$$

and the variance

$$\sigma_X^2 = E(X^2) - \mu_X^2$$
$$= e^{2\mu + \sigma^2}\left(e^{\sigma^2} - 1\right).$$

It is important to note that parameters μ and σ^2 are not the mean and variance of the Lognormal(μ, σ^2) distribution.

Although all moments exist, the MGF does not exist for a lognormal distribution. To see this, we consider

$$M_X(t) = E(e^{tX})$$
$$= \int_0^\infty e^{tx} \frac{1}{\sqrt{2\pi}\sigma} \frac{1}{x} e^{-\frac{1}{2\sigma^2}(\ln x - \mu)^2} dx$$
$$= \int_0^\infty \frac{1}{\sqrt{2\pi}\sigma} e^{te^{\ln x} - \frac{1}{2\sigma^2}(\ln x - \mu)^2} d\ln x$$
$$= \int_{-\infty}^\infty \frac{1}{\sqrt{2\pi}\sigma} e^{te^y - \frac{1}{2\sigma^2}(y - \mu)^2} dy \quad \text{(by setting } y = \ln x)$$
$$\geq \int_c^{c+1} \frac{1}{\sqrt{2\pi}\sigma} e^{te^y - \frac{1}{2\sigma^2}(y - \mu)^2} dy \quad \text{for any } c > 0$$
$$\geq \frac{1}{\sqrt{2\pi}\sigma} e^{te^c - \frac{1}{2\sigma^2}(c - \mu)^2}(c + 1 - c) \quad \text{for } c \text{ sufficiently large}$$
$$= \frac{1}{\sqrt{2\pi}\sigma} e^{te^c - \frac{1}{2\sigma^2}(c - \mu)^2}$$
$$\to \infty \text{ as } c \to \infty$$

because $te^c - (c - \mu)^2/2\sigma^2 \to \infty$ as $c \to \infty$ and $t > 0$.

Intuitively, the lognormal distribution has a long right-tail. When $t > 0$, the exponential function e^{tX} takes rather large values with a relatively large probability, rendering the non-existence of the expectation of e^{tX}.

Question: What is the characteristic function of the Lognormal(μ, σ^2) distribution?

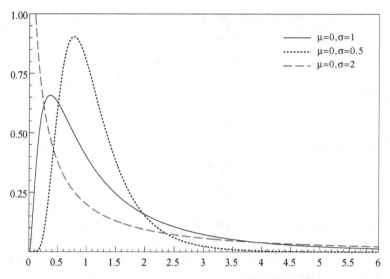

Figure 4.8: PDF's of lognormal distribution

The lognormal distribution is very popular in modeling applications when the variable of interest is nonnegative and skewed to the right. In particular, it has been widely used to model the distribution of asset prices, commodity prices, incomes and populations. To appreciate this, we consider following nonnegative economic variable

$$X_t = X_{t-1}(1 + Y_t),$$

where $\{Y_t\}$ are a sequence of IID random variables such that Y_t is independent of X_{t-1}.

The recursive relationship leads to

$$X_n = X_0 \prod_{t=1}^{n} (1 + Y_t)$$

and so

$$\ln X_n = \ln X_0 + \sum_{t=1}^{n} \ln(1 + Y_t).$$

By the CLT, for large n, $\sum_{t=1}^{n} \ln(1+Y_t)$ will be approximately normally distributed, after suitable standardization. Thus, its exponential, $\Pi_{t=1}^{n} (1 + Y_t)$ is approximately lognormally distributed.

Among other things, the lognormality assumption offers a great deal of convenience in analysis. To see this, suppose a stock price $P_t \sim$

Lognormal($\mu t, \sigma^2 t$), where the time t changes continuously. Then $\ln P_t \sim N(\mu t, \sigma^2 t)$, and the log-return

$$R_t = \ln(P_t/P_{t-1})$$
$$= \ln(P_t) - \ln(P_{t-1}),$$

which is approximately equal to the relative price change from time $t-1$ to time t, is also normally distributed. Furthermore, the cumulative return $\sum_{t=1}^{m} R_t$ over m time periods from $t = 1$ to $t = m$ is also normally distributed. Black and Scholes (1973) use the lognormal distribution for the underlying stock price in deriving the European options prices.

Lognormal distributions have been also found useful in representing the distribution of size for varied kinds of natural economic units (e.g., Gibrat 1930, 1931). The relationship between leaving a company and employees tenure has been described by lognormal distributions with great success (Young 1971, McClean 1976). O'Neill and Wells (1972) point out that the lognormal distribution can be effectively shed to fit the distribution for individual insurance claim payments. The lognormal distribution is also a serious competitor to the Weibull distribution to be introduced below in modeling lifetime distributions.

4.3.6 Gamma and Generalized Gamma Distributions

A nonnegative CRV X follows a Gamma(α, β) distribution if its PDF

$$f_X(x) = \begin{cases} \frac{1}{\Gamma(\alpha)\beta^\alpha} x^{\alpha-1} e^{-x/\beta}, & x > 0, \\ 0, & x \leq 0, \end{cases}$$

where $\alpha, \beta > 0$, and $\Gamma(\alpha) = \int_0^\infty t^{\alpha-1} e^{-t} dt$ is the Gamma function.

The Gamma(α, β) distribution is a flexible family of distributions for a nonnegative random variable on $[0, \infty)$. Here, α is a shape parameter, and β is a scale parameter controlling the spread of the distribution. When $\beta = 1$, the Gamma($\alpha, 1$) distribution is called a standard Gamma distribution. Figure 4.9 shows the shape of the Gamma distribution with different values of parameters (α, β).

Question: How to verify that for any given parameters $\alpha, \beta > 0$, the integral

$$\int_0^\infty \frac{1}{\Gamma(\alpha)\beta^\alpha} x^{\alpha-1} e^{-x/\beta} dx = 1?$$

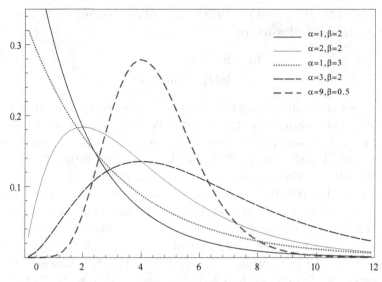

Figure 4.9: PDF's of Gamma distribution

By change of variable and the definition of the Gamma function, we have

$$\int_0^\infty \frac{1}{\Gamma(\alpha)\beta^\alpha} x^{\alpha-1} e^{-x/\beta} dx = \frac{1}{\Gamma(\alpha)} \int_0^\infty (x/\beta)^{\alpha-1} e^{-(x/\beta)} d(x/\beta)$$

$$= \frac{1}{\Gamma(\alpha)} \int_0^\infty y^{\alpha-1} e^{-y} dy \text{ (setting } y = x/\beta)$$

$$= \frac{1}{\Gamma(\alpha)} \Gamma(\alpha)$$

$$= 1.$$

We now derive the mean, variance and MGF of X. The mean

$$\mu_X = \int_{-\infty}^\infty x f_X(x) dx$$

$$= \int_0^\infty \frac{1}{\Gamma(\alpha)\beta^\alpha} x^\alpha e^{-x/\beta} dx$$

$$= \int_0^\infty \frac{\alpha\beta}{\alpha\Gamma(\alpha)\beta^{(\alpha+1)}} x^{(\alpha+1)-1} e^{-x/\beta} dx \quad \begin{array}{l} \text{(setting } \alpha^* = \alpha + 1 \text{ \&} \\ \text{using } \alpha\Gamma(\alpha) = \Gamma(\alpha + 1)) \end{array}$$

$$= \alpha\beta \int_0^\infty \frac{1}{\Gamma(\alpha^*)\beta^{\alpha^*}} x^{\alpha^*-1} e^{-x/\beta} dx$$

$$= \alpha\beta,$$

where the last integral can be viewed as the integral of the PDF of a Gamma(α^*, β) random variable.

The second moment

$$
\begin{aligned}
E(X^2) &= \int_0^\infty \frac{1}{\Gamma(\alpha)\beta^\alpha} x^{\alpha+2-1} e^{-x/\beta} dx \\
&= \int_0^\infty \frac{\alpha(\alpha+1)\beta^2}{\Gamma(\alpha+2)\beta^{\alpha+2}} x^{\alpha+2-1} e^{-x/\beta} dx \\
&= \alpha(\alpha+1)\beta^2 \int_0^\infty \frac{1}{\Gamma(\alpha+2)\beta^{\alpha+2}} x^{\alpha+2-1} e^{-x/\beta} dx \\
&= \alpha(\alpha+1)\beta^2,
\end{aligned}
$$

where the last integral can be viewed as the integral of the PDF of a Gamma($\alpha + 2, \beta$) random variable. It follows that the variance

$$
\sigma_X^2 = E(X^2) - \mu_X^2 = \alpha\beta^2.
$$

The MGF

$$
\begin{aligned}
M_X(t) &= \int_0^\infty e^{tx} \frac{1}{\Gamma(\alpha)\beta^\alpha} x^{\alpha-1} e^{-x/\beta} dx \\
&= \int_0^\infty \frac{1}{\Gamma(\alpha)\beta^\alpha} x^{\alpha-1} e^{-x(1/\beta-t)} dx \quad \left(\text{setting } \beta^* = \frac{1}{1/\beta - t} \right) \\
&= \int_0^\infty \frac{1}{\Gamma(\alpha)\beta^\alpha} x^{\alpha-1} e^{-x/\beta^*} dx \\
&= \frac{(\beta^*)^\alpha}{\beta^\alpha} \int_0^\infty \frac{1}{\Gamma(\alpha)(\beta^*)^\alpha} x^{\alpha-1} e^{-x/\beta^*} dx \\
&= \frac{(\beta^*)^\alpha}{\beta^\alpha} \\
&= (1 - \beta t)^{-\alpha}, \quad t < 1/\beta,
\end{aligned}
$$

where the last integral can be viewed as the integral of the PDF of a Gamma(α, β^*) random variable.

The Gamma distribution has a similar shape to the lognormal distribution. It has been used to model the distribution of the continuous waiting time of economic events (e.g., unemployment duration, price duration, poverty duration, etc.). It can also be used to model distributions of non-negative random variables, such as income, population, and range. Cox, Ingersoll and Ross (1985) propose an equilibrium model for the term structure of the spot interest rate. They assume that the spot interest rate process follows a square root process, that is,

$$dr_t = k(\theta - r_t)dt + \sigma\sqrt{r_t}dB_t,$$

where B_t is a Brownian motion process. For k, $\theta > 0$, the interest rate is elastically pulled toward a central location or long-term value θ, the parameter k determines the speed of adjustment.

This specification precludes negative interest rates as may occur in Vasicek's (1977) model which assumes

$$dr_t = k(\theta - r_t)dt + \sigma dB_t.$$

It can be shown that the Cox, Ingersoll and Ross (1985) model admits a steady state Gamma distribution (stationary distribution) with PDF:

$$f(r) = \frac{1}{\Gamma(\alpha)\beta^\alpha}r^{\alpha-1}e^{-r/\beta},$$

where $\alpha = \frac{2k\theta}{\sigma^2}$ and $\beta = \frac{\sigma^2}{2k}$. As a result, the steady state mean and variance of the short-term interest rate r_t are θ and $\sigma^2\theta/2k$ respectively.

There is a closely related distribution called Generalized Gamma distribution. Suppose the random variable

$$Y = \left(\frac{X - \gamma}{\beta}\right)^c$$

follows a standard Gamma distribution, i.e., $Y \sim \text{Gamma}(\alpha, 1)$ or equivalently it has the PDF

$$f_Y(y) = \frac{y^{\alpha-1}e^{-y}}{\Gamma(\alpha)}, \qquad y \geq 0.$$

Then the random variable X is said to follow a Generalized Gamma distribution with parameters $(\alpha, \beta, \gamma, c)$, where α and c are shape parameters, β is a scale parameter, and γ is a location parameter. With the univariate transformation method (Theorem 3.13), it can be shown that the PDF of X

$$f_X(x) = \frac{c}{\Gamma(\alpha)\beta^{c\alpha}}(x - \gamma)^{c\alpha-1}e^{-\left(\frac{x-\gamma}{\beta}\right)^c}, \qquad x \geq \gamma.$$

Often one can set $\gamma = 0$ in practice. Figure 4.10 plots the shape of the PDF of the Generalized Gamma distribution with various choices of parameters, particularly the choices of parameter c.

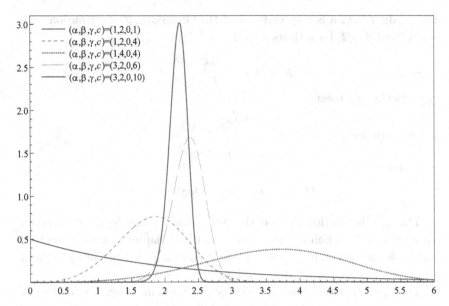

Figure 4.10: PDF's of Generalized Gamma distribution with various choices of parameter values

We note that the moments of the Generalized Gamma distribution can be obtained from the moments of the standard Gamma($\alpha, 1$) distribution by observing the following relationship:

$$
E\left[\left(\frac{X-\gamma}{\beta}\right)^k\right] = E\left[\left(\frac{X-\gamma}{\beta}\right)^{c(k/c)}\right]
$$

$$
= E\left(Y^{(k/c)}\right)
$$

$$
= \frac{\Gamma(\alpha+k/c)}{\Gamma(\alpha)}.
$$

4.3.7 Chi-Square Distribution

We now consider a special case of the Gamma distribution.

A nonnegative CRV X follows a Chi-square distribution with ν degrees of freedom, denoted as χ_ν^2, if its PDF

$$
f_X(x) = \begin{cases} \frac{1}{\Gamma(\frac{\nu}{2})\sqrt{2^\nu}}x^{\frac{\nu}{2}-1}e^{-\frac{x}{2}}, & x > 0, \\ 0, & x \le 0. \end{cases}
$$

The χ^2_ν distribution is a special case of the Gamma(α, β) distribution with $\alpha = \frac{\nu}{2}$, and $\beta = 2$. Its k-th moment is

$$E(X^k) = \frac{2^k \Gamma(\frac{\nu}{2} + k)}{\Gamma(\frac{\nu}{2})}.$$

In particular, its mean

$$E(X) = \nu$$

and its variance

$$\text{var}(X) = 2\nu.$$

The MGF

$$M_X(t) = (1 - 2t)^{-\frac{\nu}{2}} \text{ for } t < \frac{1}{2}.$$

The χ^2_ν distribution defined this way allow ν, the degree of freedom parameter, to be a non-integer value. Later, in Chapter 5, we will see that when ν is an integer, a χ^2_ν distribution is equivalent to that of the sum of ν squared independent $N(0, 1)$ random variables.

The χ^2_ν distribution is a right-skewed distribution. When the number of degrees of freedom $\nu \to \infty$, χ^2_ν becomes an approximately normal distribution with with mean ν and variance 2ν.

Like the normal distribution, the χ^2_ν distribution is one of the most important distributions in statistics and has occupied a central position in econometrics. Many popular test statistics in econometrics are in a quadratic form and have an asymptotic χ^2_ν distribution.

4.3.8 *Exponential and Weibull Distributions*

Another special case of the Gamma distribution is called an Exponential distribution.

A nonnegative CRV X follows an Exponential(β) distribution, if its PDF

$$f_X(x) = \begin{cases} \frac{1}{\beta} e^{-x/\beta}, & x > 0, \\ 0, & x \leq 0, \end{cases}$$

where $\beta > 0$. The parameter β is a scale parameter. When $\beta = 1$, X is called to follow the standard exponential distribution, denoted as EXP(1).

The Exponential(β) distribution is a special case of the Gamma$(1, \beta)$ distribution. The MGF

$$M_X(t) = E(e^{tX}) = \frac{1}{1 - \beta t}, \qquad t < \frac{1}{\beta}.$$

The mean
$$E(X) = \beta$$
and the variance
$$\text{var}(X) = \beta^2.$$

Like the geometric distribution, the exponential distribution also has the so-called "memoryless" property in the sense that for any positive numbers x and y, $x > y$,

$$P(X > x | X > y) = P(X > x - y).$$

To see this, note that for an Exponential(β) distribution, the CDF

$$F_X(x) = 1 - e^{-x/\beta} \text{ for } x \geq 0.$$

It follows that when $x > y$,

$$
\begin{aligned}
P(X > x | X > y) &= \frac{P(X > x, X > y)}{P(X > y)} \\
&= \frac{P(X > x)}{P(X > y)} \\
&= \frac{1 - F_X(x)}{1 - F_Y(y)} \\
&= \frac{e^{-x/\beta}}{e^{-y/\beta}} \\
&= e^{-(x-y)/\beta} \\
&= P(X > x - y).
\end{aligned}
$$

The exponential distribution can be viewed as a continuous analog of the discrete geometric distribution introduced in Section 4.2.4. In fact, the exponential distribution is the only continuous distribution with the memoryless property. (**Question**: How to show this?)

In fact, the Poisson process provides an interesting unified framework to link the exponential distribution and the Gamma distribution. Suppose $\{N(t) : t \geq 0\}$ is a Poisson process with rate λ. Let X_1 be the time of the first event, and for $n \geq 2$, let X_n be the time between the $(n-1)$-th and the n-th events. Then it can be shown that $\{X_1, X_2, \ldots\}$ is a sequence of identically distributed expential random variables; that is, $X_i \sim \text{EXP}(1/\lambda)$. Next, let $X = \Sigma_{i=1}^n X_i$ be the time of the n-th event. Then it can be shown that X follows a Gamma($n, 1/\lambda$) distribution. Of course, we note that the general definition of a Gamma(α, β) distribution does not require that α be an integer.

Question: Why is the exponential distribution useful in economics and finance?

The exponential distribution is of considerable importance and widely used in statistics and econometrics. Like the Gamma distribution, the exponential distribution has been used to model time durations of economic events, such as the unemployment spell of a worker, time before a credit default, the time between two trades or price changes, etc. There are many situations where one would expect an exponential distribution to give a useful description of observed phenomena. Below is an example in labor economics: Let X be the unemployment duration of a worker which has a PDF $f_X(x)$. Then the so-called hazard rate or hazard function is defined as

$$
\begin{aligned}
\lambda(x) &= \lim_{\Delta x \to 0^+} \frac{P(X \le x + \Delta x | X \ge x)}{\Delta x} \\
&= \lim_{\Delta x \to 0^+} \frac{P(x \le X \le x + \Delta x)}{P(X \ge x)\Delta x} \\
&= \left[\lim_{\Delta x \to 0^+} \frac{\int_x^{x+\Delta x} f_X(u)du}{\Delta x} \right] \cdot \frac{1}{P(X \ge x)} \\
&= \frac{f_X(x)}{P(X \ge x)} \\
&= \frac{f_X(x)}{1 - F_X(x)} \\
&= -\frac{d}{dx} \ln[1 - F_X(x)].
\end{aligned}
$$

Intuitively, the hazard function $\lambda(x)$ is the instantaneous probability that the unemployed worker will find a job right after an unemployment duration of x. Duration analysis is to model $\lambda(x)$ using economic explanatory variables. A simplest example is to assume that the hazard rate is a constant of the unemployment duration x, that is,

$$
\lambda(x) = \lambda_0 \text{ for all } x.
$$

Then the corresponding distribution of the unemployment duration X follows an Exponential($1/\lambda_0$) distribution:

$$
f_X(x) = \lambda_0 e^{-\lambda_0 x} \text{ for } x > 0.
$$

In financial econometrics, an empirical stylized fact of high-frequency stock return $\{X_t\}$ is that the absolute value of the stock return $|X_t|$ approximately follows a standard exponential distribution (Ding, Granger and Engle 1993). Here, X_t is the standardized financial return in time period t.

There are many important distributions closely related to the exponential distribution. For example, suppose $Y = (X - \alpha)^c$ follows an exponential distribution with parameter λ. Then X is said to have a Weibull distribution. Its PDF is

$$f_X(x) = \frac{c}{\lambda}(x - \alpha)^{c-1} e^{-\frac{(x-\alpha)^c}{\lambda}}, \qquad x > \alpha,$$

where α is a location parameter, λ is a scale parameter, and c is a shape parameter. It is necessary that c be greater than 1 (why?). Often, one can set $\alpha = 0$ in many applications.

The Weibull distribution is more flexible than the exponential distribution. For example, the associated hazard function is no longer a constant function. Thus, it is very useful to model hazard functions. Figure 4.11 plots the shapes of the Weibull distribution with various choices of parameters (α, λ, c).

The Weibull distribution is used in Engle and Russell (1998) to model the time duration between trades or price changes in finance.

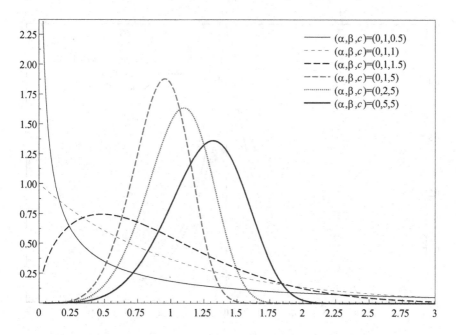

Figure 4.11: PDF's of Weibull distribution with various choices of parameter values

On the other hand, if $Y = e^{-X}$ follows an Exponential(λ) distribution, then X is said to follow an extreme value distribution. Such a distribution is used in Nelson (1991) to model the asymmetric volatility dynamics of stock returns in a time series context.

4.3.9 *Double Exponential Distribution*

A continuous random variable X follows a Double Exponential(α, λ) distribution if its PDF

$$f_X(x) = \frac{1}{2\lambda} e^{-\frac{|x-\alpha|}{\lambda}}, \qquad -\infty < x < \infty,$$

where $\lambda > 0$.

The Double Exponential(α, λ) distribution is a symmetric distribution about α, but has a fatter tail than a normal distribution. It has a peak at $x = \alpha$ where the derivative does not exist.

The mean of X,

$$E(X) = \alpha,$$

the variance

$$\mathrm{var}(X) = 2\lambda^2,$$

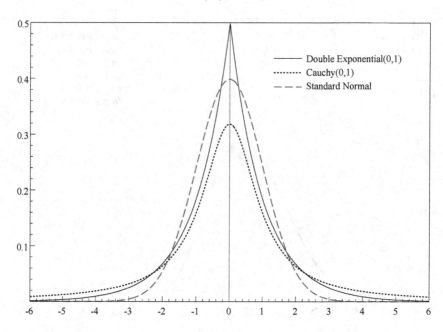

Figure 4.12: PDF's of Double Exponential$(0, 1)$, Cauchy$(0, 1)$ and $N(0, 1)$ distributions

and the MGF

$$M_X(t) = \frac{e^{\alpha t}}{1 - \lambda^2 t^2} \text{ for } |t| < \frac{1}{\lambda}.$$

The double exponential distribution is also called the Laplace distribution. When $\alpha = 0$, the absolute value of X, $Y = |X|$, follows an Exponential(λ) distribution (see Example 3.24 in Chapter 3).

Although the Double Exponential(α, λ) distribution has heavier tails than the normal distribution, all its moments exist, and has lighter tails than the Cauchy distribution. Figure 4.12 shows a Cauchy(0, 1) distribution, a standard normal distribution $N(0, 1)$, together with the Double Exponential(0, 1) distribution.

4.4 Conclusion

In this chapter we have introduced a variety of important parametric discrete and continuous probability distributions, and investigated their properties and their relationships. The flexibility of these parametric distributions depends on the functional form of the PDF, and the number of parameters. We have paid a closed attention to the interpretation of the parameters and their relationships. These distributions are widely used in modeling economic and financial data.

EXERCISE 4

4.1. Suppose $X_i \sim$ Bernoulli(p), and $X_1, \cdots, X_n$ are jointly independent. Define $X = \Sigma_{i=1}^n X_i$. Show $X \sim$ Binomial(n, p).

4.2. Let X have the standard normal PDF $f_X(x) = \left(1/\sqrt{2\pi}\right) e^{-x^2/2}$ for $-\infty < x < \infty$.

(1) Find $E(X^2)$ directly, and then by using the PDF of $Y = X^2$ and calculating $E(Y)$.

(2) Find the PDF of $Y = |X|$, and find its mean and variance. This distribution is sometimes called a *folded normal*.

4.3. The Gamma function is defined as

$$\Gamma(\alpha) = \int_0^\infty t^{\alpha-1} e^{-t} dt.$$

Verify these two identities regarding the Gamma function:

(1) $\Gamma(\alpha + 1) = \alpha \Gamma(\alpha)$;

(2) $\Gamma(\frac{1}{2}) = \sqrt{\pi}$.

4.4. Let the random variable X have the PDF $f(x) = \frac{2}{\sqrt{2\pi}} e^{-x^2/2}$, $0 < x < \infty$. Find the transformation $Y = g(X)$ and values of α and λ so that $Y \sim$ Gamma(α, λ). The PDF of a Gamma(α, λ) is given by $f_X(x) = \frac{1}{\Gamma(\alpha)\lambda^\alpha} x^{\alpha-1} e^{-x/\lambda}$, for $x > 0$.

4.5. The Pareto distribution, with parameters α and λ, has PDF $f(x) = \frac{\lambda \alpha^\lambda}{x^{\lambda+1}}$, $\alpha < x < \infty, \lambda > 0, \alpha > 0$.

(1) Verify that $f(x)$ is a PDF.

(2) Derive the mean and variance of this distribution.

(3) Prove that the variance does not exist if $\lambda \leq 2$.

4.6. Suppose X follows a binomial distribution, $B(n, p)$, where $p \in (0, 1)$, with PMF

$$f(x) = \binom{n}{x} p^x (1 - p)^{n-x} \text{ for } x = 0, 1, \cdots, n.$$

Using the uniqueness theorem of the moment generating function, show that when $n \to \infty$ but $np \to \lambda \in (0, \infty)$, the binomial distribution can be approximated by a Poisson(λ) distribution. [Hint: Show the MGF of the binomial distribution converges to the MGF of the Poisson distribution.]

4.7. Find the mean and variance of a geometric distribution with parameter $p \in (0, 1)$.

4.8. Show that the geometric distribution is characterized by the Markovian property

$$P(X = x + y | X \geq y) = P(X = x) \text{ for all positive integers } x, y.$$

4.9. (1) A electronic product can withstand a number of external shocks. However, when the K-th shock arrives, the product fails. That is, the lifetime of the product is the arrival time of the K-th shock. Suppose during the time $[0, t)$, the number of shocks follow a Poisson(λt) distribution:

$$P(X = x) = \frac{(\lambda t)^x}{x!} e^{-x}, \quad x = 0, 1, \cdots.$$

Prove that the lifetime of the product follows a Gamma(K, λ) distribution.

(2) Generally, if $X \sim$ Gamma(α, λ) and α is an integer, then

$$P(X \leq x) = P(Y \geq \alpha), \text{ for any } x,$$

where $Y \sim$ Poisson(x/λ).

4.10. Find the mean and variance of the geometric distribution with parameter p.

4.11. A series system is built in such a way that it operates only when all its components operate (so it fails when at least one component fails). Assuming that the lifetime of each components has an Exp(1) distribution and that the components operate independently, find the distribution and survival function of the system's life time T.

4.12. Suppose X is exponentially distributed. Show

$$P(X > x + y | X > x) = P(X > y) \text{ for all } 0 < x, y < \infty.$$

4.13. Many well-known distributions are special cases of the more common distributions. For each of the following named distributions derive the form of the PDF and verify that it is a PDF:

(1) if $X \sim$ Exponential(λ), then $Y = X^{1/\gamma}$ has a Weibull(γ, λ) distribution, where $\gamma > 0$ is a constant;

(2) if $X \sim$ Exponential(λ), then $Y = (2X/\lambda)^{1/2}$ has a Rayleigh distribution;

(3) if $X \sim$ Gamma(a, b), then $Y = 1/X$ has an Inverted Gamma $IG(a, b)$ distribution;

(4) if $X \sim$ Gamma($\frac{3}{2}, \lambda$), then $Y = (X/\lambda)^{1/2}$ has a Maxwell distribution;

(5) if $X \sim$ Exponential(1), then $Y = \alpha - \gamma \log X$ has a Gumbel(α, γ) distribution, where $-\infty < \alpha < \infty$ and $\gamma > 0$.

4.14. Prove the following analogs to Stein's Lemma, assuming appropriate conditions on the function $g(\cdot)$.

(1) If $X \sim \text{Gamma}(\alpha, \beta)$, then

$$E\left[g(X)(X - \alpha\beta)\right] = \lambda E\left[Xg'(X)\right].$$

(2) If $X \sim \text{Beta}(\alpha, \beta)$, then

$$E\left\{g(X)\left[\beta - (\alpha - 1)\frac{(1 - X)}{X}\right]\right\} = E\left[(1 - X)g'(X)\right].$$

4.15. Suppose X follows a double exponential distribution with PDF $f_X(x) = \frac{1}{2\sigma}e^{-|x-\mu|/\sigma}$. Show that when $\mu = 0$, the absolute value of X, $Y = |X|$, follows an Exponential(σ) distribution.

4.16. Suppose $Y = e^{-X}$ follows an Exponential(λ) distribution. Find the PDF of X. The distribution of X is called an extreme value distribution.

4.17. A random variable is said to have a two-piece normal distribution with parameters μ, σ_1, σ_2 if it has its PDF

$$f_X(x) = \begin{cases} Ae^{-(x-\mu)^2/(2\sigma_1^2)}, & x \leq \mu, \\ Ae^{-(x-\mu)/(2\sigma_2^2)}, & x > \mu. \end{cases}$$

Find:

(1) the value of constant A;

(2) the mean of X;

(3) the variance of X.

4.18. Suppose $X \sim N(\mu, \sigma^2)$. Find the PDF of $Y = 1/X$.

4.19. Suppose $X \sim \chi_\nu^2$. Then $Y = \sqrt{X}$ is called a Chi-distribution with ν degrees of freedom and is denoted as χ_ν. Show:

(1) the PDF of Y is $f_Y(y) = \frac{1}{2^{(\nu/2)-1}\Gamma(\nu/2)}e^{-y^2/2}y^{\nu-1}$, $y > 0$;

(2) $E(Y^k) = \frac{2^{k/2}\Gamma[(\nu+k)/2]}{\Gamma(\nu/2)}$.

Chapter 5

Multivariate Probability Distributions

Abstract: In this chapter we study the relationships among random variables, which will be characterized by the joint probability distribution of random variables. Most insights into multivariate distributions can be gained by focusing on bivariate distributions. We first introduce the joint probability distribution of a bivariate random vector (X, Y) via the characterization of the joint cumulative distribution function, the joint probability mass function (when (X, Y) are discrete), and the joint probability density function (when (X, Y) are continuous) respectively. We then characterize various aspects of the relationship between X and Y using the conditional distributions, correlation, and conditional expectations. The concept of independence and its implications on the joint distributions, conditional distributions and correlation are also discussed. We also introduce a class of bivariate normal distributions.

Key words: Bivariate normal distribution, Bivariate transformation, Conditional distribution, Conditional mean, Conditional variance, Correlation, Independence, Joint moment generating function, Joint probability distribution, Law of iterated expectations, Marginal distribution.

5.1 Random Vectors and Joint Probability Distributions

Any economy is a system which consists of different units (e.g., households, assets, sectors, markets, etc). These units are generally related to each other. As a consequence, economic variables are interrelated. The most important goal of economic analysis and econometric analysis is to identify the relationships between economic events or economic variables. As discussed in Chapter 2, for two events A and B, the joint probability $P(A \cap B)$ and conditional probability $P(A|B)$ describe the joint and predictive relationships between them. Such a relationship can be exploited to predict one using the other.

Definition 5.1. [Random Vector]: A n-dimensional random vector, denoted as $Z = (Z_1, \cdots, Z_n)'$, is a function from a sample space S into $\mathbb{R}^n$, the n-dimensional Euclidean space. For each outcome $s \in S$, $Z(s)$ is a n-dimensional real-valued vector and is called a realization of the random vector Z.

In this chapter, we will mainly focus on bivariate probability distributions, which can illustrate most (but not all) essentials of multivariate probability distributions. We now consider two random variables (X, Y) in most of the subsequent discussion, where both X and Y are defined on the same probability space $(S, \mathbb{B}, P)$. A realization of (X, Y) will be a pair $(x, y) \in \mathbb{R}^2$.

Having defined a bivariate random vector (X, Y), we can now discuss probabilities of events that are defined in terms of (X, Y).

Like in the univariate case, we can use the CDF, now called the joint CDF, of X and Y, to characterize their joint distribution.

Definition 5.2. [Joint CDF]: The joint CDF of X and Y is defined as

$$F_{XY}(x, y) = P(X \leq x, Y \leq y)$$
$$= P(X \leq x \cap Y \leq y)$$

for any pair $(x, y) \in \mathbb{R}^2$.

We first examine the properties of the joint CDF.

Lemma 5.1. *[Properties of $F_{XY}(x, y)$]:*
   (1) $F_{XY}(-\infty, y) = F_{XY}(x, -\infty) = 0, F_{XY}(\infty, \infty) = 1$;
   (2) $F_{XY}(x, y)$ is non-decreasing in both x and y;
   (3) $F_{XY}(x, y)$ is right-continuous in both x and y.

Below we state a useful result.

Theorem 5.1. $F_X(x) = F_{XY}(x, +\infty)$ and $F_Y(y) = F_{XY}(+\infty, y)$.

Proof: Define two events: $A = \{X \leq x\}$, $B = \{Y \leq \infty\}$. Since B always holds, we have $A \cap B = A$. It follows that $P(A) = P(A \cap B)$, that is, $F_X(x) = F_{XY}(x, \infty)$.

Theorem 5.1 implies that individual CDF's of X and Y can be obtained from the joint CDF $F_{XY}(x, y)$. These individual CDF's are called the marginal CDF's of X and Y respectively.

The joint and marginal CDF's are widely used in statistics and econometrics. Below we provide an important example called copula, which can be used to model dependence between random variables.

Example 5.1. [Bivariate Copula]: Put $U = F_X(X), V = F_Y(Y)$. Then both the probability integral transforms U and V are $U[0,1]$ random variables. The joint CDF of (U, V)

$$C(u, v) = P(U \leq u, V \leq v)$$

is called the copula associated with the joint probability distribution of (X, Y). The copula $C(u, v)$ is closely related to the joint CDF $F_{XY}(x, y)$. Suppose $F_X(\cdot)$ and $F_Y(\cdot)$ are strictly increasing functions. Then

$$
\begin{aligned}
F_{XY}(x, y) &= P(X \leq x, Y \leq y) \\
&= P[F_X(X) \leq F_X(x), F_Y(Y) \leq F_Y(y)] \\
&= P[U \leq F_X(x), V \leq F_Y(y)] \\
&= C[F_X(x), F_Y(y)].
\end{aligned}
$$

This suggests that the joint distribution $F_{XY}(x, y)$ can be decomposed into two separate components: the marginal distributions $F_X(\cdot)$ and $F_Y(\cdot)$ on one hand, and the "pure" dependence function $C(\cdot, \cdot)$ between X and Y on the other hand. In other words, the copula $C(\cdot, \cdot)$ only specifies the dependence or association between X and Y regardless of their marginal distributions. It separates the marginal behaviors from the association between X and Y. Given the functional form $C(\cdot, \cdot)$, different marginal distributions will yield different joint distributions of (X, Y). The copula has been widely used in financial econometrics and financial industries, to model comovements among markets or among assets (Cherubini, Luciano and Vecchiato 2004).

5.1.1 The Discrete Case

We now consider the case when both X and Y are DRV's, where we can introduce a joint PMF to characterize their joint probability distribution.

Definition 5.3. [Joint PMF]: Let X and Y be two DRV's. Then their joint PMF is defined as

$$f_{XY}(x, y) = P(X = x \cap Y = y) = P(X = x, Y = y)$$

for any point $(x, y) \in \mathbb{R}^2$.

We now examine the properties of the joint PMF.

Lemma 5.2. *[Properties of $f_{XY}(x, y)$]:*
(1) $f_{XY}(x, y) \geq 0$ for all $(x, y) \in \mathbb{R}^2$;
(2) $\sum_{x \in \Omega_X} \sum_{y \in \Omega_Y} f_{XY}(x, y) = 1$, where Ω_X and Ω_Y are the supports of X and Y respectively.

Definition 5.4. [Support]: The support of a bivariate random vector (X, Y) is defined as the set of all possible pairs of (x, y) which (X, Y) will take with strictly positive probability. That is,

$$\text{Support}(X, Y) = \Omega_{XY} = \{(x, y) \in \mathbb{R}^2 : f_{XY}(x, y) > 0\}.$$

It is convenient to work on the support of (X, Y) only.

Question: Suppose Ω_X and Ω_Y are the supports of X, Y respectively. Is it true that

$$\Omega_{XY} = \Omega_X \times \Omega_Y$$
$$= \{(x, y) \in \mathbb{R}^2 : f_X(x) > 0, f_Y(y) > 0\}?$$

The answer is generally no. Consider an example of bivariate distribution for which both X and Y take nonnegative integers but with the restriction that $X \leq Y$. Then we have

$$\Omega_X = \Omega_Y = \{0, 1, 2, \cdots\}$$

while

$$\Omega_{XY} = \{(x, y) : 0 \leq x \leq y < \infty, \ x, y \text{ are integers}\}.$$

Obviously, Ω_{XY} is a subset of $\Omega_X \times \Omega_Y$.

Question: Is it true that

$$\sum_{(x,y) \in \Omega_{XY}} f_{XY}(x, y) = 1?$$

Note that in general, Ω_{XY} is a subset of $\Omega_X \times \Omega_Y$.

The joint PMF $f_{XY}(x, y)$ can be used to calculate the probability of any event defined in terms of (X, Y). For any subset $A \in \mathbb{R}^2$, we have

$$P[(X, Y) \in A] = \sum_{(x,y) \in A} f_{XY}(x, y).$$

Example 5.2. Suppose X and Y have the joint PMF

$$f_{XY}(x, y) = c|x + y| \text{ for } x = -1, 0, 1 \text{ and } y = 0, 1,$$

where c is an unknown constant.

Find: (1) the supports of X, Y, and (X, Y) respectively; (2) the value of c; (3) $P(X = 0 \text{ and } Y = 1)$; (4) $P(X = 1)$; and (5) $P(|X - Y| \leq 1)$.

Solution: (1) We have $\Omega_X = \{-1, 0, 1\}$, $\Omega_Y = \{0, 1\}$, and $\Omega_{XY} = \{(-1, 0), (0, 1), (1, 0), (1, 1)\}$. Note that Ω_{XY} is a subset of $\Omega_X \times \Omega_Y$ because $(0, 0) \in \Omega_X \times \Omega_Y$ but $(0, 0) \notin \Omega_{XY}$.

(2) Using the property that $\sum_{x \in \Omega_X} \sum_{y \in \Omega_Y} f_{XY}(x, y) = 1$, we have

$$c\left[| -1 + 0| + |-1 + 1| + |0 + 0| + |0 + 1| + |1 + 0| + |1 + 1|\right] = 1.$$

Thus, $c = \frac{1}{5}$.

(3) By the definition of the joint PMF, we have

$$P(X = 0, Y = 1) = f_{XY}(0, 1) = \frac{1}{5}|0 + 1| = \frac{1}{5}.$$

(4) Noting that the event $\{X = 1\} = \{X = 1\} \cap \{Y \in \Omega_Y\}$, we have

$$
\begin{aligned}
P(X = 1) \\
&= \sum_{y \in \Omega_Y} f_{XY}(1, y) \\
&= f_{XY}(1, 0) + f_{XY}(1, 1) \\
&= \frac{3}{5}.
\end{aligned}
$$

(5)

$$
\begin{aligned}
P(|X - Y| \leq 1) &= \sum_{(x,y) \in \Omega_{XY} : |x - y| \leq 1} f_{XY}(x, y) \\
&= f_{XY}(-1, 0) + f_{XY}(0, 1) + f_{XY}(1, 0) + f_{XY}(1, 1) \\
&= 1.
\end{aligned}
$$

We now investigate the relationship between $f_{XY}(x, y)$ and $F_{XY}(x, y)$, which is very similar to the relationship between $f_X(x)$ and $F_X(x)$ in the univariate case.

For DRV's X and Y, we can obtain the joint CDF

$$
\begin{aligned}
F_{XY}(x, y) &= P(X \leq x, Y \leq y) \\
&= \sum_{(u,v) \in \Omega_{XY}(x,y)} f_{XY}(u, v),
\end{aligned}
$$

where $\Omega_{XY}(x, y)$ is the set of all possible pairs of (u, v) in the support Ω_{XY} of (X, Y) such that $u \leq x, v \leq y$, namely,

$$\Omega_{XY}(x, y) = \{(u, v) \in \Omega_{XY}, u \leq x, v \leq y\}.$$

Thus, we can obtain $F_{XY}(x, y)$ from $f_{XY}(x, y)$.

On the other hand, by taking the differences of $F_{XY}(x, y)$ with respect to x and y, we can also recover $f_{XY}(x, y)$ from $F_{XY}(x, y)$. Without loss of generality, we assume that the possible values X can take are arranged in an increasing order: $x_1 < x_2 < x_3 < \cdots$, and the possible values Y can take are also arranged in an increasing order: $y_1 < y_2 < y_3 < \cdots$. Then, for $i > 1, j > 1$,

$$
\begin{aligned}
f_{XY}&(x_i, y_j) \\
&= \Delta_Y \Delta_X F_{XY}(x_i, y_j) \\
&= \Delta_Y [F_{XY}(x_i, y_j) - F_{XY}(x_{i-1}, y_j)] \\
&= [F_{XY}(x_i, y_j) - F_{XY}(x_i, y_{j-1})] - [F_{XY}(x_{i-1}, y_j) - F_{XY}(x_{i-1}, y_{j-1})] \\
&= F_{XY}(x_i, y_j) - F_{XY}(x_i, y_{j-1}) - F_{XY}(x_{i-1}, y_j) + F_{XY}(x_{i-1}, y_{j-1}),
\end{aligned}
$$

where Δ_X and Δ_Y are the difference operators with respect to x and y respectively. For example, for any given y, we have $\Delta_X F_{XY}(x_i, y) = F_{XY}(x_i, y) - F_{XY}(x_{i-1}, y)$.

The above formula does not cover the cases where $i = 1$ or $j = 1$. For these cases, we have

$$
f_{XY}(x_i, y_j) = \begin{cases}
F_{XY}(x_i, y_j) - F_{XY}(x_i, y_{j-1}), & \text{if } i = 1, j > 1, \\
F_{XY}(x_i, y_j) - F_{XY}(x_{i-1}, y_j), & \text{if } i > 1, j = 1, \\
F_{XY}(x_i, y_j), & \text{if } i = 1, j = 1.
\end{cases}
$$

The above bivariate concepts can be generalized to the multivariate case, where there are n random variables. For example, the joint PMF of n discrete random variables $X_1, \cdots, X_n$ is given by

$$f_{\mathbf{X}^n}(\mathbf{x}^n) = P(X_1 = x_1, \cdots, X_n = x_n)$$

for each n tuple $\mathbf{x}^n = (x_1, \cdots, x_n) \in \mathbb{R}^n$, and the joint CDF of $\mathbf{X}^n = (X_1, \cdots, X_n)$ is given by

$$F_{\mathbf{X}^n}(\mathbf{x}^n) = P(X_1 \leq x_1, \cdots, X_n \leq x_n)$$

for all $\mathbf{x}^n \in \mathbb{R}^n$.

5.1.2 *The Continuous Case*

Next, we consider the case of CRV's:

Definition 5.5. [Joint PDF]: Two random variables X and Y are said to have a joint continuous distribution if their joint CDF $F_{XY}(x, y)$ is absolutely continuous in both x and y. In this case, there exists a nonnegative function $f_{XY}(x, y)$ such that for any subset $A \subset \mathbb{R}^2$,

$$P[(X, Y) \in A] = \int \int_{(x,y) \in A} f_{XY}(x, y) dx dy.$$

In particular,

$$F_{XY}(x, y) = \int_{-\infty}^{x} \int_{-\infty}^{y} f_{XY}(u, v) du dv.$$

The function $f_{XY}(x, y)$ is called a joint PDF of (X, Y).

Lemma 5.3. [*Properties of Joint PDF* $f_{XY}(x, y)$]:
(1) $f_{XY}(x, y) \geq 0$ for all (x, y) in the xy plane;
(2) $\int_{-\infty}^{\infty} \int_{-\infty}^{\infty} f_{XY}(x, y) dx dy = 1$.

Proof: (1) Denoting

$$A(x, y) = \{(u, v) : u \leq x, v \leq y\} \text{ for any given pair } (x, y) \in \mathbb{R}^2,$$

we have

$$P[(X, Y) \in A(x, y)] = P(X \leq x, Y \leq y)$$
$$= F_{XY}(x, y)$$
$$= \int_{-\infty}^{x} \int_{-\infty}^{y} f_{XY}(u, v) dv du.$$

This formula is analogous to the double sum of the joint PMF in the discrete case. It indicates that $F_{XY}(x, y)$ can be obtained from the joint PDF $f_{XY}(x, y)$ by double integrations.

On the other hand, at the points of (x, y) where $F_{XY}(x, y)$ is differentiable, we have

$$f_{XY}(x, y) = \frac{\partial^2 F_{XY}(x, y)}{\partial x \partial y} \geq 0,$$

where the equality follows from the fundamental theorem of calculus, and the inequality follows from the fact that $F_{XY}(x, y)$ is non-decreasing in (x, y). This formula is analogous to the double differences of the joint CDF $F_{XY}(x, y)$ with respect to (x, y) in the discrete case. It indicates that one can recover the joint PDF $f_{XY}(x,y)$ by differentiating the joint CDF $F_{XY}(x,y)$ with respect to (x,y).

(2) The integral $\int_{-\infty}^{\infty}\int_{-\infty}^{\infty}f_{XY}(x,y)dxdy = 1$ follows immediately from the fact that $F_{XY}(\infty,\infty) = 1$. This completes the proof.

Question: What is the interpretation of the joint PDF $f_{XY}(x,y)$?

For any given pair (x,y) in the xy-plane, consider the event

$$A(x,y) = \left\{ x - \frac{\epsilon}{2} < X \le x + \frac{\epsilon}{2} \text{ and } y - \frac{\epsilon}{2} < Y \le y + \frac{\epsilon}{2} \right\},$$

where $\epsilon > 0$ is a small constant. That is, $A(x,y)$ is the event that (X,Y) takes values in a small rectangular area centered at point (x,y) and with each side equal to ϵ. Suppose $f_{XY}(x,y)$ is continuous at the point (x,y). Then

$$\begin{aligned}
P[A(x,y)] &= \int_{y-\epsilon/2}^{y+\epsilon/2}\int_{x-\epsilon/2}^{x+\epsilon/2} f_{XY}(u,v)dudv \\
&= f_{XY}(\bar{x},\bar{y}) \cdot \epsilon \cdot \epsilon \text{ for some } (\bar{x},\bar{y}) \\
&\approx f_{XY}(x,y)\epsilon^2 \text{ when } \epsilon \text{ is small.}
\end{aligned}$$

Thus, although $f_{XY}(x,y)$ is not a probability measure, it is proportional to the probability that (X,Y) takes values in a small rectangular area centered at point (x,y). In other words, $f_{XY}(x,y)$ is proportional to the probability that (X,Y) takes values in the area centered at point (x,y) in the xy plane.

The probability $P[(X,Y) \in A]$ has a 3-dimensional geometric interpretation. Recall that in the univariate case, when A is an interval on the real line (e.g., $A = \{x \in R : a < x \le b\}$), the probability $P(X \in A)$ is equal to the area under the curve $f_X(x)$ over the interval A, as is shown in Figure 5.1(a). Now, for the bivariate case, suppose A is an area on the xy plane. Then the probability $P[(X,Y) \in A]$ is a volume under the surface of $f_{XY}(x,y)$ over the area A in the xy plane, as shown in Figure 5.1(b).

The geometric interpretation for $P[(X,Y) \in A]$ has important implications: (1) an event that (X,Y) takes a value at any individual point (x,y), or values at any finite number of points in the xy plane, has probability zero; (2) an event that (X,Y) takes values on any one-dimensional curve in the xy plane has probability zero.

Because $F_{XY}(x,y)$ and $f_{XY}(x,y)$ can be recovered from each other, they are equivalent in the sense that they contain the same information about the joint distribution of (X,Y). However, it is often more convenient to use $f_{XY}(x,y)$ in practice. Also, like the univariate case, for each joint CDF $F_{XY}(x,y)$, there may exist some degree of arbitrariness in defining the joint PDF $f_{XY}(x,y)$ over a set of countable points (x,y) on the xy-plane, which will not alter the joint CDF $F_{XY}(x,y)$ in any way. However, there is always a most natural joint PDF which is as smooth as possible over the xy plane.

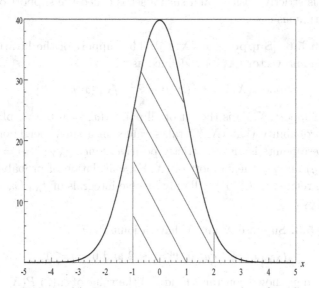

Figure 5.1(a): Geometric interpretation for $P(X \in [-1,2])$

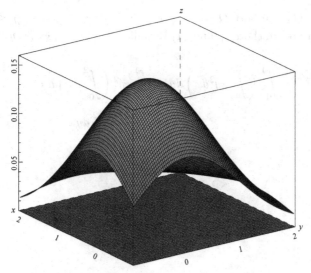

Figure 5.1(b): Geometric interpretation for $P((X,Y) \in [-1,2]^2)$

It is important to note that the joint PDF $f_{XY}(x,y)$ is defined for all pairs of (x,y) in $\mathbb{R}^2$. The PDF $f_{XY}(x,y)$ may be zero for a large set A where $P[(X,Y) \in A] = 0$, although the PDF is defined for points in A. For this reason, it is convenient to focus on the set of pairs of (x,y) at which $f_{XY}(x,y)$ is strictly positive and such a set is called the support of (X,Y) in the continuous case.

Definition 5.6. [Support of (X,Y)]: The support of the bivariate continuous random vector (X,Y) is defined as

$$\text{Support}(X,Y) = \{(x,y) \in \mathbb{R}^2 : f_{XY}(x,y) > 0\}.$$

Thus, $\text{Support}(X,Y)$ is the set of all points (x,y) in the xy plane such that the probability that (X,Y) takes values in a small neighborhood of each of these points is always strictly positive. Since $f_{XY}(x,y) = 0$ for all points (x,y) outside the support of (X,Y), calculation of probabilities for any events related to (X,Y) will only involve integrals of $f_{XY}(x,y)$ on the support Ω_{XY}.

Example 5.3. Suppose X and Y have a joint PDF

$$f_{XY}(x,y) = cy^2 \text{ if } 0 \le x \le 2 \text{ and } 0 \le y \le 1,$$

where c is an unknown constant. Find: (1) the value of c; (2) $P(X+Y > 2)$; (3) $P(X < 0.5)$; (4) $P(X = 3Y)$.

Solution: The support $\Omega_{XY} = \{(x,y) : 0 \le x \le 2, 0 \le y \le 1\}$ is a rectangular area in the xy plane. (1) Recalling $\int_{-\infty}^{\infty} \int_{-\infty}^{\infty} f_{XY}(x,y)dxdy = 1$, we have

$$\int_0^1 \left(\int_0^2 cy^2 dx \right) dy = \int_0^1 cy^2 \left(\int_0^2 dx \right) dy$$

$$= c \int_0^1 y^2 x \Big|_0^2 dy$$

$$= 2c \int_0^1 y^2 dy$$

$$= \frac{2c}{3} y^3 \Big|_0^1$$

$$= \frac{2c}{3}$$

$$= 1.$$

Thus, $c = \frac{3}{2}$.

(2)
$$P(X + Y > 2) = \int_0^1 \left(\int_{2-y}^2 \frac{3}{2} y^2 dx \right) dy = \frac{3}{8};$$

(3)
$$P(X < 0.5) = \int_0^1 \left(\int_0^{\frac{1}{2}} \frac{3}{2} y^2 dx \right) dy = \frac{1}{4};$$

(4) Because $x = 3y$ is a line, the volume over a line is zero. Therefore, we have
$$P(X = 3Y) = 0.$$

Example 5.4. Suppose X and Y have a joint PDF
$$f_{XY}(x, y) = cx^2 y \text{ for } x^2 \le y \le 1,$$
where c is an unknown constant. Find: (1) the value of c; (2) $P(X \ge Y)$.

Solution: The support $\Omega_{XY} = \{(x, y) : x^2 \le y \le 1\}$ is plotted in Figure 5.2.

(1) Using the property that
$$\int_0^1 \left(\int_{-\sqrt{y}}^{\sqrt{y}} cx^2 y dx \right) dy = 1,$$
we can solve for $c = \frac{21}{4}$.

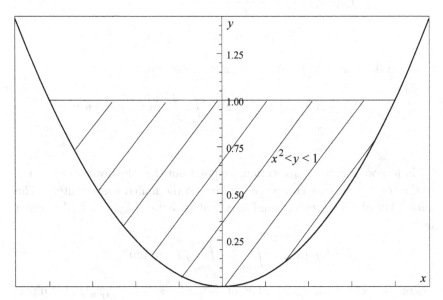

Figure 5.2: Support of (X, Y)

(2)

$$P(X \geq Y) = \int_0^1 \left(\int_y^{\sqrt{y}} cx^2 y dx \right) dy = \frac{3}{20}.$$

Example 5.5. Suppose (X, Y) has a joint CDF $F_{XY}(x, y) = \frac{1}{16}xy(x+y)$ for $0 \leq x \leq 2$ and $0 \leq y \leq 2$. Find: (1) the complete representation of the CDF $F_{XY}(x, y)$ in the entire xy-plane; (2) the joint PDF $f_{XY}(x, y)$; (3) $P(1 \leq X \leq 2, 1 \leq Y \leq 2)$.

Solution: (1) According to the properties of the joint CDF $F_{XY}(x, y)$ in Lemma 5.1, we obtain

$$F_{XY}(x, y) = \begin{cases} 0, & \text{if } x < 0 \text{ or } y < 0, \\ \frac{1}{16}xy(x+y), & \text{if } 0 \leq x \leq 2, \ 0 \leq y \leq 2, \\ \frac{1}{8}x(x+2), & \text{if } 0 \leq x \leq 2, \ y > 2, \\ \frac{1}{8}y(y+2), & \text{if } x > 2, \ 0 \leq y \leq 2, \\ 1, & \text{if } x > 2, y > 2. \end{cases}$$

(2) By partial differentiation, we have

$$f_{XY}(x, y) = \frac{\partial^2 F_{XY}(x, y)}{\partial x \partial y}$$
$$= \frac{1}{8}(x+y) \text{ for } 0 \leq x \leq 2, 0 \leq y \leq 2.$$

Note that $f_{XY}(x, y) = 0$ for all (x, y) elsewhere.

(3)

$$P(1 \leq X \leq 2, 1 \leq Y \leq 2) = \int_1^2 \int_1^2 \frac{1}{8}(x+y)dxdy$$
$$= \frac{3}{8}.$$

Before we conclude this section, we point out that the bivariate concepts in the continuous case can be generalized to the multivariate context. The joint CDF of n continuous random variables $\mathbf{X}^n = (X_1, \cdots, X_n)'$ is given by

$$F_{\mathbf{X}^n}(\mathbf{x}^n) = \int_{-\infty}^{x_1} \cdots \int_{-\infty}^{x_n} f_{\mathbf{X}^n}(\mathbf{u}^n)d\mathbf{u}^n,$$

where $f_{\mathbf{X}^n}(\mathbf{x}^n)$ is the joint PDF of $\mathbf{X}^n$, $\mathbf{x}^n = (x_1, \cdots, x_n)$, and $\mathbf{u}^n = (u_1, \cdots, u_n)$. Partial differentiation yields

$$f_{\mathbf{X}^n}(\mathbf{x}^n) = \frac{\partial^n F_{\mathbf{X}^n}(\mathbf{x}^n)}{\partial x_1 \cdots \partial x_n}$$

wherever the partial derivative exists at point $\mathbf{x}^n$ in the $\mathbb{R}^n$ space.

5.2 Marginal Distributions

Question: What information can be extracted from the joint PMF/PDF $f_{XY}(x, y)$? Intuitively, we can expect to extract the following information:

- Individual information of X, characterized by the PMF/PDF $f_X(x)$ of X;
- Individual information of Y, characterized by the PMF/PDF $f_Y(y)$ of Y;
- Predictive relationship between X and Y, characterized by suitable conditional distribution concepts.

5.2.1 *The Discrete Case*

We now investigate how to extract such information from the joint PMF/ PDF $f_{XY}(x, y)$. We first consider the case of DRV's.

Definition 5.7. [Discrete Marginal PMF's]: Suppose X and Y have a joint discrete distribution with joint PMF $f_{XY}(x, y)$. Then the marginal PMF's of X and Y are defined as

$$f_X(x) = P(X = x) = \sum_{y \in \Omega_Y} f_{XY}(x, y), \text{ where } -\infty < x < \infty,$$

$$f_Y(y) = P(Y = y) = \sum_{x \in \Omega_X} f_{XY}(x, y), \text{ where } -\infty < y < \infty.$$

To understand the definition of marginal PMF's, we define the event $\{X = x\}$ in a bivariate context. Observe that the event

$$\{X = x\} = \{X = x\} \cap \{Y \in \Omega_Y\}$$

$$= \{X = x\} \cap \left[\bigcup_{y \in \Omega_Y} \{Y = y\} \right]$$

$$= \bigcup_{y \in \Omega_Y} \{X = x\} \cap \{Y = y\},$$

where the last equality follows by the distributive laws in Theorem 2.1. This is analogous to the rule of total probability that $P(A) = \sum_{i=1}^{\infty} P(A \cap$

A_i) for a sequence of mutually exclusive and collectively exhaustive events $\{A_1, A_2, \cdots\}$ and any event A. It follows that

$$P(X = x) = P(\{X = x\} \cap \{Y \in \Omega_Y\})$$
$$= \sum_{y \in \Omega_Y} f_{XY}(x, y).$$

Intuitively, the marginal PMF of X is the probability that X takes a given value x regardless of the values taken by Y. By taking into account of all possibilities of Y, we get rid of all information of Y, and only the information of X remains. We call $f_X(x)$ the marginal PMF of X to emphasize the fact that it is the PMF of X but in the bivariate context that gives the joint distribution of the vector (X, Y). Technically, the adjective "marginal" is redundant.

The marginal PMF's have the following properties.

Lemma 5.4. *[Properties of $f_X(x)$ and $f_Y(y)$]:*
(1) $f_X(x) \geq 0$ for all $x \in (-\infty, \infty)$;
(2) $\sum_{x \in \Omega_X} f_X(x) = 1$, where Ω_X is the support of X.
Similar results for $f_Y(y)$ also hold.

Example 5.6. Suppose X and Y have the joint PMF

$$f_{XY}(x, y) = \frac{1}{5}|x + y|, \qquad x = -1, 0, 1, \quad y = 0, 1.$$

Find: (1) $f_X(x)$; and (2) $f_Y(y)$.

Solution: (1) For $x = -1$, the event $\{X = -1\}$ contains two basic outcomes: $\{X = -1, Y = 0\}$ and $\{X = -1, Y = 1\}$. These basic outcomes are mutually exclusive. Thus, it follows that

$$f_X(-1) = P(X = -1)$$
$$= f_{XY}(-1, 0) + f_{XY}(-1, 1)$$
$$= \frac{1}{5}.$$

Similarly,

$$f_X(0) = f_{XY}(0, 0) + f_{XY}(0, 1) = \frac{1}{5},$$

$$f_X(1) = f_{XY}(1, 0) + f_{XY}(1, 1) = \frac{3}{5}.$$

Then the marginal PMF of X is given by

$$f_X(x) = \begin{cases} \frac{1}{5}, & \text{if } x = -1, \\ \frac{1}{5}, & \text{if } x = 0, \\ \frac{3}{5}, & \text{if } x = 1. \end{cases}$$

(2) Similarly, we can obtain the PMF of Y

$$f_Y(y) = \begin{cases} \frac{2}{5}, & \text{if } y = 0, \\ \frac{3}{5}, & \text{if } y = 1. \end{cases}$$

We can compactly represent the joint PMF $f_{XY}(x, y)$ and the marginal PMF's of (X, Y) as a matrix form, where the marginal PMFs can be obtained from summing rows and columns respectively:

Y/X	-1	0	1
0	$\frac{1}{5}$	0	$\frac{1}{5}$
1	0	$\frac{1}{5}$	$\frac{2}{5}$

It is possible to have the same marginal PMF's, but different joint PMF's. There are many joint distributions that have the same marginal distributions. The joint PMF not only tells the marginal information but also the relationship between X and Y that is not available in the marginals. If the relationship between X and Y is changed, a different joint distribution will arise, although the marginal distributions may remain the same.

We consider a simple example.

Example 5.7. Suppose both X and Y are binary random variables, taking only value 1 or 0. Consider the following two joint PMF's for (X, Y):

Case (1):

$$f_{XY}(x, y) = \begin{cases} p, & \text{if } x = y = 1, \\ 1 - p, & \text{if } x = y = 0, \\ 0, & \text{otherwise.} \end{cases}$$

Case (2):

$$f_{XY}(x, y) = \begin{cases} p^{x+y}(1 - p)^{2-(x+y)}, & \text{if } x = 1, 0; \ y = 1, 0, \\ 0, & \text{otherwise.} \end{cases}$$

Obviously, these are two different joint distributions of (X, Y). However, it can be shown that in both cases, we have $X \sim$ Bernoulli(p), $Y \sim$ Bernoulli(p).

5.2.2   The Continuous Case

Suppose the bivariate continuous random vector (X, Y) have a joint PDF $f_{XY}(x, y)$. We first consider how to obtain the CDF of X. Observe that the event $\{X \leq x\} = \{X \leq x\} \cap \{-\infty < Y < \infty\}$, we have

$$
\begin{aligned}
F_X(x) &\equiv P(X \leq x) \\
&= P(X \leq x, -\infty < Y < \infty) \\
&= \int_{-\infty}^{x} \int_{-\infty}^{\infty} f_{XY}(u, y) du\, dy \\
&= \int_{-\infty}^{x} \left[\int_{-\infty}^{\infty} f_{XY}(u, y) dy \right] du \\
&= \int_{-\infty}^{x} f_X(u) du.
\end{aligned}
$$

Differentiating both sides of the above equation, we obtain

$$
f_X(x) = \int_{-\infty}^{\infty} f_{XY}(x, y) dy.
$$

Thus, by integrating out y, we can obtain the marginal PDF $f_X(x)$ from the joint PDF $f_{XY}(x, y)$.

Definition 5.8. [Marginal PDF's]: Suppose X and Y have a joint continuous distribution with joint PDF $f_{XY}(x, y)$. Then the marginal PDF's of X and Y are defined as follows:

$$
f_X(x) = \int_{-\infty}^{\infty} f_{XY}(x, y) dy, \quad -\infty < x < \infty,
$$

$$
f_Y(y) = \int_{-\infty}^{\infty} f_{XY}(x, y) dx, \quad -\infty < y < \infty.
$$

Unlike such marginal concepts as marginal utility and marginal productivity, the marginal PDF's are obtained by integrating the other variable, rather than by taking partial derivatives. Intuitively, by integrating out y in the joint PDF $f_{XY}(x, y)$, what remains is the information of X.

The marginal PDF's have the following properties.

Lemma 5.5. [Properties of $f_X(x)$ and $f_Y(y)$]:
(1) $f_X(x) \geq 0$ for all $x \in (-\infty, \infty)$;
(2) $\int_{-\infty}^{\infty} f_X(x) dx = 1$.
Similar results hold for $f_Y(y)$.

Example 5.8. Suppose (X, Y) have a joint PDF $f_{XY}(x, y) = 4xy$ if $0 < x < 1$ and $0 < y < 1$. Find the marginal PDF's $f_X(x)$ and $f_Y(y)$.

Solution: The support $\Omega_{XY} = \{(x, y) : 0 < x < 1, 0 < y < 1\}$ is a unit rectangular area. For $0 < x < 1$, the marginal PDF of X is

$$f_X(x) = \int_0^1 4xy \, dy = 2x.$$

For $x \leq 0$ or $x \geq 1$, we have $f_X(x) = 0$.
   For $0 < y < 1$, the marginal PDF of Y is

$$f_Y(y) = \int_0^1 4xy \, dx = 2y.$$

For $y \leq 0$ or $y \geq 1$, we have $f_Y(y) = 0$.

Example 5.9. Suppose $f_{XY}(x, y) = cy^2$ for $x^2 < y < 1$. Find: (1) $f_X(x)$; (2) $f_Y(y)$.

Solution: The support of (X, Y) is the same as the support of (X, Y) in Example 5.4, which was given in Figure 5.2, from which we can obtain the support of X as $\Omega_X = \{x \in \mathbb{R} : -1 < x < 1\}$ and the support of Y as $\Omega_Y = \{y \in \mathbb{R} : 0 < y < 1\}$.
   We first determine the value of c by using the property that

$$\int_{-\infty}^{\infty} \int_{-\infty}^{\infty} f_{XY}(x, y) dx dy = 1.$$

Given the support of (X, Y) as depicted in Figure 5.2, we have

$$c \int_{-1}^1 \left(\int_{x^2}^1 y^2 dy \right) dx = c \int_{-1}^1 \frac{1}{3} \left(1 - x^6 \right) dx$$
$$= \frac{2c}{3} \int_0^1 \left(1 - x^6 \right) dx$$
$$= \frac{4c}{7}$$
$$= 1.$$

It follows that $c = \frac{7}{4}$.
   (1) For $-1 < x < 1$, we have

$$f_X(x) = \int_{-\infty}^{\infty} f_{XY}(x,y)dy$$

$$= \int_{x^2}^{1} cy^2 dy$$

$$= c\frac{1}{3}y^3\big|_{x^2}^{1}$$

$$= \frac{7}{12}(1 - x^6).$$

Then

$$f_X(x) = \begin{cases} \frac{7}{12}(1 - x^6), & -1 < x < 1, \\ 0, & \text{otherwise.} \end{cases}$$

(2) For $0 < y < 1$, we have

$$f_Y(y) = \int_{-\infty}^{\infty} f_{XY}(x,y)dx$$

$$= \int_{-\sqrt{y}}^{\sqrt{y}} cy^2 dx$$

$$= cy^2 \cdot 2\sqrt{y}$$

$$= \frac{7}{2}y^{\frac{5}{2}}.$$

Then the marginal PDF of Y is

$$f_Y(y) = \begin{cases} \frac{7}{2}y^{\frac{5}{2}}, & 0 \le y < 1, \\ 0, & \text{otherwise.} \end{cases}$$

When there are more than two random variables, we can define not only the marginal distributions of individual random variables but also the joint distributions of any subset of random variables. For example, in the case of n discrete random variables $X_1, \cdots, X_n$, we have the marginal PMF of X_1

$$f_{X_1}(x_1) = \sum_{x_2 \in \Omega_2} \cdots \sum_{x_n \in \Omega_n} f_{\mathbf{X}^n}(\mathbf{x}^n), \qquad -\infty < x_1 < \infty,$$

and the joint PMF of the subset (X_1, X_2, X_3)

$$f_{X_1 X_2 X_3}(x_1, x_2, x_3) = \sum_{x_4 \in \Omega_4} \cdots \sum_{x_n \in \Omega_n} f_{\mathbf{X}^n}(\mathbf{x}^n), \qquad -\infty < x_1, x_2, x_3 < \infty,$$

where $\mathbf{X}^n = (X_1, \cdots, X_n), \mathbf{x}^n = (x_1, \cdots, x_n)$, and Ω_i is the support of $X_i, i = 1, \cdots, n$.

5.3 Conditional Distributions

Oftentimes when two random variables, (X, Y), are observed, the values of the two variables are related. Knowledge about the value of X gives us some information about the value of Y even if it does not tell us the value of Y exactly.

Question: How to characterize the predictive relationship between X and Y?

We can use the concept of conditional distribution of Y given X. Again, we consider the cases of discrete and continuous random variables respectively.

5.3.1 *The Discrete Case*

We first consider the case of DRV's.

Definition 5.9. [**Conditional PMF's**]: Let X and Y have a joint discrete distribution with joint PMF $f_{XY}(x, y)$ and marginal PMF's $f_X(x)$ and $f_Y(y)$. Then the conditional PMF of Y given $X = x$ is defined as

$$f_{Y|X}(y|x) = P(Y = y | X = x)$$
$$= \frac{f_{XY}(x, y)}{f_X(x)}$$

provided $f_X(x) > 0$.

Similarly, the conditional PMF of X given $Y = y$ is defined as

$$f_{X|Y}(x|y) = P(X = x | Y = y)$$
$$= \frac{f_{XY}(x, y)}{f_Y(y)}$$

provided $f_Y(y) > 0$.

Intuitively, the conditional PMF $f_{Y|X}(y|x)$ is the probability that the random variable Y will take any arbitrary value y given that the value x of the random variable X has been observed. Recall the conditional probability formula $P(A|B) = P(A \cap B)/P(B)$ for any two events in Chapter 2. Define events $A = \{X = x\}$ and $B = \{Y = y\}$. Then

$$f_{Y|X}(y|x) = P(B|A) = \frac{P(A \cap B)}{P(A)}.$$

Example 5.10. Suppose two random variables X and Y are independent, and $X \sim \text{Poisson}(\lambda_1)$, $Y \sim \text{Poisson}(\lambda_2)$. We can show that $X + Y \sim \text{Poisson}(\lambda_1 + \lambda_2)$. Find the conditional distribution of X given $X + Y = n$.

Solution: By the definition of conditional PMF and independence between X and Y, we have

$$
\begin{aligned}
P(X &= k | X + Y = n) \\
&= \frac{P(X = k, X + Y = n)}{P(X + Y = n)} \\
&= \frac{P(X = k) P(Y = n - k)}{P(X + Y = n)} \\
&= \frac{\frac{\lambda_1^k}{k!} e^{-\lambda_1} \cdot \frac{\lambda_2^{n-k}}{(n-k)!} e^{-\lambda_2}}{\frac{(\lambda_1 + \lambda_2)^n}{n!} e^{-(\lambda_1 + \lambda_2)}} \\
&= \frac{n!}{k!(n-k)!} \frac{\lambda_1^k \lambda_2^{n-k}}{(\lambda_1 + \lambda_2)^n} \\
&= \binom{n}{k} \left(\frac{\lambda_1}{\lambda_1 + \lambda_2} \right)^k \left(\frac{\lambda_2}{\lambda_1 + \lambda_2} \right)^{(n-k)} \\
&= \binom{n}{k} \left(\frac{\lambda_1}{\lambda_1 + \lambda_2} \right)^k \left(1 - \frac{\lambda_1}{\lambda_1 + \lambda_2} \right)^{(n-k)}, \quad k = 0, 1, \cdots, n.
\end{aligned}
$$

Therefore, given $X + Y = n$, X follows a binomial distribution $B(n, p)$ with $p = \frac{\lambda_1}{\lambda_1 + \lambda_2}$.

Questions: What happens to the conditional PMF $f_{Y|X}(y|x)$ if $f_X(x) = 0$?

In this case, $f_{Y|X}(y|x)$ is not well-defined, because it does not make any sense in practice to condition on something that is unlikely to occur.

It is important to emphasize that given any x with $f_X(x) > 0$, the conditional PMF $f_{Y|X}(y|x)$ is a PMF of Y. That is, given each x with $f_X(x) > 0$, we have the following:

(1) $f_{Y|X}(y|x) \geq 0$ for all $y \in (-\infty, \infty)$;

(2) $\sum_{y \in \Omega_{Y|X}(x)} f_{Y|X}(y|x) = 1$, where $\Omega_{Y|X}(x) = \{y \in \Omega_Y : f_{Y|X}(y|x) > 0\}$ is the set of all possible values of $y \in \Omega_Y$ with $f_{Y|X}(y|x) > 0$.

Different values of x can be associated with different conditional distributions of Y. For example, X can be a state variable taking two possible values: 0 and 1. When $X = 0$ (representing a bear market), the distribution for the stock return Y may have a large dispersion; when $X = 1$ (representing a bull market), the distribution for the stock return Y may have a small dispersion. Indeed, an important empirical stylized fact of financial markets is that asset returns have a larger volatility in a bear market than a bull market. For another example, suppose Y is the wage of an employee, and X is a gender dummy taking value 1 when the employee is female and taking value 0 when the employee is male. Then $f_{Y|X}(y|1)$ is the wage distribution of female employees and $f_{Y|X}(y|0)$ is the wage distribution of male employees.

With the definition of conditional PMF, we have the following multiplication rules:

$$f_{XY}(x,y) = f_{Y|X}(y|x)f_X(x) = f_{X|Y}(x|y)f_Y(y),$$

where the first equality always holds provided $f_X(x) > 0$ and the second equality holds provided $f_Y(y) > 0$. These multiplication rules suggest that joint PMFs can be equivalently characterized by specifying the conditional PMF $f_{Y|X}(y|x)$ and the marginal PMF $f_X(x)$.

Since the conditional distribution of Y given $X = x$ is possibly a different probability distribution for each value of x, we have a family of probability distributions for Y, one for each x. When we wish to describe this entire family, we will use the phrase "the distribution of $Y|X$." If, for example, X is a positive integer-valued random variable and the conditional distribution of Y given $X = x$ is Binomial(x, p), then we might say the distribution of $Y|X$ is Binomial(X, p) or write $Y|X \sim$ Binomial(X, p). Whenever we use the symbol $Y|X$ or have a random variable X as the parameter of a probability distribution of Y, we are describing the family of conditional probability distributions.

The conditional PMF $f_{Y|X}(y|x)$ is very useful because it tells how the information of X can be used to predict the probability of Y. It should be emphasized that the conditional PMF is a predictive relationship. It is not the causal relationship from X to Y. For example, it is possible that both X and Y are caused by an unobservable variable Z. In this case, X and Y are generally related to each other, and one can use the information of X to predict the probability distribution of Y, although X does not cause Y.

5.3.2 *The Continuous Case*

Next, we consider the case of CRV's. Suppose we have observed a value x for random variable X. It is desired to specify probabilities for $Y \in A$ given $X = x$, for various sets A that are of interest.

However, if X and Y are CRV's, then $P(X = x) = 0$ for every value of x. Therefore, to compute a conditional probability such as $P(Y > 5|X = 10)$, the definition $P(B|A) = P(A \cap B)/P(A)$, where $A = \{X = 10\}$, cannot be used, since the denominator $P(X = 10) = 0$. In other words, the concept of conditional PMF cannot be used for the continuous case. Yet, in reality, a value of $X = 10$ may be well observed. If, to the limit of our measurement, we observe $X = 10$, this knowledge might give us information about Y. It turns out that when X and Y are continuous, the appropriate way to define a conditional probability distribution for Y given $X = x$ is analogous to the discrete case but with PDF's replacing PMF's. In other words, the concept of the conditional PMF for the discrete case should be extended when we move to the continuous case.

Definition 5.10. [Conditional PDF's]: Let X and Y have a joint continuous distribution with joint PDF $f_{XY}(x, y)$ and marginal PDF's $f_X(x)$ and $f_Y(y)$. Then the conditional PDF of Y given $X = x$ is defined as

$$f_{Y|X}(y|x) = \frac{f_{XY}(x, y)}{f_X(x)}$$

if $f_X(x) > 0$.

Similarly, the conditional PDF of X given $Y = y$ is defined as

$$f_{X|Y}(x|y) = \frac{f_{XY}(x, y)}{f_Y(y)}$$

if $f_Y(y) > 0$.

It should be noted that for DRV's (X, Y), the conditional PMF $f_{Y|X}(y|x)$ is derived as a conditional probability for the event $\{Y = y\}$ given the event $\{X = x\}$, whereas for CRV's (X, Y), the conditional PDF $f_{Y|X}(y|x)$ is defined as the ratio of the joint PDF $f_{XY}(x, y)$ to the marginal PDF $f_X(x)$.

Example 5.11. Suppose two random variables (X, Y) follow a uniform distribution on $\{(x, y) : x^2 + y^2 \leq 1\}$. Find: (1) the conditional probability of Y given $X = x$; (2) $P(Y > 0|X = 0)$.

Solution: The joint PDF of (X, Y) is

$$f_{XY}(x, y) = \begin{cases} \frac{1}{\pi}, & x^2 + y^2 \leq 1, \\ 0, & \text{otherwise}, \end{cases}$$

and the support $\Omega_{XY} = \{(x, y) : x^2 + y^2 \leq 1\}$ is a unit circle centered at the origin $(0, 0)$ in the xy-plane. For $-1 \leq x \leq 1$, we have

$$f_X(x) = \int_{-\infty}^{\infty} f_{XY}(x, y) dy$$

$$= \int_{-\sqrt{1-x^2}}^{\sqrt{1-x^2}} \frac{1}{\pi} dy$$

$$= \frac{2}{\pi} \sqrt{1 - x^2}.$$

Then, for $-1 < x < 1$, the conditional PDF of Y given $X = x$ is

$$f_{Y|X}(y|x) = \frac{f_{XY}(x, y)}{f_X(x)}$$

$$= \begin{cases} \frac{1}{2\sqrt{1-x^2}}, & \text{if } -\sqrt{1 - x^2} \leq y \leq \sqrt{1 - x^2}, \\ 0, & \text{otherwise.} \end{cases}$$

That is, for any given $x \in (-1, 1)$, the conditional PDF of Y given $X = x$ follows a uniform distribution on the interval $[-\sqrt{1 - x^2}, \sqrt{1 - x^2}]$.

(2) The conditional distribution of Y given $X = x$ depends on the value of x. When $x = 0$, we have

$$f_{Y|X}(y|0) = \begin{cases} \frac{1}{2}, & \text{if } -1 < y < 1, \\ 0, & \text{otherwise.} \end{cases}$$

It follows that

$$P(Y > 0 | X = 0) = \int_0^{\infty} f_{Y|X}(y|0) dy$$

$$= \int_0^1 \frac{1}{2} dy$$

$$= \frac{1}{2}.$$

We now examine the properties of and provide an interpretation for the conditional PDF $f_{Y|X}(y|x)$.

Lemma 5.6. *[Properties of Conditional PDF's]: For any given $x \in \mathbb{R}$ with $f_X(x) > 0$, $f_{Y|X}(y|x)$ is a PDF of Y. That is,*

(1) $f_{Y|X}(y|x) \geq 0$ for all $y \in (-\infty, \infty)$;

(2) $\int_{-\infty}^{\infty} f_{Y|X}(y|x) dy = 1$.

The same properties hold for $f_{X|Y}(x|y)$.

Proof: Suppose $f_X(x) > 0$. Then $f_{Y|X}(y|x) = f_{XY}(x,y)/f_X(x)$ is well-defined and is nonnegative for all $y \in (-\infty, \infty)$. Moreover,

$$
\begin{aligned}
\int_{-\infty}^{\infty} f_{Y|X}(y|x)dy &= \int_{-\infty}^{\infty} \frac{f_{XY}(x,y)}{f_X(x)}dy \\
&= \frac{1}{f_X(x)} \int_{-\infty}^{\infty} f_{XY}(x,y)dy \\
&= \frac{1}{f_X(x)} f_X(x) \\
&= 1.
\end{aligned}
$$

Thus, given any value of x with $f_X(x) > 0$, $f_{Y|X}(y|x)$ is a PDF of Y. Different values of x can be associated with different distributions for Y. This implies that one can then use the information of X to predict the distribution of Y. Figure 5.3 plots a continuous family of conditional PDF's $f_{Y|X}(y|x)$ for $x \in (-1,1)$ of Example 5.11.

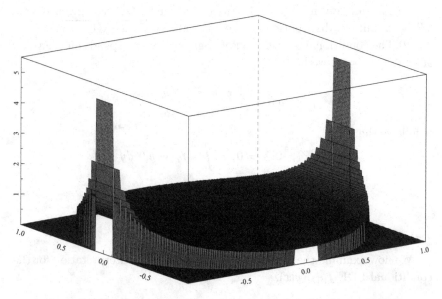

Figure 5.3: Conditional PDF's $f_{Y|X}(y|x)$ for $x \in (-1,1)$ in Example 5.11

Question: What is the interpretation for the conditional PDF $f_{Y|X}(y|x)$?

For any given pair of (x,y) with $f_{XY}(x,y) > 0$, consider two events $A(x) = \{x - \frac{\epsilon}{2} < X \le x + \frac{\epsilon}{2}\}$ and $B(y) = \{y - \frac{\epsilon}{2} < Y \le y + \frac{\epsilon}{2}\}$, where ϵ is a small positive constant. Then by the mean value theorem, we can obtain

$$P[A(x) \cap B(y))] = \int_{y-\epsilon/2}^{y+\epsilon/2} \int_{x-\epsilon/2}^{x+\epsilon/2} f_{XY}(u,v) du dv$$

$$\approx f_{XY}(x,y)\epsilon^2,$$

$$P[A(x)] = \int_{x-\epsilon/2}^{x+\epsilon/2} f_X(u) du$$

$$\approx f_X(x)\epsilon.$$

It follows that

$$P[B(y)|A(x)] = \frac{P[A(x) \cap B(y)]}{P[A(x)]}$$

$$\approx \frac{f_{XY}(x,y)\epsilon^2}{f_X(x)\epsilon} = f_{Y|X}(y|x)\epsilon.$$

This implies that the conditional PDF $f_{Y|X}(y|x)$ is proportional to

$$P[B(y)|A(x)] = P\left(y - \frac{\epsilon}{2} < Y \le y + \frac{\epsilon}{2} \,\middle|\, x - \frac{\epsilon}{2} < X \le x + \frac{\epsilon}{2} \right),$$

the conditional probability that Y takes values in the small interval $(y - \frac{\epsilon}{2}, y + \frac{\epsilon}{2}]$ given that X takes values in the small interval $(x - \frac{\epsilon}{2}, x + \frac{\epsilon}{2}]$. In other words, $f_{Y|X}(y|x)$ is proportional to the probability that Y takes values near y given that X has taken values near x.

Like in the discrete case, we also have the multiplication rules:

$$f_{XY}(x,y) = f_{Y|X}(y|x)f_X(x) = f_{X|Y}(x|y)f_Y(y),$$

where the first equality holds whenever $f_X(x) > 0$, and the second equality holds whenever $f_Y(y) > 0$.

When we are dealing with more than two random variables, we can consider various kinds of conditional distributions. For example, we can define

$$f_{X_i|\mathbf{X}^{i-1}}(x_i|\mathbf{x}^{i-1}) = \frac{f_{\mathbf{X}^i}(\mathbf{x}^i)}{f_{\mathbf{X}^{i-1}}(\mathbf{x}^{i-1})}$$

if $f_{\mathbf{X}^{i-1}}(\mathbf{x}^{i-1}) > 0$, where $\mathbf{X}^{i-1} = (X_1, \cdots, X_{i-1})'$ and $\mathbf{x}^{i-1} = (x_1, \cdots, x_{i-1})'$. We can also define

$$f_{(X_1,X_2)|(X_3,X_4)}(x_1,x_2|x_3,x_4) = \frac{f_{X_1 X_2 X_3 X_4}(x_1,x_2,x_3,x_4)}{f_{X_3 X_4}(x_3,x_4)}$$

if $f_{X_3 X_4}(x_3,x_4) > 0$.

5.4 Independence

The marginal distributions of X and Y, described by the marginal PMF's/ PDF's $f_X(x)$ and $f_Y(y)$, do not completely describe the joint distribution of X and Y. Indeed, there are many different joint distributions that have the same marginal distributions. Thus, it is hopeless to try to determine the joint PMF's/PDF's, $f_{XY}(x,y)$, from the knowledge of the marginal PMF's/PDF's, $f_X(x)$ and $f_Y(y)$. However, there is a special but important case in which we can use the marginal distributions to determine the joint distributions. This occurs when the knowledge about X gives us no information about Y. This is the case of so-called independence between X and Y. It is a characterization of no association between X and Y.

Definition 5.11. [Independence]: Two random variables X and Y are independent if

$$F_{XY}(x,y) = F_X(x)F_Y(y) \text{ for all } -\infty < x,y < \infty.$$

where $F_{XY}(\cdot), F_X(\cdot), F_Y(\cdot)$ are the joint and marginal CDF's.

The above definition of independence is equivalent to the definition that two random variables X and Y defined on the same sample space are independent if

$$P(X \in A, Y \in B) = P(X \in A)P(Y \in B)$$

for all subsets of $A \in \Omega_X$ and $B \in \Omega_Y$.

    The definition of independence allows that X and Y are discrete or continuous. In the discrete case, we can use the joint and marginal PMF's to characterize independence.

Theorem 5.2. *Two discrete random variables* (X,Y) *are independent if and only if*

$$f_{XY}(x,y) = f_X(x)f_Y(y) \text{ for all pairs of } (x,y) \in \mathbb{R}^2,$$

where $f_{XY}(x,y), f_X(x), f_Y(y)$ *are the joint and marginal PMF's.*

Proof: (1) [*Necessity*] By the definition of independence, we have

$$F_{XY}(x,y) = F_X(x)F_Y(y) \text{ for all } -\infty < x,y < \infty.$$

Without loss of generality, we assume that the possible values that X can take are arranged in an increasing order: $x_1 < x_2 < x_3 < \cdots$ and the possible values that Y can take are also arranged in an increasing order: $y_1 < y_2 < y_3 < \cdots$.

For $i > 1$, taking a difference of the above equation with respect to x, i.e., from x_{i-1} to x_i, we obtain

$$\Delta_X F_{XY}(x_i, y) = F_{XY}(x_i, y) - F_{XY}(x_{i-1}, y)$$
$$= [F_X(x_i) - F_X(x_{i-1})]F_Y(y).$$

If in addition $j > 1$, we can further take a difference of the above equation with respect to y and obtain

$$\Delta_Y \Delta_X F_{XY}(x_i, y_j) = [F_X(x_i) - F_X(x_{i-1})][F_Y(y_j) - F_Y(y_{j-1})].$$

This yields the following relationship among the joint and marginal PMF's:

$$f_{XY}(x_i, y_j) = f_X(x_i) f_Y(y_j), \qquad i, j > 1.$$

We can also obtain the same relationship when $i = 1$ or $j = 1$. (Please verify it!)

(2) [*Sufficiency*] Suppose now

$$f_{XY}(x_i, y_j) = f_X(x_i) f_Y(y_j) \text{ for all } i, j = 1, 2, \cdots.$$

Assuming $x_i \leq x < x_{i+1}, y_j \leq y < y_{j+1}$, we have

$$F_{XY}(x, y) = P(X \leq x, Y \leq y)$$
$$= \sum_{i'=1}^{i} \sum_{j'=1}^{j} f_{XY}(x_{i'}, y_{j'})$$
$$= \sum_{i'=1}^{i} f_X(x_{i'}) \sum_{j'=1}^{j} f_Y(y_{j'})$$
$$= F_X(x) F_Y(y).$$

Since i, j are arbitrary, so are x and y. Thus, $F_{XY}(x, y) = F_X(x) F_Y(y)$ for all pairs of (x, y) on the xy-plane. The proof is completed.

Next, we show that for the continuous case, we can use the joint and marginal PDF's to characterize independence.

Theorem 5.3. *Suppose X and Y are two CRV's. Then X and Y are independent if and only if*

$$f_{XY}(x, y) = f_X(x) f_Y(y) \text{ for all } (x, y) \in \mathbb{R}^2,$$

where $f_{XY}(x, y), f_X(x),$ and $f_Y(y)$ are the joint and marginal PDF's.

Proof: (1) [*Necessity*] We first show if (X, Y) are independent, then $f_{XY}(x, y) = f_X(x) f_Y(y)$ for all pairs of $(x, y) \in \mathbb{R}^2$.

Suppose (X, Y) are independent. Then by definition, we have

$$F_{XY}(x, y) = F_X(x)F_Y(y) \text{ for all } x, y.$$

Differentiating the both sides of this equation with respect to x and y respectively, we obtain

$$\frac{\partial^2 F_{XY}(x, y)}{\partial x \partial y} = \frac{\partial^2 F_X(x)F_Y(y)}{\partial x \partial y}$$

$$= \frac{\partial F_X(x)}{\partial x}\frac{\partial F_Y(y)}{\partial y}.$$

This implies

$$f(x, y) = f_X(x)f_Y(y) \text{ for all } (x, y) \in \mathbb{R}^2.$$

(2) [*Sufficiency*] Next, we show that if $f_{XY}(x, y) = f_X(x)f_Y(y)$, then (X, Y) are independent. Suppose $f_{XY}(u, v) = f_X(u)f_Y(v)$ for all $(u, v) \in \mathbb{R}^2$. Then by integration, we have

$$\int_{-\infty}^{x}\int_{-\infty}^{y} f_{XY}(u, v)dudv = \int_{-\infty}^{x}\int_{-\infty}^{y} f_X(u)f_Y(v)dudv$$

$$= \int_{-\infty}^{x} f_X(u)du \int_{-\infty}^{y} f_Y(v)dv.$$

That is,

$$F_{XY}(x, y) = F_X(x)F_Y(y), \text{ for all } -\infty < x, y < \infty,$$

which implies that X and Y are independent. The proof is completed.

Example 5.12. Suppose $f_{XY}(x, y) = 4xy$ if $0 \leq x \leq 1$ and $0 \leq y \leq 1$. Are X and Y independent?

Solution: In Example 5.8, we have obtained $f_X(x) = 2x$ for $0 \leq x \leq 1$ and $f_Y(y) = 2y$ for $0 \leq y \leq 1$. Then

$$f_X(x)f_Y(y) = 4xy = f_{XY}(x, y) \text{ for } 0 \leq x \leq, 0 \leq y \leq 1.$$

Also, we have $f_{XY}(x, y) = f_X(x)f_Y(y) = 0$ for all (x, y) outside the rectangular area defined by $0 \leq x \leq 1$ and $0 \leq y \leq 1$. It follows that $f_{XY}(x, y) = f_X(x)f_Y(y)$ for all (x, y) in the xy plane. Therefore, X and Y are independent.

Example 5.13. Suppose $f_{XY}(x, y) = 8xy$, $0 \leq x \leq y \leq 1$. Are X and Y independent?

Solution: The support $\Omega_{XY} = \{(x, y) : 0 \leq x \leq y \leq 1\}$ is an upper triangular area, as shown in Figure 5.4.

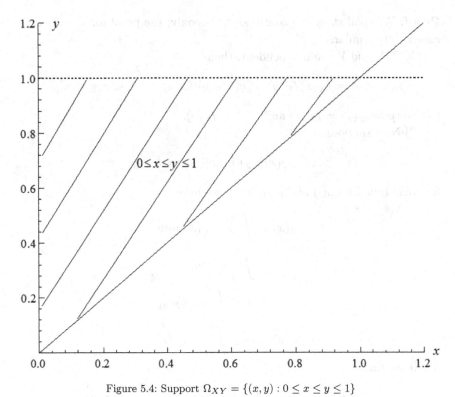

Figure 5.4: Support $\Omega_{XY} = \{(x, y) : 0 \le x \le y \le 1\}$

By integrating out y and x respectively, we can obtain the marginal PDF's:

$$f_X(x) = 4x(1 - x^2), \qquad 0 \le x \le 1,$$

and

$$f_Y(y) = 4y^3, \qquad 0 \le y \le 1.$$

Thus, X and Y are not independent, since $f_X(x)f_Y(y) \ne f_{XY}(x, y)$ when $0 \le x \le y \le 1$.

We now explore some implications of independence.

Suppose X and Y have a joint PMF or PDF $f_{XY}(x, y)$ that can be decomposed as the product of two functions $h(x)g(y)$ for all (x, y) on the xy-plane, where $h(x)$ and $g(y)$ are not necessarily a PMF or PDF of X and Y. Then the theorem below shows that X and Y are independent.

Theorem 5.4. [Factorization Theorem]: *Two random variables X and Y are independent if and only if their joint PMF/PDF can be written as*

$$f_{XY}(x, y) = g(x)h(y), \text{ for all } -\infty < x, y < \infty.$$

Proof: We shall show the continuous case only; the proof for the discrete case is very similar.

(1) If X and Y are independent, then

$$f_{XY}(x,y) = f_X(x)f_Y(y) = g(x)h(y) \text{ for all } -\infty < x, y < \infty,$$

where we set $g(x) = f_X(x)$ and $h(y) = f_Y(y)$.

(2) Now suppose

$$f_{XY}(x,y) = g(x)h(y) \text{ for all } -\infty < x, y < \infty$$

for some functions $g(\cdot)$ and $h(\cdot)$. Then we have

$$f_X(x) = \int_{-\infty}^{\infty} f_{XY}(x,y)dy$$

$$= \int_{-\infty}^{\infty} g(x)h(y)dy$$

$$= g(x)\int_{-\infty}^{\infty} h(y)dy,$$

$$f_Y(y) = h(y)\int_{-\infty}^{\infty} g(x)dx.$$

It follows that

$$f_X(x)f_Y(y) = \left[g(x)\int_{-\infty}^{\infty} h(v)dv\right]\left[h(y)\int_{-\infty}^{\infty} g(u)du\right]$$

$$= g(x)h(y)\int_{-\infty}^{\infty}\int_{-\infty}^{\infty} g(u)h(v)dudv$$

$$= f_{XY}(x,y) \text{ for all } -\infty < x, y < \infty,$$

where we have used the fact that

$$\int_{-\infty}^{\infty}\int_{-\infty}^{\infty} f_{XY}(u,v)dudv = \int_{-\infty}^{\infty}\int_{-\infty}^{\infty} g(u)h(v)dudv$$

$$= 1.$$

This completes the proof.

The factorization theorem provides a very convenient way to check independence. The key is to check whether the joint PMF/PDF can be partitioned into a product of two separate functions of x and y respectively. We note that it is important to check whether the partition holds for all points (x, y) in the whole xy-plane, rather than on a subregion of the xy-plane only.

On the other hand, for CRV's (X, Y), it is possible that $f_{XY}(x, y) \neq f_X(x)f_Y(y)$ on a set A of (x, y) values for which $\int_A dxdy = 0$. In such cases, X and Y are still called independent. This is because two PDF's that differ only on a set like A still define the same probability distribution for (X, Y).

Next, we explore the implication of independence on the conditional probability distributions.

Theorem 5.5. *Suppose two random variables X and Y are independent. Then for the conditional PMF/PDF*

$$f_{Y|X}(y|x) = f_Y(y) \text{ for all } (x, y) \in \mathbb{R}^2$$

where $f_X(x) > 0$, and

$$f_{X|Y}(x|y) = f_X(x) \text{ for all } (x, y) \in \mathbb{R}^2$$

where $f_Y(y) > 0$.

Proof: When X and Y are independent, we have $f_{XY}(x, y) = f_X(x)f_Y(y)$ for all (x, y) in the xy-plane. It follows that

$$
\begin{aligned}
f_{Y|X}(y|x) &= \frac{f_{XY}(x, y)}{f_X(x)} \\
&= \frac{f_X(x)f_Y(y)}{f_X(x)} \\
&= f_Y(y) \text{ for all } (x, y)
\end{aligned}
$$

provided $f_X(x) > 0$.

Similarly, we have

$$f_{X|Y}(x|y) = f_X(x) \text{ for all } (x, y)$$

provided $f_Y(y) > 0$. The proof is completed.

This theorem implies that when X and Y are independent, the information of X has no predictive power for the probability distribution of Y, and vice versa.

Example 5.14. Suppose two random variables X and Y have a joint PDF

$$
f_{XY}(x, y) = \begin{cases} e^{-y}, & 0 < x < y < \infty, \\ 0, & \text{otherwise.} \end{cases}
$$

(1) Find $f_X(x)$ and $f_Y(y)$; (2) find $f_{Y|X}(y|x)$ and $f_{X|Y}(x|y)$; (3) check if X and Y are independent.

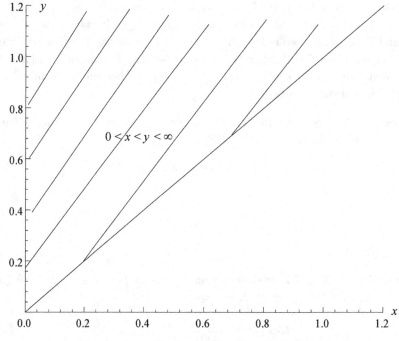

Figure 5.5: Support $\Omega_{X,Y}$ of (X, Y) in Example 5.14

Solution: The support $\Omega_{XY} = \{(x, y) : 0 < x < y < \infty\}$ is an upper triangular area as shown in Figure 5.5.

(1) The support of X is $0 < x < \infty$. By definition,

$$f_X(x) = \int_{-\infty}^{\infty} f_{XY}(x, y)dy.$$

For $x \leq 0$, $f_{XY}(x, y) = 0$. So, $f_X(x) = 0$ for $x \leq 0$.
For $x > 0$,

$$f_X(x) = \int_x^{\infty} e^{-y}dy$$

$$= e^{-x}.$$

Therefore,

$$f_X(x) = \begin{cases} e^{-x}, & x > 0, \\ 0, & \text{otherwise.} \end{cases}$$

It is important to note that no matter whether X and Y are independent, the marginal PDF $f_X(x)$ has nothing to do with y.

Next, we calculate $f_Y(y)$. The support of Y is $0 < y < \infty$. When $y \leq 0, f_{XY}(x,y) = 0$, so $f_Y(y) = 0$. For $y > 0$,

$$f_Y(y) = \int_{-\infty}^{\infty} f_{XY}(x,y)dx$$

$$= \int_0^y e^{-y}dx$$

$$= ye^{-y}.$$

It follows that

$$f_Y(y) = \begin{cases} ye^{-y}, & y > 0, \\ 0, & \text{otherwise.} \end{cases}$$

(2) First, we calculate $f_{Y|X}(y|x)$. Since $f_X(x) = 0$ for $x \leq 0$, the conditional PDF $f_{Y|X}(y|x)$ is defined for any given $x > 0$. Given any $x > 0$, we have

$$f_{Y|X}(y|x) = \frac{f_{XY}(x,y)}{f_X(x)}$$

$$= \frac{e^{-y}}{e^{-x}} \text{ for } 0 < x < y < \infty.$$

It follows that

$$f_{Y|X}(y|x) = \begin{cases} e^{-(y-x)}, & y \in (x,\infty), \\ 0, & y \in (-\infty, x]. \end{cases}$$

This implies that, conditional on $X = x$, Y follows an exponential distribution with support $y \in (x, \infty)$.

Next, we calculate $f_{X|Y}(x|y)$. Since $f_Y(y) = 0$ for $y \leq 0$, the conditional PDF $f_{X|Y}(x|y)$ is defined for any given $y > 0$. By definition,

$$f_{X|Y}(x|y) = \frac{f_{XY}(x,y)}{f_Y(y)}$$

$$= \frac{e^{-y}}{ye^{-y}}$$

$$= \frac{1}{y} \text{ for all } x \in (0, y).$$

Thus, for any given $y > 0$,

$$f_{X|Y}(x|y) = \begin{cases} \frac{1}{y}, & x \in (0, y), \\ 0, & \text{otherwise.} \end{cases}$$

This implies that conditional on $Y = y$, X follows a uniform distribution on the interval $(0, y)$.

(3) Because $f_{Y|X}(y|x) \neq f_Y(y)$ for $0 < x < y < \infty$, X and Y are not independent.

If X and Y are independent, then we have $f_{XY}(x,y) = f_X(x)f_Y(y) > 0$ on the set $\Omega_X \times \Omega_Y = \{(x,y) : x \in \Omega_X \text{ and } y \in \Omega_Y\}$, where $\Omega_X = \{x : f_X(x) > 0\}$ and $\Omega_Y = \{y : f_Y(y) > 0\}$ are the supports of X and Y respectively. Memberships in the cross-product $\Omega_X \times \Omega_Y$ can be checked by considering the x and y values separately. If $f_{XY}(x,y)$ is a joint PDF or PMF and the set where $f_{XY}(x,y) > 0$ is not a cross-product, then the random variables X and Y are not independent. An example of not being a cross-product is the set $0 < x < y < \infty$.

We now extend the definition of independence when there are more than two random variables.

Definition 5.12. The random variables $X_1, \cdots, X_n$ are mutually independent if the joint CDF is equal to the product of their marginal CDF's, namely,

$$F_{\mathbf{X}^n}(\mathbf{x}^n) = \prod_{i=1}^{n} F_{X_i}(x_i) \text{ for all } -\infty < x_1, \cdots, x_n < \infty.$$

where $\mathbf{X}^n = (X_1, \cdots, X_n)$ and $\mathbf{x}^n = (x_1, \cdots, x_n)$.

For more than two random variables, it is possible that two random variables of any pair are independent, but all of them together are not independent. This is illustrated by the example below.

Example 5.15. Suppose random variables X_1, X_2, X_3 have the following joint PDF

$$f_{X_1 X_2 X_3}(x_1, x_2, x_3) = \begin{cases} (x_1 + x_2)e^{-x_3}, & 0 < x_1 < 1, 0 < x_2 < 1, x_3 > 0 \\ 0, & \text{elsewhere.} \end{cases}$$

It can be verified that X_1, X_2 and X_3 are pairwise independent but they are not jointly independent. This is analogous to Example 2.55 in Chapter 2 where three events are not jointly independent but any two events are independent of each other.

When can we have the result that pairwise independence also implies joint independence? If this can be the case, then it will greatly simplify the verification of independence in multivariate contexts. One example is the multivariate normal distribution, in which pairwise independence implies joint independence. See Sections 5.5 and 5.8 for discussion.

The concept of independence has important applications in economics and finance. As discussed in Example 2.50, Chapter 2, a geometric random walk is often used to model asset prices, where the increments in the log-price, which are approximately equal to the relative price changes, are independent across different time periods. Thus, one can test the random walk hypothesis by checking whether relative price changes of different time periods are mutually independent. In practice, the definition of independence and various characterizations provide powerful operational ways to verify independence using observed data. For example, one could construct estimators for the joint and marginal PDF's, and then check whether the joint PDF is equal to the product of the marginal PDF's. See Hong and White (2005) and Robinson (1991) for more discussion.

5.5 Bivariate Transformation

In Chapter 3, we have discussed how to find the PMF/PDF $f_Y(y)$ of the univariate transformation $Y = g(X)$ when the PMF/PDF $f_X(x)$ of X is given. Suppose now we have a bivariate transformation

$$U = g_1(X, Y),$$
$$V = g_2(X, Y),$$

where the joint PMF/PDF $f_{XY}(x, y)$ of (X, Y) is given. How can one find $f_{UV}(u, v)$, the joint PMF/PDF of (U, V)?

We first provide some examples to show why a bivariate transformation is useful in economics.

Example 5.16. Suppose X and Y are two common factors that drive the dynamics of the Eurodollar U and Chinese Yuan V. Then both U and V are the functions of X and Y.

Example 5.17. Both demand U and supply V of some commodity are the functions of the commodity price X and the consumer income Y.

Sometimes, our interest is in finding the probability distribution of $U = g_1(X, Y)$ given the joint probability distribution of (X, Y). In this case, we can first derive the joint PMF/PDF $f_{UV}(u, v)$ of $U = g_1(X, Y)$ and $V = X$ (say), and then integrate out v to obtain the PMF/PDF $f_U(u)$.

To find the joint PMF/PDF $f_{UV}(u, v)$ of (U, V), we consider the cases of DRV's and CRV's (X, Y) respectively. For the discrete case, the new random variables (U, V) are also discrete as well. Their joint PMF can be obtained via the formula

$$f_{UV}(u,v) = P(U = u, V = v)$$
$$= \sum_{(x,y)\in A(u,v)} f_{XY}(x,y),$$

where

$$A(u,v) = \{(x,y) \in \Omega_{XY} : g_1(x,y) = u, g_2(x,y) = v\}.$$

This is the set of all possible values of (x,y) in the support of (X,Y) that satisfy the restrictions that $u = g_1(x,y)$ and $v = g_2(x,y)$.

Example 5.18. Let X and Y be independent Poisson random variables with parameters θ and λ, respectively. Thus the joint PMF of (X,Y) is:

$$f_{XY}(x,y) = \frac{\theta^x e^{-\theta}}{x!} \frac{\lambda^y e^{-\lambda}}{y!}, \qquad x = 0,1,2,\cdots, y = 0,1,2,\cdots$$

Find the PMF of $X + Y$.

Solution: Define $U = X + Y, V = Y$. The support of (U,V) is given by $\Omega_{UV} = \{(u,v) : u = v, v+1, v+2, \cdots ; v = 0, 1, \cdots\}$. For any $(u,v) \in \Omega_{UV}$, the only (x,y) value that satisfies $x + y = u, y = v$ is the pair $(x,y) = (u - v, v)$. Thus, $A(u,v) = \{(u-v, v)\}$. Then the joint PMF of (U,V) is:

$$f_{UV}(u,v) = f_{XY}(u-v, v)$$
$$= f_X(u-v)f_V(v)$$
$$= \frac{\theta^{u-v} e^{-\theta}}{(u-v)!} \frac{\lambda^v e^{-\lambda}}{v!} \qquad u = v, v+1, v+2 \cdots ; v = 0, 1, \cdots.$$

Since for any given integer $u \geq 0$, $f_{UV}(u,v) > 0$ only for $v = 0, 1, \cdots, u$. It follows that the marginal PMF of U

$$f_U(u) = \sum_{v=0}^{u} f_{UV}(u,v)$$
$$= \sum_{v=0}^{u} \frac{\theta^{u-v} e^{-\theta}}{(u-v)!} \frac{\lambda^v e^{-\lambda}}{v!}$$
$$= e^{-(\theta+\lambda)} \sum_{v=0}^{u} \frac{\theta^{u-v}}{(u-v)!} \frac{\lambda^v}{v!}$$
$$= \frac{e^{-(\theta+\lambda)}}{u!} \sum_{v=0}^{u} \binom{u}{v} \lambda^v \theta^{u-v}$$
$$= \frac{e^{-(\theta+\lambda)}}{u!} (\theta + \lambda)^u, \qquad u = 0, 1, 2, \cdots.$$

That is, $U = X + Y$ follows a Poisson$(\theta + \lambda)$ distribution. This example is related to Example 5.10.

Next, we consider the continuous case. For this purpose, we first review the concepts of the Jacobian matrix and Jacobian.

Definition 5.13. [Jacobian Matrix and Jacobian]: Consider the bivariate transformation:

$$U = g_1(X, Y),$$
$$V = g_2(X, Y),$$

where functions $g_1(\cdot, \cdot)$ and $g_2(\cdot, \cdot)$ are continuously differentiable with respect to (x, y). Then the 2×2 matrix

$$J_{UV}(x, y) = \begin{bmatrix} \frac{\partial g_1(x,y)}{\partial x} & \frac{\partial g_1(x,y)}{\partial y} \\ \frac{\partial g_2(x,y)}{\partial x} & \frac{\partial g_2(x,y)}{\partial y} \end{bmatrix}$$

is called the Jacobian matrix of (U, V), and its determinant is called the Jacobian of (U, V).

Note that the Jacobian matrix $J_{UV}(x, y)$ is not necessarily a symmetric matrix.

Definition 5.14. [Inverse Function]: Suppose $\mathbb{A}$ and $\mathbb{B}$ are subsets of $\mathbb{R}^2$, and the functions $g_1 : \mathbb{A} \to \mathbb{B}$ and $g_2 : \mathbb{A} \to \mathbb{B}$ are continuously differentiable with the determinant of $J_{UV}(x, y)$ not zero for all $(x, y) \in \mathbb{A}$. Then the following functions exist:

$$X = h_1(U, V),$$
$$Y = h_2(U, V),$$

where $h_1 : \mathbb{B} \to \mathbb{A}$ and $h_2 : \mathbb{B} \to \mathbb{A}$ are continuously differentiable on $\mathbb{B}$ and they satisfy the conditions

$$h_1[g_1(x, y), g_2(x, y)] = x,$$
$$h_2[g_1(x, y), g_2(x, y)] = y.$$

The functions $\{h_1(\cdot), h_2(\cdot)\}$ are called the inverse functions of $\{g_1(\cdot), g_2(\cdot)\}$.

Intuitively, the inverse functions $h_1(U, V)$ and $h_2(U, V)$ can be obtained by representing (X, Y) in terms of (U, V) via solving the system of equations $U = g_1(X, Y)$ and $V = g_2(X, Y)$.

The following theorem shows that the Jacobian matrix of the inverse functions is equal to the inverse of the Jacobian matrix of the original functions.

Theorem 5.6. *The Jacobian matrix of* (X, Y)

$$
J_{XY}(u, v) = \begin{bmatrix} \frac{\partial h_1(u,v)}{\partial u} & \frac{\partial h_1(u,v)}{\partial v} \\[2mm] \frac{\partial h_2(u,v)}{\partial u} & \frac{\partial h_2(u,v)}{\partial v} \end{bmatrix}
$$

$$
= J_{UV}(x, y)^{-1}
$$

where $x = h_1(u, v), y = h_2(u, v)$.

Proof: Recall the identities

$$
h_1[g_1(x, y), g_2(x, y)] = x,
$$
$$
h_2[g_1(x, y), g_2(x, y)] = y.
$$

Put $u = g_1(x, y), v = g_2(x, y)$. Differentiating the first identity with respect to x, y respectively, we obtain the following results:

$$
\frac{\partial h_1(u, v)}{\partial u} \frac{\partial g_1(x, y)}{\partial x} + \frac{\partial h_1(u, v)}{\partial v} \frac{\partial g_2(x, y)}{\partial x} = 1,
$$
$$
\frac{\partial h_1(u, v)}{\partial u} \frac{\partial g_1(x, y)}{\partial y} + \frac{\partial h_1(u, v)}{\partial v} \frac{\partial g_2(x, y)}{\partial y} = 0.
$$

Similarly, differentiating the second identity with respect to x, y respectively, we obtain the following results:

$$
\frac{\partial h_2(u, v)}{\partial u} \frac{\partial g_1(x, y)}{\partial x} + \frac{\partial h_2(u, v)}{\partial v} \frac{\partial g_2(x, y)}{\partial x} = 0,
$$
$$
\frac{\partial h_2(u, v)}{\partial u} \frac{\partial g_1(x, y)}{\partial y} + \frac{\partial h_2(u, v)}{\partial v} \frac{\partial g_2(x, y)}{\partial y} = 1.
$$

Representing these four derivative equations in a matrix form, we obtain

$$
\begin{bmatrix} \frac{\partial h_1(u,v)}{\partial u} & \frac{\partial h_1(u,v)}{\partial v} \\[2mm] \frac{\partial h_2(u,v)}{\partial u} & \frac{\partial h_2(u,v)}{\partial v} \end{bmatrix} \begin{bmatrix} \frac{\partial g_1(x,y)}{\partial x} & \frac{\partial g_1(x,y)}{\partial y} \\[2mm] \frac{\partial g_2(x,y)}{\partial x} & \frac{\partial g_2(x,y)}{\partial y} \end{bmatrix} = \begin{bmatrix} 1 & 0 \\ 0 & 1 \end{bmatrix}
$$

or

$$
J_{XY}(u, v) J_{UV}(x, y) = I_2,
$$

where I_2 is a 2×2 identity matrix. Because $J_{UV}(x, y)$ is nonsingular for all $(x, y) \in \mathbb{A}$, by multiplying the inverse of $J_{UV}(x, y)$ from the right hand side, we obtain

$$
J_{XY}(u, v) = J_{UV}^{-1}(x, y).
$$

This completes the proof.

Example 5.19. Two random variables $U, V : \mathbb{R}^2 \to \mathbb{R}^2$ are defined by $U = XY$ and $V = X$. Then $g : \mathbb{R}^2 \to \mathbb{R}^2$ is a one-to-one mapping and has an inverse function $h : \mathbb{R}^2 \to \mathbb{R}^2$. (1) Find the inverse function; (2) verify that $J_{XY}(u, v) = J_{UV}^{-1}(x, y)$, where $u = xy$ and $v = x$.

Solution: (1) Given

$$U = XY = g_1(X, Y),$$
$$V = X = g_2(X, Y),$$

we have the inverse functions:

$$X = h_1(U, V) = V,$$
$$Y = h_2(U, V) = \frac{U}{V}.$$

(2) By definition, the Jacobian matrix of the inverse function is given by

$$J_{XY}(u, v) = \begin{bmatrix} \frac{\partial h_1(u,v)}{\partial u} & \frac{\partial h_1(u,v)}{\partial v} \\ \frac{\partial h_2(u,v)}{\partial u} & \frac{\partial h_2(u,v)}{\partial v} \end{bmatrix}$$

$$= \begin{bmatrix} 0 & 1 \\ \frac{1}{v} & -\frac{u}{v^2} \end{bmatrix}.$$

Furthermore, the Jacobian matrix of the original functions $U = g_1(X, Y) = XY$, $V = g_2(X, Y) = X$ is given by

$$J_{UV}(x, y) = \begin{bmatrix} \frac{\partial g_1(x,y)}{\partial x} & \frac{\partial g_1(x,y)}{\partial y} \\ \frac{\partial g_2(x,y)}{\partial x} & \frac{\partial g_2(x,y)}{\partial y} \end{bmatrix}$$

$$= \begin{bmatrix} y & x \\ 1 & 0 \end{bmatrix}.$$

The inverse

$$J_{UV}(x, y)^{-1} = -\frac{1}{x} \begin{bmatrix} 0 & -x \\ -1 & y \end{bmatrix}$$

$$= \begin{bmatrix} 0 & 1 \\ \frac{1}{x} & -\frac{y}{x} \end{bmatrix}$$

$$= \begin{bmatrix} 0 & 1 \\ \frac{1}{v} & -\frac{u}{v^2} \end{bmatrix}$$

$$= J_{XY}(u, v),$$

where $u = xy$ and $v = x$.

Now, we consider the following general problem: suppose (X, Y) have a joint PDF $f_{XY}(x, y)$, and

$$U = g_1(X, Y)$$
$$V = g_2(X, Y).$$

Then how can we find the joint PDF $f_{UV}(u, v)$ of (U, V)?

The following bivariate transformation theorem provides a powerful method to calculate the joint PDF $f_{UV}(u, v)$.

Theorem 5.7. *[Bivariate Transformation]: Let (X, Y) be a bivariate continuous random vector with joint PDF $f_{XY}(x, y)$, and let $\Omega_{XY} = \{(x, y) \in \mathbb{R}^2 : f_{XY}(x, y) > 0\}$ be the support of (X, Y). Define*

$$U = g_1(X, Y),$$
$$V = g_2(X, Y),$$

where $g : \Omega_{XY} \to \mathbb{R}^2$ is a one-to-one and continuously differentiable function on Ω_{XY}, with $\det[J_{UV}(x, y)] \neq 0$ for all $(x, y) \in \Omega_{XY}$. Then the joint PDF of (U, V) is given by

$$f_{UV}(u, v) = f_{XY}(x, y) |\det [J_{XY}(u, v)]| \text{ for all } (u, v) \in \Omega_{UV},$$

where $x = h_1(u, v)$ and $y = h_2(u, v)$, and

$$\Omega_{UV} = \{(u, v) \in \mathbb{R}^2 : u = g_1(x, y), v = g_2(x, y) \text{ for all } (x, y) \in \Omega_{XY}\}$$

is the the support of (U, V).

Proof: For (u, v) in the support Ω_{UV} of (U, V), we have

$$
\begin{aligned}
F_{UV}(u, v) &= P(U \leq u, V \leq v) \\
&= P\left[g_1(X, Y) \leq u, g_2(X, Y) \leq v\right] \\
&= \int\int_{\mathbb{A}(u,v)} f_{XY}(x', y') dx' dy',
\end{aligned}
$$

where the double integration is taken over the set $\mathbb{A}(u, v) = \{(x, y) \in \mathbb{R}^2 : g_1(x, y) \leq u \text{ and } g_2(x, y) \leq v\}$. Making the transformation $s = g_1(x', y')$ and $t = g_2(x', y')$ and applying the change of variable formula for double integrals from calculus, we obtain

$$F_{UV}(u, v) = \int_{-\infty}^{u} \int_{-\infty}^{v} \frac{1}{|\det[J_{UV}(x', y')]|} f_{XY}(x', y') ds dt,$$

where (x', y') satisfies the restrictions that $g_1(x', y') = s$ and $g_2(x', y') = t$. By taking the partial derivatives of both sides with respect to (u, v), we can obtain

$$f_{UV}(u, v) = \frac{\partial^2 F_{UV}(u, v)}{\partial u \partial v}$$

$$= f_{XY}(x, y) \frac{1}{|\det[J_{UV}(x, y)]|},$$

where $x = h_1(u, v), y = h_2(u, v)$. Since $J_{UV}(x, y) = J_{XY}^{-1}(u, v)$, we also have

$$f_{UV}(u, v) = f_{XY}(x, y) |\det[J_{XY}(u, v)]|.$$

This completes the proof.

As discussed in Chapter 3, for the univariate transformation $Y = g(X)$, where $g(\cdot)$ is a continuously differentiable monotonic function, the PDF

$$f_Y(y) = f_X(x)|h'(y)| = f_X(x) \frac{1}{|g'(x)|},$$

where $x = h(y)$ is the inverse function of $y = g(x)$. See Theorem 3.12 in Chapter 3 for details. Thus, we can view the bivariate transformation theorem as a generalization of the univariate transformation theorem. In fact, we can also derive a similar multivariate transformation involving more than two random variables.

It is difficult to overemphasize the importance of the bivariate transformation theorem. One can use it to efficiently derive results, obtain univariate distributions, and, of course, determine bivariate distributions. We note that the one-to-one mapping is a crucial condition in applying the bivariate transformation theorem. This is similar to the univariate transformation theorem where it is crucial that $g(\cdot)$ is monotonic.

We first apply the bivariate transformation theorem to show that independence between X and Y is equivalent to independence between $g_1(X)$ and $g_2(Y)$ for any continuously differentiable one-to-one transformations $g_1(\cdot)$ and $g_2(\cdot)$.

Theorem 5.8. *Suppose $U = g_1(X)$ and $V = g_2(Y)$ are some one-to-one continuously differentiable measurable functions. Then X and Y are independent if and only if U and V are independent.*

Proof: (1) [*Necessity*] We first show that independence between X and Y implies independence between U and V. By definition, the Jacobian matrix

$$J_{UV}(x, y) = \begin{bmatrix} g_1'(x) & 0 \\ 0 & g_2'(y) \end{bmatrix}$$

and the Jacobian

$$\det[J_{UV}(x,y)] = g_1'(x)g_2'(y).$$

It follows from the bivariate transformation theorem that

$$
\begin{aligned}
f_{UV}(u,v) &= f_{XY}(x,y)\,|\det[J_{UV}(x,y)]|^{-1} \\
&= f_X(x)f_Y(y)|g_1'(x)g_2'(y)|^{-1} \\
&= \left[f_X(x)|g_1'(x)|^{-1}\right] \cdot \left[f_Y(y)|g_2'(y)|^{-1}\right] \\
&= f_U(u)f_V(v),
\end{aligned}
$$

where $x = g_1^{-1}(u)$ and $y = g_2^{-1}(v)$, which are the inverse functions of $g_1(u)$ and $g_2(v)$. Because (u,v) is arbitrary, we have that U and V are independent.

(2) [*Sufficiency*] Next, we show that independence between U and V implies independence between X and Y. The proof is analogous to the reasoning in (1) by applying the bivariate transformation theorem to the bivariate transformation $X = g_1^{-1}(U)$ and $Y = g_2^{-1}(V)$. This completes the proof.

Now we consider a variety of examples on applications of the bivariate transformation.

Example 5.20. Suppose the random variable $X \sim \text{Beta}(\alpha,\beta)$, namely its PDF

$$f_X(x) = \frac{\Gamma(\alpha+\beta)}{\Gamma(\alpha)\Gamma(\beta)}x^{\alpha-1}(1-x)^{\beta-1}, \qquad 0 < x < 1,$$

$Y \sim \text{Beta}(\alpha+\beta,\gamma)$, namely its PDF

$$f_Y(y) = \frac{\Gamma(\alpha+\beta+\gamma)}{\Gamma(\alpha+\beta)\Gamma(\gamma)}y^{\alpha+\beta-1}(1-y)^{\gamma-1}, \qquad 0 < y < 1,$$

and X and Y are independent. Define $U = XY$, $V = X$. Find the joint PDF $f_{UV}(u,v)$.

Solution: (1) We first note that the support of (X,Y)

$$
\begin{aligned}
\Omega_{XY} &= \{(x,y) \in \mathbb{R}^2 : f_{XY}(x,y) = f_X(x)f_Y(y) > 0\} \\
&= \{(x,y) \in \mathbb{R}^2 : 0 < x < 1, 0 < y < 1\}.
\end{aligned}
$$

This is a rectangular area.

Given $U = XY$ and $V = X$, we obtain the support of (U, V)

$$\Omega_{UV} = \{(u, v) \in \mathbb{R}^2 : f_{UV}(u, v) > 0\}$$
$$= \{(u, v) \in \mathbb{R}^2 : 0 < u < v, 0 < v < 1\}$$
$$= \{(u, v) \in \mathbb{R}^2 : 0 < u < v < 1\}.$$

(2) We now solve for the inverse functions $x = h_1(u, v)$ and $y = h_2(u, v)$. Given $u = xy, v = x$, we have

$$x = h_1(u, v) = v,$$
$$y = h_2(u, v) = \frac{u}{v}.$$

(3) The Jacobian matrix of (X, Y)

$$J_{XY}(u, v) = \begin{bmatrix} \frac{\partial h_1(u,v)}{\partial u} & \frac{\partial h_1(u,v)}{\partial v} \\ \frac{\partial h_2(u,v)}{\partial u} & \frac{\partial h_2(u,v)}{\partial v} \end{bmatrix}$$
$$= \begin{bmatrix} 0 & 1 \\ \frac{1}{v} & -\frac{u}{v^2} \end{bmatrix}.$$

Therefore, the Jacobian

$$\det[J_{XY}(u, v)] = \det\left(\begin{bmatrix} 0 & 1 \\ v^{-1} & -u/v^2 \end{bmatrix}\right) = -\frac{1}{v}.$$

It follows from the bivariate transformation theorem that

$$f_{UV}(u, v) = f_{XY}(x, y)|\det J_{XY}(u, v)|$$
$$= f_X(x)f_Y(y)\left|-\frac{1}{v}\right|$$
$$= \frac{\Gamma(\alpha + \beta)}{\Gamma(\alpha)\Gamma(\beta)}x^{\alpha-1}(1 - x)^{\beta-1}\frac{\Gamma(\alpha + \beta + \gamma)}{\Gamma(\alpha + \beta)\Gamma(\gamma)}y^{\alpha+\beta-1}(1 - y)^{\gamma-1}\frac{1}{v}$$
$$= \frac{\Gamma(\alpha + \beta + \gamma)}{\Gamma(\alpha)\Gamma(\beta)\Gamma(\gamma)}v^{\alpha-1}(1 - v)^{\beta-1}\left(\frac{u}{v}\right)^{\alpha+\beta-1}\left(1 - \frac{u}{v}\right)^{\gamma-1}\frac{1}{v}$$

for $0 < u < v < 1$, where we have $x = v, y = u/v$.

Example 5.21. Suppose X and Y are independent $N(0, \sigma^2)$. Define $U = X^2 + Y^2, V = X/\sqrt{U} = X/\sqrt{X^2 + Y^2}$. (1) Find $f_{UV}(u, v)$; (2) show that U and V are independent.

Solution: (1) From $U = X^2 + Y^2$, $V = X/\sqrt{X^2 + Y^2}$, and $X \sim N(0, \sigma^2)$, $Y \sim N(0, \sigma^2)$, we have the support of (U, V)

$$\Omega_{UV} = \{(u, v) \in \mathbb{R}^2 : 0 < u < \infty, -1 < v < 1\}.$$

We first note that (U, V) is not a one-to-one mapping, because there exist two pairs, (x, y) and $(x, -y)$, that correspond to the same value of (u, v). As a result, the bivariate transformation theorem cannot be applied directly. Putting $Z = Y^2$, we will consider a transformation from (X, Z) to (U, V), which is a one-to-one mapping. Then the bivariate transformation theory is applicable.

For the distribution of $Z = Y^2$, we have for any $z \geq 0$,

$$\begin{aligned}
F_Z(z) &= P(Y^2 \leq z) \\
&= P(-\sqrt{z} \leq Y \leq \sqrt{z}) \\
&= F_Y(\sqrt{z}) - F_Y(-\sqrt{z}), \\
f_Z(z) &= F_Z'(z) \\
&= f_Y(\sqrt{z}) \frac{1}{2\sqrt{z}} + f_Y(-\sqrt{z}) \frac{1}{2\sqrt{z}} \\
&= \frac{1}{\sqrt{2\pi z}\sigma} e^{-z/2\sigma^2}.
\end{aligned}$$

Since X is independent of Y, it is also independent of $Z = Y^2$. The joint PDF of (X, Z) is

$$f_{XZ}(x, z) = \frac{1}{2\pi\sigma^2\sqrt{z}} e^{-x^2/2\sigma^2} e^{-z/2\sigma^2}, \qquad -\infty < x < \infty, 0 \leq z < \infty.$$

Also, the support of (X, Z) is

$$\Omega_{XZ} = \{(x, z) \in \mathbb{R}^2 : -\infty < x < \infty, 0 \leq z < \infty\}$$

and the joint support of $U = X^2 + Z$ and $V = X/\sqrt{X^2 + Z}$ is

$$\Omega_{UV} = \{(u, v) \in \mathbb{R}^2 : 0 < u < \infty, -1 < v < 1\}.$$

We now find the inverse functions $X = h_1(U, V)$ and $Z = h_2(U, V)$. Given

$$U = X^2 + Z,$$
$$V = \frac{X}{\sqrt{X^2 + Z}},$$

we have

$$X = h_1(U, V) = V\sqrt{U},$$
$$Z = h_2(U, V) = U(1 - V^2).$$

It follows that the Jacobian matrix of (X, Z)

$$J_{XZ}(u, v) = \begin{bmatrix} \frac{\partial h_1(u,v)}{\partial u} & \frac{\partial h_1(u,v)}{\partial v} \\ \frac{\partial h_2(u,v)}{\partial u} & \frac{\partial h_2(u,v)}{\partial v} \end{bmatrix}$$

$$= \begin{bmatrix} \frac{v}{2\sqrt{u}} & \sqrt{u} \\ 1 - v^2 & -2uv \end{bmatrix}.$$

By the bivariate transformation theorem, we have

$$f_{UV}(u, v) = f_{XZ}(x, z) |\det J_{XZ}(u, v)|$$

$$= \frac{1}{2\pi\sigma^2\sqrt{z}} e^{-\frac{x^2}{2\sigma^2}} e^{-\frac{z}{2\sigma^2}} |\det J_{XZ}(u, v)|$$

$$= \frac{1}{2\pi\sigma^2\sqrt{u(1-v^2)}} e^{-\frac{u}{2\sigma^2}} \sqrt{u}$$

$$= \frac{1}{2\pi\sigma^2\sqrt{1-v^2}} e^{-\frac{u}{2\sigma^2}}, \text{ for } u > 0, -1 < v < 1.$$

It follows that

$$f_{UV}(u, v) = \begin{cases} \frac{1}{2\pi\sigma^2\sqrt{1-v^2}} e^{-\frac{u}{2\sigma^2}}, & 0 < u < \infty, -1 < v < 1, \\ 0, & \text{otherwise.} \end{cases}$$

(2) Although both U, V are functions of (X, Y), U and V are independent because their joint PDF $f_{UV}(u, v)$ can be partitioned into the product of two separate functions of u and v respectively for all $(u, v) \in \mathbb{R}^2$, with

$$g(u) = \begin{cases} e^{-\frac{u^2}{2\sigma^2}u}, & \text{if } u > 0, \\ 0, & \text{if } u \leq 0, \end{cases}$$

and

$$h(v) = \begin{cases} \frac{1}{2\pi\sigma^2\sqrt{1-v^2}}, & \text{if } -1 < v < 1, \\ 0, & \text{otherwise.} \end{cases}$$

The independence between U and V follows from the factorization theorem (Theorem 5.4).

Example 5.22. Suppose $X \sim N(\mu, \sigma^2), Y \sim N(\mu, \sigma^2)$, and X, Y are independent. Put $U = X + Y, V = X - Y$. (1) Find the joint PDF of (U, V); (2) show that U and V are independent.

Solution: The support of (U, V) is the entire xy plane. From $U = X + Y$, $V = X - Y$, we have

$$X = h_1(U, V) = \frac{1}{2}(U + V),$$

$$Y = h_2(U, V) = \frac{1}{2}(U - V).$$

The Jacobian matrix of (X, Y) is

$$J_{XY}(u, v) = \begin{bmatrix} \frac{1}{2} & \frac{1}{2} \\ \frac{1}{2} & -\frac{1}{2} \end{bmatrix}.$$

It follows from the bivariate transformation theorem (Theorem 5.7) and $x = \frac{1}{2}(u + v), y = \frac{1}{2}(u - v)$ that

$$f_{UV}(u, v) = f_{XY}(x, y)|\det J_{XY}(u, v)|$$

$$= \frac{1}{2\pi\sigma^2} e^{-\frac{1}{2\sigma^2}[(x-\mu)^2 + (y-\mu)^2]} \cdot \frac{1}{2}$$

$$= \frac{1}{4\pi\sigma^2} e^{-\frac{1}{8\sigma^2}(u+v-2\mu)^2} e^{-\frac{1}{8\sigma^2}(u-v-2\mu)^2}$$

$$= \frac{1}{4\pi\sigma^2} e^{-\frac{1}{4\sigma^2}(u-2\mu)^2} e^{-\frac{1}{4\sigma^2}v^2}$$

$$= \frac{1}{\sqrt{2\pi 2\sigma^2}} e^{-\frac{1}{4\sigma^2}(u-2\mu)^2} \frac{1}{\sqrt{2\pi 2\sigma^2}} e^{-\frac{1}{4\sigma^2}v^2}, \quad -\infty < u, v < \infty.$$

Hence, $U \sim N(2\mu, 2\sigma^2)$, $V \sim N(0, 2\sigma^2)$, and U and V are independent by the factorization theorem (Theorem 5.4), because we can partition the joint PDF

$$f_{UV}(u, v) = g(u)h(v), \quad -\infty < u, v < \infty,$$

where

$$g(u) = \frac{1}{\sqrt{2\pi 2\sigma^2}} e^{-\frac{1}{4\sigma^2}(u-2\mu)^2}, \quad -\infty < u < \infty$$

is the PDF of a $N(2\mu, 2\sigma^2)$ distribution, and

$$h(v) = \frac{1}{\sqrt{2\pi 2\sigma^2}} e^{-\frac{1}{4\sigma^2}v^2}, \quad -\infty < v < \infty$$

is the PDF of a $N(0, 2\sigma^2)$ distribution.

Example 5.23. Find the PDF of $X + Y$, where $X \sim U[0, 1]$ and $Y \sim U[0, 1]$, and X and Y are independent.

Solution: Put $U = X + Y$ and $V = X$. We shall first find the joint PDF $f_{UV}(u, v)$ of (U, V) and then integrate out v to obtain the PDF $f_U(u)$ of U. (1) From $U = X + Y, V = X$, we find the support of (U, V) as

$$\Omega_{UV} = \{(u, v) \in \mathbb{R}^2 : v \le u \le 1 + v, 0 \le v \le 1\}.$$

Figure 5.6 gives the plot of the support Ω_{UV}.

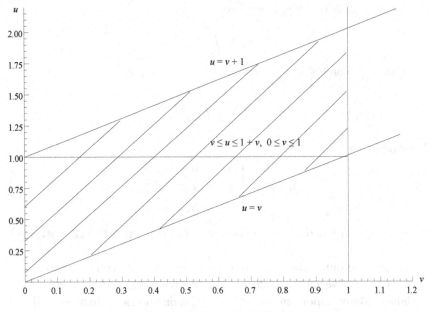

Figure 5.6: Support of (U, V)

(2) The Jacobian matrix $J_{UV}(x, y)$ of (U, V) is

$$J_{UV}(x, y) = \begin{bmatrix} \dfrac{\partial g_1(x,y)}{\partial x} & \dfrac{\partial g_1(x,y)}{\partial y} \\[2mm] \dfrac{\partial g_2(x,y)}{\partial x} & \dfrac{\partial g_2(x,y)}{\partial y} \end{bmatrix}$$

$$= \begin{bmatrix} 1 & 1 \\ 1 & 0 \end{bmatrix}.$$

Hence, the Jacobian of (U, V) is

$$\det [J_{UV}(x, y)] = -1.$$

(3) By the bivariate transformation theorem, we obtain

$$f_{UV}(u,v) = f_{XY}(x,y)| \det [J_{UV}(x,y)] |^{-1}$$
$$= f_X(x)f_Y(y)| - 1|^{-1}$$
$$= 1, \quad 0 \le v \le 1, v \le u \le 1 + v.$$

We now integrate out v to obtain the PDF $f_U(u)$ of U. The support of U is $0 \le u \le 2$. We consider two sub-intervals: $0 \le u \le 1$ and $1 \le u \le 2$.

Case (1): For $0 \le u \le 1$, we have

$$f_U(u) = \int_0^u dv = u.$$

Case (2): For $1 \le u \le 2$, we have

$$f_U(u) = \int_{u-1}^1 dv$$
$$= 2 - u.$$

It follows that

$$f_U(u) = \begin{cases} u, & 0 \le u \le 1, \\ 2 - u, & 1 \le u < 2, \\ 0, & \text{otherwise.} \end{cases}$$

This is a triangular density function over the support $u \in [0, 2]$ with peak at $u = 1$.

As pointed out earlier, we can extend the bivariate transformation further to a multivariate transformation involving more than two random variables. Many important statistics (e.g., econometric estimators and test statistics) are functions of more than two random variables. Thus, the multivariate transformation is very useful to obtain or understand the distribution of these important statistics.

Nevertheless, the multivariate transformation is often very tedious to use in practice, especially when the number of random variables involved is large. Fortunately, there exist other methods, such as those based on the moment generating function, that are convenient to use to obtain the distribution of interest. See examples in the subsequent sections of this chapter.

5.6   Bivariate Normal Distribution

We now consider a very important bivariate joint distribution called bivariate normal distribution.

Definition 5.15. [**Bivariate Normal Distribution**]: (X, Y) are jointly normally distributed, denoted as $BN(\mu_1, \mu_2, \sigma_1^2, \sigma_2^2, \rho)$, where $|\rho| \leq 1$, if their joint PDF

$$f_{XY}(x, y) = \frac{1}{2\pi\sigma_1\sigma_2\sqrt{1 - \rho^2}} e^{-\frac{1}{2(1-\rho^2)}\left[\left(\frac{x-\mu_1}{\sigma_1}\right)^2 + \left(\frac{y-\mu_2}{\sigma_2}\right)^2 - 2\rho\left(\frac{x-\mu_1}{\sigma_1}\right)\left(\frac{y-\mu_2}{\sigma_2}\right)\right]}.$$

When $(\mu_1, \mu_2, \sigma_1, \sigma_2) = (0, 0, 1, 1)$, $BN(0, 0, 1, 1, \rho)$ is called a standard bivariate normal distribution.

An alternative representation of $f_{XY}(x, y)$ is

$$f_{XY}(x, y) = \frac{1}{\sqrt{(2\pi)^2 \det(\Sigma)}} e^{-\frac{1}{2}(z-\mu)'\Sigma^{-1}(z-\mu)},$$

where $z = (x, y)'$, $\mu = (\mu_1, \mu_2)'$, and

$$\Sigma = \begin{bmatrix} \sigma_1^2 & \rho\sigma_1\sigma_2 \\ \rho\sigma_1\sigma_2 & \sigma_2^2 \end{bmatrix}$$

$$= \begin{bmatrix} \sigma_1 & 0 \\ 0 & \sigma_2 \end{bmatrix} \begin{bmatrix} 1 & \rho \\ \rho & 1 \end{bmatrix} \begin{bmatrix} \sigma_1 & 0 \\ 0 & \sigma_2 \end{bmatrix}.$$

Figure 5.7 plots the joint PDF $f_{XY}(x, y)$ of $(X, Y) \sim BN(\mu_1, \mu_2, \sigma_1^2, \sigma_2^2, \rho)$, with various combinations of parameter values.

We now calculate the marginal PDF's of X and Y and the conditional PDF's of Y given X and of X given Y when $(X, Y) \sim N(\mu_1, \mu_2, \sigma_1^2, \sigma_2^2, \rho)$. Put

$$q(x, y) = \frac{1}{1 - \rho^2} \left[\left(\frac{x - \mu_1}{\sigma_1}\right)^2 - 2\rho\left(\frac{x - \mu_1}{\sigma_1}\right)\left(\frac{y - \mu_2}{\sigma_2}\right) + \left(\frac{y - \mu_2}{\sigma_2}\right)^2 \right].$$

Then we can write the joint PDF of a bivariate normal distribution as

$$f_{XY}(x, y) = \frac{1}{2\pi\sigma_1\sigma_2\sqrt{1 - \rho^2}} e^{-\frac{1}{2}q(x,y)}.$$

By straightforward algebra, we have

$$(1 - \rho^2)q(x, y) = (1 - \rho^2)\left(\frac{x - \mu_1}{\sigma_1}\right)^2 + \left[\left(\frac{y - \mu_2}{\sigma_2}\right) - \rho\left(\frac{x - \mu_1}{\sigma_1}\right)\right]^2$$

$$= (1 - \rho^2)\left(\frac{x - \mu_1}{\sigma_1}\right)^2 + \left(\frac{y - \mu}{\sigma_2}\right)^2,$$

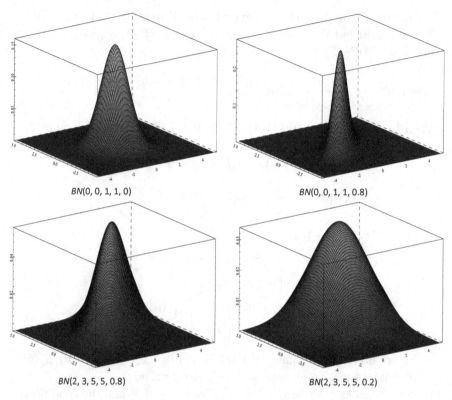

BN(0, 0, 1, 1, 0)   BN(0, 0, 1, 1, 0.8)

BN(2, 3, 5, 5, 0.8)   BN(2, 3, 5, 5, 0.2)

Figure 5.7: Joint PDF $f_{XY}(x,y)$ of $(X,Y) \sim BN(\mu_1,\mu_2,\sigma_1^2,\sigma_2^2,\rho)$

where

$$\mu = \mu_2 + \frac{\rho\sigma_2}{\sigma_1}(x - \mu_1).$$

Thus,

$$f_X(x) = \frac{1}{\sqrt{2\pi\sigma_1^2}}e^{-\frac{(x-\mu_1)^2}{2\sigma_1^2}}\int_{-\infty}^{\infty}\frac{1}{\sqrt{2\pi\sigma_2^2(1-\rho^2)}}e^{-\frac{(y-\mu)^2}{2\sigma_2^2(1-\rho^2)}}\,dy$$

$$= \frac{1}{\sqrt{2\pi\sigma_1^2}}e^{-\frac{(x-\mu_1)^2}{2\sigma_1^2}}, \qquad -\infty < x < \infty,$$

where the integral is equal to unity because it is the integral of the PDF of a $N[\mu, \sigma_2^2(1 - \rho^2)]$ distribution.

By symmetry, we can also obtain

$$f_Y(y) = \frac{1}{\sqrt{2\pi\sigma_2^2}}e^{-\frac{(y-\mu_2)^2}{2\sigma_2^2}}, \qquad -\infty < y < \infty.$$

These results imply that when X and Y are jointly normally distributed, then X and Y have marginal normal distributions respectively. In particular, $X \sim N(\mu_1, \sigma_1^2)$, and $Y \sim N(\mu_2, \sigma_2^2)$.

With $f_X(x) > 0$, the conditional PDF of Y given $X = x$ is

$$f_{Y|X}(y|x) = \frac{f_{XY}(x,y)}{f_X(x)}$$

$$= \frac{1}{\sqrt{2\pi\sigma_2^2(1-\rho^2)}} e^{-\frac{(y-\mu)^2}{2\sigma_2^2(1-\rho^2)}}, \qquad -\infty < y < \infty,$$

where, as before, $\mu = \mu_2 + \frac{\rho\sigma_2}{\sigma_1}(x - \mu_1)$. Therefore, the conditional distribution of Y given $X = x$ is also a normal distribution, $N[\mu_2 + \frac{\rho\sigma_2}{\sigma_1}(x - \mu_1), \sigma_2^2(1 - \rho^2)]$. Similarly, the conditional distribution of X given $Y = y$ is a normal distribution, $N[\mu_1 + \frac{\rho\sigma_1}{\sigma_2}(y - \mu_2), \sigma_1^2(1 - \rho^2)]$. Figure 5.8 plots the conditional PDF $f_{Y|X}(y|x)$ when $(X, Y) \sim BN(\mu_1, \mu_2, \sigma_1^2, \sigma_2^2, \rho)$.

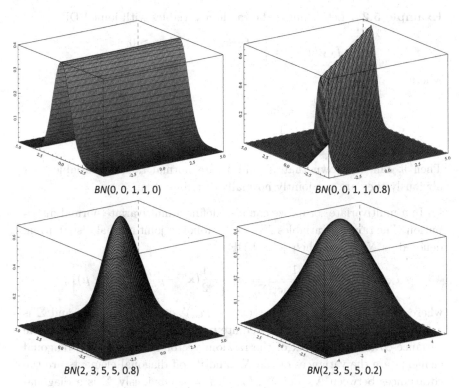

BN(0, 0, 1, 1, 0)   BN(0, 0, 1, 1, 0.8)

BN(2, 3, 5, 5, 0.8)   BN(2, 3, 5, 5, 0.2)

Figure 5.8: Conditional PDF $f_{Y|X}(y|x)$ when $(X, Y) \sim BN(\mu_1, \mu_2, \sigma_1^2, \sigma_2^2, \rho)$

The constant ρ characterizes dependence between X and Y. When $\rho = 0$, the joint PDF of (X, Y) becomes

$$f_{XY}(x,y) = \frac{1}{2\pi\sigma_1\sigma_2} e^{-\frac{1}{2}\left[\left(\frac{x-\mu_1}{\sigma_1}\right)^2 + \left(\frac{y-\mu_2}{\sigma_2}\right)^2\right]}$$

$$= \frac{1}{\sqrt{2\pi\sigma_1^2}} e^{-\frac{1}{2}\left(\frac{x-\mu_1}{\sigma_1}\right)^2} \frac{1}{\sqrt{2\pi\sigma_2^2}} e^{-\frac{1}{2}\left(\frac{y-\mu_2}{\sigma_2}\right)^2}$$

$$= f_X(x) f_Y(y) \text{ for all } (x,y) \in \mathbb{R}^2.$$

Therefore, two jointly normally distributed random variables X and Y are independent if and only if $\rho = 0$.

We have shown that two jointly normally distributed random variables have marginal normal distributions. Is the converse true? That is, suppose the marginal PDFs of X and Y are a normal PDF respectively, are X and Y jointly normal? The following example shows that this is not necessarily the case.

Example 5.24. Let X and Y be random variables with joint PDF

$$f_{XY}(x,y) = \begin{cases} 2f_X(x)f_Y(y), & \text{if } xy > 0, \\ 0, & \text{if } xy \leq 0, \end{cases}$$

where

$$f_X(x) = \frac{1}{\sqrt{2\pi}} e^{-\frac{1}{2}x^2}, \qquad -\infty < x < \infty,$$

$$f_Y(y) = \frac{1}{\sqrt{2\pi}} e^{-\frac{1}{2}y^2}, \qquad -\infty < y < \infty.$$

Then it can be shown that X and Y are normal random variables but obviously they are not jointly normally distributed.

In a multivariate setup, we can also define a multivariate normal distribution. The random variables $X_1, \cdots, X_n$ follow a joint normal distribution, denoted as $N(\mu, \Sigma)$ if their joint PDF

$$f_{\mathbf{X}^n}(\mathbf{x}^n) = \frac{1}{\sqrt{(2\pi)^n \det(\Sigma)}} \exp\left[-\frac{1}{2}(\mathbf{x}^n - \mu)'\Sigma^{-1}(\mathbf{x}^n - \mu)\right],$$

where $\mathbf{X}^n = (X_1, \cdots, X_n)', \mathbf{x}^n = (x_1, \cdots, x_n)', \mu = (\mu_1, \cdots, \mu_n)'$, and Σ is a $n \times n$ symmetric and positive definite matrix. The matrix Σ is called the variance-covariance matrix of the random vector $\mathbf{X}^n$ because its diagonal elements are the variances of the X_i, and its off-diagonal elements are the covariances between X_i and X_j, for all $i \neq j$. Obviously, Σ is a diagonal matrix if all covariances between X_i and X_j for $i \neq j$ are zero.

The assumption of multivariate normality offers a great deal of convenience in practice when calculating probabilities. For example, consider the return on a portfolio, which is a weighted average of n assets. Suppose the asset returns, $X_1, \cdots, X_n$, follow a multivariate normal distribution, it can be shown via a multivariate transformation that any linear combination of them, namely $X \equiv \sum_{i=1}^{n} c_i X_i$, will follow a normal distribution. It follows that the portfolio return also follows a normal distribution. Thus, calculation of the probabilities (e.g., $P(X < -V_{0.01}) = 0.01$) of the portfolio return will be rather convenient, under the assumption of the joint normal distribution for asset returns. Here, $V_{0.01}$ is the value at risk at the 1% significance level, which is widely used in financial risk management (see Section 3.7 in Chapter 3 for more discussion on value at risk).

5.7 Expectations and Covariance

Question: What information can we extract from a bivariate distribution?

We first define the mathematical expectation under a bivariate joint distribution.

Definition 5.16. [Expectation Under Bivariate Joint Distribution]:
Let $g : \Omega_{XY} \to \mathbb{R}$ be a real-valued measurable function, where Ω_{XY} is the support of (X, Y). Then the expectation of function $g(X, Y)$ is defined as

$$
E[g(X, Y)] = \int_{-\infty}^{\infty} \int_{-\infty}^{\infty} g(x, y) dF_{XY}(x, y)
$$

$$
= \begin{cases} \sum \sum_{(x,y) \in \Omega_{XY}} g(x, y) f_{X,Y}(x, y) \text{ if } (X, Y) \text{ are DRV's,} \\ \int_{-\infty}^{\infty} \int_{-\infty}^{\infty} g(x, y) f_{XY}(x, y) dx dy \text{ if } (X, Y) \text{ are CRV's,} \end{cases}
$$

provided the double sum or the double integral exists. Like the univariate case, we say that $E[g(X, Y)]$ exists if $E|g(X, Y)| < \infty$.

We now consider expectations of some important functions $g(X, Y)$. We first choose $g(X, Y) = X^r Y^s$. This will yield various joint product moments of (X, Y).

Definition 5.17. [Product Moments]: The r-th and s-th order product moment of (X, Y) about the origin is defined as

$$
E(X^r Y^s) = \begin{cases} \sum_{x \in \Omega_X} \sum_{y \in \Omega_Y} x^r y^s f_{XY}(x, y), \text{ if } (X, Y) \text{ are DRV's,} \\ \int_{-\infty}^{\infty} \int_{-\infty}^{\infty} x^r y^s f_{XY}(x, y) dx dy, \text{ if } (X, Y) \text{ are CRV's.} \end{cases}
$$

Similarly, the rth and sth central product moment is defined as

$$E\left\{[X - E(X)]^r[Y - E(Y)]^s\right\}$$
$$= \begin{cases} \sum_{x \in \Omega_X} \sum_{y \in \Omega_Y} (x - \mu_X)^r (y - \mu_Y)^s f_{XY}(x,y), & \text{if } (X,Y) \text{ are DRV's,} \\ \int_{-\infty}^{\infty} \int_{-\infty}^{\infty} (x - \mu_X)^r (y - \mu_Y)^s f_{XY}(x,y) dx dy, & \text{if } (X,Y) \text{ are CRV's.} \end{cases}$$

When X and Y are not independent, we say that there exists a relationship or an association between them. However, if there is a relationship or an association, the relationship or association may be weak or strong. How to measure the strength of a relationship between X and Y? Among other things, the cross-products $E(X^r Y^s)$ and $E[(X - \mu_X)^r (Y - \mu_Y)^s]$ provide a way to characterize the relationship or association between X and Y. Different choices of order (r, s) will capture different types of association between X and Y. To see this, we now examine a special case where $(r, s) = (1, 1)$. This corresponds to the expectation of function $g(X, Y) = (X - \mu_X)(Y - \mu_Y)$. This function yields a positive value if X and Y tend to move above or below their means in the same direction, and yields a negative value if X and Y tend to move above or below their means in opposite directions. Thus, it can characterize the direction of the co-movement between X and Y.

Definition 5.18. [Covariance]: Suppose $E(X^2) < \infty$ and $E(Y^2) < \infty$. Then the covariance between two variables X and Y is defined as

$$\text{cov}(X, Y) = E\left[(X - \mu_X)(Y - \mu_Y)\right]$$
$$= \int_{-\infty}^{\infty} \int_{-\infty}^{\infty} (x - \mu_X)(y - \mu_Y) dF_{XY}(x, y).$$

The covariance is a measure of the co-movement between X and Y. Suppose there is a high probability that large values of X tend to be observed with large values of Y, and small values of X with small values of Y, then $\text{cov}(X, Y) > 0$. In this case, we say that X and Y are positively correlated. Like in the univariate case, the expectation operator in the definition of $\text{cov}(X, Y)$ can be interpreted as the average value of the product $(X - \mu_X)(Y - \mu_Y)$ in infinitely many repeated independent experiments. On the other hand, suppose there is a high probability that large values of X tend to be observed with small values of Y, and small values of X tend to be observed with large values of Y, then $\text{cov}(X, Y) < 0$. In this case, we say that X and Y are negatively correlated. If their changes are irrelevant, then $\text{cov}(X, Y) = 0$. In this case, we call X and Y are uncorrelated. Thus, the signs of $\text{cov}(X, Y)$ provide information regarding the relationship between X and Y. Note that $\text{cov}(X, X) = \text{var}(X)$.

A nonzero covariance implies that there exists an association between X and Y, but it does not necessarily imply causality between X and Y. If one finds a positive correlation between oil price increase and economic slowdown, for example, it does not necessarily mean that oil price increase causes economic slowdown. For another example, if one finds that there exists a positive correlation between smoking and cancer, it does not necessarily imply that smoking causes cancer.

The following theorem provides a convenient formula to calculate covariance.

Theorem 5.9. *Suppose* (X, Y) *have finite second moments. Then*

$$cov(X, Y) = E(XY) - \mu_X \mu_Y.$$

Proof: Noting that the expectation operator $E(\cdot)$ is a linear operator, we have

$$\begin{aligned} \text{cov}(X, Y) &= E\left[(X - \mu_X)(Y - \mu_Y)\right] \\ &= E(XY - X\mu_Y - \mu_X Y + \mu_X \mu_Y) \\ &= E(XY) - \mu_X \mu_Y. \end{aligned}$$

The magnitude of $\text{cov}(X, Y)$ does not provide useful information about the strength of the relationship between X and Y, because it depends on the scales of X and Y. We now provide a normalized measure which is robust to the scales of X and Y.

Definition 5.19. [Correlation]: The correlation coefficient between X and Y is defined as

$$\rho_{XY} = \frac{\text{cov}(X, Y)}{\sigma_X \sigma_Y}.$$

The correlation coefficient ρ_{XY} is the standardized covariance. This is analogous to the definitions of skewness and kurtosis. Now the magnitude of ρ_{XY} can indicate the strength of the association which ρ_{XY} is capturing. Before we examine the nature of the association which ρ_{XY} captures, we first show that its absolute value is always less than or at most equal to unity.

Theorem 5.10. $|\rho_{XY}| \leq 1$.

Proof: This follows immediately from the Cauchy-Schwarz inequality: for any measurable functions $g(X)$ and $h(Y)$, we have

$$E|g(X)h(Y)| \leq \{E[g^2(X)]E[h^2(Y)]\}^{1/2}.$$

Setting $g(X) = X - \mu_X$ and $h(Y) = Y - \mu_Y$, we obtain

$$|\text{cov}(X, Y)| \leq E|(X - \mu_X)(Y - \mu_Y)|$$
$$\leq (\sigma_X^2 \sigma_Y^2)^{1/2}$$
$$= \sigma_X \sigma_Y.$$

It follows that $|\rho_{XY}| \leq 1$. This completes the proof.

The correlation coefficient ρ_{XY} is a measure of linear association between X and Y. To gain an insight into the nature of linear association, we first state a result for ρ_{XY} when X is a linear function of Y.

Theorem 5.11. *Suppose $Y = a + bX$, $b \neq 0$, where $\sigma_X^2 = var(X)$ exists. Then $\rho_{XY} = 1$ if $b > 0$, and $\rho_{XY} = -1$ if $b < 0$.*

Proof: Since $\mu_Y = a + b\mu_X$ and $\sigma_Y^2 = b^2\sigma_X^2$, the covariance

$$\text{cov}(X, Y) = E\left[(X - \mu_X)(Y - \mu_Y)\right]$$
$$= E\left[(X - \mu_X)(a + bX - a - b\mu_X)\right]$$
$$= bE(X - \mu_X)^2$$
$$= b\sigma_X^2.$$

It follows that

$$\rho_{XY} = \frac{\text{cov}(X, Y)}{\sigma_X \sigma_Y}$$
$$= \frac{b\sigma_X^2}{|b|\sigma_X^2}$$
$$= \frac{b}{|b|}$$
$$= \begin{cases} 1, & \text{if } b > 0, \\ -1, & \text{if } b < 0. \end{cases}$$

This completes the proof.

Thus, when there is a perfect linear relationship between X and Y, we always have $\rho_{XY} = \pm 1$, that is, ρ_{XY} achieves the maximum value of unity in absolute value. In this sense, it is a measure of linear association between X and Y.

For certain kinds of joint distributions of X and Y, the correlation coefficient ρ_{XY} proves to be a very useful characteristic of the joint distribution.

Unfortunately, the formal definition of ρ_{XY} does not reveal this fact. We have seen that if a joint distribution of two variables has a correlation coefficient, then $-1 \leq \rho_{XY} \leq 1$. If $\rho_{XY} = 1$, then there is a line with equation $Y = a + bX$, where $b > 0$, the graph of which contains all information of the probability distribution of X and Y. In this extreme case, we have $P(Y = a + bX) = 1$. If $\rho_{XY} = -1$, we have the same state of affairs except $b < 0$. This suggests the following interesting question: when ρ_{XY} does not have the one of its extreme values, is there a line in the xy-plane such that there is a high probability that X and Y tend to be concentrated in a band about this line? Under certain restrictive conditions this is in fact the case, and under those conditions we can look upon ρ_{XY} as a measure of the intensity of the concentration for X and Y to take values around the straight line.

We now illustrate this by an example.

Example 5.25. Suppose the joint PDF

$$f_{XY}(x, y) = \begin{cases} \frac{1}{4h\varepsilon}, & -\varepsilon + a + bx < y < \varepsilon + a + bx, \quad -h < x < h, \\ 0, & \text{otherwise,} \end{cases}$$

where $\varepsilon > 0$. The support Ω_{XY} is depicted in Figure 5.9.

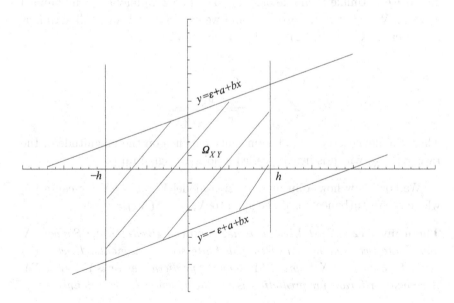

Figure 5.9: Support Ω_{XY} of Example 5.25

Here, $|Y - (a + bX)|$ is bounded within a band of ε. If and only if $\varepsilon = 0$, there exists an exact linear relationship: $Y = a + bX$. It can be shown that

$$\rho_{XY} = \frac{bh}{\sqrt{\varepsilon^2 + b^2 h^2}}.$$

Obviously, suppose the band ε shrinks to zero, then ρ_{XY} will converge to 1 when $b > 0$.

In Example 5.25, the difference $|Y - (a + bX)|$ is bounded by a constant $\varepsilon > 0$. Below we provide an example where the difference $|Y - (a + bX)|$ is equal to a random variable ε which may have an unbounded support but a finite variance. In this case, ρ_{XY} can still measure the degree of linear association.

Example 5.26. **[Linear Regression Model]:** Suppose we have

$$Y = a + bX + \varepsilon,$$

where ε is a random variable with $E(\varepsilon) = 0$, $\text{var}(\varepsilon) = \sigma_\varepsilon^2 > 0$, and it is orthogonal to X in the sense $E(X\varepsilon) = 0$. This is usually called a linear regression model in statistics and econometrics. The random variable ε can be viewed as a disturbance to an otherwise perfect linear relationship $Y = a + bX$. Unlike Example 5.25, $\varepsilon = Y - (a + bX)$ may have unbounded support. When $\sigma_\varepsilon^2 > 0$, $|\rho_{XY}| < 1$ and we expect that $|\rho_{XY}|$ will shrink as σ_ε^2 increases. Given $E(X\varepsilon) = 0$, it can be shown that

$$\rho_{XY} = \frac{\text{cov}(X, Y)}{\sigma_X \sigma_Y}$$

$$= \frac{b}{\sqrt{b^2 + \sigma_\varepsilon^2 / \sigma_X^2}}.$$

Thus, the deviation of $|\rho_{XY}|$ from unity depends on the magnitude of the ratio $\sigma_\varepsilon^2 / \sigma_X^2$, which is usually called a noise-to-signal ratio.

We now show how a linear regression model $Y = a + bX + \varepsilon$ can arise, where the disturbance ε is orthogonal to X, i.e., $E(X\varepsilon) = 0$.

Theorem 5.12. *[Best Linear Least Squares Prediction]: Suppose X and Y are two random variables with finite second moments. If we use a linear function $\alpha + \beta X$ to predict Y, then the prediction error is $Y - (\alpha + \beta X)$. A popular criterion for prediction is the mean squared error, defined as*

$$MSE(\alpha, \beta) = E[Y - (\alpha + \beta X)]^2.$$

Then the optimal coefficients (α^*, β^*) *that minimize* $MSE(\alpha, \beta)$ *are given by*

$$\alpha^* = \mu_Y - \frac{cov(X,Y)}{var(X)}\mu_X,$$

$$\beta^* = \frac{cov(X,Y)}{var(X)} = \rho_{XY}\sqrt{\frac{var(Y)}{var(X)}}.$$

Proof: We solve for the optimal coefficients (α^*, β^*) using the first order conditions(FOC):

$$\frac{\partial MSE(\alpha, \beta)}{\partial \alpha}\Big|_{(\alpha^*,\beta^*)} = -2E[Y - (\alpha^* + \beta^* X)] = 0,$$

$$\frac{\partial MSE(\alpha, \beta)}{\partial \beta}\Big|_{(\alpha^*,\beta^*)} = -2E\{X[Y - (\alpha^* + \beta^* X)]\} = 0.$$

Solving these two equations, we can obtain the desired expressions for α^* and β^*.

Define the prediction error of the best linear least squares predictor $\alpha^* + \beta^* X$,

$$\varepsilon = Y - (\alpha^* + \beta^* X).$$

Then we can write

$$Y = \alpha^* + \beta^* X + \varepsilon,$$

where ε is orthogonal to X, namely,

$$E(X\varepsilon) = 0.$$

The FOC of the minimization of $MSE(\alpha, \beta)$ ensures $E(X\varepsilon) = 0$. That is, it is the very nature of the best linear least squares prediction that ensures that ε is orthogonal to X, implying that ε contains no linear component of X that is useful to predict Y. This provides a justification for Example 5.26. It is very important to note that the optimal slope coefficient β^* is proportional to covariance $cov(X,Y)$. If $cov(X,Y) = 0$, then the linear predictor $\alpha^* + \beta^* X$ has no predictive power for Y in terms of the MSE criterion. When $cov(X,Y) \neq 0$, one can use X to predict Y using the linear predictor $\alpha^* + \beta^* X$. Nevertheless, this does not necessarily means that there exists a causality from X to Y.

Covariance has been quite useful in economic analysis. A most important empirical stylized fact in macroeconomics is called the Phillips curve, which states that unemployment and inflation tend to move in opposite directions, namely, they are negatively correlated. We now provide an important example in finance called capital asset pricing model (CAPM) to illustrate the usefulness of covariance.

Example 5.27. [Capital Asset Pricing Model (CAPM)]: Let R_{pt} be the return on a portfolio during a certain holding period t, r_{ft} is the risk-free interest rate, R_{mt} is the return on the market portfolio (often proxied by the return on the Standard & Poor 500 price index) in the same holding period. Then CAPM asserts that

$$R_{pt} - r_{ft} = \beta_p(R_{mt} - r_{ft}) + \varepsilon_{pt},$$

where $R_{pt} - r_{ft}$ is the excess return on the portfolio in the time period t, $R_{mt} - r_{ft}$ is the excess return on the market portfolio in the same period which represents the unavoidable systematic risk, and the random variable ε_{pt} represents the idiosyncratic risk peculiar to the portfolio, which could be eliminated by diversification (i.e., by forming a portfolio with a large number of assets; see Example 6.6 in Chapter 6). For simplicity, it is assumed here that $R_{mt} - r_{ft}$ and ε_{pt} are independent. Hence, the expected excess return on the portfolio is β_p times the expected excess return on the market portfolio. This implies that in the equilibrium, any asset or portfolio should be paid only for its exposure to the market risk and not for its idiosyncratic risk; that is, only the unavoidable systematic market risk should be compensated. It can be shown that the slope coefficient

$$\beta_p = \frac{\text{cov}(R_{pt} - r_{ft}, R_{mt} - r_{ft})}{\text{var}(R_{mt} - r_{ft})}$$

$$= \rho_{pm} \cdot \frac{\sqrt{\text{var}(R_{pt} - r_{ft})}}{\sqrt{\text{var}(R_{mt} - r_{ft})}},$$

where ρ_{pm} is the correlation between $R_{pt} - r_{ft}$ and $R_{mt} - r_{ft}$. The coefficient β_p is called the "beta" factor of the portfolio and has interesting economic interpretation: it measures both the riskiness of the portfolio and the reward for assuming that risk. The magnitude of β_p depends on the correlation strength between the portfolio and the market portfolio and the riskiness of the portfolio relative to the market portfolio. It indicates how aggressive the portfolio is. By definition, the beta for the market portfolio is unity, i.e., $\beta_m = 1$. It follows that the portfolio is more aggressive than the market portfolio if $\beta_p > 1$, and is less aggressive than the market portfolio if $\beta_p < 1$.

We note that under the above assumption for CAPM, we have

$$\text{var}\,(R_{pt} - r_{ft}) = \beta_p^2 \text{var}(R_{mt} - r_{ft}) + \text{var}(\varepsilon_{pt}),$$

where $\text{var}(R_{pt} - r_{ft})$ measures the total risk of the portfolio, $\text{var}(\varepsilon_{pt})$ measures the idiosyncratic risk peculiar to the portfolio, and $\beta_p^2 \text{var}(R_{mt} - r_{ft})$ is the unavoidable market risk tied to the portfolio.

To further investigate the properties of $\text{cov}(X, Y)$, we now consider the mean and variance of the linear transformation

$$Z = a + bX + cY.$$

In economics, there are many examples of $Z = a + bX + cY$. As an example, Z is the return on a portfolio which consists of two risky assets and a risk-free asset, where X and Y are the return on the two risky assets respectively, and a, b, c are the portfolio weights on the risky assets and risk-free asset (assuming the return on the risk-free asset is unity). Then one may be interested in calculating the expected return and risk of such a portfolio. For another example, Z can be the total cost of production, where bX and cY are the costs of inputs (e.g., labor and capital), and a is a setup cost.

Theorem 5.13. *Suppose $Z = a + bX + cY$. Then*
(1) $E(Z) = a + b\mu_X + c\mu_Y$;
(2) $var(Z) = b^2\sigma_X^2 + c^2\sigma_Y^2 + 2bc \cdot cov(X, Y)$.

For simplicity, we set $a = 0, b = c = 1$. Then Theorem 5.13 implies

$$\text{var}(X + Y) = \sigma_X^2 + \sigma_Y^2 + 2\text{cov}(X, Y).$$

When $\text{cov}(X, Y) > 0$, we have $\text{var}(X + Y) > \sigma_X^2 + \sigma_Y^2$. That is, the variation of the sum is greater than the sum of individual variations. This follows because both X and Y move in the same direction with a high probability, and so it is more likely that $X + Y$ tends to be extremely large or small.

On the other hand, when $\text{cov}(X, Y) < 0$, the variation of the sum is less than the sum of individual variations. This follows because X and Y move in opposite directions with a high probability so that $X + Y$ does not vary much. These results have important implication on the risk of the portfolio investment: other things being equal, positive correlation between risky assets increases the risk of a portfolio, whereas negative correlation between risky assets decreases the risk of the portfolio.

More generally, we have the following multivariate result:

Theorem 5.14. *Suppose* $X_1, \cdots, X_n$ *are a sequence of n random variables, and $Y = a_0 + \sum_{i=1}^{n} a_i X_i$, where the a_i are constants. Then we have*

(1) $E(Y) = a_0 + \sum_{i=1}^{n} a_i E(X_i)$;

(2) $var(Y) = \sum_{i=1}^{n} a_i^2 var(X_i) + 2 \sum_{i=2}^{n} \sum_{j=1}^{i-1} a_i a_j cov(X_i, X_j)$.

We now provide an interpretation for the constant ρ in a bivariate normal distribution $BN(\mu_1, \mu_2, \sigma_1^2, \sigma_2^2, \rho)$ introduced in Section 5.6. That is, ρ is the correlation coefficient between X and Y.

Theorem 5.15. *Suppose two random variables (X, Y) follow a bivariate normal distribution $BN(\mu_1, \mu_2, \sigma_1^2, \sigma_2^2, \rho)$. Then the correlation coefficient $\rho_{XY} = \rho$.*

Proof:

$$\rho_{XY} = \frac{cov(X, Y)}{\sigma_1 \sigma_2}$$

$$= \int_{-\infty}^{\infty} \int_{-\infty}^{\infty} \left(\frac{x - \mu_1}{\sigma_1}\right) \left(\frac{y - \mu_2}{\sigma_2}\right) f(x, y) dx dy$$

$$\left(\text{setting } u = \frac{x - \mu_1}{\sigma_1}, v = \frac{y - \mu_2}{\sigma_2}\right)$$

$$= \int_{-\infty}^{\infty} \int_{-\infty}^{\infty} uv \frac{1}{2\pi\sqrt{1 - \rho^2}} e^{-\frac{1}{2(1-\rho^2)}(u^2 - 2u\rho v + (\rho v)^2 - (\rho v)^2 + v^2)} du dv$$

$$= \frac{1}{2\pi\sqrt{1 - \rho^2}} \int_{-\infty}^{\infty} \int_{-\infty}^{\infty} uv e^{-\frac{1}{2(1-\rho^2)}[(u - \rho v)^2 + (1 - \rho^2)v^2]} du dv$$

(now setting $w = u - \rho v, u = w + \rho v$)

$$= \frac{1}{2\pi\sqrt{1 - \rho^2}} \int_{-\infty}^{\infty} \int_{-\infty}^{\infty} wv e^{-\frac{1}{2(1-\rho^2)}[w^2 + (1 - \rho^2)v^2]} dw dv$$

$$+ \rho \frac{1}{2\pi\sqrt{1 - \rho^2}} \int_{-\infty}^{\infty} \int_{-\infty}^{\infty} v^2 e^{-\frac{1}{2(1-\rho^2)}[w^2 + (1 - \rho^2)v^2]} dw dv$$

$$= 0 + \rho$$

$$= \rho.$$

Thus, the constant ρ is the correlation coefficient between X and Y.

Example 5.28. How can one obtain a bivariate normal distribution $BN(\mu_1, \mu_2, \sigma_1^2, \sigma_2^2, \rho)$ if two independent $N(0, 1)$ random variables Z_1 and Z_2 are given?

Solution: Define

$$X = \mu_1 + aZ_1 + bZ_2,$$
$$Y = \mu_2 + cZ_1 + dZ_2,$$

where constants a, b, c, d satisfy the restrictions that

$$a^2 + b^2 = \sigma_1^2,$$
$$c^2 + d^2 = \sigma_2^2,$$
$$ac + bd = \rho\sigma_1\sigma_2.$$

Then using the bivariate transformation theorem, it can be shown that $(X, Y) \sim BN(\mu_1, \mu_2, \sigma_1^2, \sigma_2^2, \rho)$.

In statistics, there is a so-called independent component analysis (ICA), where random variables $X_1, \cdots, X_n$ can be represented as a linear combination of a set of independent components $Z_1, \cdots, Z_m$, and n and m need not be equal to each other. Example 5.28 is a special case of independent component analysis. Here, independent components may be interpreted as independent random shocks to the system of variables $X_1, \cdots, X_n$. In this framework, it is straightforward to assess the impact of each shock to the system in terms of variance and the expected marginal effect.

Because ρ_{XY} is a measure of the linear association between X and Y, it is often referred to as the linear correlation coefficient. And, to avoid confusion, the word "linearity" is often used to describe the types of correlation with respect to ρ_{XY}. For instance, the term "positively linearly correlated" is often used in place of "positively correlated".

The correlation coefficient ρ_{XY} is only a particular kind of linear association. Two random variables X and Y may have a strong dependence but their correlation is zero, because their relationship is nonlinear. In other words, ρ_{XY} may not capture certain nonlinear associations. This is illustrated by the example below.

Example 5.29. Suppose $X \sim N(0, \sigma^2)$ and $Y = X^2$. Then

$$
\begin{aligned}
\mathrm{cov}(X, Y) &= E(XY) - \mu_X\mu_Y \\
&= E(X^3) \\
&= \int_{-\infty}^{\infty} x^3 \frac{1}{\sqrt{2\pi\sigma^2}} e^{-\frac{x^2}{2\sigma^2}}\, dx \\
&= 0,
\end{aligned}
$$

where the integral is zero because the integrand is an odd function (i.e., $g(-x) = -g(x)$ for all x). Thus, although Y is a deterministic (nonlinear) function of X, they are uncorrelated. This indicates that $\mathrm{cov}(X, Y)$ cannot capture some important nonlinear relationships between X and Y.

5.8   Joint Moment Generating Function

In this section, we first define the joint MGF of (X, Y) and then discuss its properties.

Definition 5.20. [Joint MGF]: The joint MGF of (X, Y) is defined as

$$M_{XY}(t_1, t_2) = E\left(e^{t_1 X + t_2 Y}\right), \qquad -\infty < t_1, t_2 < \infty,$$

provided the expectation exists for all (t_1, t_2) in some neighborhood of $(0, 0)$.

The joint MGF $M_{XY}(t_1, t_2)$ is the bivariate generalization of the MGF $M_X(t)$ introduced in Section 3.8, Chapter 3. It contains rich information about the joint probability distribution of (X, Y). The marginal MGF's of X and Y can be obtained from the joint MGF $M_{XY}(t_1, t_2)$:

$$M_X(t_1) = M_{XY}(t_1, 0),$$
$$M_Y(t_2) = M_{XY}(0, t_2).$$

When the joint MGF $M_{XY}(t_1, t_2)$ exists for all (t_1, t_2) in some neighborhood of $(0, 0)$, the product moment $E(X^r Y^s)$ exists for all orders (r, s), and $M_{XY}(t_1, t_2)$ can be used to generate various product moments $E(X^r Y^s)$.

Theorem 5.16. *Suppose the joint MGF $M_{XY}(t_1, t_2)$ exists for all (t_1, t_2) in some neighborhood of $(0, 0)$. Then for all nonnegative integers $r, s \geq 0$,*

$$E(X^r Y^s) = M_{XY}^{(r,s)}(0, 0),$$

and

$$\mathrm{cov}(X^r, Y^s) = M_{XY}^{(r,s)}(0, 0) - M_X^{(r)}(0) M_Y^{(s)}(0).$$

In particular,

$$\mathrm{cov}(X, Y) = M_{XY}^{(1,1)}(0, 0) - M_X^{(1)}(0) M_Y^{(1)}(0).$$

Proof: Given $M_{XY}(t_1, t_2) = \int_{-\infty}^{\infty} \int_{-\infty}^{\infty} e^{t_1 x + t_2 y} dF_{XY}(x, y)$, we have

$$M_{XY}^{(r,s)}(t_1, t_2) = \frac{\partial^{r+s}}{\partial t_1^r \partial t_2^s} \int_{-\infty}^{\infty} \int_{-\infty}^{\infty} e^{t_1 x + t_2 y} dF_{XY}(x, y)$$

$$= \int_{-\infty}^{\infty} \int_{-\infty}^{\infty} \frac{\partial^{r+s} e^{t_1 x + t_2 y}}{\partial t_1^r \partial t_2^s} dF_{XY}(x, y)$$

$$= \int_{-\infty}^{\infty} \int_{-\infty}^{\infty} x^r y^s e^{t_1 x + t_2 y} dF_{XY}(x, y).$$

It follows that

$$M_{XY}^{(r,s)}(0,0) = \int_{-\infty}^{\infty} \int_{-\infty}^{\infty} x^r y^s dF_{XY}(x,y) = E(X^r Y^s).$$

Moreover,

$$M_{XY}^{(r,s)}(0,0) - M_X^{(r)}(0)M_Y^{(s)}(0) = E(X^r Y^s) - E(X^r)E(Y^s)$$
$$= \text{cov}(X^r, Y^s).$$

It is interesting to note that with the choice of $(r,s) = (1,1)$, $\text{cov}(X,Y)$ can be obtained by differentiating the joint MGF $M_{XY}(t_1, t_2)$ and the marginal MGF's $M_X(t_1), M_Y(t_2)$. This completes the proof.

Like in the univariate case, the joint MGF $M_{XY}(t_1, t_2)$, when it exists for all (t_1, t_2) in some neighborhood of $(0,0)$, can be used to uniquely characterize the joint probability distribution of (X, Y).

Example 5.30. Suppose X, Y follow a bivariate normal distribution $BN(\mu_1, \mu_2, \sigma_1^2, \sigma_2^2, \rho)$. Then their joint MGF

$$M_{XY}(t_1, t_2) = e^{\mu_1 t + \mu_2 t + \frac{\sigma_1^2 t_1^2 + \sigma_2^2 t_2^2 + 2\rho\sigma_1\sigma_2 t_1 t_2}{2}}$$
$$= e^{\mu' t + \frac{1}{2}t'\Sigma t},$$

where $t = (t_1, t_2)'$, $\mu = (\mu_1, \mu_2)'$ and

$$\Sigma = \begin{bmatrix} \sigma_1^2 & \rho\sigma_1\sigma_2 \\ \rho\sigma_1\sigma_2 & \sigma_2^2 \end{bmatrix}.$$

In a multivariate setup, we can define the joint MGF of random variables $X_1, \cdots, X_n$:

$$M_{\mathbf{X}^n}(\mathbf{t}) = E\left(e^{\mathbf{t}'\mathbf{X}^n}\right) = E\left(e^{\sum_{i=1}^n t_i X_i}\right),$$

provided the expectation exists for $\mathbf{t} = (t_1 \cdots, t_n)'$ in some neighborhood of the origin $(0, \cdots, 0)'$, where $\mathbf{X}^n = (X_1, \cdots, X_n)'$, $\mathbf{t} = (t_1, \cdots, t_n)'$.

The joint MGF may not exist for some joint distributions. However, we can always define a joint characteristic function, which exists for any joint distribution and has similar properties to the joint MGF. For space, we will not discuss the joint characteristic function. Interested readers are referred to Hong (1999) in a time series context.

5.9 Implications of Independence on Expectations

To examine the impact of independence on expectations under a joint probability distribution, we first state an important result of independence.

Theorem 5.17. *Suppose (X, Y) are independent. Then for any measurable and integrable functions $h(X)$ and $q(Y)$,*

$$E\left[h(X)q(Y)\right] = E[h(X)]E[q(Y)]$$

or equivalently,

$$cov[h(X), q(Y)] = 0.$$

Proof:

$$
\begin{aligned}
E[h(X)q(Y)] &= \int_{-\infty}^{\infty} \int_{-\infty}^{\infty} h(x)q(y) dF_{XY}(x, y) \\
&= \int_{-\infty}^{\infty} \int_{-\infty}^{\infty} h(x)q(y) dF_X(x) dF_Y(y) \quad \text{(by independence)} \\
&= \int_{-\infty}^{\infty} h(x) dF_X(x) \int_{-\infty}^{\infty} q(y) dF_Y(y) \\
&= E[h(X)]E[q(Y)].
\end{aligned}
$$

Thus, when X and Y are independent, any measurable transformation of X and Y, no matter linear or nonlinear, will be uncorrelated.

5.9.1 *Independence and Moment Generating Functions*

To illustrate the usefulness of the above theorem, we first state a corollary about the impact of independence on the MGF.

Corollary 5.1. *Suppose X and Y are independent and their marginal MGF's $M_X(t)$ and $M_Y(t)$ exist for all t in a neighborhood of 0. Then $M_{X+Y}(t)$ exists for all t in a small neighborhood of 0, and*

$$M_{X+Y}(t) = M_X(t)M_Y(t) \text{ for all } t \text{ in a neighborhood of 0.}$$

Note that $M_{X+Y}(t)$ is the MGF of random variable $X + Y$. It is not the joint MGF $M_{XY}(t_1, t_2)$ of (X, Y).

Proof: Applying Theorem 5.17, we choose

$$g(X, Y) = e^{t(X+Y)} = e^{tX}e^{tY} = h(X)q(Y).$$

It follows that if X and Y are independent, then $M_{X+Y}(t)$ exists for all t in a small neighborhood of 0, and

$$E[e^{t(X+Y)}] = E(e^{tX})E(e^{tY}).$$

That is,

$$M_{X+Y}(t) = M_X(t)M_Y(t).$$

This property of the MGF for the sum of independent random variables is rather useful in characterizing the distributions for some random variables which themselves are the sum of other independent random variables. We now illustrate this via a variety of examples.

Example 5.31. [Reproductive Property of Normal Distribution]:
Suppose $X \sim N(\mu_1, \sigma_1^2), Y \sim N(\mu_2, \sigma_2^2)$ and X, Y are independent. Then show

$$X \pm Y \sim N(\mu_1 \pm \mu_2, \sigma_1^2 + \sigma_2^2).$$

Solution: The MGF's of X and Y are

$$M_X(t) = Ee^{tX} = e^{\mu_1 t + \frac{\sigma_1^2}{2}t^2},$$
$$M_Y(t) = Ee^{tY} = e^{\mu_2 t + \frac{\sigma_2^2}{2}t^2}.$$

By independence,

$$\begin{aligned}
M_{X+Y}(t) &= M_X(t)M_Y(t) \\
&= e^{\mu_1 t + \frac{1}{2}\sigma_1^2 t^2} \cdot e^{\mu_2 t + \frac{1}{2}\sigma_2^2 t^2} \\
&= e^{\mu t + \frac{1}{2}\sigma^2 t^2},
\end{aligned}$$

where $\mu = \mu_1 + \mu_2, \sigma^2 = \sigma_1^2 + \sigma_2^2$. It follows that

$$X + Y \sim N(\mu_1 + \mu_2, \sigma_1^2 + \sigma_2^2).$$

That is, the sum of two independent normal variables is also a normal variable. In fact, this result can be extended to a linear combination of many independent normal random variables. This property is called the reproductive property of the normal distribution.

Example 5.32. [Reproductive Property of Poisson Distribution]:
Suppose $X_1, \cdots, X_n$ are n independent random variables, with X_i having a Poisson(λ_i) distribution, $i = 1, \cdots, 2$. Show that $\Sigma_{i=1}^n X_i$ follows a Poisson(λ) distribution, where $\lambda = \Sigma_{i=1}^n \lambda_i$.

Solution: For a random variable X_i that follows a Poisson(λ_i) distribution, the MGF

$$M_i(t) = e^{\lambda_i(e^t - 1)}, \qquad -\infty < t < \infty.$$

Put $X = \Sigma_{i=1}^n X_i$. Then its MGF

$$
\begin{aligned}
M_X(t) &= E\left(e^{t\Sigma_{i=1}^n X_i}\right) \\
&= \prod_{i=1}^n M_i(t) \\
&= \prod_{i=1}^n e^{\lambda_i(e^t - 1)} \\
&= e^{\lambda(e^t - 1)},
\end{aligned}
$$

where $\lambda = \Sigma_{i=1}^n \lambda_i$. Therefore, $X \sim$ Poisson(λ). In other words, the sum of n independent Poisson(λ_i) random variables is a Poisson(λ) random variable with $\lambda = \sum_{i=1}^n \lambda_i$. This is called the reproductive property of the Poisson distribution.

Example 5.33. [Reproductive Property of χ^2 Distribution]: Suppose $X_1, X_2, \cdots$ are independent $\chi^2_{\nu_i}$ respectively. Show $\Sigma_{i=1}^n X_i$ follows a χ^2_ν, where $\nu = \Sigma_{i=1}^n \nu_i$.

Solution: From Chapter 4, we have that the MGF of a $\chi^2_{\nu_i}$ distribution

$$M_i(t) = (1 - 2t)^{-\nu_i/2}, \qquad t < \frac{1}{2}.$$

Put $X = \sum_{i=1}^n X_i$. Then its MGF

$$
\begin{aligned}
M_X(t) &= E\left(e^{tX}\right) \\
&= \prod_{i=1}^n M_i(t) \\
&= \prod_{i=1}^n (1 - 2t)^{-\nu_i/2} \\
&= (1 - 2t)^{-\frac{1}{2}\Sigma_{i=1}^n \nu_i}.
\end{aligned}
$$

It follows that $X \sim \chi^2_\nu$, where $\nu = \sum_{i=1}^n \nu_i$. That is, the sum of independent χ^2 random variables is still a χ^2 random variable with the degrees of freedom equal to the sum of individual degrees of freedom. This is called the reproductive property of the χ^2 distribution.

Dykstra and Hewett (1972) give two examples that shed light on some properties of the Chi-square distribution. One example shows that the sum of two random variables can be distributed as a χ^2, where one of the variables is χ^2-distributed but the other variable is positive but not necessarily χ^2-distributed. The other example shows the fact that the sum of two χ^2 random variables can have a χ^2-distribution, but these two χ^2 random variables need not be independent.

Example 5.34. Suppose $X_1, X_2, \cdots$ are independent random variables having exponential distributions with the same parameter $\beta > 0$. Show that $\Sigma_{i=1}^{n} X_i$ follows a Gamma(n, β) distribution.

Solution: For X_i having an EXP(β) distribution, we have

$$M_i(t) = (1 - \beta t)^{-1}, \quad t < \frac{1}{\beta}.$$

Put $X = \sum_{i=1}^{n} X_i$. Then its MGF

$$
\begin{aligned}
M_X(t) &= \prod_{i=1}^{n} M_i(t) \\
&= (1 - \beta t)^{-n}.
\end{aligned}
$$

This implies that $X \sim$ Gamma(n, β). That is, the sum of n independent exponential random variables with same parameter β follows a Gamma(n, β) distribution.

5.9.2   *Independence and Uncorrelatedness*

Applying Theorem 5.17, we immediately obtain an important result that independence implies uncorrelatedness.

Corollary 5.2. *Suppose (X, Y) are independent. Then* $cov(X, Y) = 0$.

Proof: Set $g(X) = X - \mu_X$ and $h(Y) = Y - \mu_Y$, and then apply Theorem 5.17.

Since independence rules out all possible types of association between X and Y, whereas uncorrelatedness only implies the absence of a linear association, independence is clearly a stronger condition than uncorrelatedness. Therefore, independence implies uncorrelatedness but uncorrelatedness does not necessarily imply independence. This is illustrated by the following examples.

Example 5.35. Put $Y = X^2$, where X follows a continuous probability distribution that is symmetric about 0 (i.e., $f_X(x) = f_X(-x)$ for all $-\infty < x < \infty$). Then

$$\text{cov}(X, Y) = 0.$$

Solution: Because the distribution of X is symmetric about 0, we have $E(X) = 0$ and $E(X^3) = 0$. It follows that

$$
\begin{aligned}
\text{cov}(X, Y) &= E(XY) - E(X)E(Y) \\
&= E(X^3) - 0 \cdot E(X^2) \\
&= E(X^3) \\
&= \int_{-\infty}^{\infty} x^3 f_X(x) dx \\
&= 0.
\end{aligned}
$$

Therefore, $\text{cov}(X, Y)$ cannot capture the quadratic association between X and $Y = X^2$ when X is symmetrically distributed about zero.

Example 5.36. [Uncorrelated Bivariate Student's t-distribution]: Random variables X and Y are called to follow a standard bivariate Student t-distribution if their joint PDF is given by

$$f_{XY}(x, y) = \frac{1}{2\pi} \left[1 + \frac{1}{\nu}(x^2 + y^2) \right]^{-\frac{\nu}{2}}, \qquad \nu \geq 3,$$

where the shape parameter ν is called the degree of freedom. It can be shown that $\text{cov}(X, Y) = 0$ (verify it!). However, for any given $\nu < \infty$, X and Y are not independent because the joint PDF $f_{XY}(x, y)$ cannot be partitioned into the product of two separate functions of x and y respectively.

Note that when $\nu \to \infty$, we have

$$f_{XY}(x, y) \to \frac{1}{2\pi} e^{-\frac{1}{2}(x^2 + y^2)}$$

given the fact that $(1 + a/\nu)^\nu \to e^a$ as $\nu \to \infty$. In this case, X and Y are independent $N(0, 1)$ random variables. Figure 5.10 plots the joint PDF $f_{XY}(x, y)$ of the bivariate standard Student t-distribution with various choices of degree of freedom ν.

Example 5.37. Suppose X has a PDF $f(x - \theta)$, where $f(\cdot)$ is symmetric about 0 and $E|X| < \infty$. Define $Y = \mathbf{1}(|X - \theta| < 2)$, where $\mathbf{1}(\cdot)$ is the indicator function, taking value 1 if $|X - \theta| < 2$ and 0 if $|X - \theta| \geq 2$. Check if $\text{cov}(X, Y) = 0$.

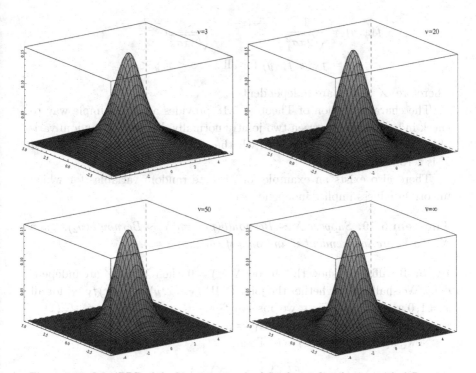

Figure 5.10: Joint PDF of the bivariate standard Student t-distribution with different ν

Solution: This is left as an exercise.

The above examples show that while independence implies uncorrelatedness, the converse is not true. However, there exist some special cases under which $\text{cov}(X, Y) = 0$ if and only if (X, Y) are independent. We consider two examples.

Theorem 5.18. *Suppose (X, Y) are jointly normally distributed. Then $\text{cov}(X, Y) = 0$ if and only if X and Y are independent.*

Proof: (1) [*Necessity*] By Corollary 5.2, if X, Y are independent, then we have $\text{cov}(X, Y) = 0$.

(2) [*Sufficiency*] Next, we show $\text{cov}(X, Y) = 0$ implies independence between X and Y. For a bivariate normal distribution $BN(\mu_1, \mu_2, \sigma_1^2, \sigma_2^2, \rho)$, we have shown in Theorem 5.15 that the correlation coefficient $\rho_{XY} = \rho$. Thus, $\text{cov}(X, Y) = 0$ implies $\rho = 0$. It follows that the joint PDF

$$f(x,y) = \frac{1}{\sqrt{2\pi\sigma_1^2}} e^{-\frac{1}{2\sigma_1^2}(x-\mu_1)^2} \frac{1}{\sqrt{2\pi\sigma_2^2}} e^{-\frac{1}{2\sigma_2^2}(y-\mu_2)^2}$$

$$= f_X(x)f_Y(y) \text{ for all } -\infty < x, y < \infty.$$

Therefore, X and Y are independent.

The characterization of Theorem 5.18 provides a rather simple way to check independence between two jointly normally distributed random variables. That is, it suffices to check whether the correlation between X and Y is zero.

There also exists an example for discrete random variables for which uncorrelatedness implies independence.

Theorem 5.19. *Suppose $X \sim Bernoulli(p_1)$ and $Y \sim Bernoulli(p_2)$. Then X and Y are independent if and only if $cov(X, Y) = 0$.*

Proof: It suffices to show that if $cov(X, Y) = 0$, then X and Y are independent. We shall check whether the joint PMF $f_{XY}(x, y) = f_X(x)f_Y(y)$ for all $x = 1, 0$ and $y = 1, 0$. First, we observe $f_X(1) = p_1, f_X(0) = 1 - p_1, f_Y(1) = p_2, f_Y(0) = 1 - p_2, \mu_X = p_1$ and $\mu_Y = p_2$.

Now suppose $cov(X, Y) = 0$. Then

$$E(XY) = E(X)E(Y)$$

or equivalently

$$\sum_{x=0}^{1} \sum_{y=0}^{1} xy f_{XY}(x, y) = p_1 p_2.$$

Since $\sum_{x=0}^{1} \sum_{y=0}^{1} xy f_{XY}(x, y) = f_{XY}(1, 1)$. We have

$$f_{XY}(1, 1) = p_1 p_2 = f_X(1)f_Y(1).$$

Next, given $f_X(1) = \sum_{y=0}^{1} f_{XY}(1, y)$, we have

$$f_{XY}(1, 0) = f_X(1) - f_{XY}(1, 1)$$
$$= p_1(1 - p_2)$$
$$= f_X(1)f_Y(0).$$

Similarly, we can show

$$f_{XY}(0, 1) = (1 - p_1)p_2 = f_X(0)f_Y(1)$$

and

$$f_{XY}(0, 0) = (1 - p_1)(1 - p_2) = f_X(0)f_Y(0).$$

It follows that $f_{XY}(x, y) = f_X(x)f_Y(y)$ for all $x = 1, 0$ and all $y = 1, 0$. Therefore, X and Y are independent. The proof is completed.

Question: Theorem 5.17 shows that if X and Y are independent, then

$$\text{cov}[h(X), q(Y)] = 0$$

for any measurable functions $h(X)$ and $q(Y)$. Now, suppose $\text{cov}[h(X), q(Y)]$ $= 0$ for any measurable functions $h(X)$ and $q(Y)$. Are X and Y independent?

To answer this question, we consider $h_x(X) = \mathbf{1}(X \leq x), q_y(Y) = \mathbf{1}(Y \leq y)$ for any given $(x, y) \in \mathbb{R}^2$. Then

$$
\begin{aligned}
E[h_x(X)] &= E[\mathbf{1}(X \leq x)] \\
&= F_X(x), \\
E[q_y(Y)] &= E[\mathbf{1}(Y \leq y)] \\
&= F_Y(y), \\
E[h_x(X)q_y(Y)] &= E[\mathbf{1}(X \leq x)\mathbf{1}(Y \leq y)] \\
&= F_{XY}(x, y).
\end{aligned}
$$

Now, suppose we have

$$\text{cov}[h_x(X), q_y(Y)] = 0 \text{ for all } -\infty < x, y < \infty$$

or equivalently

$$E[h_x(X)q_y(Y)] = E[h_x(X)]E[q_y(Y)] \text{ for all } -\infty < x, y < \infty.$$

Then

$$F_{XY}(x, y) = F_X(x)F_Y(y) \text{ for all } -\infty < x, y < \infty.$$

It follows that X and Y must be independent.

We can also use the joint MGF to characterize independence.

Theorem 5.20. *Suppose the joint MGF $M_{XY}(t_1, t_2) = E(e^{t_1 X + t_2 Y})$ exists for all (t_1, t_2) in some neighborhood of the origin $(0, 0)$. Then (X, Y) are independent if and only if*

$$M_{XY}(t_1, t_2) = M_X(t_1)M_Y(t_2),$$

for all (t_1, t_2) in some neighborhood of $(0, 0)$.

Proof: (1) [*Necessity*] Suppose (X, Y) are independent. Then, by Theorem 5.17, we have

$$
\begin{aligned}
M_{XY}(t_1, t_2) &= E(e^{t_1 X + t_2 Y}) \\
&= E(e^{t_1 X} \cdot e^{t_2 Y}) \\
&= E\left(e^{t_1 X}\right) E\left(e^{t_2 Y}\right) \\
&= M_X(t_1) M_Y(t_2).
\end{aligned}
$$

(2) [*Sufficiency*] Next, we shall show that if $M_{XY}(t_1, t_2) = M_X(t_1) M_Y(t_2)$ for all (t_1, t_2) in some neighborhood of $(0, 0)$, then (X, Y) are independent. We apply the uniqueness theorem of the MGF. By the uniqueness theorem of the MGF, the bivariate distribution $F_{XY}(x, y)$ is the unique distribution that corresponds to the joint $M_{XY}(t_1, t_2)$, and the bivariate distribution $F_X(x) F_Y(y)$ is the unique distribution that corresponds to the joint MGF $M_X(t_1) M_Y(t_2)$. If $M_{XY}(t) = M_X(t_1) M_Y(t_2)$ for all (t_1, t_2) in some neighborhood of $(0, 0)$, then $F_X(x) F_Y(y)$ is also the unique distribution associated with $M_{XY}(t)$. It follows that $F_{XY}(x, y) = F_X(x) F_Y(y)$ for all $-\infty < x, y < \infty$. This completes the proof.

Again, it is important to emphasize the difference between $M_{XY}(t_1, t_2)$ and $M_{X+Y}(t)$: $M_{XY}(t_1, t_2) = E(e^{t_1 X + t_2 Y})$ is the joint MGF of the bivariate vector (X, Y), whereas $M_{X+Y}(t) = E(e^{t(X+Y)})$ is the MGF of the sum $X + Y$.

To sum up, we have learnt three basic approaches to characterizing independence, that is, checking whether the joint CDF, or the joint PMF/PDF, or the joint MGF, is equal to the product of their marginal counterparts, respectively.

There exists an alternative insightful representation of the MGF-based characterization of independence:

Theorem 5.21. *Suppose $M_{XY}(t_1, t_2)$ exists for all (t_1, t_2) in a neighborhood of $(0, 0)$. Then (X, Y) are independent if and only if*

$$
cov(e^{t_1 X}, e^{t_2 Y}) = 0
$$

for all (t_1, t_2) in the neighborhood of $(0, 0)$.

Proof: Using the formula $cov(U, V) = E(UV) - E(U)E(V)$, it is straightforward to show that

$$
\begin{aligned}
\mathrm{cov}(e^{t_1 X}, e^{t_2 Y}) &= E\left(e^{t_1 X} e^{t_2 Y}\right) - E\left(e^{t_1 X}\right) E\left(e^{t_2 Y}\right) \\
&= M_{XY}(t_1, t_2) - M_X(t_1) M_Y(t_2).
\end{aligned}
$$

Because $M_{XY}(t_1, t_2) = M_X(t_1)M_Y(t_2)$ for all (t_1, t_2) in a neighborhood of $(0,0)$ if and only if (X, Y) are independent, it follows immediately that $\text{cov}(e^{t_1 X}, e^{t_2 Y}) = 0$ for all t_1, t_2 in a neighborhood of 0 if and only if (X, Y) are independent. The proof is completed.

Since $\text{cov}(e^{t_1 X}, e^{t_2 Y})$ is the covariance between exponential transformations $e^{t_1 X}$ and $e^{t_2 Y}$, it can be viewed as a generalized covariance. Theorem 5.21 shows that X and Y are independent if and only if this generalized covariance $\text{cov}(e^{t_1 X}, e^{t_2 Y})$ is zero for all pairs of (t_1, t_2) in some neighborhood of the origin $(0,0)$. The next theorem shows that the generalized covariance $\text{cov}(e^{t_1 X}, e^{t_2 Y})$ can be viewed as a covariance generating function, that is, a function that can be used to generate various covariances $\text{cov}(X^r, Y^s)$, as shown below.

Theorem 5.22. *Suppose $M_{XY}(t_1, t_2)$ exists for all (t_1, t_2) in a neighborhood of $(0,0)$. Then*

$$\text{cov}(X, Y) = \frac{\partial^2 \text{cov}(e^{t_1 X}, e^{t_2 Y})}{\partial t_1 \partial t_2}\bigg|_{(t_1, t_2) = (0,0)}.$$

Moreover, for any positive integers r, s,

$$\text{cov}(X^r, Y^s) = \frac{\partial^{r+s} \text{cov}(e^{t_1 X}, e^{t_2 Y})}{\partial t_1^r \partial t_2^s}\bigg|_{(t_1, t_2) = (0,0)}.$$

When $M_{XY}(t_1, t_2)$ exists for all (t_1, t_2) in a neighborhood of $(0,0)$, all derivatives of $\text{cov}(e^{t_1 X}, e^{t_2 Y})$ at the origin $(0,0)$ exist. Therefore, using MacLaurin's series expansion $e^{tX} = \sum_{r=0}^{\infty} \frac{(tX)^r}{r!}$, we obtain

$$\text{cov}(e^{t_1 X}, e^{t_2 Y}) = \sum_{r=1}^{\infty} \sum_{s=1}^{\infty} \frac{t_1^r t_2^s}{r!s!} \frac{\partial^{r+s} \text{cov}(e^{t_1 X}, e^{t_2 Y})}{\partial t_1^r \partial t_2^s}\bigg|_{t=(0,0)}$$

$$= \sum_{r=1}^{\infty} \sum_{s=1}^{\infty} \frac{t_1^r t_2^s}{r!s!} \text{cov}(X^r, Y^s).$$

Thus, $\text{cov}(e^{t_1 X}, e^{t_2 Y})$ contains information on all covariances $\{\text{cov}(X^r, Y^s)\}$ of various orders. If X and Y are independent, then $\text{cov}(X^r, Y^s) = 0$ for all $r, s > 0$, including $\text{cov}(X, Y) = 0$. In general, $\text{cov}(X, Y) = 0$ is only one of an infinite set of implications for independence. It is possible that $\text{cov}(X, Y) = 0$ but $\text{cov}(X^r, X^s) \neq 0$ for some choice of (r, s). In this case, X and Y are uncorrelated but they are not independent. An economic example is a sequence of high-frequency asset returns $\{X_t\}$ over time. It is often found that $\{X_t\}$ is serially uncorrelated over different time periods, that is, $\text{cov}(X_t, X_{t-j}) = 0$ for all $j > 0$. However, there exists persistent volatility clustering, that is, a large volatility today tends to be followed by another large volatility tomorrow, and a small volatility today tends to be followed by another small volatility, and these patterns alternate over times. This implies $\text{cov}(X_t^2, X_{t-j}^2) > 0$ at least for some $j > 0$.

Finally, we consider the following problem: suppose all moments of X exist, and $\text{cov}(X^r, Y^s) = 0$ for all $r, s > 0$. Are X and Y independent?

This is analogous to the question in the univariate case whether the equality of all moments of two random variables X and Y imply identical distributions (see Theorem 3.25). The answer is yes, under the condition that both X and Y have bounded supports.

Theorem 5.23. *Suppose X and Y have bounded supports. Then $\text{cov}(X^r, Y^s) = 0$ for all $r, s > 0$ if and only if X and Y are independent.*

Proof: Given the bounded supports for both X and Y, there exists a constant $M < \infty$ such that $P(|X| < M) = 1$ and $P(|Y| < M) = 1$. Therefore, $E|X^r| \leq M^r$ and $E|Y^s| \leq M^s$. It follows that for any given $t_1, t_2 \in (-\infty, \infty)$, $\text{cov}(e^{t_1 X}, e^{t_2 Y})$ exists because

$$
\begin{aligned}
\left| \text{cov}(e^{t_1 X}, e^{t_2 Y}) \right| &= \left| \sum_{r=1}^{\infty} \sum_{s=1}^{\infty} \frac{t_1^r t_2^s}{r! s!} \text{cov}(X^r, Y^s) \right| \\
&\leq \sum_{r=1}^{\infty} \sum_{s=1}^{\infty} \frac{|t_1|^t |t_2|^s}{r! s!} |\text{cov}(X^r, Y^s)| \\
&\leq \sum_{r=1}^{\infty} \sum_{s=1}^{\infty} \frac{|t_1|^r |t_2|^s}{r! s!} (2M)^r (2M)^s \\
&= e^{2(|t_1| M + |t_2| M)} < \infty,
\end{aligned}
$$

where we have made use of the fact that $|\text{cov}(X^r, Y^s)| \leq [\text{var}(X^{2r}) \text{var}(Y^{2s})]^{1/2} \leq (2M)^{r+s}$.

(1) [*Necessity*] Suppose X and Y are independent. Then applying Theorem 5.17 with the choices of $h(X) = X^r, q(Y) = Y^s$, we have $\text{cov}(X^r, Y^s) = 0$ for all $r, s > 0$.

(2) [*Sufficiency*] Now, suppose $\text{cov}(X^r, Y^s) = 0$ for all $r, s > 0$. Then

$$
\text{cov}(e^{t_1 X}, e^{t_2 Y}) = \sum_{r=1}^{\infty} \sum_{s=1}^{\infty} \frac{t_1^r t_2^s}{r! s!} \text{cov}(X^r, Y^s) = 0
$$

for all $t_1, t_2 \in (-\infty, \infty)$, which implies independence between X and Y by Theorem 5.21. This completes the proof.

Theorem 5.23 can be viewed as a generalization of Theorem 3.25 from the univariate context to the bivariate context. This implies that we can check whether all possible covariances between X^r and Y^s are zero in order to check independence between X and Y.

There are other definitions of generalized covariance. For example, we can define a generalized covariance

$$\sigma_{XY}(x,y) \equiv \text{cov}[\mathbf{1}(X \leq x), \mathbf{1}(Y \leq y)], \qquad -\infty < x, y < \theta,$$

where $\mathbf{1}(\cdot)$ is the indicator function that takes value 1 if the condition inside holds and takes 0 otherwise. Straightforward algebra shows that

$$\text{cov}[\mathbf{1}(X \leq x), \mathbf{1}(Y \leq y)] = F_{XY}(x,y) - F_X(x)F_Y(y).$$

It follows immediately that X and Y are independent if and only if the generalized covariance $\sigma_{XY}(x,y) = 0$ for all pairs of (x,y). Interestingly, the population version of Spearman's correlation or Spearman's rho, which is the correlation between rank orders of the observations in a sample, is defined as

$$S_{XY} = \frac{\text{cov}[F_X(X), F_Y(Y)]}{\sqrt{\text{var}[F_X(X)]\text{var}[F_Y(Y)]}}.$$

It can be shown that

$$\text{cov}[F_X(X), F_Y(Y)] = E\left[\sigma_{XY}(X,Y)\right].$$

Intuitively, the population Spearman's rho can be interpreted as the correlation between the probability integral transforms of X and Y. Spearman's correlation does not have the desired property that it is zero if and only if X and Y are independent. However, it can capture certain nonlinear dependences between X and Y.

The aforementioned discussion investigates the relationship between zero covariance and independence. Before concluding this section, we discuss the relationship between correlation and causality by asking the following questions:

Questions: What is the relationship between correlation and causality? In particular, if $\text{cov}(X,Y) \neq 0$, does this imply that there exists causality between X and Y? On other hand, if there exists causality between X and Y, does this imply $\text{cov}(X,Y) \neq 0$.

5.10 Conditional Expectations

Question: What information can we extract from a conditional distribution $f_{Y|X}(y|x)$?

We first define the expectation under a conditional distribution.

Definition 5.21. [Conditional Expectation]: The conditional expectation of $g(X, Y)$ given $X = x$ is defined as

$$E[g(X,Y)|X = x] = E[g(X,Y)|x]$$

$$= \begin{cases} \sum_{y \in \Omega_Y(x)} g(x,y) f_{Y|X}(y|x), & \text{if } (X,Y) \text{ are DRV's,} \\ \int_{-\infty}^{\infty} g(x,y) f_{Y|X}(y|x) dy, & \text{if } (X,Y) \text{ are CRV's,} \end{cases}$$

where $\Omega_Y(x) = \{y \in \Omega_Y : f_{Y|X}(y|x) > 0\}$ is the support of Y conditional on the event that $X = x$.

When taking a conditional expectation, one treats x as fixed, and conditional expectation is the expectation with respect to the conditional distribution of Y given $X = x$ instead of an unconditional distribution of Y.

Since y is integrated out, the conditional expectation $E[g(X,Y)|X = x]$ is a function of x only. Thus, $E[g(X,Y)|X]$ is a function of random variable X.

With the conditional expectation, $E[g(X,Y)|X]$, we can state the law of iterated expectations.

Theorem 5.24. [Law of Iterated Expectations]: *Suppose $g(X,Y)$ is a measurable function and $E[g(X,Y)]$ exists. Then*

$$E[g(X,Y)] = E_X\{E[g(X,Y)|X]\}$$
$$= E_Y\{E[g(X,Y)|Y]\}.$$

Proof: We focus on the continuous case. Suppose (X,Y) have a joint PDF $f_{XY}(x,y)$. By the multiplication rule, we have $f_{XY}(x,y) = f_{Y|X}(y|x) f_X(x)$ if $f_X(x) > 0$. It follows that

$$E[g(X,Y)] = \int_{-\infty}^{\infty} \int_{-\infty}^{\infty} g(x,y) f_{XY}(x,y) dx dy$$

$$= \int_{-\infty}^{\infty} \int_{-\infty}^{\infty} g(x,y) f_{Y|X}(y|x) f_X(x) dx dy$$

$$= \int_{-\infty}^{\infty} \left[\int_{-\infty}^{\infty} g(x,y) f_{Y|X}(y|x) dy \right] f_X(x) dx$$

$$= \int_{-\infty}^{\infty} E[g(X,Y)|X = x] f_X(x) dx$$

$$= E_X\{E[g(X,Y)|X]\}.$$

Similarly it can be shown for the case of discrete random variables. This completes the proof.

The law of iterated expectations provides a two-stage procedure to compute an unconditional expectation. Thus, it is also called the law of total expectations. We shall provide an economic interpretation for this law of iterated expectations when we consider a special function of $g(X, Y)$ below.

In practice, we often write $E[g(X, Y)] = E\{E[g(X, Y)|X]\}$, by abusing the notation "E." The same notation "E" stands for different expectations in the same equation. The inside notation E is the expectation with respect to the conditional distribution of $Y|X$, and the outside notation E is the expectation with respect to the marginal distribution of X.

We now introduce various conditional moments of Y given X and examine their properties.

Definition 5.22. [Conditional Mean]: The conditional mean of Y given $X = x$ is defined as

$$E(Y|x) = E(Y|X = x)$$
$$= \begin{cases} \sum_{y \in \Omega_Y(x)} y f_{Y|X}(y|x) & \text{if } (X, Y) \text{ are DRV's,} \\ \int_{-\infty}^{\infty} y f_{Y|X}(y|x) dy & \text{if } (X, Y) \text{ are CRV's.} \end{cases}$$

The conditional mean $E(Y|X = x)$ is the average value of Y given that X has taken value x. We now provide an economic interpretation for $E(Y|X)$ and the law of iterated expectations via an example below.

Example 5.38. [Average Wage and Law of Iterated Expectations]: Suppose Y is the wage of an employee, and X is a gender variable of the employee, taking value 0 if the employee is female, and taking value 1 if the employee is male. Then $E(Y|X = 0)$ is the average wage of female employees and $E(Y|X = 1)$ is the average wage of male employees.

Applying the law of iterated expectations, we obtain the overall average wage of employees

$$E(Y) = E[E(Y|X)]$$
$$= P(X = 0) \cdot E(Y|X = 0) + P(X = 1) \cdot E(Y|X = 1),$$

where $P(X = 0)$ is the proportion of female employees in the labor force, and $P(X = 1)$ is the proportion of male employees in the labor force. The law of iterated expectations thus provides insight into the income distribution of the labor market.

The conditional mean $E(Y|X)$ is a function of X only. It is also called the regression function of Y on X to signify the predictive relationship of X for Y. It is the primary interest in econometrics and economics.

Why?

Suppose we use a function (or model) $g(X)$ of X to predict Y. Then the prediction error is $Y - g(X)$. A criterion to evaluate the predictive ability of model $g(X)$ is the mean squared error of $g(X)$, which is defined as

$$\text{MSE}(g) = E[Y - g(X)]^2.$$

$\text{MSE}(g)$ is the average of the squared prediction errors. The smaller $\text{MSE}(g)$, the better the predictor $g(X)$. The theorem below shows that the conditional mean is the optimal predictor for Y in terms of mean squared error.

Theorem 5.25. *[Mean Squared Error Criterion]: Let X and Y be random variables defined on the same sample space and suppose Y has a finite variance. Then the conditional mean $E(Y|X)$ is the optimal minimizer for the minimization problem of $E[Y - g(X)]^2$; that is,*

$$E(Y|X) = \arg \min_{g(\cdot)} E[Y - g(X)]^2,$$

where the minimization is over all measurable and square-integrable functions.

Proof: Put $g_0(X) = E(Y|X)$. Then we have

$$
\begin{aligned}
MSE(g) &= E[Y - g(X)]^2 \\
&= E\left\{[Y - g_0(X)] + [g_0(X) - g(X)]\right\}^2 \\
&= E\{[Y - g_0(X)]^2\} + E\{[g_0(X) - g(X)]^2\} \\
&\quad + 2E\{[Y - g_0(X)][g_0(X) - g(X)]\} \\
&= E\{[Y - g_0(X)]^2\} + E\{[g_0(X) - g(X)]^2\},
\end{aligned}
$$

where the last term, namely the cross product term, is identically zero:

$$
\begin{aligned}
E\{[Y - g_0(X)][g_0(X) - g(X)]\} &= E_X\{E[(Y - g_0(X))(g_0(X) - g(X))|X]\} \\
&= E_X\{(g_0(X) - g(X))E[(Y - g_0(X))|X]\} \\
&= E_X[(g_0(X) - g(X)) \cdot 0] \\
&= 0
\end{aligned}
$$

by the law of iterated expectations and $E[(Y - g_0(X))|X] = 0$.

In the decomposition of MSE(g), the first term $E[Y - g_0(X)]^2$ does not depend on $g(\cdot)$. Thus, minimizing MSE(g) with respect to $g(\cdot)$ is equivalent to minimizing the second term $E[g_0(X) - g(X)]^2$, which achieves the minimum if and only if we choose $g(X) = g_0(X)$. This completes the proof.

Theorem 5.26. *[Regression Identity]: Suppose $E(Y|X)$ exists. Then there is a random variable ε such that*

$$Y = E(Y|X) + \varepsilon,$$

where ε satisfies the condition

$$E(\varepsilon|X) = 0.$$

Proof: Define $\varepsilon = Y - E(Y|X)$. Then we have $Y = E(Y|X) + \varepsilon$, where

$$\begin{aligned}
E(\varepsilon|X) &= E\left\{[Y - E(Y|X)]X\right\} \\
&= E(Y|X) - E[E(Y|X)|X] \\
&= E(Y|X) - E(Y|X) \\
&= 0.
\end{aligned}$$

The random variable ε is called a disturbance to the regression function $E(Y|X)$. It reflects the degree of uncertainty about the relationship between Y and X. When $\varepsilon = 0$, we have a perfect deterministic relationship between Y and X. An example is $Y = g_0(X)$ for some measurable function $g_0(\cdot)$.

What does it mean by the condition of $E(\varepsilon|X) = 0$? Put it simply, it implies that ε contains no systematic information of X that can be used to predict the expected value of Y with respect to stochastic factors rather than X. All systematic information of X that can be used to predict the expected value of Y has been incorporated in $E(Y|X)$.

The regression function $E(Y|X)$ can be a linear or nonlinear function of X. When $E(Y|X) = a + bX$, we call it a linear regression function.

We now provide two examples for which the regression function is a linear function of X.

Example 5.39. Let the joint PDF $f_{XY}(x, y) = e^{-y}$ for $0 < x < y < \infty$ be given as in Example 5.14 of this chapter. Find the regression function $E(Y|X = x)$.

Solution: In Example 5.14 of this chapter, we have shown that for any given $x > 0$,

$$f_{Y|X}(y|x) = e^{-(y-x)} \text{ for } y \in (x, \infty).$$

It follows that

$$E(Y|x) = \int_{-\infty}^{\infty} y f_{Y|X}(y|x) dy$$

$$= \int_{x}^{\infty} y e^{-(y-x)} dy$$

$$= e^{x} \int_{x}^{\infty} y e^{-y} dy$$

$$= -e^{x} \int_{x}^{\infty} y d e^{-y}$$

$$= 1 + x.$$

Example 5.40. [**Bivariate Normal Distribution**]: Suppose (X, Y) follow a bivariate normal distribution, $BN(\mu_1, \mu_2, \sigma_1^2, \sigma_2^2, \rho)$. Then the conditional mean

$$E(Y|X) = \mu_2 + \rho \frac{\sigma_2}{\sigma_1}(X - \mu_1)$$

which is a linear function of X.

When $E(Y|X) = a + bX$, we have a linear regression equation

$$Y = a + bX + \varepsilon,$$

where $E(\varepsilon|X) = 0$.

Lemma 5.7. *Suppose* $Y = a + bX + \varepsilon$, *where* $E(\varepsilon|X) = 0$. *Then* $E(X\varepsilon) = 0$.

Proof: By the law of interacted expectations in Theorem 5.24, we have

$$E(X\varepsilon) = E[E(X\varepsilon|X)]$$

$$= E[XE(\varepsilon|X)]$$

$$= E(X \cdot 0)$$

$$= 0.$$

This implies $\text{cov}(X, \varepsilon) = 0$. (Why?)

Since ε has no information on X that can be used to predict the expected value of Y, it should be orthogonal to X. By orthogonality, we mean that ε and X are uncorrelated, i.e., $\text{cov}(X, \varepsilon) = 0$. In fact, by the law of iterated expectations, $E(\varepsilon|X) = 0$ implies that ε is orthogonal to any measurable function $h(X)$ of X, that is, $E[\varepsilon h(X)] = 0$.

There is an important difference between $E(\varepsilon|X) = 0$ and $E(X\varepsilon) = 0$. Although $E(\varepsilon|X) = 0$ implies $E(X\varepsilon) = 0$, the converse is not true. This can be seen from the example below.

Example 5.41. Suppose $\varepsilon = (X^2 - 1) + u$, where X and u are independent $N(0,1)$ random variables. Then

$$E(\varepsilon|X) = X^2 - 1 + E(u|X)$$
$$= X^2 - 1$$

where $E(u|X) = E(u) = 0$. On the other hand,

$$E(X\varepsilon) = E(X^3 - X + Xu)$$
$$= 0.$$

In this example, ε contains a predictable (nonlinear) component $X^2 - 1$ so that $E(\varepsilon|X) \neq 0$ but ε is orthogonal to X or any linear function of X. This result has a profound implication. Suppose

$$Y = a + bX + \varepsilon,$$

where $E(X\varepsilon) = 0$. Then it does not necessarily imply that the linear model $g(X) = a + bX$ is the optimal predictor for Y in terms of the mean squared error criterion, because $E(X\varepsilon) = 0$ does not imply $E(\varepsilon|X) = 0$.

The concept of the conditional mean or regression function has wide applications in economics and finance. Below are a few examples in economics and finance.

Example 5.42. [Consumption Function and Marginal Propensity to Consume]: Suppose Y is consumption, X is income, and

$$Y = \alpha + \beta X + \varepsilon,$$

where ε represents other stochastic factors that affect consumption, with $E(\varepsilon|X) = 0$. The conditional expectation

$$E(Y|X) = \alpha + \beta X$$

is called a consumption function and the derivative

$$\frac{dE(Y|X)}{dX} = \beta$$

is the expected marginal propensity to consume; that is, the expected increase in consumption when income is increased by one unit. This is the most important concept in the Keynesian macroeconomic theory.

Example 5.43. [Conditional Mean and Efficient Market Hypothesis (EMH)]: Suppose Y_t is the asset return in time period t, and I_{t-1} is the information available at time $t - 1$. Suppose one attempts to use the available information I_{t-1} to predict the asset return Y_t, but the expected return $E(Y_t|I_{t-1})$ using the information I_{t-1} is the same as the long-run market average return $E(Y_t)$. Then the information I_{t-1} has no predictive power for the future return Y_t, and we say that the asset market is efficient with respect to the information set I_{t-1}. Formally, such an hypothesis can be stated as follows:

$$E(Y_t|I_{t-1}) = E(Y_t).$$

There are three forms of the efficient market hypothesis:

- The weak form EMH, where I_{t-1} only contains the information of historical asset returns available at time $t - 1$;
- The semi-strong form EMH, where I_{t-1} contains all publicly available information at time $t - 1$;
- The strong form EMH, where I_{t-1} contains not only publicly available information but also some inside information at time $t - 1$.

Example 5.44. [Expected Shortfall and Financial Risk Management]: In Example 3.33 of Chapter 3, we have introduced the concept of value at risk of a portfolio over a certain time period. The value of risk $V_t(\alpha)$ at level α is defined as

$$P[X_t < -V_t(\alpha)|I_{t-1}] = \alpha,$$

where X_t is the return on the portfolio in time period t, and I_{t-1} is the information available at time $t - 1$. The value at risk $V_t(\alpha)$ is the threshold level which actual loss will exceed with probability α. This has been used to determine the level of capital to prepare for the occurrence of extreme adverse market events.

In practice it has been known that the risk capital level based on the value at risk may not be prudent enough. Instead, some practitioners have advocated and used the so-called expected shortfall, defined as $ES_t(\alpha) = -E[X_t|X_t < -V_t(\alpha)]$, to set the level of risk capital. The expected shortfall at level α is the expected loss given that a crisis has occurred (a crisis occurs when actual loss exceeds the value at risk). When α is small, the expected shortfall is larger than the value at risk, thus providing more prudent risk management. See Figure 5.11.

Example 5.45. [Dynamic Asset Pricing Model and Euler Equation]: In the standard consumer based asset pricing model, the represen-

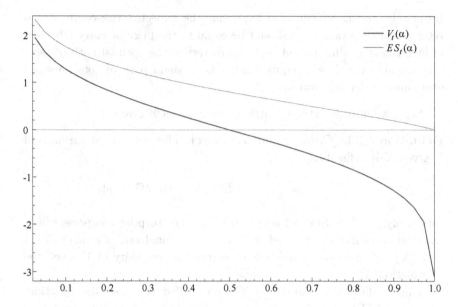

Figure 5.11: $ES_t(\alpha)$ and $V_t(\alpha)$ for $\alpha \in (0,1)$

tative agent (investor) maximizes his expected lifetime utility

$$\max_{\{C_t\}} E \sum_{j=0}^{\infty} \beta^j u(C_{t+j})$$

by choosing an optimal path of consumptions $\{C_t, t = 1, 2, \cdots\}$ subject to an intertemporal budget constraint, where β is a time discount factor, and $u(\cdot)$ is the utility function of the economic agent in each time period. The first order condition for this intertemporal maximization problem is given by

$$E(M_{t+1}R_{t+1}|I_t) = 1,$$

where $M_{t+1} = \beta u'(C_{t+1})/u'(C_t)$ is called the stochastic discount factor which reflects the risk attitude of the representative economic agent, and R_{t+1} is the gross asset return from time t to time $t+1$. This FOC is called the Euler equation, which is a conditional mean characterization. Define an expectational error

$$\varepsilon_{t+1} = M_{t+1}R_{t+1} - 1.$$

Then the Euler equation can be equivalently represented as follows:

$$E(\varepsilon_{t+1}|I_t) = 0.$$

Intuitively, the Euler equation says that the expected risk-compensated return in time period $t + 1$ should be equal to the price of unity (the cost of investment in time period t). It characterizes the optimal path of asset investments over time or equivalently, the optimal path of consumptions over times under uncertainty.

Next, we introduce the conditional variance of Y given X.

Definition 5.23. [**Conditional Variance**]**:** The conditional variance of Y given X is defined as

$$\text{var}(Y|X = x) = \int_{-\infty}^{\infty} [y - E(Y|X = x)]^2 dF_{Y|X}(y|x).$$

Intuitively, $\varepsilon = Y - E(Y|X)$ is the unexpected or surprise component of Y when one uses $E(Y|X)$ to predict Y. The conditional variance $\text{var}(Y|X) = \text{var}(\varepsilon|X)$ measures the magnitude of surprise or volatility of Y given the information X.

The conditional variance $\text{var}(Y|X)$ is called the *scedastic* function. When $\text{var}(Y|X) = \sigma^2$ is a constant, i.e., $\text{var}(Y|X)$ does not depend on X, we say that there exists conditional homoskedasticity (same variance). This is an important assumption in the classical regression analysis (see, for example, Chapter 10).

In general, the conditional variance of Y given X is a function of X, i.e., it depends on X. In this case, $\text{var}(Y|X) \neq \sigma^2$, and we say that there exists conditional heteroskedasticity. Conditional heteroskedasticity is a rule rather than exception. For example, large firms usually have large variations in output.

We now provide two important examples for conditional variance modeling.

Example 5.46. [**Interest Rate Volatility and Level Effect**]**:** It is well known that there exists an important empirical stylized fact about the interest rate volatility: the interest rate volatility depends on the interest rate level. That is, the higher the interest rate level, the higher the interest rate volatility, as displayed in Figure 5.12 for the U.S. short-term interest rates. This phenomenon is called the level effect of interest rate volatility and it is often modeled as

$$\text{var}(r_t|I_{t-1}) = \alpha r_{t-1}^{\rho},$$

where r_t is the short-term interest rate in time t, and I_{t-1} is the information available at time $t - 1$. See Chen *et al.* (1992) for more discussion.

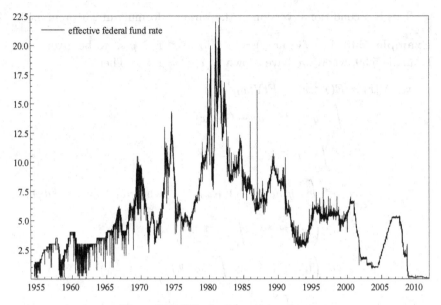

Figure 5.12: U.S. short-term interest rates

Example 5.47. [ARCH Models and Volatility Clustering]: It is often observed in financial markets that a large volatility of an asset price today tends to be followed by another large volatility tomorrow, and a small volatility today tends to be followed by another small volatility tomorrow, as shown in Figure 1.2 for S&P500 daily returns. This phenomenon is called volatility clustering (Mandelbrot 1963). To explain this stylized fact, Engle (1982) proposes a class of AutoRegressive Conditional Heteroskedasticity (ARCH) models to predict volatility. Suppose Y_t is the asset return in time t. Then the ARCH(1) model assumes

$$\text{var}(Y_t | I_{t-1}) = \alpha + \beta Y_{t-1}^2,$$

where $\alpha, \beta > 0$, and I_{t-1} contains information on all past asset returns.

We now state a formula which provides a convenient way to calculate the conditional variance.

Lemma 5.8.

$$var(Y|X) = E(Y^2|X) - [E(Y|X)]^2.$$

Proof: This is left as an exercise.

This is a conditional version of the variance formula in Theorem 3.13.

Example 5.48. Let $f_{XY}(x, y) = e^{-y}$ for $0 < x < y < \infty$ be given as in Example 5.39, where we have shown $E(Y|x) = 1 + x$. Then

$$
\begin{aligned}
\operatorname{var}(Y|x) &= E(Y^2|x) - [E(Y|x)]^2 \\
&= \int_x^\infty y^2 e^{-(y-x)} dy - (1+x)^2 \\
&= e^x \int_x^\infty y^2 e^{-y} dy - (1+x)^2 \\
&= -e^x \int_x^\infty y^2 de^{-y} - (1+x)^2 \qquad \text{(where } de^{-y} = -e^{-y} dy\text{)} \\
&= -e^x \left(y^2 e^{-y}\big|_x^\infty - \int_x^\infty e^{-y} dy^2 \right) - (1+x)^2 \\
&= -e^x \left(0 - x^2 e^{-x} - 2\int_x^\infty y e^{-y} dy \right) - (1+x)^2 \\
&= x^2 + 2e^x \int_x^\infty y e^{-y} dy - (1+x)^2 \\
&= x^2 + 2\int_x^\infty y e^{-(y-x)} dy - (1+x)^2 \\
&= x^2 + 2(1+x) - (1+x)^2 \\
&= 1.
\end{aligned}
$$

There exists conditional homoskedasticity in this example.

Example 5.49. [Bivariate Normal Distribution and Conditional Homoskedasticity]: Suppose X and Y follow a bivariate normal distribution $BN(\mu_1, \mu_2, \sigma_1^2, \sigma_2^2, \rho)$. Show the conditional variance $\operatorname{var}(Y|X) = \sigma_2^2(1 - \rho^2)$.

Solution: In Section 5.6, we have shown that for the bivariate normal distribution, the conditional distribution of Y given X is a normal distribution with mean $\mu_2 + \frac{\rho\sigma_2}{\sigma_1}(X - \mu_1)$ and variance $\sigma_2^2(1 - \rho^2)$. Therefore, $\operatorname{var}(Y|X) = \sigma_2^2(1 - \rho^2)$, which does not depend on X.

We now provide some examples of conditional heteroskedasticity.

Example 5.50. Suppose $Y = Z\sqrt{1 + X^2}$, where Z is a random variable with mean 0 and variance 1, and is independent of X. Find: (1) $E(Y|X)$; (2) $\operatorname{var}(Y|X)$.

Solution: (1)
$$E(Y|X) = E(Z\sqrt{1+X^2}|X)$$
$$= \sqrt{1+X^2}E(Z|X)$$
$$= \sqrt{1+X^2}E(Z)$$
$$= 0 = E(Y).$$

(2)
$$\text{var}(Y|X) = E(Y^2|X) - [E(Y|X)]^2$$
$$= E(Y^2|X)$$
$$= E[Z^2(1+X^2)|X]$$
$$= (1+X^2)E(Z^2|X)$$
$$= (1+X^2)E(Z^2)$$
$$= 1+X^2.$$

Example 5.51. [Random Coefficient Model]: Suppose
$$Y = \alpha + \beta X + u_3$$
$$= (\alpha_0 + u_1) + (\beta_0 + u_2)X + u_3,$$

where u_1, u_2, u_2, X are jointly independent, and $E(u_i) = 0$ for $i = 1, 2, 3$. This is called a random coefficient model. Find: (1) $E(Y|X)$; (2) $\text{var}(Y|X)$.

Solution: (1) We first write
$$Y = \alpha_0 + \beta_0 X + \varepsilon,$$
where
$$\varepsilon = u_1 + u_2 X + u_3.$$

Since $E(\varepsilon|X) = 0$, we have
$$E(Y|X) = \alpha_0 + \beta_0 X.$$

(2)
$$\text{var}(Y|X) = \text{var}(\varepsilon|X)$$
$$= \sigma_{u_1}^2 + \sigma_{u_2}^2 X^2 + \sigma_{u_3}^2$$
$$= (\sigma_{u_1}^2 + \sigma_{u_3}^2) + \sigma_{u_2}^2 X^2.$$

Theorem 5.27. [*Variance Decomposition*]: *For any two random variables X and Y with finite second moments,*
$$\text{var}(Y) = \text{var}[E(Y|X)] + E[\text{var}(Y|X)].$$

Proof: Putting $g_0(X) = E(Y|X)$, we have $Y = g_0(X) + \varepsilon$. It follows that

$$
\begin{aligned}
\text{var}(Y) &= \text{var}[g_0(X) + \varepsilon] \\
&= \text{var}[g_0(X)] + \text{var}(\varepsilon) + 2\text{cov}[g_0(X), \varepsilon] \\
&= \text{var}[g_0(X)] + \text{var}(\varepsilon),
\end{aligned}
$$

where

$$
\begin{aligned}
\text{cov}[g_0(X), \varepsilon] &= E[g_0(X)\varepsilon] - E[g_0(X)]E(\varepsilon) \\
&= 0
\end{aligned}
$$

by the law of iterated expectations in Theorem 5.24 and $E(\varepsilon|X) = 0$. Since the first term $\text{var}[g_0(X)] = \text{var}[E(Y|X)]$, it suffices to show $\text{var}(\varepsilon) = E[\text{var}(Y|X)]$. By $E(\varepsilon|X) = 0$ and the law of iterated expectations, we have $E(\varepsilon) = 0$ and

$$
\begin{aligned}
\text{var}(\varepsilon) &= E(\varepsilon^2) \\
&= E[E(\varepsilon^2|X)] \\
&= E[\text{var}(Y|X)],
\end{aligned}
$$

where $\text{var}(Y|X) = E(\varepsilon^2|X)$ by definition.

Intuitively, the variance decomposition theorem states that the total variation of Y, $\text{var}(Y)$, is equal to the sum of two components: the first is the variability of the best MSE predictor $E(Y|X)$, which measures how well $E(Y|X)$ can predict Y. The more $E(Y|X)$ can vary, the better it can predict Y. The second component, $E[\text{var}(Y|X)]$, is the averaged squared prediction error $Y - E(Y|X)$. Since $\text{var}(Y)$ is a constant, increasing the variability of the best predictor $E(Y|X)$ will decrease the average of squared prediction errors. In the best case, when $Y = g(X)$ for some measurable function $g(\cdot)$, we have $E(Y|X) = Y$. In this case, $E(Y|X)$ can perfectly predict the variation of Y, and there is no prediction error. On the other hand, if X is independent of Y, then we have $E(Y|X) = E(Y)$, a constant. In this case, there is no variation in $E(Y|X)$, and the mean squared prediction error achieves its maximum value $\text{var}(Y)$.

In econometrics, the first two conditional moments are most important. However, there has been increasing interest in higher order conditional moments in economic and financial applications. Below are two examples:

- Conditional Skewness:

$$
S(Y|X) = \frac{E(\varepsilon^3|X)}{[\text{var}(\varepsilon|X)]^{3/2}}.
$$

- Conditional Kurtosis:

$$K(Y|X) = \frac{E(\varepsilon^4|X)}{[\text{var}(\varepsilon|X)]^2}.$$

Example 5.52. Suppose (X, Y) follow a bivariate normal distribution $N(\mu_1, \mu_2, \sigma_1^2, \sigma_2^2, \rho)$. Find $E(Y|X)$, $\text{var}(Y|X)$, $S(Y|X)$ and $K(Y|X)$.

Solution: We have shown that the conditional distribution of Y given X is a normal distribution

$$Y|X \sim N\left[\mu_2 + \frac{\rho\sigma_2}{\sigma_1}(X - \mu_1), \sigma_2^2(1 - \rho^2)\right].$$

It follows that the conditional mean

$$E(Y|X) = \mu_2 + \frac{\sigma_2}{\sigma_1}\rho(X - \mu_1),$$

the conditional variance

$$\text{var}(Y|X) = \sigma_2^2(1 - \rho^2),$$

the conditional skewness

$$S(Y|X) = 0,$$

and the conditional kurtosis

$$K(Y|X) = 3.$$

This example shows that for the bivariate normal distribution, only the first conditional moment $E(Y|X)$ depends on X, all other conditional higher order comments are constant, i.e., they do not depend on X.

On the other hand, it may be possible that the lower order conditional moments are constant, yet the higher order conditional moments will depend on X.

Example 5.53. Suppose the conditional distribution of Y given X is a Lognormal$(0, X^2)$ distribution. Then conditional on X, $\ln(Y)$ follows a $N(0, X^2)$ distribution. It follows that

$$E(Y^k|X) = e^{\frac{k^2}{2}X^2}, \qquad k = 1, 2, \cdots.$$

Put $\mu(X) = E(Y|X)$ and $\sigma^2(X) = \text{var}(Y|X)$. Define the standardized random variable

$$Z = \frac{Y - \mu(X)}{\sigma(X)}.$$

It can be shown that $E(Z|X) = 0$, $\text{var}(Z|X) = 1$, but $E(Z^3|X)$ is a function of X (please check it!). In this example, the first two conditional moments of Z do not depend on X, but all other higher order conditional moments of Z are functions of X.

In the practice of econometric modeling, there is an important question, namely, which conditional moment should be used in practice?

This depends on the economic application we have in mind. For some applications such as study on the market efficiency hypothesis and dynamic asset pricing, we need to model the conditional mean of economic variables. For applications such as studies on volatility spillover, we need to model the conditional variance. For applications such as financial risk management, hedging, and derivatives pricing, we need to model higher order conditional moments or even the entire conditional distribution.

More generally, we can consider the problem of decision-making under uncertainty. Suppose an economic agent has a loss function $l(e)$ and he is making an optimal decision a to minimize the expected loss conditional on the available information $X = x$ he has. The minimization problem is

$$\min_a E\left[l(Y - a)|X = x\right] = \min_a \int_{-\infty}^{\infty} l(y - a) f_{Y|X}(y|x) dx,$$

where Y is a random payoff which is unknown when the economic agent is making a decision, and $f_{Y|X}(y|x)$ is the conditional PDF of Y given $X = x$. In practice, the conditional PDF $f_{Y|X}(y|x)$ is usually unknown and need to be modeled. When the loss function is a quadratic loss, i.e.,

$$l(e) = e^2,$$

the optimal decision is the conditional mean:

$$a^*(X) = E(Y|X).$$

When the loss function is a so-called linexp function

$$l(e) = \frac{1}{\alpha^2}\left[e^{\alpha e} - (1 + \alpha e)\right],$$

and the conditional distribution of Y given X is a normal distribution, the optimal decision is a linear combination of the conditional mean and conditional variance:

$$a^*(X) = E(Y|X) + \frac{\alpha}{2}\text{var}(Y|X).$$

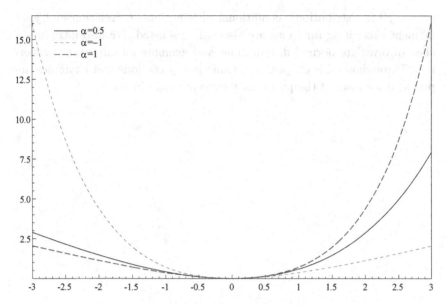

Figure 5.13: The linexp loss function for various values of α

The linexp loss function is asymmetric except for $\alpha = 0$, which yields the squared loss function. See Figure 5.13 for the shapes of the linexp loss function for various values of α.

More generally, if the loss function $l(e)$ is a generic loss function, then the optimal decision $a^*(X)$ will depend on the entire conditional distribution of Y given X. In this case, one need to model the conditional PDF $f_{Y|X}(y|x)$; the first few conditional moments are not sufficient. For more discussion, see Granger (1999).

5.11 Conclusion

The most important goal in economic analysis and econometric analysis is to identify various economic relationships. In this chapter, we first characterize the joint probability distribution of two random variables X and Y by using the joint CDF, the joint PMF when (X, Y) are discrete, the joint PDF when (X, Y) are continuous, as well as their conditional counterparts. We also consider the methods of deriving the joint distribution of a bivariate transformation. We then examine the predictive relationships between X and Y using the conditional distributions, correlation, and conditional expectations. The concept of independence and its implications on the joint distributions, conditional distributions, correlation and joint moment generating functions are also fully discussed. We also introduce a class of bivariate normal distributions and examine its important properties. Throughout this chapter, economic interpretations and examples are provided for most of the probability concepts introduced.

EXERCISE 5

5.1. A joint PDF is defined by

$$f_{XY}(x,y) = \begin{cases} c(x+2y), & \text{if } 0 < y < 1 \text{ and } 0 < x < 2, \\ 0, & \text{otherwise.} \end{cases}$$

Find: (1) the value of c;
     (2) the marginal PDF of X;
     (3) the joint CDF of X and Y;
     (4) the PDF of the random variable $Z = 9/(X+1)^2$.

5.2. Suppose (X, Y) has a joint PDF

$$f_{XY}(x,y) = \begin{cases} 1 + \theta x, & \text{if } -y < x < y, 0 < y < 1, \\ 0, & \text{otherwise,} \end{cases}$$

where θ is a constant.

(1) Determine the possible value(s) of θ so that $f_{XY}(x,y)$ is a joint PDF. Give your reasoning.

(2) Let $\theta = 0$. Check if X and Y are independent. Give your reasoning.

5.3. For m random variables $X_i, i = 1, \cdots, m$, denote $F_{X_i}(x_i)$ as the marginal distribution of X_i and $F(x_1, \cdots, x_m)$ as the joint distribution of $(X_1, \cdots, X_m)$. Then the function $C : [0,1]^m \to [0,1]$ that satisfies

$$F(x_1, \cdots, x_m) = C[F_{X_1}(x_1), \cdots, F_{X_m}(x_m)], \quad x_j \in (-\infty, \infty),$$

is called the copula associated with $F(x_1, \cdots, x_m)$. Copula contains all information about dependence among components of a random vector but no information about the marginal distributions. Suppose $X = (X_1, \cdots, X_m)$ with joint CDF $F_X(x) = P(X_1 \leq x_1, \cdots, X_m \leq x_m)$ and marginal CDFs $F_{X_i}(x_i) = P(X_i \leq x_i)$, where $x = (x_1, \cdots, x_m)$. Assume that $F_{X_i}(\cdot)$ is strictly increasing for all $i = 1, \cdots, m$. Show:

(1) the copula of X is given by $C_X(u) = F_X[F_{X_1}^{-1}(u_1), \cdots, F_{X_m}^{-1}(u_m)]$;

(2) suppose $Y_i = g_i(X_i)$, where $g_i(\cdot)$ is strictly monotonically increasing, $i = 1, \cdots, m$. Show $C_X(u) = C_Y(u)$ for all u. That is, the copula is invariant to any strictly increasing transformation of the random variables.

5.4. (1) Find $P(X > \sqrt{Y})$ if X and Y are jointly distributed with PDF $f_{XY}(x,y) = x + y$ for $0 \leq x \leq 1, 0 \leq y \leq 1$.

(2) Find $P(X^2 < Y < X)$ if X and Y are jointly distributed with PDF $f_{XY}(x,y) = 2x$ for $0 \leq x \leq 1, 0 \leq y \leq 1$.

5.5. Prove that if the joint CDF of X and Y satisfies $F_{XY}(x,y) = F_X(x)F_Y(y)$, that is, if X and Y are independent, then for any pair of intervals (a,b) and (c,d), $P(a \leq X \leq b, c \leq Y \leq d) = P(a \leq X \leq b)P(c \leq Y \leq d)$.

5.6. The random pair (X,Y) has the joint probability distribution

			X	
		1	2	3
	2	$\frac{1}{12}$	$\frac{1}{6}$	$\frac{1}{12}$
Y	3	$\frac{1}{6}$	0	$\frac{1}{6}$
	4	0	$\frac{1}{3}$	0

(1) Show that X and Y are dependent.

(2) Give a probability table for random variables U and V that have the same marginals as X and Y but are independent.

5.7. Suppose X and Y are independent $N(0,1)$ random variables. Find:
  (1) $P(X^2 + Y^2 < 1)$;
  (2) $P(X^2 < 1)$, after verifying that X^2 is distributed as χ_1^2.

5.8. Let X be an Exponential(1) random variable, and define Y to be the integer part of $X + 1$, that is, $Y = i + 1$ if and only if $i \leq X < i + 1$, where $i = 0, 1, \cdots$.

  (1) Find the distribution of Y. What well-known distribution does Y have?

  (2) Find the conditional distribution of $X - 4$ given $Y \geq 5$.

5.9. Suppose $g(x) \geq 0$ and $\int_0^\infty g(x)dx = 1$, show that $f(x,y) = \frac{2g(\sqrt{x^2+y^2})}{\pi\sqrt{x^2+y^2}}$, for $x, y > 0$, is a joint PDF.

5.10. Suppose (X,Y) has a joint PDF

$$f(x,y) = e^{-y} \text{ for } 0 < x < y < \infty.$$

Find: (1) $f_X(x)$; (2) $f_Y(y)$; (3) $f_{X|Y}(x|y)$; (4) $f_{Y|X}(y|x)$; (5) are X and Y independent?

5.11. (X,Y) follow a bivariate normal distribution if their joint PDF

$$f_{XY}(x,y) = \frac{1}{2\pi\sigma_1\sigma_2\sqrt{1-\rho^2}}e^{-\frac{1}{2(1-\rho^2)}\left[\left(\frac{x-\mu_1}{\sigma_1}\right)^2 - 2\rho\left(\frac{x-\mu_1}{\sigma_1}\right)\left(\frac{y-\mu_2}{\sigma_2}\right) + \left(\frac{y-\mu_2}{\sigma_2}\right)^2\right]},$$

where $-\infty < \mu_1, \mu_2 < \infty, 0 < \sigma_1, \sigma_2 < \infty, -1 \leq \rho \leq 1$. Find: (1) $f_X(x)$; (2) $f_Y(y)$; (3) $f_{Y|X}(y|x)$; (4) $f_{X|Y}(x|y)$; (5) the conditions on parameters

$(\mu_1, \mu_2, \sigma_1, \sigma_2, \rho)$ under which X and Y will be independent. [Hint: When finding $f_X(x)$, you can form a term with form

$$z^2 = \left[\left(\frac{y - \mu_2}{\sigma_2} \right) - \rho \left(\frac{x - \mu_1}{\sigma_1} \right) \right]^2$$

and integrate it out first.]

5.12. Let X be $N(0, \sigma^2)$. Show that the CDF of the conditional distribution of X given $X > c$ is

$$F_{X|X>c}(x) = \frac{\Phi(x/\sigma) - \Phi(c/\sigma)}{1 - \Phi(c/\sigma)}, \; x > c$$

and the PDF of this distribution is

$$f_{X|X>c}(x) = \frac{\phi(x/\sigma)}{\sigma[1 - \Phi(c/\sigma)]}, \qquad x > c,$$

where $\phi(x)$ and $\Phi(x)$ are the PDF and CDF of $N(0, 1)$. Such a distribution is called a truncated distribution.

5.13. Suppose the random variables X and Y have the following joint PDF

$$f(x, y) = \begin{cases} 8xy & \text{for } 0 \le x \le y \le 1, \\ 0 & \text{otherwise.} \end{cases}$$

Also, let $U = X/Y$ and $V = Y$. Determine the joint PDF of U and V.

5.14. (1) Let X_1 and X_2 be independent $N(0,1)$ random variables. Find the PDF of $(X_1 - X_2)^2/2$.

(2) If $X_i, i = 1, 2$, are independent Gamma$(\alpha_i, 1)$ random variables, find the marginal distributions of $X_1/(X_1 + X_2)$ and $X_2/(X_1 + X_2)$.

5.15. Suppose X_1, X_2 are independent standard Gamma random variables, with possibly different parameters α_1, α_2. Show:

(1) the random variables

$$X_1 + X_2 \text{ and } \frac{X_1}{X_1 + X_2}$$

are mutually independent;

(2) the distribution of $X_1 + X_2$ is a standard Gamma with $\alpha = \alpha_1 + \alpha_2$;

(3) the distribution of $X_1/(X_1 + X_2)$ is a Beta distribution with parameters α_1, α_2.

5.16. Suppose X_1 and X_2 are independent $N(0, \sigma^2)$ random variables.

(1) Find the joint distribution of Y_1 and Y_2, where $Y_1 = X_1^2 + X_2^2$ and $Y_2 = X_1/\sqrt{Y_1}$.

(2) Show that Y_1 and Y_2 are independent, and interpret this result geometrically.

5.17. Let $X \sim \text{Beta}(\alpha, \beta)$, and $Y \sim \text{Beta}(\alpha + \beta, \gamma)$ be independent random variables. Find the distribution of XY by making the transformation given in (1) and (2) and integrating out V:
  (1) $U = XY, V = Y$;
  (2) $U = XY, V = X/Y$.

5.18. Let $X \sim N(\mu, \sigma^2)$, and let $Y \sim N(\gamma, \sigma^2)$. Suppose X and Y are independent. Define $U = X + Y$ and $V = X - Y$. Show that U and V are independent normal random variables. Find the distribution of each of them.

5.19. Show that (1) if $X_1 \sim N(0, \sigma_1^2), X_2 \sim N(0, \sigma_2^2)$, and X_1 and X_2 are independent, then $X_1 X_2 / \sqrt{X_1^2 + X_2^2}$ is normally distributed; (2) if in addition $\sigma_1^2 = \sigma_2^2$, then $(X_1^2 - X_2^2)/(X_1^2 + X_2^2)$ is normally distributed.

5.20. Suppose $X_1 \sim N(0, 1), X_2 \sim N(0, 1)$ and X_1 and X_2 are independent. Find the distribution of (1) X_1/X_2; and (2) $X_1/|X_2|$ respectively.

5.21. Let Z_1, Z_2 be independent standard normal random variables. Define
$$X = \mu_1 + aZ_1 + bZ_2,$$
$$Y = \mu_2 + cZ_1 + dZ_2,$$
where constants a, b, c, d satisfy the restrictions that
$$a^2 + b^2 = \sigma_1^2,$$
$$c^2 + d^2 = \sigma_2^2,$$
$$ac + bd = \rho\sigma_1\sigma_2.$$
Show that $(X, Y) \sim BN(\mu_1, \mu_2, \sigma_1^2, \sigma_2^2, \rho)$.

5.22. Let X and Y be two independent uniform random variables on $[0, 1]$. Show that the random variables $U = \cos(2\pi X)\sqrt{-2\ln Y}$ and $V = \sin(2\pi X)\sqrt{-2\ln Y}$ are independent standard normal random variables.

5.23. Find the PDF of $X - Y$, where $X \sim U[0, 1], Y \sim U[0, 1]$, and X and Y are independent.

5.24. Suppose $X_1 \sim \text{Cauchy}(0, 1), X_2 \sim \text{Cauchy}(0, 1)$, and X_1 and X_2 are independent. Show that $aX_1 + bX_2$ has a Cauchy distribution.

5.25. Suppose $X_1 \sim \text{Gamma}(\alpha_1, 1), X_2 \sim \text{Gamma}(\alpha_2, 1)$, and X_1 and X_2 are independent. Show that $X_1 + X_2$ and $X_1/(X_1 + X_2)$ are independent. Also, find the marginal distributions of $X_1 + X_2$ and $X_1/(X_1 + X_2)$, respectively.

5.26. Let $U = a_1 + b_1 X$ and $V = a_2 + b_2 Y$, where (X, Y) have the joint PDF $f_{XY}(x, y)$, and $a_1, a_2, b_1 > 0, b_2 > 0$ are given constants. Find the joint PDF of (U, V).

5.27. Suppose the PDF of X_i is

$$\frac{1}{\sigma_i} f\left(\frac{x - \theta_i}{\sigma_i}\right), \qquad i = 1, 2,$$

and X_1, X_2 are independent. Show that the PDF of $X_1 + X_2$ is of the form of

$$\frac{1}{\sigma} f\left(\frac{x - \theta}{\sigma}\right)$$

for some σ and θ.

5.28. Suppose X_1, X_2, X_3 have a continuous joint PDF $f(x, y, z)$. Define $Y_1 = F_1(X_1), Y_2 = F_2(X_1, X_2)$ and $Y_3 = F_3(X_1, X_2, X_3)$, where

$$F_1(x) = P(X_1 \leq x),$$
$$F_2(x_1, x_2) = P(X_2 \leq x_2 | X_1 = x_1),$$
$$F_3(x_1, x_2, x_3) = P(X_3 \leq x_3 | X_2 = x_2, X_1 = x_1).$$

Show that Y_1, Y_2, Y_3 are mutually independent and each of them is uniformly distributed over [0,1]. [Hint: Define $f_{X_2 | X_1}(x_2 | x_1) = \frac{f_{X_1 X_2}(x_1, x_2)}{f_{X_1}(x_1)}$. Then $F_2(x_1, x_2) = \int_{-\infty}^{x_2} f_{X_2 | X_1}(y | x_1) dy$.]

5.29. Suppose the distribution of Y, conditional on $X = x$, is $N(x, x^2)$ and that the marginal distribution of X is uniform(0,1).

(1) Find $E(Y)$, $\mathrm{var}(Y)$, and $\mathrm{cov}(X, Y)$.

(2) Prove that Y/X and X are independent.

5.30. Consider two random variables (X, Y). Suppose X is uniformly distributed over $(-1, 1)$, that is, the PDF of X is

$$f_X(x) = \begin{cases} \frac{1}{2}, & -1 < x < 1, \\ 0, & \text{otherwise.} \end{cases}$$

Also, the conditional PDF of Y give $X = x$ is

$$f_{Y|X}(y|x) = \frac{1}{\sqrt{2\pi}} e^{-\frac{(y - \alpha - \beta x)^2}{2}} \qquad \text{for } -\infty < y < \infty \text{ and } -1 < x < 1.$$

Find: (1) $E(Y)$; (2) $\mathrm{cov}(X, Y)$.

5.31. Let X and Y be random variables with finite expectations. Show that if $P(X \leq Y) = 1$, then $E(X) \leq E(Y)$.

5.32. Let X and Y be two independent standard exponential random variables. Find $E[\max(X, Y)]$.

5.33. Let X and Y be nonnegative random variables with an arbitrary joint probability distribution function. Let $\mathbf{1}(x, y)$ be 1 if $X > x$ and $Y > y$, and 0 otherwise.

(1) Show $\int_0^\infty \int_0^\infty \mathbf{1}(x, y)dxdy = XY$.

(2) By calculating the expected values of both sides in part (1), show that $E(XY) = \int_0^\infty \int_0^\infty P(X > x, Y > y)dxdy$.

5.34. A generalization of the Beta distribution is the Dirichlet distribution. In its bivariate version, (X, Y) have a joint PDF $f_{XY}(x, y) = kx^{a-1}y^{b-1}(1 - x - y)^{c-1}$, $0 < x < 1$, $0 < y < 1$, $0 < y < 1 - x < 1$, where $a > 0$, $b > 0$, and $c > 0$ are constants.

(1) Show that $k = \frac{\Gamma(a+b+c)}{\Gamma(a)\Gamma(b)\Gamma(c)}$.

(2) Show that, marginally, both X and Y follow a Beta distribution.

(3) Find the conditional distribution of $Y|X = x$, and show that $Y|(1 - X)$ follows a Beta(b, c) distribution.

(4) Show that $E(XY) = \frac{ab}{(a+b+c+1)(a+b+c)}$, and find the covariance $\text{cov}(X, Y)$.

5.35. Let X_1, X_2 and X_3 be uncorrelated random variables, each with mean μ and variance σ^2. Find, in terms of μ and σ^2, $\text{cov}(X_1 + X_2, X_2 + X_3)$ and $\text{cov}(X_1 + X_2, X_1 - X_2)$.

5.36. Suppose (X, Y) is a bivariate random vector with means μ_X and μ_Y, and variances σ_X^2 and σ_Y^2. Let $U = X + Y$ and $V = X - Y$. Show that U and V are uncorrelated if and only if $\sigma_X^2 = \sigma_Y^2$.

5.37. Let $g(\cdot)$ and $h(\cdot)$ be two PDFs with CDFs $G(\cdot)$ and $H(\cdot)$, respectively. Show that for $-1 \le \alpha \le 1$, the function

$$f(x, y) = g(x)h(y)\left\{1 + \alpha\left[2G(x) - 1\right]\left[2H(y) - 1\right]\right\}$$

is a joint PDF of two random variables. Moreover, show that $g(\cdot)$ and $h(\cdot)$ are the marginal PDFs of the joint PDF $f(x, y)$.

5.38. Suppose X, Y have a joint PMF

$$f_{XY}(x, y) = \begin{cases} \frac{1}{3}, & (x, y) = (-1, 1), (0, 0), (1, 1), \\ 0, & \text{otherwise.} \end{cases}$$

(1) Find $\text{cov}(X, Y)$. (2) Are X, Y independent? Give your reasoning.

5.39. Suppose f_1, f_2, g_1, g_2 are univariate densities with means $\mu_1, \mu_2, \zeta_1,$ ζ_2, respectively, and random variables (X, Y) have a joint PDF

$$f_{X,Y}(x, y) = af_1(x)g_1(y) + (1 - a)f_2(x)g_2(y),$$

where $0 < a < 1$ is known.

(1) Show that the marginal distributions are given by $f_X(x) = af_1(x) +$ $(1 - a)f_2(x)$ and $f_Y(y) = ag_1(y) + (1 - a)g_2(y)$.

(2) Show that X and Y are independent if and only if $[f_1(x) - f_2(x)][g_1(y) - g_2(y)] = 0$.

(3) Show that $\text{cov}(X, Y) = a(1 - a)(\mu_1 - \mu_2)(\zeta_1 - \zeta_2)$, and thus explain how to construct dependent uncorrelated random variables.

(4) Letting f_1, f_2, g_1, g_2 be binomial PMFs, give examples of combinations of parameters that lead to independent (X, Y) pairs, correlated (X, Y) pairs, and uncorrelated but dependent (X, Y) pairs.

5.40. Let Z_1 and Z_2 are two independent $N(0, 1)$ random variables. Define

$$X = \mu_1 + aZ_1 + bZ_2,$$
$$Y = \mu_2 + cZ_1 + dZ_2,$$

where constants a, b, c, d satisfy the restrictions that

$$a^2 + b^2 = \sigma_1^2,$$
$$c^2 + d^2 = \sigma_2^2,$$
$$ac + bd = \rho\sigma_1\sigma_2.$$

Show that (X, Y) follow a bivariate normal distribution $BN(\mu_1, \mu_2, \sigma_1^2, \sigma_2^2, \rho)$.

5.41. Suppose (X, Y) have a bivariate normal PDF

$$f_{XY}(x, y) = \frac{1}{2\pi\sqrt{1 - \rho^2}}e^{-\frac{1}{2(1-\rho^2)}(x^2 - 2\rho xy + y^2)}.$$

Show that $\text{corr}(X, Y) = \rho$, and $\text{corr}(X^2, Y^2) = \rho^2$. [Hint: Conditional expectation will simplify calculations.]

5.42. Suppose (X, Y) follow a standard bivariate normal distribution with correlation coefficient ρ. Define $U = (Y - \rho X)/\sqrt{1 - \rho^2}$. Show that U is normally distributed and independent of X.

5.43. Show $\text{var}(Y|X) = E(Y^2|X) - [E(Y|X)]^2$.

5.44. Suppose the joint PDF of (X, Y) is a uniform PDF on the circle $x^2 + y^2 \leq 1$. Find: (1) $E(Y|X)$; (2) $\text{var}(Y|X)$.

5.45. Suppose (X, Y) have a joint normal distribution $N(\mu_1, \mu_2, \sigma_1^2, \sigma_2^2, \rho)$. Find: (1) $E(Y|X)$; (2) $\text{var}(Y|X)$.

5.46. Show that if $X = (X_1, \cdots, X_m)'$ follows a multivariate normal distribution with mean vector $\mu = E(X) = (\mu_1, \cdots, \mu_m)'$ and variance-covariance matrix $\Sigma = \text{cov}(X, X)$, then for any $\lambda = (\lambda_1, \cdots, \lambda_m)'$ with $\lambda'\lambda = 1$, $\lambda'X$, has a normal distribution with mean $\lambda'\mu$ and variance $\lambda'\Sigma\lambda$.

5.47. Suppose X and Y are random variables such that $E(Y|X) = 7 - \frac{1}{4}X$ and $E(X|Y) = 10 - Y$. Determine the correlation between X and Y.

5.48. Suppose $E(Y|X) = 1 + 2X$ and $\text{var}(X) = 2$. Find $\text{cov}(X, Y)$.

5.49. Suppose $Y = \alpha_0 + \alpha_1 X + \varepsilon\sqrt{\beta_0 + \beta_1 X^2}$, where ε and X are mutually independent random variables, with $E(\varepsilon) = 0$ and $\text{var}(\varepsilon) = 1$. Find: (1) $E(Y|X)$; (2) $\text{var}(Y|X)$.

5.50. Suppose (X, Y) have a joint distribution such that $E(Y^2) < \infty$ and $0 < \text{var}(X) < \infty$. Let

$$\mathbb{A} = \{g : \mathbb{R} \to \mathbb{R} | g(x) = \alpha + \beta x, \quad -\infty < \alpha, \beta < \infty\}$$

be a class of linear functions.

(1) Show that $g^*(X) = \alpha^* + \beta^* X$ is the optimal solution to $\min_{g \in \mathbb{A}} E\left[Y - g(x)\right]^2$ if and only if $E(u^*) = 0$ and $E(Xu^*) = 0$, where $u^* = Y - g^*(X)$.

(2) Find the expressions for α^* and β^* in terms of $\mu_X, \sigma_X^2, \mu_Y, \sigma_Y^2$ and $\text{cov}(X, Y)$.

5.51. Let X and Y be two random variables and $0 < \sigma_X^2 < \infty$. Show that if $E(Y|X) = \alpha_o + \alpha_1 X$, then $\alpha_1 = \text{cov}(X, Y)/\sigma_X^2$.

5.52. Suppose $E(Y|X)$ is a linear function of X, i.e., $E(Y|X) = a + bX$ for some constants a, b. Find the expressions of a and b in terms of $\mu_X, \sigma_X^2, \mu_Y, \sigma_Y^2$ and $\text{cov}(X, Y)$. Is it true that when $E(Y|X)$ is linear in X, then $E(Y|X)$ does not depend on X if and only if $\text{cov}(X, Y) = 0$? Give your reasoning.

5.53. Two random variables X and Y have a joint PDF

$$f_{XY}(x, y) = \frac{1}{\alpha + \beta x} e^{-\frac{y}{\alpha + \beta x}} \text{ for } 0 < y < \infty, 0 < x < 1,$$

where $0 < \alpha < \infty, 0 < \beta < \infty$ are two given constants.

(1) Find the conditional PDF $f_{Y|X}(y|x)$.

(2) Find the conditional mean $E(Y|X)$.

(3) Are (X, Y) independent? Give your reasoning.

5.54. Suppose (X, Y) have a joint PDF

$$f_{XY}(x, y) = \begin{cases} xe^{-y}, & \text{if } 0 < x < y < \infty, \\ 0, & \text{otherwise.} \end{cases}$$

(1) Find the conditional PDF $f_{Y|X}(y|x)$ of Y given $X = x$.
(2) Find the conditional mean $E(Y|x)$.
(3) Find the conditional variance $\text{var}(Y|x)$.
(4) Are X and Y independent? Give your reasoning.

5.55. Suppose (X, Y) is a bivariate random vector whose conditional mean $E(Y|X)$ has the form

$$E(Y|X) = \alpha_0 + \alpha_1 X + \alpha_2 X^2,$$

where $X \sim N(0, 1)$, and $\alpha_0, \alpha_1, \alpha_2$ are some constants.

(1) Find the mean of Y.
(2) If $\alpha_1 = \alpha_2 = 0$, is it always true that $\text{cov}(X, Y) = 0$? Give your reasoning.
(3) If $\text{cov}(X, Y) = 0$, is it always true that $\alpha_1 = \alpha_2 = 0$? Give your reasoning.

5.56. Suppose (X, Y) is a bivariate vector whose conditional mean $E(Y|X)$ has the form

$$E(Y|x) = \alpha_0 + \alpha_1 x + \alpha_2 x^2,$$

where $E(X) = 0$, $V(X) > 0$, and $\alpha_o, \alpha_1, \alpha_2$ are some constants.

(1) If $E(Y|x) = \alpha_0$ for all x, is it always true that $\text{cov}(X, Y) = 0$? Give your reasoning.
(2) If $\text{cov}(X, Y) = 0$, is it always true that $E(Y|x) = \alpha_0$ for all x? Give your reasoning.

5.57. For any two random variables X and Y with finite variances, show

(1) $\text{cov}(X, Y) = \text{cov}(X, E(Y|X))$;
(2) X and $Y - E(Y|X)$ are uncorrelated;
(3) $\text{var}[Y - E(Y|X)] = E[\text{var}(Y|X)]$.

5.58. (1) Suppose $E(Y|X) = E(Y)$. Show $\text{cov}(X, Y) = 0$. (2) Does $\text{cov}(X, Y) = 0$ imply $E(Y|X) = E(Y)$. If yes, prove it. If not, provide an example.

5.59. Suppose Y is a Bernoulli random variable, and X is a random variable. Show that Y and X are independent if and only if $E(Y|X) = E(Y)$.

5.60. Suppose $E(Y|X) = \alpha + \beta X$ and $\text{var}(Y|X) = \sigma^2$. Show $\text{var}(Y|X) = \sigma^2(1 - \rho_{XY}^2)$.

5.61. Suppose $X_1, \cdots, X_n$ are n IID random variables with variance σ^2. Denote $\bar{X}_n = n^{-1}\Sigma_{i=1}^n X_i$. Prove that for any integers i and j, $1 \le i < j \le n$, the correlation coefficient between $X_i - \bar{X}_n$ and $X_j - \bar{X}_n$ is $-\frac{1}{n-1}$.

5.62. Suppose X has a PDF

$$f(x) = \begin{cases} |x|, & -1 < x < 1, \\ 0, & \text{otherwise.} \end{cases}$$

Let $Y = X^2$.

　(1) Find $\text{cov}(X, Y)$.

　(2) Are X and Y independent? Explain.

5.63. Suppose X_1, X_2, X_3 are Bernoulli(p_i) random variables respectively, where $i = 1, 2, 3$, and they are pairwise uncorrelated, i.e., $\text{cov}(X_i, X_j) = 0$ for $i \ne j$. Are X_1, X_2, X_3 jointly independent? Provide your reasoning.

5.64. Suppose $\{X_i\}_{i=1}^n$ is an IID sequence from a $N(\mu, \sigma^2)$ distribution. Define $\bar{X}_n = n^{-1}\Sigma_{i=1}^n X_i$. Show that for all n, $\bar{X}_n$ and $g(X_1 - \bar{X}_n, \cdots, X_n - \bar{X}_n)$ are independent, where $g(\cdot, \cdots, \cdot)$ is any measurable function.

5.65. Suppose $X \sim \text{Poisson}(\lambda_1), Y \sim \text{Poisson}(\lambda_2)$, and X and Y are independent. Show:

　(1) $X + Y \sim \text{Poisson}(\lambda_1 + \lambda_2)$;

　(2) the conditional distribution of X given $X + Y = n$ is a Binomial distribution $B(n, \frac{\lambda_1}{\lambda_1 + \lambda_2})$.

5.66. We have one unit of capital to invest in two bonds A and B. If we invest w_1 in bond A and the remaining $w_2 = 1 - w_1$ in bond B, then (w_1, w_2) constitute a portfolio. Denote the return of bond A and B as random variable X with mean μ_X, variance σ_X^2 and random variable Y with mean μ_Y, variance σ_Y^2 respectively. The correlation coefficient between X and Y is ρ. Find:

(1) the average return and risk of portfolio (w_1, w_2);

(2) the portfolio weights (w_1^*, w_2^*) which minimize the investment risk.

Chapter 6

Introduction to Sampling Theory

Abstract: In the previous chapters, we have discussed probability theory. In this chapter, we will introduce some basic concepts in statistics. The basic idea of statistical inference is to assume that the observed data is generated from some unknown probability distribution, which is often assumed to have a known functional form up to some unknown parameters. The purpose of statistical inference is to develop theory and methods to make inference on the unknown parameters based on observed data.

Key words: Degree of freedom, F-distribution, F-test, Minimal sufficient statistic, Normality, Parameter, Population, Random sample, Sampling distribution, Sample mean, Sample variance, Statistics, Student t-distribution, Sufficient statistic, t-test.

6.1 Population and Random Sample

Statistical analysis is based on outcomes of a large number of repeated experiments of same or similar kind. Suppose a random variable X_i denotes the outcome of the i-th experiment. Then we will obtain a sequence of outcomes, $X_1, \cdots, X_n$, if n experiments are implemented. This sequence of outcomes then constitutes a so-called random sample. From the sample information, one can make inference of the underlying probability distribution which has generated the observed data.

Definition 6.1. [Random Sample]: A random sample, denoted as $\mathbf{X}^n = (X_1, \cdots, X_n)$, is a sequence of n random variables $X_1, \cdots, X_n$. A realization of the random sample $\mathbf{X}^n$, denoted as $\mathbf{x}^n = (x_1, \cdots, x_n)$, is called a data set generated from $\mathbf{X}^n$ or a sample point of $\mathbf{X}^n$. A random sample $\mathbf{X}^n$ can generate many different data sets. The collection of all possible sample points of $\mathbf{X}^n$ constitutes the sample space of the random sample $\mathbf{X}^n$.

We now consider some examples.

Example 6.1. [Throwing n Coins]: Suppose we throw n coins. Let X_i denote the outcome of throwing the i-th coin, with $X_i = 1$ if the head shows up, and $X_i = 0$ if the tail shows up. Then $\mathbf{X}^n = (X_1, \cdots, X_n)'$ constitutes a random sample. If we throw n coins, we will obtain a sequence of real numbers, such as

$$\mathbf{x}^n = (1, 1, 0, 0, 1, 0, \cdots, 1).$$

This sequence is a data set of size n from the random sample $\mathbf{X}^n$.

Obviously, if we throw the n coins again, we will get a different sequence, such as

$$\mathbf{x}^n = (1, 0, 0, 1, 1, 1, \cdots, 0).$$

This is another data set from the random sample $\mathbf{X}^n$. Apparently, the random sample $\mathbf{X}^n$ can generate a total of 2^n data sets, each with size n.

Example 6.2. [Chinese GDP Annual Growth Rate]: Let X_i denote the Chinese GDP growth rate in year i, from 1953 to 2010. Then $\mathbf{X}^n = (X_1, \cdots, X_n)$ constitutes a random sample with sample size $n = 58$. The observed data $\mathbf{x}^n = (x_1, \cdots, x_n)$, depicted in Figure 6.1, is a realization of $\mathbf{X}^n$.

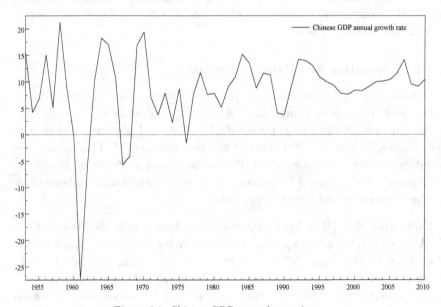

Figure 6.1: Chinese GDP annual growth rate

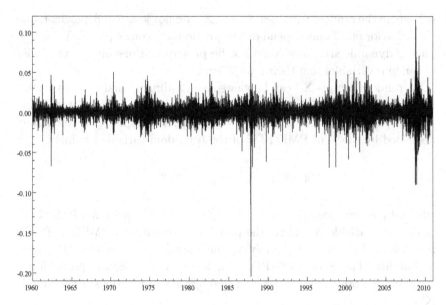

Figure 6.2: Daily returns on S&P500 closing price index

Example 6.3. Let X_i be the return on the S&P500 price index at day i, from January 4, 1960 to December 31, 2010. Then $\mathbf{X}^n = (X_1, \cdots, X_n)$ forms a random sample with size $n = 12839$. The observed data set $\mathbf{x}^n = (x_1, \cdots, x_n)$, depicted in Figure 6.2, is a realization of the random sample $\mathbf{X}^n$.

While in theory a random sample $\mathbf{X}^n$ can generate many different data sets $\mathbf{x}^n$, each with size n, one may only observe or obtain one data set $\mathbf{x}^n$ in practice. This is the case with Examples 6.2 and 6.3. For example, if we would like to obtain another data set (i.e., a different sequence of realizations) for the Chinese GDP growth rate, we would have to let the Chinese economy repeat again back from 1953, and this is simply impossible due to the non-experimental nature of a real economy. Even it were possible to repeat some social-economic experiments, it may be extremely costly. However, in statistical analysis, we still assume that the only observed data in Example 6.2 or Example 6.3 is one of many possible realizations from the random sample $\mathbf{X}^n$.

On the other hand, for some random samples, the order of the random variables $X_1, \cdots, X_n$ in the sample, together with their realizations, may not be altered freely. An example is the time series random sample of Example

6.2, where the random variables $X_1, \cdots, X_n$ are not jointly independent, and the behavior of X_i may depend on the previous outcomes $\{X_{i-1}, X_{i-2}, \cdots\}$. Such a dynamic structure could not be preserved if one altered the order of random variables and their realizations.

A random sample $\mathbf{X}^n$ can be viewed as a n-dimensional random vector, namely, $\mathbf{X}^n : S \to \mathbb{R}^n$, where S is the sample space of the underlying random experiment. The information of a random sample $\mathbf{X}^n$ is completely described by the joint PMF/PDF of the n random variables, namely,

$$f_{\mathbf{X}^n}(\mathbf{x}^n) = \prod_{i=1}^{n} f_{X_i | \mathbf{X}^{i-1}}(x_i | \mathbf{x}^{i-1}),$$

where, by convention, $f_{X_1 | \mathbf{X}^0}(x_1 | x^0) = f_{X_1}(x_1)$ is the marginal PMF/PDF of random variable X_1. Here, the product of conditional PMF's or PDF's is obtained by repeatedly applying the multiplication rule for the joint probability. The joint PMF/PDF can be used to calculate probabilities involving the random sample $\mathbf{X}^n$ (including various functions of the sample $\mathbf{X}^n$).

The above definition of a random sample covers both independent samples and time series samples. For the former, the random variables $X_1, \cdots, X_n$ in the sample are jointly independent, and for the latter, the random variables $X_1, \cdots, X_n$ in the sample are not jointly independent. In this book, to focus on fundamental ideas in statistics, we will mainly consider independent samples only.

Definition 6.2. [IID Random Sample]: The sequence of random variables $X_1, \cdots, X_n$ is called an independent and identically distributed (IID) random sample of size n from the population distribution $F_X(x)$ if
    (1) random variables $X_1, \cdots, X_n$ are mutually *independent*;
    (2) each random variable X_i has the same marginal distribution $F_X(x)$.

Question: What is the interpretation and implication of an IID random sample?

Suppose we have an experiment in which the variable of interest X has a probability distribution described by $F_X(x)$. Suppose the random experiment is repeated n times. Then we observe n outcomes for the variable of interest, denoted as $\mathbf{x}^n = (x_1, \cdots, x_n)$. If we denote X_i as the variable of interest associated with the i-th experiment, then X_i has the probability distribution $F_X(x)$ and x_i can be viewed as an observed value (i.e., a realization) of X_i. Furthermore, the experiments are taken in such a way that the outcomes of n experiments are irrelevant to each other; that is, $X_1, \cdots, X_n$ are mutually independent. This ensures that the n random variables will generally have different realizations. In short, identical distribution means repeated experiments of same kind, and independence means that experiments are implemented independently so that new information can be obtained from each experiment. (If $X_1, \cdots, X_n$ are highly correlated, their realizations will be similar and have little variations, thus containing not much new information.) The main purpose of statistical analysis is to infer the population distribution $F_X(x)$ based on an observed data set generated from a large number of repeated experiments of the same kind.

For an IID random sample, the population is represented by the common marginal CDF $F_X(x)$ of each X_i. In many applications, one often assumes that the corresponding PMF/PDF $f_X(x) = f(x, \theta)$, where the functional form $f(\cdot, \cdot)$ is known, but the parameter θ is unknown. For example, suppose $\mathbf{X}^n$ is a random sample from a $N(\mu, \sigma^2)$ population. Then

$$
\begin{aligned}
f_X(x) &= f(x, \theta) \\
&= \frac{1}{\sqrt{2\pi\sigma^2}} e^{-\frac{(x-\mu)^2}{2\sigma^2}}, \qquad -\infty < x < \infty,
\end{aligned}
$$

where $\theta = (\mu, \sigma^2)$. Given an observed data $\mathbf{x}^n$ from a random sample $\mathbf{X}^n$, one can make inference of the parameter $\theta = (\mu, \sigma^2)$.

The positive integer n is called the sample size. In general, $\mathbf{X}^n = (X_1, \cdots, X_n)$ is obtained by repeating the experiment n times. The random variable X_i represents the possible outcome of the i-th experiment. By repeating the experiment, one can ensure that each realization or outcome is generated from the same population. Moreover, the experiments are implemented independently. This ensures that $\mathbf{X}^n = (X_1, \cdots, X_n)$ is an IID random sample. An independent experiment will provide new information.

Question: How to define the population if the random variables $X_1, \cdots, X_n$ in the sample are not identically distributed?

The random variables $X_1, \cdots, X_n$ in a random sample may not have identical probability distributions, due to the existence of heterogeneity among economic agents or structural changes of economic relationships over times. Although each X_i has a different distribution, we may assume that they may still share certain common features (e.g., common parameter values) in their probability distributions or their economic relationships, and these common features of distributions can be defined as the population.

Question: How to extract information from a data set $\mathbf{x}^n$?

When a random sample is drawn, we observe a realization $\mathbf{x}^n$ of $\mathbf{X}^n$, and some summary of the value $\mathbf{x}^n$ is usually computed. Any well-defined summary of the data set $\mathbf{x}^n$ may be expressed mathematically as a function $T(\mathbf{x}^n)$ whose domain includes the sample space (i.e., the support) of $\mathbf{X}^n$. The function $T(\cdot)$ may be scalar- or vector-valued and $T(\mathbf{X}^n)$ is called a statistic.

Definition 6.3. [Statistic]: Let $\mathbf{X}^n = (X_1, \cdots, X_n)$ be a random sample of size n from a population. A *statistic* $T(\mathbf{X}^n) = T(X_1, \cdots, X_n)$ is a real-valued or vector-valued function of a random sample $\mathbf{X}^n$.

The function $T(\cdot)$ is a mapping from the n-dimensional sample space of $\mathbf{X}^n$ to a low-dimensional Euclidean space. For notational simplicity, we have suppressed the possible dependence of the functional form $T(\cdot)$ on the sample size n. A statistic $T(\mathbf{X}^n)$ does not involve any unknown parameter. It is entirely a function of random sample $\mathbf{X}^n$. Given any data set $\mathbf{x}^n$, we can obtain a real-valued number or vector for the statistic $T(\mathbf{X}^n)$.

A statistic $T(\mathbf{X}^n)$ can be used to effectively summarize some features of data (e.g., maximum and minimum values, median, mean, standard deviation, etc), to estimate unknown parameters, to do hypothesis testing, etc. We now provide some examples for statistic $T(\mathbf{X}^n)$.

Example 6.4. Let $\mathbf{X}^n = (X_1, \cdots, X_n)$ be a random sample. Then the sample mean

$$\bar{X}_n = n^{-1} \sum_{i=1}^{n} X_i$$

and the sample variance

$$S_n^2 = (n-1)^{-1} \sum_{i=1}^{n} (X_i - \bar{X}_n)^2$$

are two statistics.

The sample mean $\bar{X}_n$ and sample variance S_n^2 are used to estimate μ_X and σ_X^2 of the population distribution $F_X(x)$. Later in this chapter, we will provide justification that $\bar{X}_n$ and S_n^2 are good estimators of μ_X and σ_X^2 respectively. More formally, we will develop various concepts to measure the closeness of an estimator to the parameter of interest in Chapters 7 and 8.

Example 6.5. Let $\mathbf{X}^n = (X_1, \cdots, X_n)$ be an IID random sample from the population $f(x, \theta)$, where θ is some unknown parameter. Then the logarithm of the joint PMF/PDF of $\mathbf{X}^n$

$$\hat{L}(\theta|\mathbf{X}^n) = \ln \prod_{i=1}^{n} f(X_i, \theta) = \sum_{i=1}^{n} \ln f(X_i, \theta)$$

is called the log-likelihood function of θ, conditional on the random sample $\mathbf{X}^n$. Note that $\hat{L}(\theta|\mathbf{X}^n)$ depends on the random sample $\mathbf{X}^n$, but it is not a statistic, because it also depends on the unknown parameter θ.

Next, we define the sampling distribution for statistic $T(\mathbf{X}^n)$.

Definition 6.4. [Sampling Distribution]: The probability distribution of a statistic $T(\mathbf{X}^n)$ is called the *sampling distribution* of $T(\mathbf{X}^n)$.

Since $T(\mathbf{X}^n)$ is a function of n random variables, $T(\mathbf{X}^n)$ itself is a random variable or random vector. The distribution of $T(\mathbf{X}^n)$ is called the sampling distribution because this distribution is usually derived from the joint distribution of the variables $X_1, \cdots, X_n$ in the random sample. The sampling distribution of $T(\mathbf{X}^n)$ is different from the population distribution $F_X(x)$. The latter is the marginal distribution of each X_i in an IID random sample $\mathbf{X}^n$.

Given that $T(\mathbf{X}^n)$ is a function of random vector $\mathbf{X}^n$, we can in principle derive the sampling distribution of $T(\mathbf{X}^n)$ by using the transformation technique, as discussed in Chapter 5. This is particularly tractable when $\mathbf{X}^n$ is an IID random sample. As will be seen later, the sampling distribution of of a statistic $T(\mathbf{X}^n)$ plays a vital role in statistical inference. For example, it is needed to obtain critical values when constructing a confidence interval estimator and a hypothesis test statistic.

As a measure of summary for the information contained in the random sample $\mathbf{X}^n$, a statistic $T(\mathbf{X}^n)$ can be viewed as a partition of the sample space of $\mathbf{X}^n$. A random sample $\mathbf{X}^n$ can generate many data sets $\mathbf{x}^n$, each of which is called a sample point in the sample space of $\mathbf{X}^n$. Let

$$A(t) = \{\mathbf{x}^n : T(\mathbf{x}^n) = t\}$$

be the collection of all sample points $\mathbf{x}^n$ that satisfy the restriction $T(\mathbf{x}^n) = t$. Then a single value of $T(\mathbf{x}^n) = t$ summarizes all sample points in $A(t)$. Note that $A(t)$ is a subset in the sample space of $\mathbf{X}^n$.

It is a common practice to apply the terms "sample", "statistic", "sample mean", "sample variance", to the values of random variables/vectors $\mathbf{X}^n, T(\mathbf{X}^n), \bar{X}_n$, and S_n^2 rather than the random variables/vectors themselves. Intuitively, this makes more sense and it conforms with colloquial usage.

In the subsequent sections of this chapter, we will use the sample mean $\bar{X}_n$ and sample variance S_n^2 as leading examples to introduce basic concepts, theory, and methods in statistical inference. The main reason to focus on the sample mean $\bar{X}_n$ and sample variance S_n^2 is that classical statistical theory often assumes a normal population $N(\mu, \sigma^2)$, where it suffices to estimate the mean μ and variance σ^2, and where we can derive the exact sampling distributions for $\bar{X}_n$ and S_n^2 for each finite sample size n. It should be understood that there are many other statistics to be introduced later. For example, $T(\mathbf{X}^n) = \max_{1 \leq i \leq n}(X_i)$ and $T(\mathbf{X}^n) = \min_{1 \leq i \leq n}(X_i)$ are the maximum and minimum statistics respectively. They are the special cases of the so-called order statistics.

6.2 The Sampling Distribution of the Sample Mean

We first consider the sample mean.

Definition 6.5. [Sample Mean]: Suppose $\mathbf{X}^n = (X_1, \cdots, X_n)$ is a random sample from a population with mean μ and variance σ^2. Then

$$T(\mathbf{X}^n) \equiv \bar{X}_n = \frac{1}{n} \sum_{i=1}^{n} X_i$$

is the sample mean for the random sample $\mathbf{X}^n$.

Because $X_1, \cdots, X_n$ are random variables, so is $\bar{X}_n$. The distribution of $\bar{X}_n$ is called the sampling distribution of $\bar{X}_n$. When one has only a single observed sample $\mathbf{x}^n$, the sample mean $\bar{x}_n$ does not appear random. However, if we realize that the observed sample $\mathbf{x}^n$ is only one of many possible samples that could have been drawn, and that each sample has a different sample mean, then we can see that the sample mean is in fact random.

Recall that we have solved the minimization problem

$$\mu = \arg \min_{-\infty < a < \infty} E(X - a)^2$$

in Chapter 3. We have a similar result by minimizing the sample analog of $E(X - a)^2$:

Theorem 6.1. *Suppose $\mathbf{X}^n$ is a random sample. Then*

$$\bar{X}_n = \arg \min_{-\infty < a < \infty} \sum_{i=1}^{n} (X_i - a)^2.$$

The objective function $\sum_{i=1}^{n}(X_i-a)^2$ is called the sum of squared residuals. This theorem shows that the sample mean $\bar{X}_n$ is the best solution for minimizing the sum of squared residuals $\sum_{i=1}^{n}(X_i-a)^2$. It is essentially the Ordinary Least Squares (OLS) estimator for a very simple linear regression model

$$X_i = a + \varepsilon_i,$$

where $\{\varepsilon_i\}$ is an IID sequence with $E(\varepsilon_i) = 0$ and $\mathrm{var}(\varepsilon_i) = \mathrm{var}(X_i) = \sigma^2$.

We now investigate the statistical properties of the sample mean estimator $\bar{X}_n$, which are important for making inference of the unknown population mean μ. In particular, we shall ask the following questions:

- What is the mean of $\bar{X}_n$?
- What is the variance of $\bar{X}_n$?
- What is the sampling distribution of $\bar{X}_n$?

We first examine the mean of $\bar{X}_n$.

Theorem 6.2. *Suppose $X_1, \cdots, X_n$ are a sequence of n identically distributed random variables with the same population mean μ. Then for all $n \geq 1$,*

$$E(\bar{X}_n) = \mu.$$

Proof:

$$E(\bar{X}_n) = \frac{1}{n}\sum_{i=1}^{n}E(X_i)$$

$$= \frac{1}{n}\sum_{i=1}^{n}\mu$$

$$= \mu.$$

Thus, for any given sample size $n \geq 1$, the mean of the sample mean $\bar{X}_n$ is the same as the population mean μ. Note that this result does not require that the random variables $X_1, \cdots, X_n$ be mutually independent.

Intuitively, the result of $E(\bar{X}_n) = \mu$ implies that the sample mean estimator $\bar{X}_n$ does not make a systematic mistake in estimating the population mean μ. If one generates a large number of data sets $\mathbf{x}^n$, each of which gives a value $\bar{x}_n$ for $\bar{X}_n$, then the average of these sample mean values will be arbitrarily close to the population mean μ for any given sample size n. There is neither a systematic upward bias nor a downward bias away from the population mean μ. This is analogous to shooting a target for a large number of times, where the hits spread more or less evenly to the left and right hand sides of the target.

Next, we derive the variance of $\bar{X}_n$.

Theorem 6.3. *Suppose* $\mathbf{X}^n$ *is an IID random sample from a population with mean* μ *and variance* σ^2. *Then for all* $n \geq 1$,

$$var(\bar{X}_n) = \frac{\sigma^2}{n}.$$

Proof: Recall that when X and Y are mutually independent, we have

$$\begin{aligned} var(a + bX + cY) &= b^2\sigma_X^2 + c^2\sigma_Y^2 + 2bc \cdot cov(X, Y) \\ &= b^2\sigma_X^2 + c^2\sigma_Y^2. \end{aligned}$$

That is, the variance of the sum of independent random variables is equal to the sum of individual variances.

Similarly, for an IID random sample $\mathbf{X}^n$, we have

$$\begin{aligned} var(\bar{X}_n) &= var\left(\sum_{i=1}^{n} n^{-1}X_i\right) \\ &= \sum_{i=1}^{n} n^{-2}var(X_i) \\ &= \sum_{i=1}^{n} n^{-2}\sigma^2 \\ &= \frac{\sigma^2}{n}. \end{aligned}$$

It should be emphasized that the variance σ^2/n of $\bar{X}_n$ is different from the population variance σ^2 of each random variable X_i. It measures the precision of the sample mean $\bar{X}_n$ from its center $E(\bar{X}_n)$. The result that $var(\bar{X}_n) = \sigma^2/n$ implies that the dispersion of $\bar{X}_n$ from its center $E(\bar{X}_n)$ shrinks to zero as $n \to \infty$. Since $E(\bar{X}_n) = \mu$, we have the mean squared error of $\bar{X}_n$

$$\begin{aligned} E(\bar{X}_n - \mu)^2 &= var(\bar{X}_n) \\ &= \frac{\sigma^2}{n} \to 0 \text{ as } n \to \infty. \end{aligned}$$

That is, the average of the squared distance between $\bar{X}_n$ and μ shrinks to zero as $n \to \infty$. Thus, the sample mean estimator $\bar{X}_n$ becomes closer and closer to μ as $n \to \infty$.

The result $\text{var}(\bar{X}_n) = \sigma^2/n$ looks rather simple in statistics but it has profound implication on risk diversification in finance.

Example 6.6. [**Idiosyncratic Risk Elimination via Diversification**]: According to the standard capital asset pricing model (CAPM), the return of asset i over certain holding period is given as follows:

$$R_i = \alpha + \beta_i R_m + \varepsilon_i,$$

where α is a constant representing the return on the risk-free asset, R_m is the market risk factor common to all individual assets, β_i is a factor loading coefficient, and ε_i represents an idiosyncratic risk associated with asset i. It is further assumed here that the sequence of $(\varepsilon_1, \cdots, \varepsilon_n)$ is IID with mean 0 and variance σ^2, and is uncorrelated with the market risk factor R_m. The risk of asset i, as measured by its variance, is given by

$$\text{var}(R_i) = \beta_i^2 \text{var}(R_m) + \sigma^2,$$

where $\beta_i^2 \text{var}(R_m)$ is a systematic risk which cannot be avoided, and σ^2 is the idiosyncratic risk which can be eliminated by forming a portfolio with a large number of assets. To see this, we consider the return on an equal-weighting portfolio with n assets:

$$\bar{R}_n = \sum_{i=1}^{n} \frac{1}{n} R_i$$
$$= \alpha + \bar{\beta}_n R_m + \bar{\varepsilon}_n,$$

where the average beta $\bar{\beta}_n = n^{-1} \sum_{i=1}^{n} \beta_i \to \beta \neq 0$ as $n \to \infty$, and $\bar{\varepsilon}_n = n^{-1} \sum_{i=1}^{n} \varepsilon_i$ is the sample mean of the individual risk sample $(\varepsilon_1, \cdots, \varepsilon_n)$. It follows that

$$\text{var}(\bar{R}_n) = \bar{\beta}_n^2 \text{var}(R_m) + \frac{\sigma^2}{n}$$
$$\to \beta^2 \text{var}(R_m) \text{ as } n \to \infty.$$

Thus, the idiosyncratic risks associated with individual assets can be eliminated by including a very large number n of assets contained in the portfolio.

Next, we derive the sampling distribution of $\bar{X}_n$. We consider the case where $\mathbf{X}^n$ is an IID random sample from a $N(\mu, \sigma^2)$ population.

Theorem 6.4. *Suppose $\mathbf{X}^n = (X_1, \cdots, X_n)$ is an IID normally distributed random sample with population mean μ and population variance $\sigma^2 < \infty$.*

Define the standardized sample mean

$$Z_n = \frac{\bar{X}_n - E(\bar{X}_n)}{\sqrt{var(\bar{X}_n)}}$$

$$= \frac{\bar{X}_n - \mu}{\sigma/\sqrt{n}}$$

$$= \frac{\sqrt{n}(\bar{X}_n - \mu)}{\sigma}.$$

Then

$$Z_n \sim N(0,1) \text{ for all } n \geq 1.$$

Proof: Put $Y_i = (X_i - \mu)/\sigma$. Then $Y_i \sim N(0,1)$ and has MGF

$$M_{Y_i}(t) = e^{\frac{1}{2}t^2} \text{ for all } i$$

by Theorem 4.2.

Now consider the MGF of $Z_n = n^{-1/2} \sum_{i=1}^{n} Y_i$:

$$M_{Z_n}(t) = E\left(e^{tZ_n}\right)$$

$$= E\left(e^{tn^{-\frac{1}{2}}\Sigma_{i=1}^{n}Y_i}\right)$$

$$= E\left(\prod_{i=1}^{n} e^{tn^{-\frac{1}{2}}Y_i}\right)$$

$$= \prod_{i=1}^{n} E\left(e^{tn^{-\frac{1}{2}}Y_i}\right)$$

$$= \prod_{i=1}^{n} M_{Y_i}\left(tn^{-\frac{1}{2}}\right)$$

$$= \left[e^{\frac{1}{2}(tn^{-\frac{1}{2}})^2}\right]^n$$

$$= e^{\frac{1}{2}t^2}.$$

It follows that $Z_n \sim N(0,1)$ for all $n \geq 1$. This completes the proof.

Theorem 6.4 implies that the sum of n independent normal random variables is still a normal variable. This is called the reproductive property of the normal distribution.

When the random sample X^n is not from a normal population, $\bar{X}_n$ and Z_n no longer follow a normal distribution. For example, in Example 6.1, $n\bar{X}_n$ follows a Binomial(n, p) distribution for any given n.

Figure 6.3 plots the sampling distribution of the sample mean $\bar{X}_n$ for various sample sizes ($n = 1, 2, 5, 10, 30$) under following population distributions: (1) $N(0,1)$; (2) Bernoulli$(\frac{1}{2})$; (3) $U[0,1]$; (4) EXP(1), all standardized

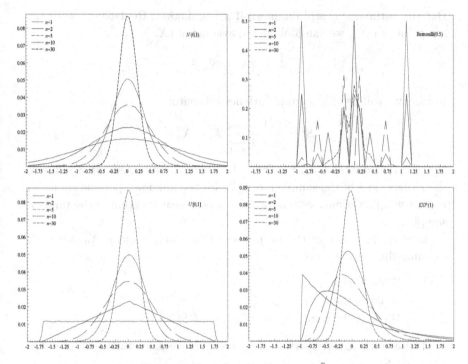

Figure 6.3: Sampling distribution of $\bar{X}_n$

with mean 0 and variance 1. As can be seen, the sampling distributions of $\bar{X}_n$ are all centered at the population mean (i.e., $E(\bar{X}_n) = \mu$) for each sample size n, and the dispersion shrinks as the sample size n increases (i.e., $\text{var}(\bar{X}_n) = \sigma^2/n$).

6.3 The Sampling Distribution of the Sample Variance

We have known how to estimate population mean μ, by using the sample mean $\bar{X}_n$, and have shown that $\bar{X}_n$ is a good estimator for μ in the sense that $E(\bar{X}_n - \mu)^2 = \sigma^2/n \to 0$ as $n \to \infty$. The next question is how to estimate the population variance σ^2.

Recall the variance formula

$$\sigma^2 = E(X_i - \mu)^2,$$

one plausible estimator for σ^2 might be the sample average

$$n^{-1} \sum_{i=1}^{n} (X_i - \mu)^2.$$

But this estimator is still unknown if μ is unknown. Replacing μ with the sample mean $\bar{X}_n$, we can look at the average of $(X_i - \bar{X}_n)^2$:

$$\frac{1}{n} \sum_{i=1}^{n} (X_i - \bar{X}_n)^2.$$

In fact, we will use the sample variance estimator

$$S_n^2 = \frac{1}{n-1} \sum_{i=1}^{n} (X_i - \bar{X}_n)^2.$$

Note that we use the scale factor $n-1$ rather than n in defining S_n^2, due to the replacement of the sample mean $\bar{X}_n$ for the unknown population mean μ. The sample variance estimator S_n^2 is a nonlinear function of the random sample $\mathbf{X}^n$.

Like in the study of the properties of the sample mean $\bar{X}_n$, we shall examine the following properties for S_n^2:

- What is the mean of S_n^2?
- What is the variance of S_n^2?
- What is the sampling distribution of S_n^2?

These statistical properties of S_n^2 are important in statistical inference involving S_n^2.

We first consider the mean of S_n^2.

Theorem 6.5. *Suppose* $\mathbf{X}^n = (X_1, \cdots, X_n)$ *is an IID random sample from a population with* (μ, σ^2). *Then for all* $n > 1$,

$$E(S_n^2) = \sigma^2.$$

Proof: Using the formula $(a - b)^2 = a^2 - 2ab + b^2$, we have

$$\sum_{i=1}^{n} (X_i - \bar{X}_n)^2 = \sum_{i=1}^{n} [(X_i - \mu) - (\bar{X}_n - \mu)]^2$$

$$= \sum_{i=1}^{n} (X_i - \mu)^2 - 2\sum_{i=1}^{n} (X_i - \mu)(\bar{X}_n - \mu) + \sum_{i=1}^{n} (\bar{X}_n - \mu)^2$$

$$= \sum_{i=1}^{n} (X_i - \mu)^2 - 2(\bar{X}_n - \mu)\sum_{i=1}^{n} (X_i - \mu) + n(\bar{X}_n - \mu)^2$$

$$= \sum_{i=1}^{n} (X_i - \mu)^2 - 2n(\bar{X}_n - \mu)^2 + n(\bar{X}_n - \mu)^2$$

$$= \sum_{i=1}^{n} (X_i - \mu)^2 - n(\bar{X}_n - \mu)^2,$$

where we have used the fact

$$\sum_{i=1}^{n}(X_i - \mu) = n(\bar{X}_n - \mu).$$

Taking the expectations for both sides, we have

$$E\sum_{i=1}^{n}(X_i - \bar{X}_n)^2 = \sum_{i=1}^{n}E(X_i - \mu)^2 - nE[(\bar{X}_n - \mu)^2]$$

$$= n\sigma^2 - n \cdot \frac{\sigma^2}{n} = (n-1)\sigma^2,$$

where we have used the fact that $E(\bar{X}_n - \mu)^2 = \frac{\sigma^2}{n}$ from Section 6.2. It follows that

$$E(S_n^2) = E\left[\frac{1}{n-1}\sum_{i=1}^{n}(X_i - \bar{X}_n)^2\right] = \sigma^2.$$

It is important to assume independence among the n random variables $X_1, \ldots, X_n$ here, because we have used the fact that $E(\bar{X}_n - \mu)^2 = \sigma^2/n$.

As seen above, the reason of using $n-1$ instead of n is to ensure that S_n^2 is unbiased for σ^2, that is, $E(S_n^2) = \sigma^2$. This is because that μ is unknown and we have to replace it with $\bar{X}_n$. This leads to the loss of one degree of freedom, from n to $n-1$, in the observations used in computing the sample variance S_n^2. For the concept of the loss of degrees of freedom, see later discussion in this section.

Since S_n^2 is a nonlinear function of the random sample $\mathbf{X}^n$, the variance and the sampling distribution of S_n^2 depend on the distribution of each random variable X_i, and are generally rather complicated. There exists, however, a special case from which the sampling distribution and the variance of S_n^2 are relatively easy to obtain. This occurs when $\mathbf{X}^n$ is an IID random sample from a $N(\mu, \sigma^2)$ population.

Before we state the sampling distribution of S_n^2, we first review the properties of the χ^2 distribution, which plays important roles in problems of sampling from normally distributed populations.

Recall from Chapter 4 that a nonnegative random variable follows a χ_ν^2 distribution if its PDF

$$f_X(x) = \frac{1}{\sqrt{2^\nu}\Gamma(\frac{\nu}{2})}x^{\frac{\nu}{2}-1}e^{-\frac{x}{2}}, \qquad x > 0,$$

where ν is called the degree of freedom for the Chi-square distribution.

The degree of freedom ν of a χ_ν^2 need not be an integer. When ν is an integer, however, there exists an intuitive representation for a χ_ν^2 random variable.

Lemma 6.1. *Let $Z_1, \cdots, Z_\nu$ be IID $N(0,1)$ random variables, where ν is a positive integer. Then*

$$\sum_{i=1}^{\nu} Z_i^2 \sim \chi_\nu^2.$$

That is, the sum of squares of ν independent N(0,1) random variable follows a χ_ν^2 distribution.

Proof: Recall that when $Z_i \sim N(0,1)$, we have $Z_i^2 \sim \chi_1^2$, whose MGF

$$M_{Z_i^2}(t) = (1 - 2t)^{-\frac{1}{2}}.$$

Put $X = \Sigma_{i=1}^{\nu} Z_i^2$. Then given the independence among $Z_1, \cdots, Z_\nu$, we have

$$
\begin{aligned}
M_X(t) &= E\left(e^{tX}\right) \\
&= E\left(e^{t\Sigma_{i=1}^{\nu} Z_i^2}\right) \\
&= \prod_{i=1}^{\nu} E\left(e^{iZ_i^2}\right) \\
&= \left[(1 - 2t)^{-\frac{1}{2}}\right]^{\nu} \\
&= (1 - 2t)^{-\frac{\nu}{2}}.
\end{aligned}
$$

It follows that $X \sim \chi_\nu^2$ by the uniqueness of the MGF. This is called the reproductive property of the χ^2 distribution.

The χ_ν^2 distribution is a special case of the Gamma distribution, Gamma$(\frac{\nu}{2}, 2)$. Its mean and variance are, respectively,

$$E\left(\chi_\nu^2\right) = \nu$$

and

$$\text{var}\left(\chi_\nu^2\right) = 2\nu.$$

The shape of the χ_ν^2 is not symmetric about the mean, and it is skewed to the right. The skewness vanishes to zero as $\nu \to \infty$.

We now show that the sampling distribution of the sample variance S_n^2, after suitably standardized, is a χ_{n-1}^2 distribution.

Theorem 6.6. *Suppose $\mathbf{X}^n = (X_1, \cdots, X_n)$ is an IID $N(\mu, \sigma^2)$ random sample. Then for each $n > 1$,*

$$\frac{(n-1)S_n^2}{\sigma^2} = \frac{\sum_{i=1}^{n}(X_i - \bar{X}_n)^2}{\sigma^2} \sim \chi_{n-1}^2,$$

where χ_{n-1}^2 is a Chi-square distribution with $n-1$ degrees of freedom.

Proof: It is straightforward to establish the recursive relation

$$(n-1)S_n^2 = (n-2)S_{n-1}^2 + \frac{n-1}{n}(X_n - \bar{X}_{n-1})^2.$$

We shall show the theorem by induction, which consists of the following two steps.

(1) We first consider $n = 2$. We have

$$\frac{(2-1)S_2^2}{\sigma^2} = \frac{1}{2\sigma^2}(X_2 - X_1)^2$$

$$= \left(\frac{X_2 - X_1}{\sqrt{2}\sigma}\right)^2$$

$$\sim \chi_1^2$$

because $(X_2 - X_1)/\sqrt{2}\sigma \sim N(0,1)$.

(2) Next, suppose that for $n = \nu$, an arbitrary positive integer with $\nu > 1$, we have $(\nu - 1)S_\nu^2/\sigma^2 \sim \chi_{\nu-1}^2$. Then we shall show $\nu S_{\nu+1}^2/\sigma^2 \sim \chi_\nu^2$.

For $n = \nu + 1$, we have

$$\frac{\nu S_{\nu+1}^2}{\sigma^2} = \frac{(\nu-1)S_\nu^2}{\sigma^2} + \frac{\nu}{(\nu+1)\sigma^2}(X_{\nu+1} - \bar{X}_\nu)^2.$$

Here, $X_{\nu+1} \sim N(\mu, \sigma^2)$, $\bar{X}_\nu \sim N(\mu, \frac{1}{\nu}\sigma^2)$, and $X_{\nu+1}$ and $\bar{X}_\nu$ are independent. It follows that

$$X_{\nu+1} - \bar{X}_\nu \sim N\left(0, \sigma^2 + \frac{\sigma^2}{\nu}\right)$$

or equivalently

$$\sqrt{\frac{\nu}{(\nu+1)\sigma^2}}(X_{\nu+1} - \bar{X}_\nu) \sim N(0,1).$$

Hence, $\frac{\nu}{\nu+1}(X_{\nu+1} - \bar{X}_\nu)^2/\sigma^2 \sim \chi_1^2$. Suppose this term is independent of S_ν^2. Then, given $(\nu-1)S_\nu^2/\sigma^2 \sim \chi_{\nu-1}^2$ and the fact that the sum of two independent χ^2 random variables follow a χ^2 distribution, we have $\nu S_{\nu+1}^2/\sigma^2 \sim \chi_\nu^2$ by the reproductive property of the χ^2 distribution (see Example 5.33). The theorem will thus be proved provided the following result is shown:

Theorem 6.7. *Suppose* $\mathbf{X}^n$ *is an IID* $N(\mu, \sigma^2)$ *random sample. Then for any* $n > 1$, S_n^2 *and* $\bar{X}_n$ *are mutually independent.*

Both S_n^2 and $\bar{X}_n$ are functions of the same random variables $\{X_i\}_{i=1}^n$, but they are independent. The independence between S_n^2 and $\bar{X}_n$ is solely due to the normality assumption for the random sample $\mathbf{X}^n$. See a similar example in Example 5.21, Chapter 5. To show Theorem 6.7, we will use the following lemma.

Lemma 6.2. *Let* $X_i \sim$ *IID* $N(\mu, \sigma^2)$, $i = 1, \cdots, n$. *For constants* a_{ij}, *and* b_{rj}, *define*

$$U_i = \sum_{j=1}^{n} a_{ij} X_j, \ i = 1, \cdots, \nu,$$

$$V_r = \sum_{j=1}^{n} b_{rj} X_j, \ r = 1, \cdots, m,$$

where $\nu + m \leq n$. *Then*
(1) the random variables U_i *and* V_r *are independent if and only if* $cov(U_i, V_r) = 0$;
(2) the random vectors $(U_1, \cdots, U_\nu)$ *and* $(V_1, \cdots, V_m)$ *are independent if and only if* U_i *is independent of* V_r *for all* i, r, *where* $i = 1, \cdots, \nu, r = 1, \cdots, m$.

Intuitively, the U_i random variables and the V_r random variables follow a joint normal distribution. Thus, the U_i random variables and the V_r random variables are independent if and only if their covariances are zero for all pairs of i, r.

Proof of Theorem 6.8: Note that $S_n^2 = (n-1)^{-1} \sum_{i=1}^{n} (X_i - \bar{X}_n)^2$ is a function of n random variables $(X_1 - \bar{X}_n), \cdots, (X_n - \bar{X}_n)$. It suffices to show that $\bar{X}_n$ and $(X_1 - \bar{X}_n, \cdots, X_n - \bar{X}_n)$ are mutually independent.

We apply Lemma 6.2. Put $U_1 = \bar{X}_n - \mu$, and $V_r = X_r - \bar{X}_n, r = 1, \cdots, n$. We first show that U_1 and V_r are mutually independent for all $r = 1, \cdots, n$. Because for any given $r = 1, \cdots, n$, we have

$$\begin{aligned}
cov(U_1, V_r) &= E(U_1 V_r) \\
&= E\left[(\bar{X}_n - \mu)(X_t - \mu)) \right] - E\left(\bar{X}_n - \mu \right)^2 \\
&= \frac{\sigma^2}{n} - \frac{\sigma^2}{n} \\
&= 0.
\end{aligned}$$

It follows from Lemma 6.2(1) that U_1 and V_r are independent. We have immediately from Lemma 6.2(2) that U_1 and $(V_1, \cdots, V_n)$ are mutually independent.

Now, put $g(U_1) = U_1 + \mu$, and $h(V_1, \cdots, V_n) = (n-1)^{-1} \sum_{r=1}^{n} V_r^2$. Then $g(U_1)$ and $h(V_1, \cdots, V_n)$ are independent, i.e., $\bar{X}_n$ and S_n^2 are independent. This completes the proof.

There exists an alternative way to prove independence between $\bar{X}_n$ and S_n^2. We now provide a heuristic proof.

Let $\mathbf{X} = (X_1, \cdots, X_n)'$ be a $n \times 1$ vector, $\mathbf{i} = (1, \cdots, 1)'$ be a $n \times 1$ vector of ones, and $\mathbf{I}$ be a $n \times n$ identity matrix, where $\mathbf{A}'$ denotes the transpose of a vector or matrix $\mathbf{A}$. Define a $n \times n$ matrix

$$\mathbf{M} = \mathbf{I} - \frac{1}{n}\mathbf{i}\mathbf{i}'.$$

Note that $\mathbf{M}^2 = \mathbf{M}$ and $\mathbf{M}' = \mathbf{M}$. Then we have

$$\bar{X}_n = \frac{\mathbf{i}'\mathbf{X}}{n},$$
$$(n-1)S_n^2 = (\mathbf{M}\mathbf{X})'(\mathbf{M}\mathbf{X})$$
$$= \mathbf{X}'\mathbf{M}^2\mathbf{X}$$
$$= \mathbf{X}'\mathbf{M}\mathbf{X}.$$

To show that $\bar{X}_n$ and S_n^2 are independent, it suffices to show the random variable $\mathbf{i}'\mathbf{X}$ and the $n \times 1$ random vector $\mathbf{M}\mathbf{X}$ are independent.

Put

$$\mathbf{Z} = \begin{pmatrix} \mathbf{i}'\mathbf{X} \\ \mathbf{M}\mathbf{X} \end{pmatrix}$$
$$= \begin{pmatrix} \mathbf{i}' \\ \mathbf{M} \end{pmatrix} \mathbf{X}$$
$$= \mathbf{A}\mathbf{X}, \quad \text{say},$$

where $\mathbf{A}$ is a $(n+1) \times n$ matrix. Because $\mathbf{Z}$ is a linear combination of $\mathbf{X}$, and $\mathbf{X} \sim N(0, \sigma^2\mathbf{I})$ is a vector of IID normal random variables, $\mathbf{Z}$ follows a multivariate normal distribution. Furthermore, we have the variance-covariance matrix between $\mathbf{i}'\mathbf{X}$ and $\mathbf{M}\mathbf{X}$

$$\text{cov}(\mathbf{i}'\mathbf{X}, \mathbf{M}\mathbf{X}) \equiv E\left\{[\mathbf{i}'\mathbf{X} - E(\mathbf{i}'\mathbf{X})][\mathbf{M}\mathbf{X} - E(\mathbf{M}\mathbf{X})]'\right\}$$
$$= E\left\{\mathbf{i}'[\mathbf{X} - E(\mathbf{X})][\mathbf{X} - E(\mathbf{X})]'\mathbf{M}'\right\}$$
$$= \mathbf{i}'E\left\{[\mathbf{X} - E(\mathbf{X})][\mathbf{X} - E(\mathbf{X})]'\right\}\mathbf{M}$$
$$= \mathbf{i}'\sigma^2\mathbf{I}\mathbf{M}$$
$$= \sigma^2\mathbf{i}'\mathbf{M}$$
$$= 0$$

given $\mathbf{i}'\mathbf{M} = \mathbf{0}$ (please check it!). Since $\mathbf{i}'\mathbf{X}$ and $\mathbf{M}\mathbf{X}$ follow a joint normal distribution, and they are uncorrelated, it follows that $\mathbf{i}'\mathbf{X}$ and $\mathbf{M}\mathbf{X}$ are mutually independent.

We now provide an interpretation for degrees of freedom associated with the sample variance S_n^2. Theorem 6.6 states that when $\{X_i\}_{i=1}^n$ is IID $N(\mu, \sigma^2)$, $(n-1)S_n^2/\sigma^2 \sim \chi_{n-1}^2$, where $n-1$ is called the number of degrees of freedom. This is a concept associated with sums of squares. The random sample $\mathbf{X}^n = (X_1, \cdots, X_n)$ are n independent observations, we now use them to estimate σ^2. If we knew μ, an estimator for σ^2 would be $n^{-1}\sum_{i=1}^n (X_i - \mu)^2$. Unfortunately we usually do not know the population mean μ. Therefore, we have to replace it with the sample mean $\bar{X}_n$ and use the estimator $S_n^2 = (n-1)^{-1}\sum_{t=1}^n (X_i - \bar{X}_n)^2$. Here, the n observations used are $(X_1 - \bar{X}_n, \cdots, X_n - \bar{X}_n)$. These n observations are subject to one restriction

$$\sum_{i=1}^n (X_i - \bar{X}_n) = 0.$$

Thus, given the $n-1$ observations, we can always obtain the remaining one from the above restriction. In this sense, in estimating S_n^2, we lose one degree of freedom in the original sample due to the restriction. The sum of squares $\sum_{i=1}^n (X_i - \bar{X}_n)^2$ has only $n-1$ degrees of freedom.

More generally, the number of degrees of freedom associated with a sum of squares is given by the number of observations used to compute the sum of squares minus the number of parameters that have to be replaced by their sample estimates. The number of parameters replaced is equal to the number of restrictions placed on the observations used to form the sum of squares.

Specifically, in a classical linear regression model

$$Y_i = X_i'\theta_0 + \varepsilon_i, \qquad i = 1, \ldots, n,$$

where X_i is a $p \times 1$ explanatory vector, and θ_0 is an unknown $p \times 1$ parameter vector, $\{\varepsilon_i\}_{i=1}^n$ is an IID disturbance sequence from a $N(0, \sigma_\varepsilon^2)$ population. For simplicity, we assume that $\{X_i\}_{i=1}^n$ are nonstochastic. The OLS estimator $\hat{\theta}$ for θ_0 solves the minimization problem

$$\hat{\theta} = \arg\min_\theta \sum_{i=1}^n (Y_i - X_i'\theta)^2.$$

The first order conditions are

$$\sum_{i=1}^n X_i(Y_i - X_i'\hat{\theta}) = \mathbf{0}.$$

This is a system of p equations. Solving for them, we obtain

$$\hat{\theta} = \left(\sum_{i=1}^{n} X_i X_i' \right)^{-1} \sum_{i=1}^{n} X_i Y_i$$
$$= (\mathbf{X}'\mathbf{X})^{-1}\mathbf{X}'\mathbf{Y},$$

where $\mathbf{X}'\mathbf{X} = \Sigma_{i=1}^{n} X_i X_i'$ is a $p \times p$ matrix, and $\mathbf{Y} = (Y_1, \cdots, Y_n)'$ is a $n \times 1$ vector. Under a set of regularity conditions, we have

$$\hat{\theta} - \theta_0 \sim N(0, \sigma_\varepsilon^2 (\mathbf{X}'\mathbf{X})^{-1}).$$

In order to make inference of θ_0, we have to estimate the error variance σ_ε^2. We can use the residual variance estimator

$$s^2 = \frac{1}{n-p} \sum_{i=1}^{n} (Y_i - X_i'\hat{\theta})^2.$$

The reason of using $n-p$ is that the $p \times 1$ parameter vector θ_0 is unknown, and has to be replaced by the OLS estimator $\hat{\theta}$. This leads to the loss of p degrees of freedom in the n estimated residuals $\{e_i = Y_i - X_i'\hat{\theta}\}_{i=1}^{n}$, which obey p restrictions imposed in the first order conditions. Using $n-p$ ensures that $E(s^2) = \sigma_\varepsilon^2$ under suitable regularity conditions. See Chapter 10 for more discussion.

We now come back to derive the mean and variance of S_n^2 under the normality assumption on the random sample $\mathbf{X}^n$. The result $(n-1)S_n^2/\sigma^2 \sim \chi_{n-1}^2$ provides an alternative simpler method to prove $E(S_n^2) = \sigma^2$, but under a stronger condition. Given $E(\chi_{n-1}^2) = n - 1$, we have

$$E\left[\frac{(n-1)S_n^2}{\sigma^2} \right] = n - 1$$

or

$$\frac{n-1}{\sigma^2} E(S_n^2) = n - 1.$$

It follows that

$$E(S_n^2) = \sigma^2.$$

Also, from the result $(n-1)S_n^2/\sigma^2 \sim \chi_{n-1}^2$, we can derive $\text{var}(S_n^2)$ from the χ_{n-1}^2 distribution.

Theorem 6.8. *Suppose* $\mathbf{X}^n = (X_1, \cdots, X_n)$ *is an IID* $N(\mu, \sigma^2)$ *random sample. Then for all* $n > 1$,

$$\text{var}(S_n^2) = \frac{2\sigma^4}{n-1}.$$

Proof: Because

$$\frac{(n-1)S_n^2}{\sigma^2} \sim \chi_{n-1}^2$$

and the variance of χ_{n-1}^2 is $2(n-1)$, we have

$$\text{var}\left[\frac{(n-1)S_n^2}{\sigma^2}\right] = 2(n-1),$$

or

$$\left[\frac{(n-1)^2}{\sigma^4}\right] \cdot \text{var}(S_n^2) = 2(n-1).$$

Therefore, $\text{var}(S_n^2) = 2\sigma^4/(n-1)$. This completes the proof.

The fact that $\text{var}(S_n^2) = 2\sigma^4/(n-1)$ and $E(S_n^2) = \sigma^2$ implies that

$$\begin{aligned}
\text{MSE}(S_n^2) &= E(S_n^2 - \sigma^2)^2 \\
&= \text{var}(S_n^2) \\
&= \frac{2\sigma^4}{n-1} \to 0 \text{ as } n \to \infty.
\end{aligned}$$

Thus, there is less and less variation between the sample variance S_n^2 and σ^2 as $n \to \infty$. In other words, S_n^2 becomes closer and closer to σ^2 as n increases.

Figure 6.4 plots the sampling distributions of the sample variance S_n^2 for various sample sizes ($n = 2, 5, 10, 30$) under following population distributions: (1) $N(0,1)$; (2) Bernoulli($\frac{1}{2}$); (3) $U[0,1]$; (4) EXP(1), all standardized with mean 0 and variance 1. As can be seen, the sampling distributions of S_n^2 are all centered at the population variance (i.e., $E(S_n^2) = \sigma^2$) for all sample sizes, and the dispersion of S_n^2 shrinks as the sample size n increases.

6.4　Student's t-Distribution

We now introduce a distribution called Student's t-distribution which plays a rather important role in classical statistical inference.

Definition 6.6. [Student's t-Distribution]: Let $U \sim N(0,1), V \sim \chi_\nu^2$, and U and V are independent. Then the random variable

$$T = \frac{U}{\sqrt{V/\nu}} \sim \frac{N(0,1)}{\sqrt{\chi_\nu^2/\nu}}$$

follows a Student's t distribution with ν degrees of freedom, denoted as $T \sim t_\nu$.

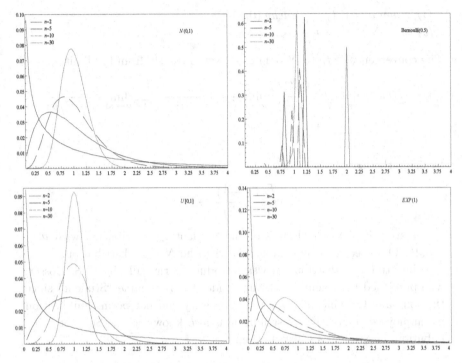

Figure 6.4: Sampling distribution of sample variance S_n^2

The PDF of a Student's t_ν distribution is

$$f_T(t) = \frac{\Gamma(\frac{\nu+1}{2})}{\Gamma(\frac{\nu}{2})} \frac{1}{(\nu\pi)^{1/2}} \frac{1}{(1 + t^2/\nu)^{(\nu+1)/2}}, \qquad -\infty < t < \infty.$$

This could be obtained by first finding the PDF $f_{TR}(t, r)$ of the bivariate transformation

$$T = U/\sqrt{V/\nu},$$
$$R = U,$$

and then integrating out R.

We first examine the properties of the Student t-distribution.

Lemma 6.3. *[Properties of Student t_ν Distribution]*:
(1) The PDF of t_ν is symmetric about 0.
(2) t_ν has a heavier distributional tail than $N(0, 1)$ (see Figure 6.5 below).
(3) Only the first $\nu - 1$ moments exist. In particular, the mean $\mu = 0$, and the variance $\sigma^2 = \nu/(\nu - 2)$ when $\nu > 2$. The MGF does not exist for any given ν.

(4) When $\nu = 1$, $t_1 \sim$ Cauchy $(0,1)$.
(5) $t_\nu \to N(0,1)$ as $\nu \to \infty$.

The convergence of t_ν to $N(0,1)$ can be seen clearly from the limit

$$\lim_{\nu \to \infty} f_T(t) = \lim_{\nu \to \infty} \sqrt{\frac{2}{\nu}} \frac{\Gamma(\frac{\nu+1}{2})}{\Gamma(\frac{\nu}{2})} \lim_{\nu \to \infty} \frac{1}{(1+t^2/\nu)^{1/2}} \frac{1}{\sqrt{2\pi}} \lim_{\nu \to \infty} \frac{1}{(1+t^2/\nu)^{\nu/2}}$$

$$= \frac{1}{\sqrt{2\pi}} e^{-\frac{1}{2}t^2}$$

by using the fact that $(1+a/\nu)^\nu \to e^a$ as $\nu \to \infty$. Here, as $\nu \to \infty$,

$$\sqrt{\frac{2}{\nu}} \frac{\Gamma\left(\frac{\nu+1}{2}\right)}{\Gamma\left(\frac{\nu}{2}\right)} \to 1.$$

Figure 6.5 plots the PDF's of the Student t_ν distribution with $\nu = 1, 5, 10, 30$ respectively, in comparison with the $N(0,1)$ distribution.

The Student t distribution was introduced originally by W.S. Gosset, who published her scientific writings under the pen name "Student" since the company for which she worked, a brewery, did not permit publication by employees. Thus, the t distribution is also known as the Student t distribution, or Student's t distribution.

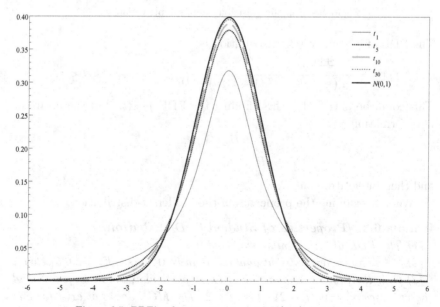

Figure 6.5: PDF's of t_1, t_5, t_{10}, t_{30} and $N(0,1)$ distributions

The Student t distribution has a heavier tail than the normal distribution, and so it is more appropriate in modeling high frequency financial returns. For example, Bollerslev (1987) suggests using a t distribution with degree of freedom larger than 2 to replace the original normal innovation to catch the leptokurtosis and fat tail of stock return.

The Student t-distribution has classical importance in statistical inference. Recall that when $\mathbf{X}^n$ is an IID $N(\mu, \sigma^2)$ random sample, we have for all $n \geq 1$,

$$\frac{\bar{X}_n - \mu}{\sigma/\sqrt{n}} \sim N(0, 1).$$

This is an important result, which could be used for confidence interval estimation and hypothesis testing about μ when the population variance σ^2 is known. However, the major difficulty in applying this result is that in most realistic applications the population standard deviation σ is unknown. This makes it necessary to replace σ with an estimator, usually the sample standard deviation S_n. Thus, the theory that follows leads to the exact distribution of

$$\frac{\bar{X}_n - \mu}{S_n/\sqrt{n}}$$

for random samples from a $N(\mu, \sigma^2)$ population.

Theorem 6.9. *Let $\mathbf{X}^n = (X_1, \cdots, X_n)$ be an IID random sample from a $N(\mu, \sigma^2)$ distribution. Then for all $n > 1$, the standardized sample mean*

$$\frac{\bar{X}_n - \mu}{S_n/\sqrt{n}} = \frac{\frac{\bar{X}_n - \mu}{\sigma/\sqrt{n}}}{\sqrt{\frac{(n-1)S_n^2}{\sigma^2}/(n-1)}}$$

$$\sim \frac{N(0,1)}{\sqrt{\chi_{n-1}^2/(n-1)}}$$

$$\sim t_{n-1},$$

where t_{n-1} is the Student t-distribution with $n-1$ degrees of freedom.

Proof: Put $U = (\bar{X}_n - \mu)/(\sigma/\sqrt{n})$, and $V = (n-1)S_n^2/\sigma^2$. Then $U \sim N(0,1)$ and $V \sim \chi_{n-1}^2$. Also, $\bar{X}_n$ and S_n^2 are independent by Theorem 6.7. It follows that

$$\frac{\bar{X}_n - \mu}{S_n/\sqrt{n}} = \frac{(\bar{X}_n - \mu)/(\sigma/\sqrt{n})}{\sqrt{(n-1)S_n^2/[\sigma^2(n-1)]}}$$

$$\sim t_{n-1}.$$

This completes the proof.

To illustrate the usefulness of the Student t distribution in statistical inference, we now consider confidence interval estimation for the population mean μ and hypothesis testing about the population mean μ respectively.

Example 6.7. [Confidence Interval Estimation for Population Mean μ]: Suppose $\mathbf{X}^n = (X_1, \cdots, X_n)$ is an IID random sample from a $N(\mu, \sigma^2)$ population, where both μ and σ^2 are unknown. We are interested in constructing a confidence interval estimator for μ at the $(1 - \alpha)100\%$ confidence level.

Given $\alpha \in (0, 1)$, a $(1 - \alpha)100\%$-confidence interval estimator for μ is defined as a random interval $[\hat{L}, \hat{U}]$ such that the probability that the true population mean μ falls between the interval $[\hat{L}, \hat{U}]$ is equal to $1 - \alpha$; namely,

$$P\left(\hat{L} < \mu < \hat{U}\right) = 1 - \alpha.$$

To construct an interval estimator for μ when σ^2 is unknown, we define the upper-tailed critical value $C_{t_{n-1}, \alpha}$ of a Student's t_{n-1} distribution by

$$P\left(t_{n-1} > C_{t_{n-1}, \alpha}\right) = \alpha,$$

as shown in Figure 6.6. By Theorem 6.9 and the symmetry of the Student-t distribution, we have

$$P\left[\left|\frac{\sqrt{n}(\bar{X}_n - \mu)}{S_n}\right| > C_{n-1, \frac{\alpha}{2}}\right] = \alpha$$

or equivalently,

$$P\left[\left|\frac{\sqrt{n}(\bar{X}_n - \mu)}{S_n}\right| \leq C_{n-1, \frac{\alpha}{2}}\right] = 1 - \alpha.$$

This yields a $(1 - \alpha)100\%$ confidence interval estimator for μ when σ^2 is unknown:

$$P\left(\bar{X}_n - \frac{S_n}{\sqrt{n}} C_{t_{n-1}, \frac{\alpha}{2}} < \mu < \bar{X}_n + \frac{S_n}{\sqrt{n}} C_{t_{n-1}, \frac{\alpha}{2}}\right) = 1 - \alpha.$$

The random interval estimator

$$\left[\bar{X}_n - \frac{S_n}{\sqrt{n}} C_{t_{n-1}, \frac{\alpha}{2}}, \bar{X}_n + \frac{S_n}{\sqrt{n}} C_{t_{n-1}, \frac{\alpha}{2}}\right]$$

is computable when σ^2 is unknown. Note that here, the sampling distribution of

$$\frac{\bar{X}_n - \mu}{S_n / \sqrt{n}}$$

plays a crucial role of determining the critical value $C_{t_{n-1}, \alpha/2}$, and thus the confidence interval estimator.

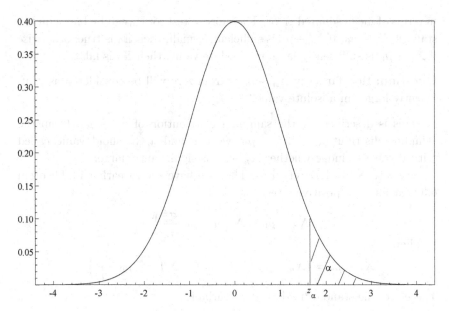

Figure 6.6: Upper-tailed critical value $C_{n-1,\alpha}$ of Student's t_{n-1} distribution

Example 6.8. [**Hypothesis Testing on Population Mean: The t-test**]: Suppose there is an IID $N(\mu, \sigma^2)$ random sample $\mathbf{X}^n = (X_1, \cdots, X_n)$ of size n, and we are interested in testing the hypothesis

$$\mathbb{H}_0 : \mu = \mu_0,$$

where μ_0 is a given (known) constant (e.g., $\mu_0 = 0$). How can we test this hypothesis?

To test the hypothesis $\mathbb{H}_0$, we consider the following statistic

$$\bar{X}_n - \mu_0 = (\bar{X}_n - \mu) + (\mu - \mu_0).$$

When $\mathbb{H}_0$ is true, $\mu = \mu_0$. It follows that

$$\bar{X}_n - \mu_0 = \bar{X}_n - \mu \to 0 \text{ as } n \to \infty$$

in terms of mean squared error. Therefore, the statistic $\bar{X}_n - \mu_0$ will be close to zero as $n \to \infty$. On the other hand, if $\mathbb{H}_0$ is false, i.e. $\mu \neq \mu_0$, then

$$\bar{X}_n - \mu_0 = (\bar{X}_n - \mu) + (\mu - \mu_0)$$
$$\to \mu - \mu_0 \neq 0 \text{ as } n \to \infty$$

in terms of mean squared error. Therefore, a test for $\mathbb{H}_0$ can be based on the statistic $\bar{X}_n - \mu_0$. If $\bar{X}_n - \mu_0$ is sufficiently small, then $\mathbb{H}_0$ is true; otherwise if $\bar{X}_n - \mu_0$ is sufficiently large in absolute value, then $\mathbb{H}_0$ is false.

Question: How far away $\bar{X}_n - \mu_0$ is from zero will be considered as "sufficiently large" in absolute value?

This is described by the sampling distribution of $\bar{X}_n - \mu_0$. From the sampling distribution of $\bar{X}_n - \mu_0$, we can find a threshold value called critical value to judge whether $\bar{X}_n - \mu_0$ is significantly large.

Suppose $\mathbf{X}^n \sim$ IID $N(\mu, \sigma^2)$. Then we have shown earlier in Theorem 6.3 that for each positive integer n,

$$\bar{X}_n - \mu_0 \sim N\left(\mu - \mu_0, \frac{\sigma^2}{n}\right).$$

It follows that

$$\bar{X}_n - \mu_0 = (\bar{X}_n - \mu) + (\mu - \mu_0) \sim N\left(\mu - \mu_0, \frac{\sigma^2}{n}\right).$$

Therefore, the standardized random variable

$$\frac{\bar{X}_n - \mu_0}{\sigma/\sqrt{n}} = \frac{\bar{X}_n - \mu}{\sigma/\sqrt{n}} + \frac{\sqrt{n}(\mu - \mu_0)}{\sigma}$$

$$\sim N\left(\frac{\sqrt{n}(\mu - \mu_0)}{\sigma}, 1\right).$$

When the hypothesis $\mathbb{H}_0$ holds,

$$\frac{\bar{X}_n - \mu_0}{\sigma/\sqrt{n}} \sim N(0,1),$$

which implies that $(\bar{X}_n - \mu_0)/(\sigma/\sqrt{n})$ will take small and finite values with a very high probability. There is only a very small probability that $(\bar{X}_n - \mu_0)/(\sigma/\sqrt{n})$ will take a large value. On the other hand, when $\mathbb{H}_0$ is false,

$$\frac{\bar{X}_n - \mu_0}{\sigma/\sqrt{n}} \to \infty \text{ as } n \to \infty$$

with high probability. Therefore, we can test $\mathbb{H}_0$ by examining whether $(\bar{X}_n - \mu_0)/(\sigma/\sqrt{n})$ is large in absolute value.

However, the quantity

$$\frac{\bar{X}_n - \mu_0}{\sigma/\sqrt{n}}$$

is not a feasible statistic, because it involves the unknown parameter σ (note that μ_0 is a given parameter value (e.g., $\mu_0 = 0$) and it causes no trouble!).

We have to replace σ with an estimator for σ, say the sample standard deviation S_n. This leads us to consider the following feasible statistic

$$T(\mathbf{X}^n) = \frac{\bar{X}_n - \mu_0}{S_n/\sqrt{n}}.$$

However, the distribution of $T(\mathbf{X}^n)$ is no longer $N(0,1)$; instead it becomes a Student t-distribution with $n-1$ degrees of freedom. Under $\mathbb{H}_0 : \mu = \mu_0$,

$$T(\mathbf{X}^n) \sim t_{n-1}$$

for all $n > 1$. This follows because under $\mathbb{H}_0$

$$
\begin{aligned}
T(\mathbf{X}^n) &= \frac{\bar{X}_n - \mu_0}{S_n/\sqrt{n}} \\
&= \frac{\bar{X}_n - \mu}{S_n/\sqrt{n}} + \frac{\sqrt{n}(\mu - \mu_0)}{S_n} \\
&= \frac{\bar{X}_n - \mu}{S_n/\sqrt{n}} \\
&\sim t_{n-1}.
\end{aligned}
$$

Thus, with a very high probability, the t-test statistic $T(\mathbf{X}^n)$ will take small and finite values, and there is only a tiny probability that the $T(\mathbf{X}^n)$ statistic will take a large value when $\mathbb{H}_0$ is true.

On the other hand, when $\mathbb{H}_0 : \mu = \mu_0$ is false, i.e., when $\mu \neq \mu_0$, we have

$$T(\mathbf{X}^n) = \frac{\bar{X}_n - \mu}{S_n/\sqrt{n}} + \frac{\sqrt{n}(\mu - \mu_0)}{S_n} \to \infty$$

as $n \to \infty$ with high probability. In other words, the statistic $T(\mathbf{X}^n)$ diverges to infinity with probability approaching 1 as $n \to \infty$ under the alternative to $\mathbb{H}_0$.

We can propose the following t-test decision rule using critical values:

- Reject the hypothesis $\mathbb{H}_0 : \mu = \mu_0$ at a prespecified significance level $\alpha \in (0,1)$ if

$$|T(\mathbf{X}^n)| > C_{t_{n-1}, \frac{\alpha}{2}},$$

  where $C_{t_{n-1}, \frac{\alpha}{2}}$ is the upper-tailed critical value of the Student t_{n-1} distribution at level $\frac{\alpha}{2}$, determined by $P(t_{n-1} > C_{t_{n-1}, \frac{\alpha}{2}}) = \frac{\alpha}{2}$.

- Accept the hypothesis $\mathbb{H}_0$ at the significance level α if $|T(\mathbf{X}^n)| \leq C_{t_{n-1}, \frac{\alpha}{2}}$.

Figure 6.7 plots the rejection region and acceptance region of the t-test decision rule based on the critical values when the sample size $n = 20$ at the 5% significance level.

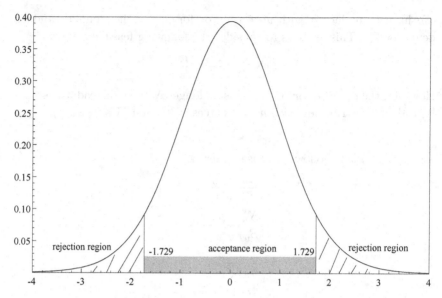

Figure 6.7: Rejection and acceptance regions of t-test for $n = 20$, $\alpha = 5\%$

Intuitively, the t-test decision rules says that when $|T(\mathbf{X}^n)| > C_{t_{n-1},\alpha/2}$, it implies that $\bar{X}_n - \mu_0$ is significantly different from 0, so $\mathbb{H}_0$ should be rejected. On the other hand, if $|T(\mathbf{X}^n)| \leq C_{t_{n-1},\alpha/2}$, then $\bar{X}_n - \mu_0$ is not significantly different from 0, and so $\mathbb{H}_0$ should be accepted.

In testing $\mathbb{H}_0$ using a data generated from the random sample $\mathbf{X}^n$ of size n, there exist two type of errors. One possibility is that $\mathbb{H}_0$ is true but we reject it. This is possible because the test statistic $T(\mathbf{X}^n)$ follows a Student t_{n-1} distribution under $\mathbb{H}_0$, which has an unbounded support. Thus, there exists a small probability that $T(\mathbf{X}^n)$ can still take a larger value than the critical value under $\mathbb{H}_0$. This is the so-called Type I error. The significance level α controls Type I error. Conventional choices of α are 10%, 5% and 1%. If

$$P\left[|T(\mathbf{X}^n)| > C_{t_{n-1},\frac{\alpha}{2}}|\mathbb{H}_0\right] = \alpha,$$

we call the decision rule a size α test or a test with size α.

On the other hand, the probability

$$P\left[|T(\mathbf{X}^n)| > C_{t_{n-1},\frac{\alpha}{2}}|\mathbb{H}_0 \text{ is false}\right]$$

is called the power function of the size α t-test. When $P[|T(\mathbf{X}^n)| > C_{t_{n-1},\frac{\alpha}{2}}|\mathbb{H}_0 \text{ is false}] < 1$, there exists a possibility that one may accept $\mathbb{H}_0$ when it is false. This is called the Type II error.

When n is finite, due to the nature of limited information offered by the random sample $\mathbf{X}^n$, both Type I and Type II errors are unavoidable and there usually exists a tradeoff between them. In practice, one usually sets a level for the Type I error and then minimizes the Type II error.

When n is large, one can use critical values of the standard normal $N(0,1)$ distribution, because they will be close to those of the t-distribution. Suppose $\mathbf{X}^n$ is an IID $N(\mu, \sigma^2)$ sequence. Then

$$\frac{\bar{X}_n - \mu}{S_n/\sqrt{n}} \sim t_{n-1} \to N(0,1) \text{ as } n \to \infty.$$

This follows because $t_{n-1} \to N(0,1)$ as $n \to \infty$. Thus, for large n, there is little difference whether one uses critical values of t_{n-1} or $N(0,1)$. In practice, the normal approximation for t_{n-1} works rather well if $n-1 \geq 30$.

The above critical value-based decision rule can be equivalently described by using the so-called P-value of the test statistic $T(\mathbf{X}^n)$. Given any observed data set $\mathbf{x}^n$, we can compute a value (a realization)

$$T(\mathbf{x}^n) = \frac{\bar{x}_n - \mu_0}{s_n/\sqrt{n}}$$

for the t-test statistic $T(\mathbf{X}^n)$. Then the probability

$$p(\mathbf{x}^n) = P\left(|t_{n-1}| > |T(\mathbf{x}^n)|\right)$$

is called the P-value of the t-test statistic $T(\mathbf{X}^n)$ when a data set $\mathbf{x}^n$ is observed. It can be viewed as the probability that the t-test statistic $T(\mathbf{X}^n)$ is larger than the observed value $T(\mathbf{x}^n)$ when $\mathbb{H}_0$ holds. When the observed value $T(\mathbf{x}^n)$ is large, $p(\mathbf{x}^n)$ will be small. Thus, a small P-value is strong evidence against the null hypothesis $\mathbb{H}_0$, while a large P-value shows that the data are consistent with $\mathbb{H}_0$. Accordingly, the above critical value-based decision rule is equivalent to the following P-value based decision rule:

- Reject the hypothesis $\mathbb{H}_0$ at the significance level α if $p(\mathbf{x}^n) < \alpha$.
- Accept the hypothesis $\mathbb{H}_0$ at the significance level α if $p(\mathbf{x}^n) \geq \alpha$.

By definition, the P-value is the smallest value of the significance level α at which $\mathbb{H}_0$ can be rejected. The P-value not only tells us whether $\mathbb{H}_0$ should be accepted or rejected at a given significance level, but it also tells us whether the decision to accept or reject $\mathbb{H}_0$ is a close call. As a matter of fact, many statistical software applications report P-values for test statistics or parameter estimators.

A rejection of the null hypothesis $\mathbb{H}_0$ based on either of the above decision rules is called a *statistically significant* effect. From a statistical perspective, for any deviation from $\mathbb{H}_0$ (i.e., any difference between $\mu - \mu_0$), no matter how small it is, a rejection decision will be made as long as the sample size n is sufficiently large. However, a small difference $\mu - \mu_0$ may not be practically significant or important from an economic perspective. For example, one may be interested in whether the expected return (μ) on a mutual fund is significantly different from a pre-specified rate (μ_0) of return. By significance here, we usually mean its economic significance. That is, the size of the difference $\mu - \mu_0$ should be large enough to consider an investment on the mutual fund, due to (e.g.,) existence of transaction costs. However, a statistic test like the t-test introduced above will reject a small difference $\mu - \mu_0$ as long as the sample size n is sufficiently large. In other words, an economically insignificant effect is likely to be statistically significant.

For more formal discussion on hypothesis testing, see Chapter 9.

6.5   Snedecor's F Distribution

Another distribution that plays an important role in connection with sampling from normal populations is the F distribution, named after Sir Ronald A. Fisher, one of the most prominent statisticians of the nineteenth century. Originally, it was studied as the sampling distribution of the ratio of two independent random variables with Chi-square distributions each divided by their respective degrees of freedom, and this is how we shall present it here.

Definition 6.7. [**The F Distribution**]: Let U and V be two independent Chi-square random variables with p and q degrees of freedom respectively. Then the random variable

$$F = \frac{U/p}{V/q} \sim F_{p,q}$$

follows a F distribution with p and q degrees of freedom.

What is the PDF of a $F_{p,q}$ distribution?

The PDF is given by

$$f_F(x) = \frac{\Gamma(\frac{p+q}{2})}{\Gamma(\frac{p}{2})\Gamma(\frac{q}{2})} \left(\frac{p}{q}\right)^{p/2} \frac{x^{(p/2)-1}}{[1 + (p/q)x]^{(p+q)/2}}, \qquad 0 < x < \infty.$$

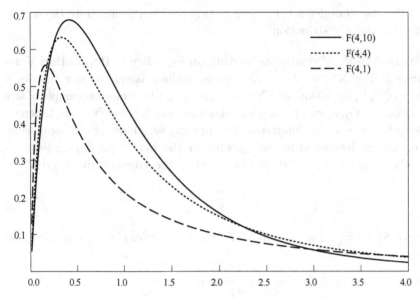

Figure 6.8: PDF's of F distribution with different freedoms (p, q)

This PDF could be obtained by using the bivariate transformation

$$F = (U/p)/(V/q),$$
$$G = U,$$

and then integrating out G.

Figure 6.8 shows the PDF of the F distribution with different freedoms (p, q).

Lemma 6.4. *[Properties of $F_{p,q}$ Distribution]:*

(1) If $X \sim F_{p,q}$, then $X^{-1} \sim F_{q,p}$.

(2) If $X \sim t_q$, then $X^2 \sim F_{1,q}$.

(3) If $q \to \infty$, then $p \cdot F_{p,q} \to \chi_p^2$.

Proof: Result (1) follows from the definition of an F random variable. For Result (2), recall that a t_q random variable is defined as

$$t_q \sim \frac{Z}{\sqrt{\chi_q^2/q}},$$

where $Z \sim N(0,1)$ and it is independent of χ_q^2. It follows that

$$t_q^2 \sim \frac{\chi_1^2/1}{\chi_q^2/q} \sim F_{1,q}.$$

We now consider a hypothesis testing problem to illustrate the importance of the F distribution.

Example 6.9. [Hypothesis Testing on Equality of Population Variances]: Let $\mathbf{X}^n = (X_1, \cdots, X_n)$ be a random sample of size n from a $N(\mu_X, \sigma_X^2)$ population, and $\mathbf{Y}^m = (Y_1, \cdots, Y_m)$ be a random sample of size m from a $N(\mu_Y, \sigma_Y^2)$ population. Assume that $\mathbf{X}^n$ and $\mathbf{Y}^m$ are independent. Suppose we are interested in comparing variability of the population, i.e., we are interested in testing whether the hypothesis $\mathbb{H}_0 : \sigma_X^2 = \sigma_Y^2$ holds. Then a test statistic can be based on the sample variance ratio

$$\frac{S_X^2}{S_Y^2}.$$

Since $S_X^2 \to \sigma_X^2$ as $n \to \infty$ in MSE, and $S_Y^2 \to \sigma_Y^2$ as $m \to \infty$ in MSE, we have

$$\frac{S_X^2}{S_Y^2} \to \frac{\sigma_X^2}{\sigma_Y^2} \text{ as } n, m \to \infty.$$

Under $\mathbb{H}_0 : \sigma_X^2 = \sigma_Y^2$, we have

$$\frac{S_X^2}{S_Y^2} = \frac{S_X^2/\sigma_X^2}{S_Y^2/\sigma_Y^2}$$

$$= \frac{\frac{(n-1)S_X^2/\sigma_X^2}{n-1}}{\frac{(m-1)S_Y^2/\sigma_Y^2}{m-1}}$$

$$\sim \frac{\chi_{n-1}^2/(n-1)}{\chi_{m-1}^2/(m-1)}$$

$$\sim F_{n-1,m-1}.$$

If $\mathbb{H}_0$ is false, and so $\sigma_X^2 \neq \sigma_Y^2$, then $\frac{S_X^2}{S_Y^2} \neq \frac{S_X^2/\sigma_X^2}{S_Y^2/\sigma_Y^2} \sim F_{n-1,m-1}$. Therefore, by checking whether S_X^2/S_Y^2 follows $F_{n-1,m-1}$, we can test whether the variances are equal. Because the F-distribution is closely related to the ratio of sample variances, it is sometime called the variance ratio distribution. However, a random variable that follows an F-distribution need not be a ratio of sample variances.

The F-test is an important testing principle in classical statistics and econometrics, where S_X^2 and S_Y^2 are generalized to the sums of squared residuals of a restricted regression model and an unrestricted regression model respectively. For example, consider the classical linear regression model

$$Y_i = X_i'\beta + Z_i'\gamma + \varepsilon_i,$$

where β is a $p \times 1$ parameter vector, γ is a $q \times 1$ parameter vector, $\{\varepsilon_i\}$ is a sequence of IID $N(0, \sigma_\varepsilon^2)$ random variables and is independent of $\mathbf{X} = (X_1, \cdots, X_n)'$ and $\mathbf{Z} = (Z_1, \cdots, Z_n)'$. Suppose we are interested in testing the null hypothesis $\mathbb{H}_0 : \gamma = \mathbf{0}$. Under $\mathbb{H}_0$, Z_i has no impact on the conditional mean $E(Y_i|X_i, Z_i)$. Under the alternative to $\mathbb{H}_0$, Z_i has impact on the conditional mean $E(Y_i|X_i, Z_i)$, and as a result, a restricted linear regression model

$$Y_i = X_i'\beta + \varepsilon_i$$

is called suffering from the omitted variables problem.

To test the hypothesis $\mathbb{H}_0$, we can run two OLS regressions. One is the unrestricted regression model

$$Y_i = X_i'\beta + Z_i'\gamma + \varepsilon_i,$$

with the OLS estimator

$$(\hat{\beta}, \hat{\gamma}) = \arg \min_{\beta \in \mathbb{R}^p, \gamma \in \mathbb{R}^q} \sum_{i=1}^{n} (Y_i - X_i'\beta - Z_i'\gamma)^2.$$

The residual variance estimator is

$$S_U^2 = \frac{1}{n-p-q} \sum_{i=1}^{n} (Y_i - X_i'\hat{\beta} - Z_i'\hat{\gamma})^2,$$

where the subscript U denotes the unrestricted model. Another is the restricted regression model

$$Y_i = X_i'\beta + u_i,$$

with the OLS estimator

$$\tilde{\beta} = \arg \min_{\beta \in \mathbb{R}^p} \sum_{i=1}^{n} (Y_i - X_i'\beta)^2.$$

The residual variance estimator is

$$S_R^2 = \frac{1}{n-p} \sum_{i=1}^{n} (Y_i - X_i'\tilde{\beta})^2,$$

where R denotes the restricted model. To test $\mathbb{H}_0$, one can compare the residual variance estimators, S_U^2 and S_R^2. Under $\mathbb{H}_0$, they will converge to the same limit. Under the alternative, the restricted model is misspecified in the sense that the omitted variables Z_i have impact on Y_i, and S_U^2 and S_R^2 converge to different limits with $\lim_{n\to\infty}(S_R^2/S_U^2) > 1$. A test statistic for $\mathbb{H}_0$ can then be constructed as

$$F = \frac{[(n-p)S_R^2 - (n-p-q)S_U^2]/q}{(n-p-q)S_U^2/(n-p-q)}$$
$$\sim F_{q,n-p-q}.$$

The F statistic is nonnegative because the sum of squared residuals of the restricted model is always larger than that of the unrestricted model. It can be shown that under suitable regularity conditions and the null hypothesis,

$$F \sim \frac{\chi_q^2/q}{\chi_{n-p-q}^2/(n-p-q)} \sim F_{q,n-p-q}.$$

Because $\lim_{n\to\infty}(S_R^2/S_U^2) > 1$ under the alternative, the upper-tailed critical value of the $F_{p,q}$ distribution should be used. See Chapter 10 for more discussion.

To sum up the discussion of Sections 6.4 and 6.5, the t and F distributions are important in applications that are based on data generated from a normal population. The normality assumption is reasonable when a random variable itself is the sum of many "small" random variables, due to the effect of the Central Limit Theorem (CLT) to be introduced in Chapter 7. Some economic and financial data, however, are highly non-normally distributed. As a consequence, the theory of the sampling distribution for normal populations will not be applicable, and we need to use asymptotic theory or other tools to investigate the sampling distribution of statistics of interest. See Chapter 7 for an introduction to asymptotic theory.

6.6 Sufficient Statistics

There is a so-called KISS principle in econometrics and statistics, namely, *"Keep It Sophisticatedly Simple"*. We now introduce a concept called sufficient statistic which reflects such a principle.

Suppose we are interested in making inference of parameter θ using a set of data generated from a random sample $\mathbf{X}^n$ from a population $f_X(x) = f(x, \theta)$. Under what conditions, can the information about θ that is contained in the random sample $\mathbf{X}^n$ be completely summarized by some low-dimensional functions of $\mathbf{X}^n$, say, some statistic $T(\mathbf{X}^n)$?

Suppose a random experiment gives rise to a realization $\mathbf{x}^n$ for the random sample $\mathbf{X}^n$, and suppose Person A observes $\mathbf{x}^n$ while Person B only observes the value of $t = T(\mathbf{x}^n)$. Generally, Person A knows better than Person B about the unknown parameter value of θ.

However, there may exist situations in which Person B can do just as well as Person A. This occurs when the statistic $T(\mathbf{X}^n)$ summarizes all information about θ that is contained in $\mathbf{X}^n$, so that individual values of $\mathbf{x}^n$ are irrelevant in search for a good estimator of θ. A statistic $T(\mathbf{X}^n)$ that has this desired property is called a sufficient statistic for parameter θ. An important implication of a sufficient statistic for parameter θ is that one can then just keep the sufficient statistic $T(\mathbf{X}^n)$, which is very low dimensional. This is rather convenient in light of the fact that the original random sample $\mathbf{X}^n$ has a very high dimension equal to the sample size n.

When a random sample $\mathbf{X}^n = (X_1, \cdots, X_n)$ is given, there might be too much irrelevant or redundant information if we are only interested in knowing a specific aspect of the population, say the unknown value of parameter θ. There exist some methods of data reduction that "throw away" irrelevant information and maintain only the essential information about the unknown parameter θ. The sufficiency principle is such a method.

For example, suppose the random sample $\mathbf{X}^n \sim$ IID $N(\mu, \sigma^2)$, where $\theta = (\mu, \sigma^2)$. Then for inference of θ, only the sample mean $\bar{X}_n$ and the sample variance S_n^2 should be retained, because they are sufficient statistics for (μ, σ^2).

A natural question follows immediately: how can one check $(\bar{X}_n, S_n^2)$ are sufficient for $\theta = (\mu, \sigma^2)$ for a random sample $\mathbf{X}^n$ from a normal population? More generally, how can one find a sufficient statistic for parameter θ associated with some population?

We first give a formal definition of a sufficient statistic.

Definition 6.8. [Sufficient Statistic]: Let $\mathbf{X}^n$ be a random sample from some population with parameter θ. A statistic $T(\mathbf{X}^n)$ is a sufficient statistic for θ if the conditional probability of the sample $\mathbf{X}^n = \mathbf{x}^n$ given that the value of the statistic $T(\mathbf{X}^n) = T(\mathbf{x}^n)$ does not depend on θ; that is,

$$f_{\mathbf{X}^n|T(\mathbf{X}^n)}[\mathbf{x}^n|T(\mathbf{x}^n), \theta] = h(\mathbf{x}^n) \text{ for all possible } \theta,$$

where the left hand side is the conditional PMF/PDF of $\mathbf{X}^n = \mathbf{x}^n$ given $T(\mathbf{X}^n) = T(\mathbf{x}^n)$, which generally depends on θ. The right hand side $h(\mathbf{x}^n)$ does not depend on θ; it is a function of $\mathbf{x}^n$ only.

Suppose $f_{\mathbf{X}^n|T(\mathbf{X}^n)}[\mathbf{x}^n|T(\mathbf{x}^n), \theta]$, the conditional probability of $\mathbf{X}^n = \mathbf{x}^n$ given $T(\mathbf{X}^n) = T(\mathbf{x}^n)$, does not depend on θ. Then all sample points $\{\mathbf{x}^n\}$ which yield the same value of $T(\mathbf{x}^n) = t$ for $T(\mathbf{X}^n)$, will be just equally likely to occur for any value of θ. In other words, since the conditional probability of $\mathbf{X}^n = \mathbf{x}^n$ given $T(\mathbf{X}^n) = T(\mathbf{x}^n)$ does not depend on θ, the data $\mathbf{x}^n$ beyond the value of $T(\mathbf{x}^n) = t$ does not provide any additional useful information about θ. As a consequence, the knowledge of the data set $\mathbf{x}^n$ except the value of $T(\mathbf{x}^n) = t$ will not help in the inference of θ. The sufficient statistic $T(\mathbf{X}^n)$ for θ thus captures all information in the sample $\mathbf{X}^n$ that is relevant to θ. All knowledge about θ that can be gained from the observed value $\mathbf{x}^n$ of the sample $\mathbf{X}^n$ can just as well be gained from the value of $T(\mathbf{x}^n)$ alone.

To gain insight into a sufficient statistic $T(\mathbf{X}^n)$, we consider discrete cases. First of all, sufficiency implies that the conditional PMF

$$f_{\mathbf{X}^n|T(\mathbf{X}^n)}\left[\mathbf{x}^n|T(\mathbf{x}^n),\theta\right] \equiv P_\theta[\mathbf{X}^n = \mathbf{x}^n|T(\mathbf{X}^n) = T(\mathbf{x}^n)]$$
$$= h(\mathbf{x}^n)$$

for all θ, where $P_\theta(\cdot)$ is the probability measure under the probability distribution of $\mathbf{X}^n$ which is usually indexed by θ. Recall that the full information of a random sample $\mathbf{X}^n$ is described by the joint probability of $\mathbf{X}^n = \mathbf{x}^n$, denoted by $P(\mathbf{X}^n = \mathbf{x}^n) = f_{\mathbf{X}^n}(\mathbf{x}^n,\theta)$. This joint probability depends on θ in general. For example, when $\mathbf{X}^n$ is an IID random sample with population PMF $f(x,\theta)$. Then

$$f_{\mathbf{X}^n}(\mathbf{x}^n,\theta) = \prod_{i=1}^{n} f(x_i,\theta).$$

Because $\mathbf{X}^n = \mathbf{x}^n$ implies $T(\mathbf{X}^n) = T(\mathbf{x}^n)$ but not vice versa given $T(\cdot)$ is a function, we have the event $A = \{\mathbf{X}^n = \mathbf{x}^n\} \subseteq B = \{T(\mathbf{X}^n) = T(\mathbf{x}^n)\}$. Therefore, $A = A \cap B$, and as a result, the joint PMF of the random sample $\mathbf{X}^n$

$$f_{\mathbf{X}^n}(\mathbf{x}^n,\theta) = P(\mathbf{X}^n = \mathbf{x}^n)$$
$$= P(A \cap B)$$
$$= P(A|B)P(B)$$
$$= P[\mathbf{X}^n = \mathbf{x}^n|T(\mathbf{X}^n) = T(\mathbf{x}^n)]P[T(\mathbf{X}^n) = T(\mathbf{x}^n)]$$
$$= h(\mathbf{x}^n)f_{T(\mathbf{X}^n)}[T(\mathbf{x}^n),\theta]$$

by sufficiency, where $f_{T(\mathbf{X}^n)}[T(\mathbf{x}^n),\theta] \equiv P[T(\mathbf{X}^n) = T(\mathbf{x}^n)]$ depends on θ but $h(\mathbf{x}^n)$ does not depend on θ.

We see that only the marginal probability $P[T(\mathbf{X}^n) = T(\mathbf{x}^n)]$ of the sufficient statistic $T(\mathbf{X}^n)$ is related to θ and the other part $h(\mathbf{x}^n)$ is irrelevant to θ. Therefore, if we are interested in making inference of θ, then we can only retain the information of $T(\mathbf{X}^n)$; all other information in the random sample $\mathbf{X}^n$ is redundant or useless in making inference of θ. In other words, using the information of the low-dimensional $T(\mathbf{x}^n)$ can do just as well as using the high-dimensional sample point $\mathbf{x}^n$ for the purpose of making inference of parameter θ.

For example, the so-called Maximum Likelihood Estimation (MLE) for θ, to be introduced in Chapter 8, is to maximize the objective function — the log-likelihood function

$$\ln f_{\mathbf{X}^n}(\mathbf{x}^n, \theta) = \ln h(\mathbf{x}^n) + \ln f_{T(\mathbf{X}^n)}[T(\mathbf{x}^n), \theta].$$

Because the first part is irrelevant to θ, we have

$$\max_{\theta \in \Theta} \ln f_{\mathbf{X}^n}(\mathbf{x}^n, \theta) = \max_{\theta \in \Theta} \ln f_{T(\mathbf{X}^n)}[T(\mathbf{x}^n), \theta],$$

where Θ is a parameter space. In other words, it suffices to maximize the log-likelihood function $\ln f_{T(\mathbf{X}^n)}[T(\mathbf{x}^n), \theta]$ of the sufficient statistic $T(\mathbf{X}^n)$ for the MLE of θ.

We now address an important question: how can one check if a statistic $T(\mathbf{X}^n)$ is sufficient for parameter θ?

It is often tedious to check whether a statistic is sufficient for θ using the definition of a sufficient statistic directly. We now provide a factorization theorem which makes it easier to check the sufficiency of a statistic.

Theorem 6.10. *[Factorization Theorem]: Let* $f_{\mathbf{X}^n}(\mathbf{x}^n, \theta)$ *denote the joint PDF (or PMF) of a random sample* $\mathbf{X}^n$. *A statistic* $T(\mathbf{X}^n)$ *is a sufficient statistic for* θ *if and only if there exist functions* $g(t, \theta)$ *and* $h(\mathbf{x}^n)$ *such that for any sample point* $\mathbf{x}^n$ *in the sample space of* $\mathbf{X}^n$ *and for any parameter value* $\theta \in \Theta$,

$$f_{\mathbf{X}^n}(\mathbf{x}^n, \theta) = g[T(\mathbf{x}^n), \theta] h(\mathbf{x}^n),$$

where $g(t, \theta)$ *depends on parameter* θ *but* $h(\mathbf{x}^n)$ *does not depend on parameter* θ.

Proof: We shall show only the discrete case, where $f_{\mathbf{X}^n}(\mathbf{x}^n, \theta) = P(\mathbf{X}^n = \mathbf{x}^n)$.

(1) [*Necessity*]: When $T(\mathbf{X}^n)$ is sufficient, noting that $\{\mathbf{X}^n = \mathbf{x}^n\} \subseteq \{T(\mathbf{X}^n) = T(\mathbf{x}^n)\}$, we have

$$\{\mathbf{X}^n = \mathbf{x}^n\} = \{\mathbf{X}^n = \mathbf{x}^n\} \cap \{T(\mathbf{X}^n) = T(\mathbf{x}^n)\}.$$

It follows that

$$f_{\mathbf{X}^n}(\mathbf{x}^n, \theta) = P(\mathbf{X}^n = \mathbf{x}^n)$$
$$= P[\mathbf{X}^n = \mathbf{x}^n, T(\mathbf{X}^n) = T(\mathbf{x}^n)]$$
$$= P[\mathbf{X}^n = \mathbf{x}^n | T(\mathbf{X}^n) = T(\mathbf{x}^n)] P[T(\mathbf{X}^n) = T(\mathbf{x}^n)]$$
$$= h(\mathbf{x}^n) P[T(\mathbf{X}^n) = T(\mathbf{x}^n)]$$
$$= h(\mathbf{x}^n) g[T(\mathbf{x}^n), \theta],$$

where $g[T(\mathbf{x}^n), \theta] = P[T(\mathbf{X}^n) = T(\mathbf{x}^n)]$ and $h(\mathbf{x}^n) = P[\mathbf{X}^n = \mathbf{x}^n | T(\mathbf{X}^n) = T(\mathbf{x}^n)]$. The latter does not depend on parameter θ.

(2) [*Sufficiency*]: Now suppose we have

$$f_{\mathbf{X}^n}(\mathbf{x}^n, \theta) = g[T(\mathbf{x}^n), \theta] h(\mathbf{x}^n).$$

We shall show that the conditional probability $P[\mathbf{X}^n = \mathbf{x}^n | T(\mathbf{X}^n) = T(\mathbf{x}^n)]$ does not depend on θ.

Because

$$\{\mathbf{X}^n = \mathbf{x}^n\} = \{\mathbf{X}^n = \mathbf{x}^n\} \cap \{T(\mathbf{X}^n) = T(\mathbf{x}^n)\},$$

we have

$$P[\mathbf{X}^n = \mathbf{x}^n | T(\mathbf{X}^n) = T(\mathbf{x}^n)]$$
$$= \frac{P[\mathbf{X}^n = \mathbf{x}^n, T(\mathbf{X}^n) = T(\mathbf{x}^n)]}{P[T(\mathbf{X}^n) = T(\mathbf{x}^n)]}$$
$$= \frac{P(\mathbf{X}^n = \mathbf{x}^n)}{P[T(\mathbf{X}^n) = T(\mathbf{x}^n)]}$$
$$= \frac{g[T(\mathbf{x}^n), \theta] h(\mathbf{x}^n)}{P[T(\mathbf{X}^n) = T(\mathbf{x}^n)]}.$$

We now consider the denominator:

$$P[T(\mathbf{X}^n) = T(\mathbf{x}^n)]$$
$$= \sum_{\{\mathbf{y}^n : T(\mathbf{y}^n) = T(\mathbf{x}^n)\}} f_{\mathbf{X}^n}(\mathbf{y}^n, \theta)$$
$$= \sum_{\{\mathbf{y}^n : T(\mathbf{y}^n) = T(\mathbf{x}^n)\}} g[T(\mathbf{y}^n), \theta] h(\mathbf{y}^n)$$
$$= \sum_{\{\mathbf{y}^n : T(\mathbf{y}^n) = T(\mathbf{x}^n)\}} g[T(\mathbf{x}^n), \theta] h(\mathbf{y}^n)$$
$$= g[T(\mathbf{x}^n), \theta] \sum_{\{\mathbf{y}^n : T(\mathbf{y}^n) = T(\mathbf{x}^n)\}} h(\mathbf{y}^n)$$

where the sum is taken over all possible sample points $\{\mathbf{y}^n\}$ in the sample space (i.e., the support) of $\mathbf{X}^n$ that yield the same value of $T(\mathbf{y}^n) = T(\mathbf{x}^n)$. It follows that the conditional probability

$$P\left[\mathbf{X}^n = \mathbf{x}^n | T(\mathbf{X}^n) = T(\mathbf{x}^n)\right] = \frac{g[T(\mathbf{x}^n), \theta] h(\mathbf{x}^n)}{P[T(\mathbf{X}^n) = T(\mathbf{x}^n)]}$$

$$= \frac{g[T(\mathbf{x}^n), \theta] h(\mathbf{x}^n)}{g[T(\mathbf{x}^n), \theta] \sum_{\{\mathbf{y}^n : T(\mathbf{y}^n) = T(\mathbf{x}^n)\}} h(\mathbf{y}^n)}$$

$$= \frac{h(\mathbf{x}^n)}{\sum_{\{\mathbf{y}^n : T(\mathbf{y}^n) = T(\mathbf{x}^n)\}} h(\mathbf{y}^n)},$$

which does not depend on θ. This completes the proof.

We now consider a few examples to illustrate the use of the factorization theorem.

Example 6.10. Suppose $\mathbf{X}^n \sim$ IID Bernoulli(θ), where $0 < \theta < 1$. Show that the sample proportion $T(\mathbf{X}^n) = n^{-1} \sum_{i=1}^{n} X_i$ is a sufficient statistic for θ. Note that $\theta = E(X_i)$.

Solution: The PMF of a Bernoulli(θ) random variable X_i is

$$f(x_i, \theta) = \theta^{x_i} (1 - \theta)^{1 - x_i},$$

where x_i takes value 0 or 1. Suppose $\mathbf{x}^n$ is a realization (i.e., a data set) of the random sample $\mathbf{X}^n$. We have

$$P(\mathbf{X}^n = \mathbf{x}^n) = \prod_{i=1}^{n} f(x_i, \theta)$$

$$= \prod_{i=1}^{n} \theta^{x_i} (1 - \theta)^{1 - x_i}$$

$$= \theta^{\sum_{i=1}^{n} x_i} (1 - \theta)^{n - \sum_{i=1}^{n} x_i}$$

$$= \theta^{n T(\mathbf{x}^n)} (1 - \theta)^{n - n T(\mathbf{x}^n)}$$

$$= g[T(\mathbf{x}^n), \theta] h(\mathbf{x}^n),$$

where $T(\mathbf{X}^n) = n^{-1} \sum_{i=1}^{n} X_i$, $h(\mathbf{x}^n) = 1$, and $g[T(\mathbf{x}^n), \theta] = \theta^{n T(\mathbf{x}^n)} (1 - \theta)^{n - n T(\mathbf{x}^n)}$. Note that $G_n = n T(\mathbf{X}^n) = \sum_{i=1}^{n} X_i$ is also a sufficient statistic for θ.

Example 6.11. Let $\mathbf{X}^n \sim$ IID $N(\mu, \sigma^2)$, where σ^2 is a known value. Then show $T(\mathbf{X}^n) = \bar{X}_n$ is a sufficient statistic for μ.

Solution: In this example, the (unknown) parameter $\theta = \mu$. Since σ^2 is a given (known) number, it is no longer a parameter. The joint PDF of $\mathbf{X}^n$

$$f_{\mathbf{X}^n}(\mathbf{x}^n, \mu) = \prod_{i=1}^{n} f(x_i, \theta)$$

$$= \prod_{i=1}^{n} \frac{1}{\sqrt{2\pi\sigma^2}} e^{-\frac{(x_i - \mu)^2}{2\sigma^2}}$$

$$= \frac{1}{(2\pi\sigma^2)^{n/2}} e^{-\frac{\sum_{i=1}^{n}(x_i - \bar{x}_n + \bar{x}_n - \mu)^2}{2\sigma^2}}$$

$$= \frac{1}{(2\pi\sigma^2)^{n/2}} e^{-\frac{\sum_{i=1}^{n}(x_i - \bar{x}_n)^2 + n(\bar{x}_n - \mu)^2}{2\sigma^2}}$$

$$= \left[\frac{1}{(2\pi\sigma^2)^{n/2}} e^{-\frac{\sum_{i=1}^{n}(x_i - \bar{x}_n)^2}{2\sigma^2}} \right] e^{-\frac{n(\bar{x}_n - \mu)^2}{2\sigma^2}}$$

$$= h(\mathbf{x}^n) g(\bar{x}_n, \mu),$$

where

$$h(\mathbf{x}^n) = \frac{1}{(2\pi\sigma^2)^{n/2}} e^{-\frac{\sum_{i=1}^{n}(x_i - \bar{x}_n)^2}{2\sigma^2}},$$

$$g[T(\mathbf{x}^n), \theta] = e^{-\frac{n(\bar{x}_n - \mu)^2}{2\sigma^2}}.$$

It follows that $T(\mathbf{X}^n) = \bar{X}_n$ is a sufficient statistic for μ.

Example 6.12. Let $\mathbf{X}^n \sim$ IID $N(\mu, \sigma^2)$, where μ, σ^2 are unknown parameters. Then $T(\mathbf{X}^n) = (\bar{X}_n, S_n^2)$ is a sufficient statistic for (μ, σ^2).

Solution: In this example, the unknown parameter $\theta = (\mu, \sigma^2)$ is a two-dimensional vector. Because the joint PDF of the random sample $\mathbf{X}^n$

$$f_{\mathbf{X}^n}(\mathbf{x}^n, \mu, \sigma^2) = \prod_{i=1}^{n} \frac{1}{\sqrt{2\pi}\sigma} e^{-\frac{(x_i - \mu)^2}{2\sigma^2}}$$

$$= \frac{1}{(\sqrt{2\pi\sigma^2})^n} e^{-\frac{\sum_{i=1}^{n}(x_i - \mu)^2}{2\sigma^2}}$$

$$= \frac{1}{(2\pi\sigma^2)^{n/2}} e^{-\frac{(n-1)[(n-1)^{-1}\sum_{i=1}^{n}(x_i - \bar{x}_n)^2]}{2\sigma^2} - \frac{n(\bar{x}_n - \mu)^2}{2\sigma^2}}$$

$$= \frac{1}{(2\pi\sigma^2)^{n/2}} e^{-\frac{(n-1)s_n^2 + n(\bar{x}_n - \mu)^2}{2\sigma^2}}$$

$$= g[T(\mathbf{x}^n), \theta] h(\mathbf{x}^n),$$

where $h(\mathbf{x}^n) = 1$ for all $\mathbf{x}^n$, it follows that the two-dimensional statistic $T(\mathbf{X}^n) = (\bar{X}_n, S_n^2)$ is a sufficient statistic for $\theta = (\mu, \sigma^2)$.

For a normally distributed random sample $\mathbf{X}^n$ with unknown μ and σ^2, it suffices to summarize the data by reporting the sample mean and sample variance, because $(\bar{X}_n, S_n^2)$ is a sufficient statistic for (μ, σ^2). However, suppose it is not normal. Then $(\bar{X}_n, S_n^2)$ may not be sufficient statistics. In other words, a sufficient statistic $T(\mathbf{X}^n)$ is generally model-dependent or population distribution dependent. It is a sufficient statistic for parameter θ under some population distribution but may not be a sufficient statistic under other population distributions.

Question: Can you provide an example of population distribution for which $(\bar{X}_n, S_n^2)$ are not sufficient statistics for $\theta = (\mu, \sigma^2)$?

Theorem 6.11. *[Invariance Principle]: If $T(\mathbf{X}^n)$ is a sufficient statistic for θ, then any one-to-one function $R(\mathbf{X}^n) = r[T(\mathbf{X}^n)]$ is also a sufficient statistic for θ, and a sufficient statistic for the transformed parameter $r(\theta)$.*

Proof: Because $T(\mathbf{X}^n)$ is a sufficient statistic for θ, the joint PMF/PDF of the random sample $\mathbf{X}^n$

$$f_{\mathbf{X}^n}(\mathbf{x}^n, \theta) = g[T(\mathbf{x}^n), \theta]h(\mathbf{x}^n)$$

for some functions $g(\cdot, \cdot)$ and $h(\cdot)$. Next, because the function $r(\cdot)$ is a one-to-one mapping, its inverse function $r^{-1}(\cdot)$ exists and $T(\mathbf{x}^n) = r^{-1}[R(\mathbf{x}^n)]$. It follows that

$$f_{\mathbf{X}^n}(\mathbf{x}^n, \theta) = g\{r^{-1}[R(\mathbf{x}^n)], \theta\}h(\mathbf{x}^n)$$
$$= \tilde{g}[R(\mathbf{x}^n), \theta]h(\mathbf{x}^n)$$

where $\tilde{g}(\cdot, \theta) = g[r^{-1}(\cdot), \theta]$ depends on parameter θ. Hence, $R(\mathbf{X}^n)$ is a sufficient statistic for θ by the definition of sufficient statistic.

Similarly, because $\theta = r^{-1}[r(\theta)] = r^{-1}(\beta)$, where $\beta = r(\theta)$ is a transformed parameter, we have

$$f_{\mathbf{X}^n}(\mathbf{x}^n, \theta) = g\{r^{-1}[R(\mathbf{x}^n)], r^{-1}(\beta)\}h(\mathbf{x}^n)$$
$$= g^*[R(\mathbf{x}^n), \beta]h(\mathbf{x}^n),$$

where the function $g^*(\cdot, \beta) = g[r^{-1}(\cdot), r^{-1}(\beta)]$ depends on parameter β. It follows that $R(\mathbf{X}^n)$ is also a sufficient statistic for β. This completes the proof.

Next, we discuss the sufficient statistics for a family of distributions called exponential family, which includes many important distributions as special cases.

Definition 6.9. [Exponential Family]: A family of probability distributions is called an exponential family if their population PMF/PDF can be expressed as

$$f(x,\theta) = h(x)c(\theta)e^{\sum_{j=1}^{k} w_j(\theta)t_j(x)}.$$

Most important distributions introduced in Chapter 4 — both discrete and continuous — belong to the exponential family. An example is the normal $N(\mu,\sigma^2)$ distribution, whose PDF

$$f(x,\theta) = \frac{1}{\sqrt{2\pi}\sigma}e^{-\frac{1}{2\sigma^2}(x-\mu)^2}$$

$$= \frac{1}{\sqrt{2\pi}\sigma}e^{-\frac{x^2}{2\sigma^2}+\frac{\mu}{\sigma^2}x-\frac{\mu^2}{2\sigma^2}},$$

where

$$h(x) = 1,$$

$$c(\theta) = \frac{1}{\sqrt{2\pi\sigma^2}}e^{-\frac{\mu^2}{2\sigma^2}},$$

$$w_1(\theta) = -\frac{1}{2\sigma^2},$$

$$w_2(\theta) = \frac{\mu}{\sigma^2},$$

$$t_1(x) = x^2,$$

$$t_2(x) = x.$$

Theorem 6.12. *Let* $\mathbf{X}^n = (X_1,\cdots,X_n)$ *be an IID random sample from the population* $f(x,\theta)$. *If*

$$f(x,\theta) = h(x)c(\theta)e^{\sum_{j=1}^{k} w_j(\theta)t_j(x)},$$

then the $k \times 1$ *statistic vector*

$$T(\mathbf{X}^n) = \left[\sum_{i=1}^{n} t_1(X_i),\cdots,\sum_{i=1}^{n} t_k(X_i)\right]$$

is a sufficient statistic for θ.

Proof: This is left as an exercise.

Sometimes, the information of $\mathbf{X}^n$ about a scalar parameter θ cannot be summarized by one scalar statistic, and several scalar statistics together are needed instead. In such cases, a sufficient statistic is a vector, namely $T(\mathbf{X}^n) = [T_1(\mathbf{X}^n),\cdots,T_k(\mathbf{X}^n)]'$. Thus, the dimension of a sufficient statistic $T(\mathbf{X}^n)$ may not be the same as the dimension of θ.

It is always true that the random sample $\mathbf{X}^n$ itself is a sufficient statistic for parameter θ. This is because we can always partition the joint PMF/PDF of $\mathbf{X}^n$ as

$$f_{\mathbf{X}^n}(\mathbf{x}^n, \theta) = g[T(\mathbf{x}^n), \theta]h(\mathbf{x}^n),$$

where $T(\mathbf{x}^n) = \mathbf{x}^n$, $h(\mathbf{x}^n) = 1$, and $g[T(\mathbf{x}^n), \theta] = f_{\mathbf{X}^n}(\mathbf{x}^n, \theta)$ for all $\mathbf{x}^n$. By the factorization theorem, $T(\mathbf{X}^n) = \mathbf{X}^n$ is always a sufficient statistic. Nevertheless, such a sufficient statistic is not effective at all in summarizing the sample information.

Generally speaking, there exist many sufficient statistics for the same parameter θ. For example, any one-to-one function of a sufficient statistic is also a sufficient statistic for parameter θ (see Theorem 6.11). Sufficient statistics for the same parameter θ may differ from each other in the degree of summarizing the sample information. What is the most efficient way to summarize information of θ that is contained in a random sample $\mathbf{X}^n$?

Definition 6.10. [Minimal Sufficient Statistic]: A sufficient statistic $T(\mathbf{X}^n)$ is called a minimal sufficient statistic for parameter θ if, for any other sufficient statistic $R(\mathbf{X}^n), T(\mathbf{X}^n)$ is a function of $R(\mathbf{X}^n)$. That is, for any sufficient statistic $R(\mathbf{X}^n)$, there always exists some function $r(\cdot)$ such that $T(\mathbf{X}^n) = r[R(\mathbf{X}^n)]$.

All sufficient statistics of θ contain all sample information that is relevant to θ, but the minimal sufficient statistic achieves the greatest possible summary of the data among all sufficient statistics for parameter θ.

Why?

To see this, suppose $T(\mathbf{X}^n) = r[R(\mathbf{X}^n)]$, and $t = r(\tau)$. Define two subsets of sample points in the sample space of $\mathbf{X}^n$:

$$A_n(\tau) = \{\mathbf{x}^n : R(\mathbf{x}^n) = \tau\},$$
$$B_n(t) = \{\mathbf{x}^n : T(\mathbf{x}^n) = t\}$$
$$= \{\mathbf{x}^n : r[R(\mathbf{x}^n)] = r(\tau)\}.$$

The first subset $A_n(t)$ is indexed by t and the second subset $B_n(\tau)$ is indexed by τ, where $t = r(\tau)$. Then $A_n(\tau) \subseteq B_n(t)$ because $R(\mathbf{x}^n) = \tau$ implies $T(\mathbf{x}^n) = r[R(\mathbf{x}^n)] = r(\tau) = t$, but $T(\mathbf{x}^n) = t$ does not imply $R(\mathbf{x}^n) = \tau$. Therefore, the sample information summarized by $T(\mathbf{x}^n) = t$ is a larger set than the sample information summarized by $R(\mathbf{x}^n) = \tau$. This implies that $T(\mathbf{X}^n)$ summarizes a larger information set of the random sample $\mathbf{X}^n$ for parameter θ.

A minimal sufficient statistic is not unique. Any one-to-one function of a minimal sufficient statistic is also a minimal sufficient statistic. That is, if $T(\mathbf{X}^n)$ is a minimal sufficient statistic, so is $g[T(\mathbf{X}^n)]$, where $g(\cdot)$ is a one-to-one function.

Question: How can one find a minimal sufficient statistic?

The following theorem provides a convenient approach to checking whether a statistic $T(\mathbf{X}^n)$ is a minimal sufficient statistic for parameter θ.

Theorem 6.13. *Let $f_{\mathbf{X}^n}(\mathbf{x}^n, \theta)$ be the PMF/PDF of a random sample $\mathbf{X}^n$. Suppose there exists a function $T(\mathbf{X}^n)$ such that, for two sample points $\mathbf{x}^n$ and $\mathbf{y}^n$ in the sample space of $\mathbf{X}^n$, the ratio of joint PMF/PDF $f_{\mathbf{X}^n}(\mathbf{x}^n, \theta)/f_{\mathbf{X}^n}(\mathbf{y}^n, \theta)$ is constant as a function of θ (i.e. is independent of θ) if and only if $T(\mathbf{x}^n) = T(\mathbf{y}^n)$. Then $T(\mathbf{X}^n)$ is a minimal sufficient statistic for parameter θ.*

Proof: (1) First we shall show that $T(\mathbf{X}^n)$ is a sufficient statistic for θ under the stated condition. Define the partition sets of the sample space of $\mathbf{X}^n$ induced by $T(\mathbf{x}^n) = t$ as $A(t) = \{\mathbf{x}^n : T(\mathbf{x}^n) = t\}$. For each $A(t)$, we choose and fix one element $\mathbf{x}_t^n \in A(t)$. In other words, for any sample point $\mathbf{x}^n$ with $T(\mathbf{x}^n) = t$, let $\mathbf{x}_t^n$ be a fixed element that is in the same set $A(t)$ as $\mathbf{x}^n$. Since $\mathbf{x}^n$ and $\mathbf{x}_t^n$ are in the same set $A(t)$, we have $T(\mathbf{x}^n) = T(\mathbf{x}_t^n)$ and hence, $f_{\mathbf{X}^n}(\mathbf{x}^n, \theta)/f_{\mathbf{X}^n}(\mathbf{x}_t^n, \theta)$ is constant as a function of θ given the condition in the theorem. Thus, we can define a function $h(\mathbf{x}^n) = f_{\mathbf{X}^n}(\mathbf{x}^n, \theta)/f_{\mathbf{X}^n}(\mathbf{x}_t^n, \theta)$, which does not depend on θ and is a function of $\mathbf{x}^n$ only (note that $t = T(\mathbf{x}^n)$ depends on $\mathbf{x}^n$). Also, we define a function $g(t, \theta) = f_{\mathbf{X}^n}(\mathbf{x}_t^n, \theta)$. Then we have

$$
\begin{aligned}
f_{\mathbf{X}^n}(\mathbf{x}^n, \theta) &= f_{\mathbf{X}^n}(\mathbf{x}_t^n, \theta) \frac{f_{\mathbf{X}^n}(\mathbf{x}^n, \theta)}{f_{\mathbf{X}^n}(\mathbf{x}_t^n, \theta)} \\
&= f_{\mathbf{X}^n}(\mathbf{x}_t^n, \theta) h(\mathbf{x}^n) \\
&= g(t, \theta) h(\mathbf{x}^n) \\
&= g[T(\mathbf{x}^n), \theta] h(\mathbf{x}^n),
\end{aligned}
$$

where the last equality follows from $t = T(\mathbf{x}^n)$. Thus, by the factorization theorem, $T(\mathbf{X}^n)$ is a sufficient statistic for θ.

(2) Now we shall show that $T(\mathbf{X}^n)$ is minimal. Let $\tilde{T}(\mathbf{X}^n)$ be any other sufficient statistic for θ. By the factorization theorem, there exist functions $\tilde{g}(\cdot, \cdot)$ and $\tilde{h}(\cdot)$ such that $f_{\mathbf{X}^n}(\mathbf{x}^n, \theta) = \tilde{g}[\tilde{T}(\mathbf{x}^n), \theta]\tilde{h}(\mathbf{x}^n)$. Let $\mathbf{x}^n$ and $\mathbf{y}^n$ be any two sample points in the sample space of $\mathbf{X}^n$ with $\tilde{T}(\mathbf{x}^n) = \tilde{T}(\mathbf{y}^n)$.

Then

$$\frac{f_{\mathbf{X}^n}(\mathbf{x}^n, \theta)}{f_{\mathbf{X}^n}(\mathbf{y}^n, \theta)} = \frac{\tilde{g}[\tilde{T}(\mathbf{x}^n), \theta]\tilde{h}(\mathbf{x}^n)}{\tilde{g}[\tilde{T}(\mathbf{y}^n), \theta]\tilde{h}(\mathbf{y}^n)} = \frac{\tilde{h}(\mathbf{x}^n)}{\tilde{h}(\mathbf{y}^n)},$$

which does not depend on θ. Since the ratio $f_{\mathbf{X}^n}(\mathbf{x}^n, \theta)/f_{\mathbf{X}^n}(\mathbf{y}^n, \theta)$ does not depend on θ, the conditions of the present theorem imply $T(\mathbf{x}^n) = T(\mathbf{y}^n)$. In other words, we have $\tilde{T}(\mathbf{x}^n) = \tilde{T}(\mathbf{y}^n)$ implies $T(\mathbf{x}^n) = T(\mathbf{y}^n)$. This means that for any given $\mathbf{x}^n$,

$$\left\{ \mathbf{y}^n : \tilde{T}(\mathbf{y}^n) = \tilde{T}(\mathbf{x}^n) \right\} \subseteq \{ \mathbf{y}^n : T(\mathbf{y}^n) = T(\mathbf{x}^n) \}.$$

Thus, $T(\mathbf{x}^n)$ is minimal. This completes the proof.

Example 6.13. Let $\mathbf{X}^n$ be an IID random sample from a $N(\mu, \sigma^2)$ population with both μ and σ^2 unknown. Let $\mathbf{x}^n$ and $\mathbf{y}^n$ denote two sample points in the sample space of $\mathbf{X}^n$, and let $(\bar{x}_n, s_X^2)$ and $(\bar{y}_n, s_Y^2)$ be the sample means and sample variances corresponding to $\mathbf{x}^n$ and $\mathbf{y}^n$ respectively. Then,

$$\frac{f_{\mathbf{X}^n}(\mathbf{x}^n, \theta)}{f_{\mathbf{X}^n}(\mathbf{y}^n, \theta)} = \frac{(2\pi\sigma^2)^{-n/2} e^{-[n(\bar{x}_n - \mu)^2 + (n-1)s_X^2]/2\sigma^2}}{(2\pi\sigma^2)^{-n/2} e^{-[n(\bar{y}_n - \mu)^2 + (n-1)s_Y^2]/2\sigma^2}}$$
$$= 1$$

if and only if $(\bar{x}_n, s_X^2) = (\bar{y}_n, s_Y^2)$. Thus, $(\bar{X}_n, S_n^2)$ is a minimal sufficient statistic for (μ, σ^2).

6.7 Conclusion

The basic idea of statistical analysis is to use a subset or sample information to infer the knowledge of the data generating process. In this chapter, we have introduced some basic concepts and ideas of statistical theory, including the concepts of population, random sample, data set, statistic, parameter and statistical inference. We examine in detail the statistical properties of two important statistics — sample mean and sample variance estimators, establishing the finite sample distribution theory for them under the assumption of an IID normal random sample. This finite sample theory highlights the importance of the Student-t and F distributions in statistical inference. Finally, we introduce the concept of sufficient statistic and discuss its role in data reduction. The sufficiency principle best captures the essential idea of statistical analysis, namely, how to most efficiently summarize the observed data in inference of the population distribution or population parameter.

EXERCISE 6

6.1. Consider an independent and identically distributed random sample $X^n = (X_1, X_2, X_3)$, where X_i follows binary distribution with $P(X_i = 0) = P(X_i = 1) = \frac{1}{2}$, for $i = 1, 2, 3$. Define the sample mean $\bar{X}_n = \frac{1}{3}(X_1 + X_2 + X_3)$. Find: (1) the sampling distribution of $\bar{X}_n$; (2) the mean of $\bar{X}_n$; (3) the variance of $\bar{X}_n$.

6.2. A community has five families whose annual incomes are 1, 2, 3, 4, and 5 respectively. Suppose a survey is to be made to two of the five families and the choice of the two is random. Find the sampling distribution of the sample mean of the family income. Give your reasoning clearly.

6.3. Suppose the return of asset i is given by

$$R_i = \alpha + \beta_i R_m + X_i, \qquad i = 1, \ldots, n,$$

where R_i is the return on asset i, α is the return on the risk-free asset, R_m is the return on the market portfolio which represents the market risk, and X_i represents an idiosyncratic risk peculiar to the characteristics of asset i. Assume $0 < \beta_i < \infty$.

We consider an equal-weighting portfolio that consists of n assets. The return on such an equal-weighting portfolio is then given by

$$\bar{R}_n = \sum_{i=1}^{n} \frac{1}{n} R_i$$

$$= \alpha + \left(\frac{1}{n} \sum_{i=1}^{n} \beta_i \right) R_m + \bar{X}_n$$

$$= \alpha + \bar{\beta} R_m + \bar{X}_n,$$

where $\bar{\beta} = n^{-1} \Sigma_{i=1}^n \beta_i$ and $\bar{X}_n = n^{-1} \Sigma_{i=1}^n X_i$ is the sample mean of the random sample $\mathbf{X}^n = (X_1, \cdots, X_n)$. Assume the random sample $\mathbf{X}^n$ is an independent and identically distributed random sample with population mean μ and population variance σ^2. Also, assume that R_m and $\mathbf{X}^n$ are mutually independent.

The total risk of the equal-weighting portfolio is measured by its variance.

(1) Show

$$\text{var}(\bar{R}_n) = \bar{\beta}^2 \text{var}(R_m) + \text{var}(\bar{X}_n).$$

That is, the risk of the portfolio contains a market risk and a component contributed by n individual risks;

(2) Show that idiosyncratic risks can be eliminated by forming a portfolio with a large number of assets, that is, by letting $n \to \infty$.

6.4. Suppose there are k IID random samples from a population Bernoulli(p) distribution, with sample sizes equal to $n_1, \cdots, n_k$ respectively. Assume these k random samples are independent of each other. Based on these k random samples, define k sample means, $\bar{X}_{n_1}, \cdots, \bar{X}_{n_k}$, respectively. Define an overall sample mean $\bar{X} = k^{-1} \Sigma_{i=1}^{n_k} \bar{X}_{n_i}$. Find: (1) the mean of $\bar{X}$; (2) the variance of $\bar{X}$.

6.5. Suppose $\mathbf{X}^n = (X_1, \cdots, X_n)$ is an IID $N(\mu_1, \sigma_1^2)$ random sample, $\mathbf{Y}^m = (Y_1, \cdots, Y_m)$ is an IID $N(\mu_2, \sigma_2^2)$ random sample, and the two random samples are mutually independent. Find the distribution of $\bar{X}_n - \bar{Y}_m$, where $\bar{X}_n$ and $\bar{Y}_m$ are the sample means of the first and second random samples respectively.

6.6. Suppose $\mathbf{X}^n = (X_1, \cdots, X_n)$ is an IID $N(\mu, \sigma^2)$ random sample, $\mathbf{Y}^n = (Y_1, \cdots, Y_n)$ is an IID $N(\mu, \sigma^2)$ random sample, and the two random samples are mutually independent. Let $\bar{X}_n$ and $\bar{Y}_n$ be the sample means of the first and second random samples respectively, and let S_X^2 and S_Y^2 be the sample variances of the first and second random samples respectively. Find:

(1) the distribution of $(\bar{X}_n - \bar{Y}_n)/\sqrt{2\sigma^2/n}$;

(2) the distribution of $(\bar{X}_n - \bar{Y}_n)/\sqrt{2S_X^2/n}$;

(3) the distribution of $(\bar{X}_n - \bar{Y}_n)/\sqrt{2S_Y^2/n}$;

(4) the distribution of $(\bar{X}_n - \bar{Y}_n)/\sqrt{(S_X^2 + S_Y^2)/n}$;

(5) the distribution of $(\bar{X}_n - \bar{Y}_n)/\sqrt{S_n^2/n}$, where S_n^2 is the sample variance of the difference sample $\mathbf{Z}^n = (Z_1, \ldots, Z_n)$, where $Z_i = X_i - Y_i, i = 1, 2, \ldots, n$.

6.7. Let $\mathbf{X}^n = (X_1, \cdots, X_n)$ be an IID $N(\mu, \sigma^2)$ random sample. Find a function of S_n^2, the sample variance, say $g(S_n^2)$, that satisfies $E[g(S_n^2)] = \sigma$. (Hint: Try $g(S_n^2) = c\sqrt{S_n^2}$, where c is a constant.)

6.8. Establish the following recursion relations for sample means and sample variances. Let $\bar{X}_n$ and S_n^2 be the mean and variance, respectively, of $(X_1, \cdots, X_n)$. Then suppose another observation, X_{n+1}, becomes available. Show that

(1) $\bar{X}_{n+1} = \frac{X_{n+1} + n\bar{X}_n}{n+1}$;

(2) $nS_{n+1}^2 = (n-1)S_n^2 + \frac{n}{n+1}(X_{n+1} - \bar{X}_n)^2$.

6.9. Let $X_i, i = 1, 2, 3$, be independent with $N(i, i^2)$ distributions. For each of the following situations, use X_1, X_2, X_3 to construct a statistic with the indicated distribution:

(1) Chi-square distribution with 3 degrees of freedom;

(2) t distribution with 2 degrees of freedom;

(3) F distribution with 1 and 2 degrees of freedom.

6.10. Let $U \sim N(0,1), V \sim \chi_\nu^2$, and U and V are independent. Then the random variable

$$T = \frac{U}{\sqrt{V/\nu}}$$

follows a Student's t distribution with ν degrees of freedom, denoted as $T \sim t_\nu$. Show that the PDF of T is

$$f(t) = \frac{\Gamma(\frac{\nu+1}{2})}{\Gamma(\frac{\nu}{2})} \frac{1}{(\nu\pi)^{1/2}} \frac{1}{(1+t^2/\nu)^{(\nu+1)/2}}, \qquad -\infty < t < \infty.$$

6.11. Show that for a Student's t_ν random variable X, (1) $E(X) = 0$; and (2) $\mathrm{var}(X) = \nu/(\nu-2)$ for $\nu > 2$.

6.12. Let U and V be two independent Chi-square random variables with p and q degrees of freedom respectively. Then the random variable

$$X = \frac{U/p}{V/q} \sim F_{p,q}$$

follows a F distribution with p and q degrees of freedom. Show that the PDF of F is

$$f(x) = \frac{\Gamma(\frac{p+q}{2})}{\Gamma(\frac{p}{2})\Gamma(\frac{q}{2})} \left(\frac{p}{q}\right)^{p/2} \frac{x^{(p/2)-1}}{[1+(p/q)x]^{(p+q)/2}}, \qquad 0 < x < \infty.$$

6.13. For a $F_{p,q}$ random variable X, find (1) $E(X)$; (2) $\mathrm{var}(X)$.

6.14. Let X be one observation from $N(0,\sigma^2)$ population. Is $|X|$ a sufficient statistic?

6.15. Let $X_1, \ldots, X_n$ be independent random variables with densities

$$f_{X_i}(x, \theta) = \begin{cases} e^{i\theta - x}, & x \geq i\theta, \\ 0, & x < i\theta. \end{cases}$$

Prove that $T = \min_{1 \leq i \leq n}(X_i/i)$ is a sufficient statistic for θ.

6.16. Prove the following theorem: Let $\mathbf{X}^n = (X_1, \cdots, X_n)$ be an IID random sample from a PDF or PMF $f(x,\theta)$ that belongs to an exponential family given by

$$f(x, \theta) = h(x)c(\theta)\exp\left[\sum_{j=1}^{k} w_j(\theta)t_j(x)\right],$$

where $\theta = (\theta_1, \ldots, \theta_d), d \leq k$. Then

$$T(\mathbf{X}^n) = \left[\sum_{i=1}^{n} t_1(X_i), \cdots, \sum_{i=1}^{n} t_k(X_i) \right]$$

is a sufficient statistic for θ.

6.17. Let $\mathbf{X}^n = (X_1, \cdots, X_n)$ be a random sample from a Gamma(α, β) distribution. Find a two-dimensional sufficient statistic for (α, β).

6.18. Let $\mathbf{X}^n = (X_1, \cdots, X_n)$ be a random sample from a population with PDF

$$f(x, \theta) = \theta x^{\theta-1}, \qquad 0 < x < 1, \theta > 0.$$

Is $\sum_{i=1}^{n} X_i$ sufficient for θ? Give your reasoning.

6.19. Let X be a random variable with an $F_{p,q}$ distribution.
   (1) Derive the PDF of X.
   (2) Derive the mean and variance of X.
   (3) Show that $1/X$ has an $F_{q,p}$ distribution.
   (4) Show that $(p/q)X/[1 + (p/q)X])$ has a Beta$(p/2, q/2)$ distribution.

6.20. Prove the following:
   (1) the statistic $(\sum_{i=1}^{n} X_i, \sum_{i=1}^{n} X_i^2)$ is sufficient, but not minimal sufficient in the $N(\mu.\mu)$ family;
   (2) the statistic $\sum_{i=1}^{n} X_i^2$ is minimal sufficient in the $N(\mu, \mu)$ family;
   (3) the statistic $(\sum_{i=1}^{n} X_i, \sum_{i=1}^{n} X_i^2)$ is minimal sufficient in the $N(\mu, \mu^2)$ family;
   (4) the statistic $(\sum_{i=1}^{n} X_i, \sum_{i=1}^{n} X_i^2)$ is minimal sufficient in the $N(\mu, \sigma^2)$ family.

6.21. Suppose $\mathbf{X}^n = (X_1, \cdots, X_n)$ is an IID random sample from a Poisson(α) distribution with probability mass function

$$f_X(x) = e^{-\alpha} \frac{\alpha^x}{x!} \text{ for } x = 0, 1, 2, \cdots,$$

where α is unknown. Find a sufficient statistics for α.

6.22. Suppose $(X_1, \cdots, X_n)$ is a random sample from a Beta distribution whose PDF is

$$f(x) = \frac{\Gamma(\alpha + \beta)}{\Gamma(\alpha)\Gamma(\beta)} x^\alpha (1 - x)^\beta, \qquad 0 < x < 1,$$

where the value of α is known but the value of β is unknown. Find a sufficient statistic for β. Is the sufficient statistic you obtain a minimal sufficient statistic?

6.23. Suppose $X^n \equiv \{X_1, \cdots, X_n\}$ is a random sample from a Gamma(α, β) distribution whose PDF is

$$f(x) = \frac{1}{\Gamma(\alpha)\beta^\alpha} x^{\alpha-1} e^{-x/\beta}, \qquad x > 0,$$

where parameters $\alpha > 0, \beta > 0$. Suppose that the value of α is unknown but the value of β is known. Find a sufficient statistic for α. Give your reasoning.

6.24. Suppose $(X_1, \cdots, X_n)$ is a random sample from a Weibull distribution whose PDF

$$f_X(x) = \begin{cases} \frac{\alpha}{\beta} x^{\alpha-1} \exp(-x^\alpha/\beta), & \text{if } x > 0, \\ 0, & \text{otherwise,} \end{cases}$$

where $\alpha > 0, \beta > 0$. Suppose α is known, so β is the only unknown parameter.

(1) Find a sufficient statistic for β.

(2) Is the sufficient statistic obtained in part (1) a minimal sufficient statistic? Give your reasoning.

6.25. Let $(X_1, \cdots, X_n)$ be an IID random sample from a $N(\theta, \theta)$ population, where θ is unknown. Find a sufficient statistic for θ and check if it is a minimum sufficient statistic.

Chapter 7

Convergences and Limit Theorems

Abstract: This chapter will introduce basic analytic tools for asymptotic analysis or large sample analysis when the sample size $n \to \infty$. We will discuss four modes of convergence — convergence in quadratic mean, convergence in probability, almost sure convergence and convergence in distribution, and two limit theorems — the law of large numbers and the central limit theorem. These are basic tools used to establish the asymptotic (i.e., large sample) properties of estimators and test statistics.

An important empirical stylized fact in economics is that most economic variables, particularly high-frequency financial time series, have distributions whose tails are heavier than those of the normal distribution. With a nonnormal random sample, the distributions of econometric estimators and test statistics usually depend on the sample size n and are generally difficult to obtain. Although various mathematical techniques could be used to determine these distributions, they are so complicated that few, if any, of us would be interested in using them to compute probabilities about econometric estimators and test statistics. Asymptotic theory is concerned with various kinds of convergence of sequences of statistics as the sample size n grows. It provides a convenient way to approximate, for large values of n (i.e., when $n \to \infty$), the exact finite sample distributions of econometric estimators and statistics. This simplifies statistical inference in practice.

Key words: Almost sure convergence, Asymptotic analysis, Central limit theorem, Chebyshev's inequality, Convergence in distribution, Convergence in quadratic mean, Convergence in probability, Delta method, Large sample theory, Markov's inequality, Slutsky's theorem, Strong law of large numbers, Weak law of large numbers.

7.1   Limits and Orders of Magnitude: A Review

We first review some elementary concepts in limit theory for nonstochastic sequences.

Definition 7.1. [Limit]: Let $\{b_n, n = 1, 2, \cdots\}$ be a sequence of nonstochastic real numbers. If there exists a real number b such that for every real number $\epsilon > 0$, there exists a finite integer $N(\epsilon)$ such that for all $n \geq N(\epsilon)$, we have $|b_n - b| < \epsilon$, then b is the limit of the sequence $\{b_n, n = 1, 2, \cdots\}$.

When $\{b_n, n = 1, 2, \cdots\}$ converges to b as $n \to \infty$, we write $b_n \to b$ as $n \to \infty$, or $\lim_{n\to\infty} b_n = b$. The constant $\epsilon > 0$ can be set to be very small. The smaller ϵ is, the larger $N(\epsilon)$ will be. One can interpret ϵ as a prespecified **tolerance level** for the discrepancy between b_n and b.

Example 7.1. $b_n = 1 - n^{-1}$. Then $b_n \to 1$ as $n \to \infty$. This is because for $b = 1$, and for every $\epsilon > 0$, there exists $N(\epsilon) = [\epsilon^{-1}] + 1$, where $[\cdot]$ denotes the integer part, such that for all $n \geq N(\epsilon)$,

$$|b_n - b| = \frac{1}{n} < \epsilon.$$

In this example, if $\epsilon = 10^{-4}$, then we need $N(\epsilon) = 1/\epsilon + 1 = 10^4 + 1$; if $\epsilon = 10^{-8}$, then we need $N(\epsilon) = 10^8 + 1$.

Example 7.2. Let $b_n = (1 + a/n)^n$, where a is a constant. Then $b_n \to e^a$ as $n \to \infty$.

Example 7.3. $b_n = (-1)^n$. Then $\{b_n, n = 1, 2, \cdots\}$ is bounded by a constant in the sense that $|b_n| \leq M$ for some constant $M > 1$ and for all $n \geq 1$. However, its limit does not exist.

Solution: Set $\epsilon = \frac{1}{2}$. Then there exist no b and $N(\epsilon)$ such that for all $n > N(\epsilon)$, we can have $|b_n - b| < \epsilon$.

Definition 7.2. [Continuity]: The function $g : \mathbb{R} \to \mathbb{R}$ is continuous at point b if for any sequence $\{b_n, n = 1, 2, \cdots\}$ such that $b_n \to b$ as $n \to \infty$, we have $g(b_n) \to g(b)$ as $n \to \infty$.

An alternative but equivalent definition of continuity is as follows: for each given $\epsilon > 0$ there exists a $\delta = \delta(\epsilon)$ such that whenever $|b_n - b| < \delta$, we have $|g(b_n) - g(b)| < \epsilon$. When $g(\cdot)$ is continuous at b, we can write $\lim_{b_n \to b} g(b_n) = g(\lim_{n\to\infty} b_n) = g(b)$. In other words, the limit of a sequence of values for a continuous function is equal to the value of the function at the limit.

Example 7.4. Suppose $a_n \to a$ and $b_n \to b$ as $n \to \infty$. Then as $n \to \infty$,
 (1) $a_n + b_n \to a + b$;
 (2) $a_n b_n \to ab$;
 (3) $a_n/b_n \to a/b$ if $b \neq 0$.

Example 7.5. Define a function

$$F(x) = \begin{cases} 0 \text{ if } x < 0, \\ \frac{1}{2} \text{ if } 0 \leq x < 1, \\ 1 \text{ if } x \geq 1. \end{cases}$$

Here, $F(x)$ is neither continuous at 0 nor at 1. This is because for at least one sequence $\{b_n\}$ such that $b_n = -\frac{1}{n}$, we have $b_n \to b = 0$, but $F(b_n) = F(-\frac{1}{n}) = 0$ for all $n \geq 1$, and $\lim_{n\to\infty} F(-\frac{1}{n}) = 0 \neq F(0) = \frac{1}{2}$.

Definition 7.3. [**Order of Magnitude**]: (1) A sequence $\{b_n, n = 1, 2, \cdots\}$ is at most of order n^λ, denoted $b_n = O(n^\lambda)$ or $n^{-\lambda} b_n = O(1)$, if for **some** (sufficiently large) real number $M < \infty$, there exists a finite integer $N(M)$ such that for all $n \geq N(M)$, we have $\left|n^{-\lambda} b_n\right| < M$.
 (2) A sequence $\{b_n\}$ is of order smaller than n^λ, denoted $b_n = o(n^\lambda)$ or $n^{-\lambda} b_n = o(1)$, if for **every** real number $\epsilon > 0$ there exists a finite integer $N(\epsilon)$ such that for all $n \geq N(\epsilon)$, we have $\left|n^{-\lambda} b_n\right| < \epsilon$.

In the definition of $b_n = O(n^\lambda)$, the constant M is usually a very big number. Note that it suffices to find one constant M only. For $\lambda > 0$, $b_n = O(n^\lambda)$ implies that b_n grows to infinity at a rate slower than or at most equal to n^λ. In particular, if

$$\lim_{n\to\infty} \frac{b_n}{n^\lambda} = C < \infty,$$

then $b_n = O(n^\lambda)$ or $n^{-\lambda} b_n = O(1)$.

Example 7.6. $b_n = 4 + 2n + 6n^2$. Then $b_n = O(n^2)$, because

$$\frac{b_n}{n^2} = \frac{4}{n^2} + \frac{2n}{n^2} + \frac{6n^2}{n^2} \to 6 < 2M = 2 \cdot 6 \text{ (say)}$$

for all n sufficiently large. Intuitively, the order of b_n is determined by the dominant term (i.e., n^2) that grows to infinity fastest.

We note that it is possible that $\lim_{n\to\infty} b_n/n^\lambda$ does not exist but $|b_n/n^\lambda|$ is bounded; in this case, we still have $b_n = O(n^\lambda)$, as is illustrated by the following example.

Example 7.7. $b_n = (-1)^n$. Then $b_n = O(1)$.

$$|b_n| = 1 < M \equiv 1.01 \text{ for all } n \geq 1.$$

In the definition of $b_n = o(n^\lambda)$, the constant ϵ can be a very small value. Then the smaller ϵ is, the bigger $N(\epsilon)$ will have to be. Intuitively, $b_n = o(n^\lambda)$ implies that b_n grows at a rate strictly slower than n^λ. That is,

$$\lim_{n \to \infty} \frac{b_n}{n^\lambda} = 0.$$

Example 7.8. $b_n = 4 + 2n + 6n^2$. Then $b_n = o(n^{2+\delta})$ for all $\delta > 0$.

Obviously, if $b_n = o(n^\lambda)$, then $b_n = O(n^\lambda)$. Intuitively, if b_n grows at a rate slower than n^λ, it will grow at most at a rate of n^λ.

Lemma 7.1. *Let a_n and b_n be scalars.*
(1) If $a_n = O(n^\lambda)$ and $b_n = O(n^\mu)$, then $a_n b_n = O(n^{\lambda+\mu})$, and $a_n + b_n = O(n^\kappa)$, where $\kappa = \max(\lambda, \mu)$.
(2) If $a_n = o(n^\lambda)$ and $b_n = o(n^\mu)$, then $a_n b_n = o(n^{\lambda+\mu})$, and $a_n + b_n = o(n^\kappa)$, where $\kappa = \max(\lambda, \mu)$.
(3) If $a_n = O(n^\lambda)$ and $b_n = o(n^\mu)$, then $a_n b_n = o(n^{\lambda+\mu})$, and $a_n + b_n = O(n^\kappa)$, where $\kappa = \max(\lambda, \mu)$.

Proof: (1) Because a_n grows at most at rate n^λ, b_n grows at most at rate n^μ, the product $a_n b_n$ will grow at most at rate $n^{\lambda+\mu}$. This is because for all n sufficiently large (i.e., for all $n \geq N(M)$),

$$\left| \frac{a_n b_n}{n^{\lambda+\mu}} \right| = \left| \frac{a_n}{n^\lambda} \frac{b_n}{n^\mu} \right| \leq M \cdot M = M^2.$$

On the other hand, the sum $a_n + b_n$ will be dominated by the term that grows to infinity faster: For all n sufficiently large,

$$\left| \frac{a_n + b_n}{n^\kappa} \right| = \left| \frac{a_n}{n^\kappa} + \frac{b_n}{n^\kappa} \right| \leq \epsilon + M \leq 2M.$$

Therefore, $a_n + b_n = O(n^\kappa)$.
(2) Similar to the proof of Result (1).
(3) The product $a_n b_n = o(n^{\lambda+\mu})$ because

$$\left| \frac{a_n b_n}{n^{\lambda+\mu}} \right| = \left| \frac{a_n}{n^\lambda} \frac{b_n}{n^\mu} \right| \leq M \cdot \left| \frac{b_n}{n^\mu} \right| \to 0 \text{ as } n \to \infty$$

given $a_n = O(n^\lambda)$ and $b_n = o(n^\mu)$.

Example 7.9. Suppose $a_n = O(1)$, and $b_n = o(1)$. Then $a_n b_n = o(1)$, and $a_n + b_n = O(1)$.

7.2 Motivation for Convergence Concepts

Why do we need convergence concepts in econometrics?

Recall that a random sample $\mathbf{X}^n = (X_1, \cdots, X_n)$ of size n is a sequence of random variables $X_1, \cdots, X_n$. It can be viewed as an n-dimensional real-valued random vector, where the dimension n may go to infinity. Its realization is an n-dimensional vector $\mathbf{x}^n = (x_1, \ldots, x_n)$. A realization $\mathbf{x}^n$ of $\mathbf{X}^n$ is usually called a sample point or a data set generated from the random sample $\mathbf{X}^n$.

Since $\mathbf{X}^n$ is a sequence of n random variables, we can use the joint probability distribution of $\mathbf{X}^n$ to characterize the random sample. Put $\mathbf{x}^{i-1} = (x_{i-1}, \cdots, x_1)$ for $i = 1, 2, \cdots$. Then repeatedly applying the multiplication rule, the joint PMF/PDF of $\mathbf{X}^n$ is

$$f_{\mathbf{X}^n}(\mathbf{x}^n) = \prod_{i=1}^{n} f_{X_i|\mathbf{X}^{i-1}}(x_i|\mathbf{x}^{i-1}),$$

where $f_{X_i|\mathbf{X}^{i-1}}(x_i|\mathbf{x}^{i-1})$ is the conditional PMF/PDF of X_i given $\mathbf{X}^{i-1} = \mathbf{x}^{i-1}$. By convention, $f_{X_1|\mathbf{X}^0}(x_1|x^0) = f_{X_1}(x_1)$ is the unconditional PMF/PDF of X_1.

When $\mathbf{X}^n$ is an IID random sample from a population PMF/PDF $f_X(\cdot)$, the joint PMF/PDF of the random sample $\mathbf{X}^n$ is given by

$$f_{\mathbf{X}^n}(\mathbf{x}^n) = \prod_{i=1}^{n} f_X(x_i).$$

This joint probability distribution is called the sampling distribution of the random sample $\mathbf{X}^n$. It completely describes the probability law of random sample $\mathbf{X}^n$.

For an IID random sample X^n, each X_i has the same PMF/PDF $f_X(x)$, the so-called population distribution. Usually, $f_X(x)$ is assumed to be a parametric model in the sense that $f_X(x) = f(x, \theta)$ for some value of a finite-dimensional parameter θ, where the functional form of $f(\cdot, \cdot)$ is known but θ is unknown. For example, if $f_X(x)$ is assumed to follow a normal $N(\mu, \sigma^2)$ distribution, we have

$$f_X(x) = f(x, \theta)$$
$$= \frac{1}{\sqrt{2\pi\sigma^2}} e^{-\frac{1}{2\sigma^2}(x-\mu)^2},$$

where $\theta = (\mu, \sigma^2)$.

One important objective of statistical analysis is to estimate the unknown parameter θ when we are given a data set $\mathbf{x}^n$, which is a realization of the random sample $\mathbf{X}^n$. An estimator for θ is a function of $\mathbf{X}^n$, and so it is a statistic. Recall that a statistic $Z_n = T(\mathbf{X}^n)$ is a function of $\mathbf{X}^n$ only and does not involve any unknown parameters. It is a random variable or vector.

To motivate the importance of various convergence concepts, we consider two simple statistics — the sample mean and sample variance. Let $\mathbf{X}^n$ be an IID random sample of size n, from a population distribution with mean μ and variance σ^2. Suppose the mean μ is unknown, so we need to use the sample information $\mathbf{X}^n$ to estimate it. It is expected that the sample mean estimator

$$\bar{X}_n = T(\mathbf{X}^n) = n^{-1} \sum_{i=1}^n X_i$$

can be used to estimate μ.

Similarly, suppose σ^2 is unknown. We can use the sample variance estimator

$$S_n^2 = (n-1)^{-1} \sum_{i=1}^n (X_i - \bar{X}_n)^2$$

to estimate σ^2.

If the sample size n is sufficiently large, then it is expected that $\bar{X}_n$ will be "close" to μ and S_n^2 will be "close" to σ^2. The larger the sample size n is, the closer $\bar{X}_n$ is to μ and the closer S_n^2 is to σ^2.

Question: How can one measure the closeness of $\bar{X}_n$ to μ and the closeness of S_n^2 to σ^2?

Since $\bar{X}_n$ and S_n^2 are random variables, the convergence concepts for nonstochastic sequences reviewed in Section 7.1 do not apply.

Recall that both $\bar{X}_n$ and S_n^2 are mappings from S to the real line; that is, $\bar{X}_n : S \to \mathbb{R}$ and $S_n^2 : S \to \mathbb{R}^+$, where S is the sample space of the underlying random experiment. Suppose a random experiment is conducted, and a basic outcome $s \in S$ occurs. Then we observe a data set $\mathbf{x}^n = (x_1, \cdots, x_n)$, where $x_i = X_i(s)$. It is a realization of the random sample $\mathbf{X}^n$. From the data set $\mathbf{x}^n$, we can compute an estimate $\bar{x}_n = \bar{X}_n(s)$ for μ and an estimate $s_n^2 = S_n^2(s)$ for σ^2. Different outcomes s will yield different estimates for μ and σ^2 respectively.

In fact, each basic outcome $s \in S$ will generate a sequence of real numbers $\{\bar{x}_n = \bar{X}_n(s), n = 1, 2, \cdots\}$ and a sequence of real numbers $\{s_n^2 = S_n^2(s), n = 1, 2, \cdots\}$ respectively. These nonstochastic sequences are called a sample path for the sample mean $\bar{X}_n$ and a sample path for the sample variance S_n^2 respectively, when a basic outcome s occurs. There are many such sample paths for both $\bar{X}_n$ and S_n^2 respectively, as illustrated in Figure 7.1. These sample paths correspond to different basic outcomes $s \in S$. We need to develop some suitable convergence concepts and distance measures between $\bar{X}_n$ and μ, or between S_n^2 and σ^2. Intuitively, a common feature of these different convergence concepts is that they define that most nonstochastic sequences $\{\bar{X}_n(s), n = 1, 2, \cdots\}$ and $\{S_n^2(s), n = 1, 2, \cdots\}$ converge to μ and σ^2 respectively.

7.3 Convergence in Quadratic Mean and L_p-Convergence

Suppose we have a sequence of random variables $Z_1, \cdots, Z_n$, and as the positive integer n becomes large, Z_n "converges" to some random variable (including a constant) Z. We emphasize that here we allow but do not restrict Z_n and Z to be statistics. For example, Z_n can be functions of random sample $\mathbf{X}^n$ and parameter θ.

We now describe the convergence of Z_n to Z using different criteria.

Definition 7.4. [**Convergence in Quadratic Mean**]: Let $\{Z_n, n = 1, 2, \cdots\}$ be a sequence of random variables and Z be a random variable. Then the sequence $\{Z_n, n = 1, 2, \cdots\}$ converges in quadratic mean (or converges in mean square) to Z if

$$E(Z_n - Z)^2 \to 0 \text{ as } n \to \infty,$$

or equivalently,

$$\lim_{n \to \infty} E(Z_n - Z)^2 = 0.$$

It is also denoted as $Z_n \overset{q.m.}{\to} Z$ or $Z_n - Z = o_{q.m}(1)$.

Intuitively, the convergence in quadratic mean means that the weighted average of the squared deviations between $Z_n(s)$ and $Z(s)$ vanishes to 0 as $n \to \infty$. Here, the average is taken over all possible basic outcomes $\{s\}$ weighted by their probabilities of occurring. When Z_n converges to Z in quadratic mean, it is possible that there exist some sample paths for which $Z_n(s)$ does not converge to $Z(s)$. However, the quadratic deviations of these sample paths all together weighted by the probabilities of their occurrings become negligible as n becomes large.

In the application to the sample mean $\bar{X}_n$, we set $Z_n = \bar{X}_n$ and $Z = \mu$.

Example 7.10. Suppose $X^n = (X_1, \cdots, X_n)$ is an IID random sample from a population with mean μ and variance σ^2. Define $Z_n = \bar{X}_n$. Show $\bar{X}_n \overset{q.m.}{\to} \mu$.

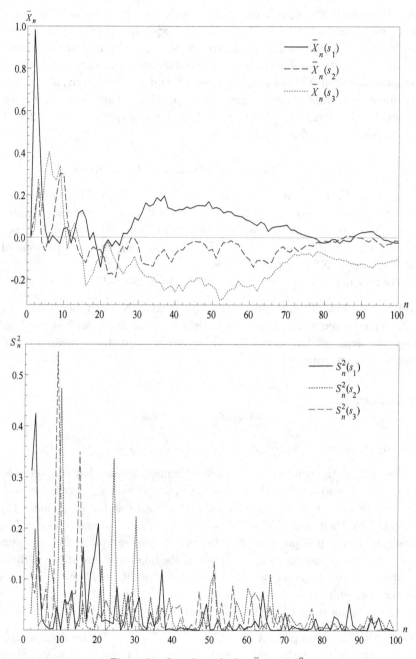

Figure 7.1: Sample paths for $\bar{X}_n$ and S_n^2

Proof: It suffices to show $\lim_{n\to\infty} E(\bar{X}_n - \mu)^2 = 0$. Noting $E(\bar{X}_n) = \mu$, we have

$$E(\bar{X}_n - \mu)^2 = \text{var}(\bar{X}_n)$$

$$= \text{var}\left(\frac{1}{n}\sum_{i=1}^{n} X_i\right)$$

$$= \frac{1}{n^2}\text{var}\left(\sum_{i=1}^{n} X_i\right)$$

$$= \frac{1}{n^2}\sum_{i=1}^{n}\text{var}(X_i)$$

$$= \frac{\sigma^2}{n}$$

$$\to 0 \text{ as } n \to \infty.$$

More generally, we can define a convergence concept called L_p-convergence of which the convergence in quadratic mean is a special case with $p = 2$.

Definition 7.5. [L_p-convergence]: Let $0 < p < \infty$, and let $\{Z_n, n = 1, 2, \cdots\}$ be a sequence of random variables with $E|Z_n|^p < \infty$, and let Z be a random variable with $E|Z|^p < \infty$. Then Z_n converges in L_p to Z if

$$\lim_{n\to\infty} E|Z_n - Z|^p = 0.$$

In connection with L_p-convergence, the following inequalities are useful in econometric analysis:

- Holder's inequality

$$E|XY| \leq (E|X|^p)^{1/p}(E|Y|^q)^{1/q},$$

where $p > 1$ and $1/p + 1/q = 1$.
- Minkowski's inequality

$$E|X + Y|^p \leq \left[(E|X|^p)^{1/p} + (E|Y|^p)^{1/p}\right]^p$$

for $p \geq 1$.

In the above analysis, both Z_n and Z are scalar. What happens if Z_n and Z are $d \times 1$ random vectors, where d is fixed (i.e., d does not change as $n \to \infty$)?

A sequence of random vectors $\{Z_n, n = 1, 2, \cdots\}$ converges to Z in L_p, if each component of the vector Z_n, $Z_{in} \xrightarrow{L_p} Z_i$ for $i = 1, \cdots, d$. In other words, component-wise convergences ensure joint convergence of the entire vector Z_n, and vice versa.

7.4   Convergence in Probability

Next, we introduce the concept of convergence in probability.

Definition 7.6. [**Convergence in Probability**]: A sequence of random variables $\{Z_n, n = 1, 2, \cdots\}$ converges in probability to a random variable Z if for every small constant $\epsilon > 0$,

$$P[|Z_n - Z| > \epsilon] \to 0 \text{ as } n \to \infty.$$

When Z_n converges in probability to Z, we write $\lim_{n\to\infty} P(|Z_n - Z| > \epsilon) = 0$ for every $\epsilon > 0$, or $p\lim_{n\to\infty} Z_n = Z$, or $Z_n \overset{p}{\to} Z$, or $Z_n - Z = o_P(1)$, or $Z_n - Z \overset{p}{\to} 0$.

Convergence in probability is widely used in statistical analysis, because it is often easy to establish. Convergence in probability is also called weak convergence.

The constant $\epsilon > 0$ could be viewed as a prespecified cutoff point such that the difference $|Z_n - Z|$ will be considered as a "large deviation" if $|Z_n - Z| > \epsilon$, and as a "small deviation" if $|Z_n - Z| \leq \epsilon$. Apparently, the smaller ϵ is, the larger n will be required to ensure that $|Z_n(s) - Z(s)| \leq \epsilon$ for any given $s \in S$.

There are other ways of expressing the definition of convergence in probability. For example, one alternative definition is: given any $\epsilon > 0$ and any $\delta > 0$, there exists a finite integer $N = N(\epsilon, \delta)$ such that for all $n > N$, we have $P(|Z_n - Z| > \epsilon) < \delta$.

Intuitively, for n sufficiently large, the probability that the difference $|Z_n - Z|$ takes large values (i.e., larger than ϵ) is rather small. In other words, if Z_n converges in probability to Z, Z_n will be arbitrarily close to Z with a very high probability when n is sufficiently large.

Recall that the random variable Z_n is a measurable mapping from the sample space S to $\mathbb{R}$. Define a set in S:

$$A_n(\epsilon) = \{s \in S : |Z_n(s) - Z(s)| \leq \epsilon\},$$

i.e. $A_n(\epsilon)$ is a subset of S that consists of all basic outcomes $s \in S$ such that the difference $|Z_n(s) - Z(s)|$ is small in the sense that $|Z_n(s) - Z(s)| \leq \epsilon$. The size of $A_n(\epsilon)$ depends on both ϵ and n. When Z_n converges in probability to Z, the probability of "small deviations"

$$P[|Z_n - Z| \leq \epsilon] = P[A_n(\epsilon)] \to 1 \text{ as } n \to \infty.$$

That is, $P[A_n(\epsilon)]$ may not be equal to 1 for any finite n, but as $n \to \infty$, it becomes arbitrarily close to 1 (alternatively, the complement $A_n^c(\epsilon)$, which is the set of basic outcomes in S with "large deviations", can occur for each n, but the probability of its occurring vanishes to zero as $n \to \infty$). For this reason, convergence in probability is also called convergence with probability approaching 1.

Because the set $A_n^c(\epsilon) = \{s : |Z_n(s) - Z(s)| > \epsilon\}$ of "large differences (i.e., larger than ϵ)" depends on n, a given basic outcome $s \in S$ might be in $A_n^c(\epsilon)$ for some n and might be outside $A_n^c(\epsilon)$ for other n. However, when Z_n converges in probability to Z, it becomes less and less likely that the difference $|Z_n - Z|$ is larger than the tolerance level ϵ as n increases.

When $Z_n \xrightarrow{p} b$, where b is a constant, we say that Z_n is consistent for b, and b is the probability limit of Z_n, denoted as $b = p\lim_{n\to\infty} Z_n$.

Example 7.11. Suppose $\mathbf{X}^n = (X_1, \cdots, X_n)$ is an IID random sample from a $U[0, \theta]$ distribution, where $\theta > 0$ is an unknown parameter. Define a statistic $Z_n = \max_{1 \leq i \leq n}(X_i)$. Is Z_n consistent for θ?

Solution: Given $\{|Z_n - Z| > \epsilon\} = \{Z_n - Z > \epsilon\} \cup \{Z_n - Z < -\epsilon\}$, we have

$$
\begin{aligned}
P(|Z_n - \theta| > \epsilon) &= P(Z_n > \theta + \epsilon) + P(Z_n < \theta - \epsilon) \\
&= P(Z_n < \theta - \epsilon) \\
&= P\left[\max_{1 \leq i \leq n}(X_i) < \theta - \epsilon\right] \\
&= P(X_1 < \theta - \epsilon, X_2 < \theta - \epsilon, \cdots, X_n < \theta - \epsilon) \\
&= \prod_{i=1}^{n} P(X_i < \theta - \epsilon) \text{ by independence} \\
&= \left(\frac{\theta - \epsilon}{\theta}\right)^n \\
&= \left(1 - \frac{\epsilon}{\theta}\right)^n \\
&\to 0 \text{ as } n \to \infty \text{ for any given } \epsilon > 0.
\end{aligned}
$$

It follows that Z_n is consistent for θ. The statistic $Z_n = \max_{1 \leq i \leq n}|X_i|$ is called an order statistic which involves some sort of ranking for the n random variables in the random sample $\mathbf{X}^n$.

In a similar way, we can define convergence in probability with order n^α, where α can be a positive or negative number:

- The sequence of random variables $\{Z_n, n = 1, 2, \cdots\}$ is said to be of order smaller than n^α in probability if $Z_n/n^\alpha \xrightarrow{p} 0$ as $n \to \infty$. This is denoted as $Z_n = o_P(n^\alpha)$.
- The sequence of random variables $\{Z_n, n = 1, 2, \cdots\}$ is said to be at most of order n^α in probability if for any given $\delta > 0$, there exist a constant $M = M(\delta) < \infty$ and a finite integer $N = N(\delta)$, such that $P(|Z_n/n^\alpha| > M) < \delta$ for all $n > N$. This is denoted as $Z_n = O_P(n^\alpha)$.

Intuitively, for $Z_n = O_P(n^\alpha)$ with $\alpha > 0$, the order n^α is the fastest growth rate at which Z_n goes to infinity with probability approaching one. When $\alpha < 0$, the order n^α is the slowest convergence rate at which Z_n vanishes to 0 with probability approaching 1. In fact, the definition of $Z_n = O_P(n^\alpha)$ involves the concept of boundedness in probability.

Definition 7.7. [Boundedness in Probability]: For every constant $\delta > 0$, there exist a constant $M = M(\delta)$ and an integer $N = N(\delta)$ such that $P(|Z_n| > M) < \delta$ for all $n \geq N$. Then $Z_n = O_P(1)$ and Z_n is called bounded in probability.

Intuitively, $Z_n = O_P(1) = O_P(n^0)$ implies that for n sufficiently large, the event that $|Z_n|$ takes any value larger than a very large constant has a tiny probability. In other words, $|Z_n|$ is bounded by a constant with a very high probability for all n sufficiently large.

Example 7.12. If $Z_n \sim N(0, 1)$ for all $n \geq 1$. Then $Z_n = O_P(1)$ because for any given $\delta > 0$, there exists a finite constant $M = \Phi^{-1}(1 - \frac{\delta}{2}) < \infty$, where $\Phi(\cdot)$ is the $N(0, 1)$ CDF, such that

$$P(|Z_n| > M) = 2[1 - \Phi(M)] = \delta < 2\delta$$

for all $n \geq 1$.

Below, we shall introduce a weak law of larger numbers (WLLN). Before this, we first provide a useful inequality that bounds the tail probability by moments.

Lemma 7.2. [Markov's Inequality]: *Suppose X is a random variable and $g(X)$ is a nonnegative function. Then for any $\epsilon > 0$, and any $k > 0$,* *we have*

$$P[g(X) \geq \epsilon] \leq \frac{E[g(X)^k]}{\epsilon^k}.$$

Proof: Let $1(\cdot)$ be the indicator function that takes values 1 and 0, depending on whether the statement is true. Then

$$
\begin{aligned}
P[g(X) > \epsilon] &= \int_{\{x:g(x)>\epsilon\}} dF_X(x) \\
&= \int_{-\infty}^{\infty} 1[g(x) > \epsilon] dF_X(x) \\
&\leq \int_{-\infty}^{\infty} 1[g(x) > \epsilon] \frac{g(x)^k}{\epsilon^k} dF_X(x) \\
&\leq \int_{-\infty}^{\infty} \frac{g(x)^k}{\epsilon^k} dF_X(x) \\
&= \frac{1}{\epsilon^k} E[g(X)^k].
\end{aligned}
$$

Markov's inequality is a standard tool for proving convergence in probability. It bounds the tail probability by a moment condition. The thickness of the tail probability depends on the magnitude of moments of the distribution. The heavier the tail probability of the distribution is, the larger the magnitude of the moments of the distribution.

When $k = 2$ and $g(x) = |x|$, Markov's inequality is called Chebyshev's inequality. The latter is widely used in econometrics and statistics because it is often easier to compute the second moment.

The following inequality provides a tighter bound on the tail probability of the sample mean $\bar{X}_n$.

Lemma 7.3. *[Bernstein's inequality]: Let $X_1, \cdots, X_n$ be independent random variables with mean zero and bounded support: $|X_i| < M$ for all $i = 1, \cdots, n$. Let $\sigma_i^2 = var(X_i)$. Suppose $V_n \geq \Sigma_{i=1}^n \sigma_i^2$. Then for each constant $\epsilon > 0$,*

$$P\left[\left|\sum_{i=1}^n X_i\right| > \epsilon\right] \leq 2e^{-\frac{1}{2}\epsilon^2/(V_n + \frac{1}{3}M\epsilon)}.$$

Bernstein's inequality gives a sharp exponentially decaying bound on the tail probability when the support of X_i is bounded. The extensions of Bernstein's inequality to unbounded supports or time series dependent random samples can be found in White and Wooldridge (1990).

We now state the weak law of large numbers.

Theorem 7.1. *[Weak Law of Large Numbers (WLLN)]: Let* $\mathbf{X}^n =$ $(X_1, \cdots, X_n)$ *be an IID random sample with* $E(X_i) = \mu$ *and* $var(X_i) =$ $\sigma^2 < \infty$. *Define* $\bar{X}_n = n^{-1} \sum_{i=1}^{n} X_i$. *Then for any given constant* $\epsilon > 0$ *and as* $n \to \infty$,

$$P[|\bar{X}_n - \mu| \leq \epsilon] \to 1,$$

$$\text{or } \bar{X}_n - \mu \xrightarrow{p} 0 \text{ or } \bar{X}_n - \mu = o_P(1).$$

Proof: First, observe $E(\bar{X}_n) = \mu$ and $var(\bar{X}_n) = \sigma^2/n$. By Chebychev's inequality (that is, Markov's inequality with $g(\bar{X}_n) = |\bar{X}_n - \mu|$ and $k = 2$), we have

$$P[|\bar{X}_n - \mu| > \epsilon] \leq \frac{E(\bar{X}_n - \mu)^2}{\epsilon^2}$$

$$= \frac{var(\bar{X}_n)}{\epsilon^2}$$

$$= \frac{\sigma^2}{n\epsilon^2}.$$

It follows that

$$P[|\bar{X}_n - \mu| \leq \epsilon] = 1 - P[|\bar{X}_n - \mu| > \epsilon]$$

$$\geq 1 - \frac{\sigma^2}{n\epsilon^2}$$

$$\to 1$$

as $n \to \infty$. Therefore, $\bar{X}_n \xrightarrow{p} \mu$, or $\bar{X}_n - \mu = o_P(1)$.

WLLN says that the sample mean $\bar{X}_n$ will approach the population mean μ with probability approaching one as the sample size $n \to \infty$. Given any constant $\epsilon > 0$, it is possible that the difference $|\bar{X}_n - \mu|$ is larger than any pre-specified constant ϵ for any finite n, but this becomes less and less likely as the sample size n increases. Intuitively, when n is sufficiently large, we can be practically certain that the error made with the sample mean $\bar{X}_n$ will be less than any pre-assigned positive constant ϵ.

In the above WLLN theorem, we have assumed a finite variance. Although such an assumption is true and desirable in most applications, it is, in fact, a stronger assumption than is needed. The only moment condition needed is that $E|X_i| < \infty$ (see Resnik 1999, Chapter 7, or Billingsley 1995, Section 22). To appreciate why a moment condition is needed for the WLLN, one can consider the example of an IID random sample from the Cauchy distribution whose moments do not exist:

Example 7.13. Suppose $\mathbf{X}^n$ is an IID random sample from a Cauchy distribution. Then for all integers $n > 0$,

$$\bar{X}_n \sim \text{Cauchy}(0, 1).$$

Thus, the sample $\bar{X}_n$ does not converge to a constant even when $n \to \infty$.

WLLN can be interpreted by considering a large number of repeated experiments, where each experiment will deliver a realization x_i and some observed realizations may repeat. The simple average value from the large number of repeated experiments can also be expressed as the weighted average of repeated values with the weights being their relative frequencies of occurring in the experiments. Thus, when the number of repeated experiments is large, then the simple average will be arbitrarily close to the population mean.

Example 7.14. [**WLLN and Buy & Hold Trading Strategy**]: We now provide an economic interpretation for WLLN via an economic example. In finance, there is a popular investment strategy called buy-and-hold trading strategy. An investor buys an asset on some day and then hold it for a long time period before he sells it out. This is called a buy-and-hold trading strategy. What is the average daily return of this trading strategy?

Suppose X_i is the return of the asset in period i, and the returns over different time periods are $\text{IID}(\mu, \sigma^2)$. Also assume that the investor holds the asset for a total of n periods from $i = 1$ to $i = n$. Then the average return in each time period is the sample mean

$$\bar{X}_n = \frac{1}{n} \sum_{i=1}^{n} X_i.$$

When the number n of holding periods is large, we have

$$\bar{X}_n \xrightarrow{p} \mu = E(X_i)$$

as $n \to \infty$. That is, the average return of the buy-and-hold trading strategy is approximately equal to the population μ when n is sufficiently large. In other words, the population mean μ can be viewed as the long-run average return for the buy-and-hold trading strategy.

Next, we investigate the relationship between L_p-convergence and convergence in probability.

Lemma 7.4. *Suppose $Z_n \to Z$ in L_p as $n \to \infty$. Then $Z_n \xrightarrow{p} Z$ as $n \to \infty$.*

Proof: By Markov's inequality, we have for all $\epsilon > 0$,

$$P[|Z_n - Z| > \epsilon] \le \frac{E|Z_n - Z|^p}{\epsilon^p} \to 0$$

if $\lim_{n\to\infty} E|Z_n - Z|^p = 0$.

Lemma 7.4 says that L_p-convergence implies convergence in probability. This is the consequence of the fact that the tail probability is bounded by moments. This lemma provides a convenient way to prove convergence in probability.

Example 7.15. Suppose $\mathbf{X}^n = (X_1, \cdots, X_n)$ is an IID $N(\mu, \sigma^2)$ random sample. Show $S_n^2 \overset{p}{\to} \sigma^2$.

Solution: We have shown Theorem 6.7 of Chapter 6 that under a normal random sample $\mathbf{X}^n$,

$$\frac{(n-1)S_n^2}{\sigma^2} \sim \chi_{n-1}^2$$

for all $n > 1$. It follows that $E(S_n^2) = \sigma^2$ and $\text{var}(S_n^2) = 2\sigma^4/(n-1)$. Hence, we have

$$E(S_n^2 - \sigma^2)^2 = \text{var}(S_n^2)$$
$$= \frac{2\sigma^4}{n-1}$$
$$\to 0 \text{ as } n \to \infty.$$

It follows from Lemma 7.4 that $S_n^2 \overset{p}{\to} \sigma^2$ as $n \to \infty$.

Now, we ask whether the converse is true. That is, does convergence in probability imply L_p-convergence? The example below show that convergence in probability does not necessarily imply L_p-convergence.

Example 7.16. Suppose a sequence of binary random variables $\{Z_n, n = 1, 2, \cdots\}$ is defined as

Z_n	$\frac{1}{n}$	n
$f_{Z_n}(z_n)$	$1 - \frac{1}{n}$	$\frac{1}{n}$

(1) Does Z_n converge in quadratic mean to 0? Give your reasoning clearly.
(2) Does Z_n converge in probability to 0? Give your reasoning clearly.

Solution: (1) Z_n does not converge in quadratic mean to 0 because

$$E(Z_n - 0)^2 = \sum (z_n - 0)^2 f_{Z_n}(z_n)$$
$$= n^{-2}(1 - n^{-1}) + n^2(n^{-1})$$
$$> n \to \infty.$$

(2) Given any $\epsilon > 0$, for all $n > N(\epsilon) = [\epsilon^{-1}] + 1$ (so $n^{-1} < \epsilon$),

$$P(|Z_n - 0| \le \epsilon) = P(Z_n = n^{-1})$$
$$= 1 - n^{-1}$$
$$\to 1 \text{ as } n \to \infty.$$

Therefore, Z_n converges to 0 in probability.

Intuitively, Z_n converges to 0 with probability $1 - n^{-1}$ approaching one, so $Z_n \overset{p}{\to} 0$ as $n \to \infty$. However, the square of Z_n grows to infinity at a rate of n^2, which is faster than the rate n^{-1} of the vanishing probability $P(Z_n = n)$. As a consequence, the second moment does not exist in this example.

More generally, convergence in probability implies that there may exist a set of basic outcomes in sample space with "large deviations" and the probability of this set shrinks to zero as $n \to \infty$ but it is nonzero for a finite n. Some large deviations may be even explosive at a rate faster than their shrinking probabilities. As a result, the L_p-convergence may not exist.

For any continuous function $g(\cdot)$, we have the following general result:

Lemma 7.5. [Continuity]: *Suppose $g(\cdot)$ is a continuous function, and Z_n converges in probability to Z. Then $g(Z_n)$ also converges in probability to $g(Z)$. That is, if $g(\cdot)$ is continuous, then $Z_n \overset{p}{\to} Z$ as $n \to \infty$ implies*

$$g(Z_n) \overset{p}{\to} g(Z) \text{ as } n \to \infty$$

or equivalently

$$p \lim g(Z_n) = g(p \lim Z_n).$$

Proof: By the continuity of function $g(\cdot)$: given any $\epsilon > 0$, there exists a constant $\delta = \delta(\epsilon)$, such that whenever $|Z_n - Z| \le \delta$, we have

$$|g(Z_n) - g(Z)| < \epsilon.$$

Now, define two events

$$A_n(\delta) \equiv \{s \in S : |Z_n(s) - Z(s)| \le \delta\},$$
$$B_n(\epsilon) \equiv \{s \in S : |g[Z_n(s)] - g[Z(s)]| \le \epsilon\}.$$

Then continuity of $g(\cdot)$ implies $A_n(\delta) \subseteq B_n(\epsilon)$, that is, $A_n(\delta)$ is a subset of $B_n(\epsilon)$. It follows that $P[A_n(\delta)] \le P[B_n(\epsilon)]$, and so

$$P[B_n(\epsilon)^c] \le P[A_n(\delta)^c] \to 0 \text{ as } n \to \infty,$$

where $B_n(\epsilon)^c$ and $A_n(\delta)^c$ are the complements of $B_n(\epsilon)$ and $A_n(\delta)$ respectively. Because ϵ is arbitrary, so is δ. It follows that $g(Z_n) \overset{p}{\to} g(Z)$ as $n \to \infty$. This completes the proof.

This is perhaps one of the most important features of convergence in probability. It shows that the $p\lim$ operator passes through nonlinear functions, provided they are continuous. This is analogous to the well-known result in calculus that the limit of a continuous function is equal to the function of the limit. The lemma stated here behaves just like regular limits when applying a continuous function to a sequence random variables which converges in probability. On the other hand, the expectation operator $E(\cdot)$, which is used in the L_p-convergence, does not have this feature, and this makes finite sample analysis difficult for many statistics.

Example 7.17. Let Z_n converge in probability to a constant $c \neq 0$. Then show that the random variable Z_n/c converges in probability to 1.

Solution: The result follows immediately from Lemma 7.5 and the fact that the function $g(z) = z/c$ is a continuous function when $c \neq 0$.

Example 7.18. Let Z_n converge in probability to a constant $c > 0$ and let $P(Z_n < 0) = 0$ for every n. Show the random variable $\sqrt{Z_n}$ converges in probability to $\sqrt{c}$.

Solution: The result follows immediately from Lemma 7.5 and the fact that the square root function $g(z) = \sqrt{z}$ is continuous.

In an application of this example, put $Z_n = S_n^2$. Then $\sqrt{Z_n} = S_n$ is the sample standard deviation. It follows that $S_n \overset{p}{\to} \sigma$ as $n \to \infty$ given $S_n^2 \overset{p}{\to} \sigma^2$ as $n \to \infty$.

7.5 Almost Sure Convergence

Recall that a random variable Z_n is a mapping defined on the sample space S, i.e., $Z_n : S \to \mathbb{R}$. Each basic outcome $s \in S$ generates a particular sequence of real numbers $\{z_n = Z_n(s), n = 1, 2, \cdots\}$. Also each outcome s generates a single value $z = Z(s)$ for the random variable Z.

Almost sure convergence is the standard notion of pointwise convergence that one studies in first year calculus with one exception: convergence is allowed to fail on a subset in S that occurs with probability zero.

Definition 7.8. [Almost Sure Convergence]: A sequence of random variables $\{Z_n, n = 1, 2, \cdots\}$ converges almost surely to a random variable

Z if for every given constant $\epsilon > 0$,

$$P\left[\lim_{n\to\infty} |Z_n - Z| > \epsilon\right] = 0$$

or equivalently,

$$P\left[s \in S : \lim_{n\to\infty} |Z_n(s) - Z(s)| \le \epsilon\right] = 1,$$

where S is the sample space. When Z_n converges almost surely to Z, we write $Z_n \overset{a.s.}{\to} Z$, or $Z_n - Z = o_{a.s.}(1)$, or $Z_n - Z \overset{a.s.}{\to} 0$.

There are several other alternative but equivalent expressions for almost sure convergence. For example, it can also be expressed as:

$$P\left(\lim_{n\to\infty} |Z_n - Z| = 0\right) = 1.$$

For this reason, almost sure convergence is also called convergence with probability 1. When $Z = b$, a constant, we say that Z_n is strongly consistent for b if Z_n converges to b with probability one.

To provide an interpretation for almost sure convergence, we define a set

$$A(\epsilon) = \left\{s \in S : \lim_{n\to\infty} |Z_n(s) - Z(s)| \le \epsilon\right\}$$
$$= \{s \in S : |Z_n(s) - Z(s)| \le \epsilon \text{ for all } n > N(\epsilon, s)\},$$

i.e. $A(\epsilon)$ is a subset of S that consists of all basic outcomes $s \in S$ such that $\lim_{n\to\infty} |Z_n(s) - Z(s)| < \epsilon$. Intuitively $A(\epsilon)$ is a convergence set in S in the sense that for each $s \in A(\epsilon)$, the sample path $\{Z_n(s), n = 1, 2, \cdots\}$ converges to $Z(s)$ as $n \to \infty$, although the speeds of convergence can be different for different s. Then

$$P\left(\lim_{n\to\infty} |Z_n - Z| \le \epsilon\right) = P[A(\epsilon)] = 1$$

when almost sure convergence holds. In other words, almost sure convergence requires that the convergence set have probability one to occur. There may exist a set, $A(\epsilon)^c$, for which the sample path $Z_n(s)$ does not converge to $Z(s)$ as $n \to \infty$, but such a set has probability zero.

When the sample space S consists of only a finite number of basic outcomes with positive probabilities, almost sure convergence implies pointwise convergence. To see this, suppose s is a basic outcome in S, which corresponds to a sequence of real numbers $\{Z_n(s), n = 1, 2, \ldots\}$. Suppose for each basic outcome $s \in S$, the sequence of real numbers $Z_n(s)$ always converges to $Z(s)$, then we have that Z_n converges to Z almost surely. In this sense, almost sure convergence is equivalent to pointwise convergence in $s \in S$.

When S consists of continuous basic outcomes, then the convergence set $A(\epsilon) = S - \Lambda$ almost covers the entire sample space S, where Λ is a subset in S with $P(\Lambda) = 0$. This arises when Λ contains a finite or an infinite but countable number of points $s \in S$. Because $A(\epsilon)$ contains "almost all points in S" except for a set with zero probability, we call such convergence "almost sure convergence".

Example 7.19. Suppose $S = [0, 1]$ is a sample space with basic outcome s following a uniform distribution on $[0, 1]$. Define two random variables

$$Z_n(s) = s + s^n \text{ and } Z(s) = s.$$

Show that $Z_n - Z \overset{a.s.}{\to} 0$.

Solution: For every $s \in [0, 1)$, $s^n \to 0$ as $n \to \infty$. It follows that for all $s \in A(\epsilon) = [0, 1)$, we have

$$Z_n(s) = s + s^n \to s = Z(s) \text{ as } n \to \infty.$$

That is, for any $\epsilon > 0$ and any s in $[0,1)$, there exists a $N(\epsilon, s) = [\ln \epsilon / \ln s] + 1$ such that for all $n > N(\epsilon, s)$

$$|Z_n(s) - Z(s)| = s^n < \epsilon.$$

Note that the dependence of $N(\epsilon, s)$ on basic outcome s indicates that the convergence rate of $Z_n(s)$ may depend on s. On the other hand, when $s = 1$, we have

$$Z_n(s) = s + s^n = 1 + 1^n = 2,$$

$$Z(s) = s = 1,$$

$$|Z_n(s) - Z(s)| = 1 > \epsilon = \frac{1}{2} \text{ (say)}.$$

There exists $\Lambda = \{1\}$ such that $Z_n(1) = 2$ for all n so that $Z_n(1) - Z(1) \nrightarrow 0$ as $n \to \infty$.

However, $P(\Lambda) = 0$ and $P[A(\epsilon)] = 1$ because s follows a continuous distribution. Thus, $Z_n - Z = o_{a.s.}(1)$.

Similarly, we can define almost sure convergence with order n^α, where α can be positive or negative:

- The sequence of random variables $\{Z_n, n = 1, 2, \cdots\}$ is said to be of order smaller than n^α with probability one if $Z_n/n^\alpha \overset{a.s.}{\to} 0$ as $n \to \infty$. This is denoted as $Z_n = o_{a.s.}(n^\alpha)$.

- The sequence of random variables $\{Z_n, n = 1, 2, \cdots\}$ is said to be at most of order n^α with probability one if there exists some constant $M < \infty$ such that $P(|Z_n/n^\alpha| > M) = 0$ as $n \to \infty$. This is denoted as $Z_n = O_{a.s.}(n^\alpha)$.

In particular, $Z_n = O_{a.s.}(1)$ implies that with probability one, Z_n is bounded by some large constant for all n sufficiently large.

We now investigate the relationship between convergence in probability and almost sure convergence. We first compare the difference in notations between almost sure convergence

$$P\left[\lim_{n\to\infty} |Z_n - Z| > \epsilon \right] = 0$$

and convergence in probability

$$\lim_{n\to\infty} P[|Z_n - Z| > \epsilon] = 0.$$

For almost sure convergence, the convergence set $A(\epsilon) = \{s \in S : \lim_{n\to\infty} |Z_n(s) - Z(s)| < \epsilon\}$ within which each sequence $Z_n(s)$ converges to $Z(s)$ as $n \to \infty$ has probability one. For convergence in probability, the set $A_n^c(\epsilon) = \{s : |Z_n(s) - Z(s)| > \epsilon\}$ for which $|Z_n(s) - Z(s)|$ is "large" (i.e., larger than ϵ) may not have zero probability for any finite n, although its probability must vanish to zero as $n \to \infty$. The fact that the set $A_n(\epsilon)$ of "large" differences for $|Z_n - Z|$ may have a nonzero probability for any finite n implies that convergence in probability is weaker than almost sure convergence. If for some $s \in S$, $Z_n(s) \to Z(s)$ as $n \to \infty$, then the difference $|Z_n(s) - Z(s)|$ will eventually become "small" (i.e., smaller than ϵ) for all n sufficiently large. Hence, almost sure convergence implies convergence in probability.

We thus have the following lemma:

Lemma 7.6. *Suppose* $Z_n \overset{a.s.}{\to} Z$. *Then* $Z_n \overset{p}{\to} Z$.

Example 7.20. Let the sample space S be the closed interval $[0,1]$ with the occurring of basic outcomes following a uniform probability distribution on $[0,1]$. Define a random variable $Z(s) = s$ for all $s \in [0,1]$. Also, for $n = 1, 2, \cdots$ define a sequence of random variables

$$Z_n(s) = \begin{cases} s + s^n & \text{if } 0 \le s \le 1 - \frac{1}{n}, \\ s + 1 & \text{if } 1 - \frac{1}{n} < s \le 1. \end{cases}$$

(1) Does Z_n converge almost surely to Z? Give your reasoning.

(2) Does Z_n converge in probability to Z? Give you reasoning.

(3) Does Z_n converges in L_p to Z? Give your reasoning.

Solution: (1) Consider the set

$$A^c(\epsilon) = \{s \in S : \lim_{n \to \infty} |Z_n(s) - Z(s)| > \epsilon\}.$$

For any given $s \in [0, 1)$, it will eventually falls into the region $[0, 1 - n^{-1}]$ if n is sufficiently large, i.e., if $n > N(s) = [1/(1 - s)] + 1$. Thus, we have $\lim_{n \to \infty} |Z_n(s) - Z(s)| = \lim_{n \to \infty} s^n = 0$ for any $s \in [0, 1)$. Hence, the complement set $A^c(\epsilon)$ at most contains one basic outcome, namely $s = 1$. Since the occurring of basic outcome s follows a continuous distribution, we have $P[A^c(\epsilon)] = 0$. It follows that Z_n converges to Z almost surely.

(2) Because almost sure convergence implies convergence in probability, we have $Z_n \xrightarrow{p} Z$ by Lemma 7.6.

Alternatively, we can consider the set

$$A_n^c(\epsilon) = \{s \in S : |Z_n(s) - Z(s)| > \epsilon\}.$$

Without loss of generality, we assume $0 < \epsilon < 1$. Then for each n, the interval $[1 - \frac{1}{n}, 1]$ is contained in $A_n^c(\epsilon)$. Furthermore, if $|Z_n(s) - Z(s)| = s^n > \epsilon$, then $s \geq e^{n^{-1} \ln \epsilon}$. It follows that

$$A_n^c(\epsilon) = [\min(1 - n^{-1}, e^{n^{-1} \ln \epsilon}), 1]$$

and

$$P[A_n^c(\epsilon)] = 1 - \min(1 - n^{-1}, e^{n^{-1} \ln \epsilon}) \to 0 \text{ as } n \to \infty.$$

(3) For $p > 0$,

$$\begin{aligned}
E|Z_n - Z|^p &= \int_0^1 |Z_n(s) - Z(s)|^p ds \\
&= \int_0^{1 - n^{-1}} s^{pn} ds + \int_{1-n^{-1}}^1 1 \cdot ds \\
&= \frac{1}{pn + 1} \left(1 - \frac{1}{n}\right)^{pn+1} + \frac{1}{n} \\
&\to 0 \text{ as } n \to \infty.
\end{aligned}$$

Hence, $Z_n \xrightarrow{L_p} Z$ as $n \to \infty$.

While almost sure convergence implies convergence in probability, the converse may not be true. Below are two examples.

Example 7.21. Suppose $\{S, \mathbb{B}, P\}$ is a probability space, where the sample space $S = [0, 1]$, $\mathbb{B}$ is a σ-field, and P is a uniform probability measure on S. Define a sequence of random variables as follow:

$$Z_1(s) = 1 \text{ for } 0 \le s \le 1,$$

$$Z_2(s) = \begin{cases} 1, & 0 \le s \le \frac{1}{2}, \\ 0, & \frac{1}{2} < s \le 1, \end{cases}$$

$$Z_3(s) = \begin{cases} 0, & 0 \le s \le \frac{1}{2}, \\ 1, & \frac{1}{2} < s \le 1, \end{cases}$$

$$Z_4(s) = \begin{cases} 1, & 0 \le s \le \frac{1}{3}, \\ 0, & \frac{1}{3} < s \le 1, \end{cases}$$

$$Z_5(s) = \begin{cases} 0, & 0 \le s \le \frac{1}{3}, \\ 1, & \frac{1}{3} < s \le \frac{2}{3}, \\ 0, & \frac{2}{3} < s \le 1, \end{cases}$$

$$Z_6(s) = \begin{cases} 0, & 0 \le s \le \frac{2}{3}, \\ 1, & \frac{2}{3} < s \le 1. \end{cases}$$

Show that $\{Z_n, n = 1, 2, \cdots\}$ converges in probability to 0 but does not converge to 0 almost surely.

Example 7.22. Suppose $\{S, \mathbb{B}, P\}$ is a probability space, where the sample space $S = [0, 1]$, $\mathbb{B}$ is a σ-field, and P is a uniform probability measure on S. Define $Z(s) = s$ and

$$Z_n(s) = \begin{cases} 1, & \text{if } s \in \left[\frac{i}{2^k}, \frac{i+1}{2^k}\right] \text{ for } i = n - 2^k, 1 \le i \le 2^k, \\ s, & \text{otherwise.} \end{cases}$$

where $k = [\log_2 n]$ is the integer part of $\log_2 n$, and $i = 1, \cdots, 2^k$. Then:
(1) for every $\epsilon > 0$, $P(|Z_n - Z| > \epsilon) \le 1/2^k \to 0$ as $n \to \infty$, so $Z_n \xrightarrow{p} Z$ as $n \to \infty$;
(2) $E|Z_n - Z|^p = 1/2^k \to 0$ as $n \to \infty$, so $Z_n \to Z$ in L_p;
(3) for any $s \in [0, 1]$, $\lim_{n \to \infty} Z_n(s)$ does not exist, so Z_n does not converge to Z almost surely.

Question: What is the relationship between almost sure convergence and L_p-convergence?

Almost sure convergence does not imply L_p-convergence and L_p-convergence does not imply almost sure convergence.

Example 7.23. Let the sample space S be the closed interval $[0,1]$ with the occurring of basic outcome s following a uniform probability distribution on $[0,1]$. For $n = 1, 2, \cdots$, define a sequence of random variables

$$Z_n(s) = \begin{cases} 0, & \text{if } s \in [0, 1 - n^{-2}], \\ e^n, & \text{if } s \in (1 - n^{-2}, 1]. \end{cases}$$

Answer the following questions and provide reasoning: (1) $Z_n \overset{q.m.}{\to} 0$?
(2) $Z_n \overset{p}{\to} 0$? (3) $Z_n \overset{a.s.}{\to} 0$?

Solution: (1) no; (2) yes; (3) yes.

Like convergence in probability, almost sure convergence also carries over to any continuous function.

Lemma 7.7. *[Continuity]: Suppose $g(Z)$ is a continuous function, and Z_n converges almost surely to Z. Then $g(Z_n)$ also converges almost surely to $g(Z)$.*

Proof: The proof is similar to the proof of Lemma 7.5 for convergence in probability of a continuous function. Let $s \in S$ be a basic outcome. Since $Z_n(s) \to Z(s)$ as $n \to \infty$ implies $g[Z_n(s)] \to g[Z(s)]$ as $n \to \infty$ by the continuity of $g(\cdot)$, we have

$$\{s \in S : Z_n(s) \to Z(s)\} \subseteq \{s \in S : g[Z_n(s)] \to g[Z(s)]\}.$$

Hence,

$$P[s \in S : g[Z_n(s)] \to g[Z(s)]] \geq P[s \in S : Z_n(s) \to Z(s)] \to 1.$$

It follows that $g(Z_n) \overset{a.s.}{\to} g(Z)$. This completes the proof.

With the concept of almost sure convergence, we are now in a position to introduce a strong law of large numbers.

Theorem 7.2. *[Kolmogorov's Strong Law of Large Numbers (SLLN)]: Suppose $\mathbf{X}^n = (X_1, \cdots, X_n)$ be an IID random sample with $E(X_i) = \mu$ and $E|X_i| < \infty$. Define $\bar{X}_n = n^{-1}\Sigma_{i=1}^n X_i$. Then*

$$\bar{X}_n \overset{a.s.}{\to} \mu \text{ as } n \to \infty.$$

Proof: See (e.g.) Gallant (1997, pp. 132-135).

SLLN states that with probability one, the limit of the sequence of $\{\bar{X}_n, n = 1, 2, \cdots\}$ is μ. Alternatively, the set in the sample space S where the sample paths of $\bar{X}_n$ do not converge as $n \to \infty$ has probability zero. Note that all that is required for the sample mean $\bar{X}_n$ to converge almost surely to the population mean μ is that the population mean μ exists.

Finally, we state a uniform strong law of large numbers, which has substantial applications in econometrics.

Theorem 7.3. *[Uniform Strong Law of Large Numbers (USLLN)]:*
Suppose (1) $\mathbf{X}^n = (X_1, \cdots, X_n)$ *is an IID random sample; (2) function* $g(x, \theta)$ *is continuous over* $\Omega \times \Theta$ *where* Ω *is the support of* X_i *and* Θ *is a compact set in* $\mathbb{R}^d$ *with d finite and fixed; and (3)* $E\left[\sup_{\theta \in \Theta} |g(X_i, \theta)|\right] < \infty$, *where the expectation* $E(\cdot)$ *is taken over the population distribution of* X_i. *Then as* $n \to \infty$,

$$\sup_{\theta \in \Theta} \left| n^{-1} \sum_{i=1}^{n} g(X_i, \theta) - E[g(X_i, \theta)] \right| \to 0 \text{ almost surely.}$$

Moreover, $E[g(X_i, \theta)]$ *is a continuous function of* θ *over* Θ.

Different from the SLLN in Theorem 7.2, $g(X_i, \theta)$ depends on both random variable X_i and parameter θ. The USLLN in Theorem 7.3 says that the worst deviation of the sample average $n^{-1}\Sigma_{i=1}^n g(X_i, \theta)$ from the population mean $E[g(X_i, \theta)]$ that one could find over all possible values in Θ converges to zero almost surely. The uniform convergence here is with respect to the parameter space Θ. USLLN is rather useful when investigating the asymptotic behavior of nonlinear econometric estimators. See applications of USLLN in Chapter 8.

7.6 Convergence in Distribution

Suppose $\mathbf{X}^n = (X_1, \cdots, X_n)$ is an IID random sample from a nonnormal distribution such as the uniform distribution on $[0, 1]$. What is the sampling distribution of the sample mean $\bar{X}_n$?

When the random variable X_i is not normally distributed, the sampling distribution of $\bar{X}_n$ is generally unknown or very complicated. In this case, one may only like to know what is the limiting distribution of $\bar{X}_n$ when the sample size $n \to \infty$.

Convergence in distribution is the tool that is used to obtain an asymptotic approximation to the exact distribution of a statistic like $\bar{X}_n$ that depends on the positive integer n and the underlying population distribution.

Definition 7.9. **[Convergence in Distribution]:** Let $\{Z_n, n = 1, 2, \cdots\}$ be a sequence of random variables with a sequence of corresponding CDF's $\{F_n(z), n = 1, 2, \cdots\}$, and let Z be a random variable with CDF $F(z)$. Then

Z_n converges in distribution to Z as $n \to \infty$ if the CDF $F_n(z)$ converges to $F(z)$ at every continuity point $z \in (-\infty, \infty)$, namely

$$\lim_{n \to \infty} F_n(z) = F(z)$$

at every point z where $F_Z(z)$ is continuous. Here, $F(z)$ is called a limiting (or asymptotic) distribution of the sequence of random variables $\{Z_n, n = 1, 2, \cdots\}$. Convergence in distribution is denoted as $Z_n \xrightarrow{d} Z$ as $n \to \infty$.

Although we refer to a sequence of random variables, $\{Z_n, n = 1, 2, \cdots\}$, converging in distribution to a random variable Z, it is actually the sequence of CDF's $\{F_n(\cdot), n = 1, 2, \ldots\}$ that converges to the CDF $F(\cdot)$. In other words, convergence in distribution means that their CDF's converge, not the random variables $\{Z_n\}$ themselves. This is quite different from the concepts of convergence in L_p, convergence in probability, and almost sure convergence. The latter all characterize the convergence or closeness of the random variable Z_n to random variable Z.

It should be emphasized that the limiting distribution $F(\cdot)$ might not be obtained by taking the limit of $F_n(\cdot)$. For example, suppose $Z_n \sim N(0, \frac{1}{n})$. Then it has the distribution function

$$F_n(z) = \int_{-\infty}^{z} \frac{1}{\sqrt{1/n}\sqrt{2\pi}} e^{-nu^2/2} du$$

$$= \int_{-\infty}^{\sqrt{n}z} \frac{1}{\sqrt{2\pi}} e^{-v^2/2} dv$$

$$= \Phi\left(\sqrt{n}z\right),$$

where $\Phi(\cdot)$ is the $N(0,1)$ CDF. Obviously, we have

$$\lim_{n \to \infty} F_n(z) = \begin{cases} 0 & \text{if } z < 0, \\ \frac{1}{2} & \text{if } z = 0, \\ 1 & \text{if } z > 0. \end{cases}$$

Now define the function

$$F(z) = \begin{cases} 0 & \text{if } z < 0 \\ 1 & \text{if } z \geq 0. \end{cases}$$

Then $F(z)$ is a CDF and $\lim_{n \to \infty} F_n(z) = F(z)$ at every continuity point of $F(z)$. (The function $F(z)$ is not continuous at point $z = 0$.) Therefore, $F(\cdot)$ is a limiting distribution of Z_n. However, $F(\cdot)$ cannot be obtained by taking the limit of $F_n(\cdot)$, because $\lim_{n \to \infty} F_n(0) \neq F(0)$ at zero. Note that in this example $\lim_{n \to \infty} F_n(z)$ is not a CDF because it is not right-continuous.

The concept of convergence in distribution is rather useful in practice. In general, $F_n(z)$ is usually unknown or very complicated for any finite n, but $F(z)$ is known and is often simple. The convergence in distribution thus permits us to approximate the CDF $F(z)$ by using the CDF $F_n(z)$. This may not be a good approximation for small n, but it becomes more and more accurate as n increases. Thus, the asymptotic distribution $F(z)$ provides a convenient tool in statistical inference.

The distribution $F(z)$ is called the limiting distribution or asymptotic distribution of Z_n. Suppose $F(\cdot)$ has mean μ and variance σ^2. Then they will be called, respectively, the asymptotic mean and asymptotic variance of the distribution $F_n(\cdot)$. Since $F(\cdot)$ is not the limit of $F_n(\cdot)$, the asymptotic mean and variance may not be the limits of the mean and variances of the distribution $F_n(\cdot)$ respectively, even if the latter exist.

Example 7.24. Suppose $\mathbf{X}^n = (X_1, \cdots, X_n)$ is an IID $U[0, \theta]$ random sample, where θ is an unknown parameter. Let $Z_n = \max_{1 \leq i \leq n}(X_i)$ be an estimator of θ. Derive the limiting distribution of $n(\theta - Z_n)$.

Solution: For any given $0 \leq u \leq \theta$, we have

$$P\left[n(\theta - Z_n) > u\right] = P\left(Z_n < \theta - \frac{u}{n}\right)$$

$$= P\left(X_1 < \theta - \frac{u}{n}, \cdots, X_n < \theta - \frac{u}{n}\right)$$

$$= \prod_{i=1}^{n} P\left(X_i < \theta - \frac{u}{n}\right)$$

$$= \left(1 - \frac{u}{n\theta}\right)^n$$

$$\to e^{-u/\theta}$$

as $n \to \infty$, where we have used the formula $(1 - \frac{a}{n})^n \to e^{-a}$ as $n \to \infty$. It follows that for $u \geq 0$,

$$F_n(u) = 1 - P[n(\theta - Z_n) > u]$$

$$\to 1 - e^{-u/\theta}$$

as $n \to \infty$. This implies that $n(\theta - Z_n)$ converges in distribution to the exponential distribution with parameter θ, denoted as $\mathrm{EXP}(\theta)$.

We now investigate the relationship between convergence in distribution and convergence in probability. We first state a result showing that convergence in distribution implies boundedness in probability.

Lemma 7.8. *Let Z_n be a random variable with CDF $F_n(\cdot)$, and let Z be a random variable with a continuous CDF $F(\cdot)$. If $Z_n \overset{d}{\to} Z$ as $n \to \infty$, then $Z_n = O_P(1)$.*

Proof: For any given constant $\epsilon > 0$, let $M = M(\epsilon)$ be a (large) constant such that $P(|Z| > M) < \epsilon$. Let $F_n(z)$ be the CDF of Z_n. Given $Z_n \overset{d}{\to} Z$, and $F(z)$ is continuous everywhere, we have $|F_n(z) - F(z)| \leq \epsilon$ for any point $z \in (-\infty, \infty)$ and for all n sufficiently large. This implies that for all n sufficiently large, we have

$$P(Z_n > M) - P(Z > M) < \epsilon,$$
$$P(Z_n \leq -M) - P(Z \leq -M) < \epsilon.$$

It follows that

$$\begin{aligned}
P(Z_n > M) + P(Z_n < -M) &\leq P(Z_n > M) + P(Z_n \leq -M) \\
&< P(Z > M) + P(Z \leq -M) + 2\epsilon \\
&= P(Z > M) + P(Z < -M) + 2\epsilon,
\end{aligned}$$

where $P(Z = -M) = 0$ given that Z follows a continuous distribution. Therefore,

$$P(|Z_n| > M) < P(|Z| > M) + 2\epsilon < 3\epsilon \equiv \delta.$$

Because ϵ is arbitrary, so is δ, and therefore $Z_n = O_P(1)$. This completes the proof.

Intuitively, if the probability distribution of Z_n converges to a well-defined continuous probability distribution as $n \to \infty$, then Z_n is bounded in probability. This result is very useful for establishing that a sequence of random variables is bounded in probability. Often, it is easier to verify that a sequence of random variables converges in distribution.

Example 7.25. Recall from Example 7.24 that for $Z_n = \max_{1 \leq i \leq n}(X_i)$, where $\mathbf{X}^n = (X_1, \cdots, X_n)$ is an IID random sample from a $U[0, \theta]$ distribution, we have shown that $n(\theta - Z_n) \overset{d}{\to} \text{EXP}(\theta)$ as $n \to \infty$. Therefore, $n(\theta - Z_n) = O_P(1)$ and $Z_n - \theta = O_P(n^{-1})$. This implies that the convergence rate of Z_n to θ in probability is n^{-1}, which is rather rapid.

Intuitively, the observations of n random variables $\{X_i\}_{i=1}^n$ will be more or less equally spread over the interval $[0, \theta]$. Therefore, the maximal observation of $\{X_i\}_{i=1}^n$ will approach the upper bound of θ at a rate equal to n, the number of observations.

The next lemma says that convergence in probability implies convergence in distribution.

Lemma 7.9. *If $Z_n \overset{p}{\to} Z$ as $n \to \infty$, then $Z_n \overset{d}{\to} Z$ as $n \to \infty$.*

Proof: This is left as an exercise.

Intuitively, when Z_n converges to Z in probability as $n \to \infty$, the random variable Z_n will be arbitrarily close to random variable Z for n sufficiently large. Therefore, the probability law of Z_n will be arbitrarily close to the probability law of Z for n sufficiently large. That is, Z_n will converge in distribution to Z as $n \to \infty$.

Example 7.26. Suppose $\{Z_n, n = 1, 2, \cdots\}$ is an IID random sample from a population distribution $F(z)$, and Z is a random variable independent of the sequence $\{Z_n, n = 1, 2, \cdots\}$ but has the same distribution $F(z)$. Assume $\text{var}(Z) = \sigma^2 < \infty$.

    (1) Does Z_n converge in distribution to Z?
    (2) Does Z_n converge in quadratic mean to Z?
    (3) Does Z_n converge in probability to Z?

Solution: (1) To show $Z_n \overset{d}{\to} Z$, it suffices to show

$$\lim_{n \to \infty} F_n(z) = F(z) \text{ for all continuity points } \{z\}.$$

(By continuity point z, it means the point z at which $F(z)$ is continuous.) Given the identical distribution assumption, we have

$$F_n(z) = F(z) \text{ for all } z \in (-\infty, \infty) \text{ and all } n > 0.$$

Hence, we trivially have

$$\lim_{n \to \infty} F_n(z) = F(z) \text{ for all continuity points } \{z\}.$$

Hence, we have $Z_n \overset{d}{\to} Z$ as $n \to \infty$.

    (2) Because Z_n and Z are independent, we have

$$
\begin{aligned}
E(Z_n - Z)^2 &= \text{var}(Z_n - Z) \\
&= \text{var}(Z_n) + \text{var}(Z) \\
&= 2\text{var}(Z) \\
&= 2\sigma^2 > 0.
\end{aligned}
$$

Thus, Z_n does not converge to Z in quadratic mean, although it converges in Z in distribution.

    (3) No. (Why?)

The following asymptotic equivalence lemma is a rather useful device to drive the asymptotic distribution of some statistics.

Lemma 7.10. *[Asymptotic Equivalence]: If $Y_n - Z_n \overset{p}{\to} 0$ and $Z_n \overset{d}{\to} Z$ as $n \to \infty$, then $Y_n \overset{d}{\to} Z$.*

Intuitively, if two random variables Y_n and Z_n are very close to each other with probability approaching one as $n \to \infty$, they will follow the same large sample probability distribution. This lemma is very useful when one is interested in deriving the asymptotic distribution of Y_n. We can first establish the asymptotic equivalence (in probability) between Y_n and Z_n in the sense that $Y_n - Z_n \overset{p}{\to} 0$ as $n \to \infty$, then the asymptotic distributions of Y_n and Z_n will be identical. For example, suppose we are interested in deriving the asymptotic distribution of the following normalized sample Sharpe ratio

$$Y_n = \frac{\sqrt{n}\bar{X}_n}{S_n}$$

under the hypothesis that $\mu = 0$, where $\bar{X}_n$ and S_n are the sample mean and sample standard deviation for an IID random sample $\mathbf{X}^n$. We can first establish the asymptotic equivalence in probability between Y_n and the random variable

$$Z_n = \frac{\sqrt{n}\bar{X}_n}{\sigma}$$

and then derive the asymptotic distribution of Z_n, which is simpler because we do not have to deal with a random denominator.

In general, as shown in Example 7.24, convergence in distribution does not imply convergence in probability. However, there is a special case where convergence in distribution implies convergence in probability. We now prove a theorem that relates a certain limiting distribution to convergence in probability to a constant. We first introduce the concept of degenerate probability distribution.

Definition 7.10. **[Degenerate Distribution]:** A random variable Z is said to have a degenerate distribution if $P(Z = c) = 1$ for some constant c.

Theorem 7.4. *Let $F_n(z)$ be the CDF of a random variable Z_n whose distribution depends on the positive integer n. Let c denote a constant which does not depend upon n. The sequence $\{Z_n, n = 1, 2, \cdots\}$ converges in probability to constant c if and only if the limiting distribution of Z_n is degenerate at $z = c$.*

Proof: (1) [*Necessity*]: First, suppose $\lim_{n\to\infty} P(|Z_n - c| < \epsilon) = 1$ for any given $\epsilon > 0$. Then we shall show

$$\lim_{n\to\infty} F_n(z) = \begin{cases} 0 \text{ if } z < c, \\ 1 \text{ if } z > c, \end{cases}$$

from which we can define an asymptotic (i.e., limiting) distribution

$$F(z) = \begin{cases} 0 \text{ if } z < c, \\ 1 \text{ if } z \geq c. \end{cases}$$

We first observe that

$$P(|Z_n - c| \leq \epsilon) = F_n(c + \epsilon) - F_n(c - \epsilon) + P(Z_n = c - \epsilon).$$

Because $0 \leq F_n(z) \leq 1$ and $\lim_{n\to\infty} P(|Z_n - c| < \epsilon) = 1$, we must have that for all $\epsilon > 0$,

$$\lim_{n\to\infty} F_n(c + \epsilon) = 1,$$

$$\lim_{n\to\infty} F_n(c - \epsilon) = 0,$$

$$\lim_{n\to\infty} P(Z_n = c - \epsilon) = 0.$$

It follows that

$$\lim_{n\to\infty} F_n(z) = \begin{cases} 0, \text{ if } z < c, \\ 1, \text{ if } z > c. \end{cases}$$

Thus, we can define an asymptotic (i.e., limiting) distribution as follows:

$$F(z) = \begin{cases} 0, \text{ if } z < c, \\ 1, \text{ if } z \geq c, \end{cases}$$

which is the CDF for Z such that $P(Z = c) = 1$. Since $\lim_{n\to\infty} F_n(z) = F(z)$ at all continuity points $\{z\}$ on the real line (only $z = 0$ is not a continuity point), we have $Z_n \xrightarrow{d} c$ as $n \to \infty$. (Note $\lim_{n\to\infty} F_n(z)$ may not be a CDF.)

(2) [*Sufficiency*]: To complete the proof of the theorem, now suppose

$$\lim_{n\to\infty} F_n(z) = \begin{cases} 0 \text{ if } z < c, \\ 1 \text{ if } z > c. \end{cases}$$

We shall prove that $\lim_{n\to\infty} P(|Z_n - c| \leq \epsilon) = 1$ for all $\epsilon > 0$. Because

$$1 \geq P(|Z_n - c| \leq \epsilon) = F_n(c + \epsilon) - F_n(c - \epsilon) + P(Z_n = c - \epsilon)$$
$$\to 1 - 0 + 0 = 1 \text{ as } n \to \infty$$

for all $\epsilon > 0$, we then obtain the desired result immediately. This completes the proof.

Like convergence in probability and almost sure convergence, the property of convergence in distribution also carries over to any continuous function. This is called the continuous mapping theorem.

Theorem 7.5. *[Continuous Mapping Theorem]:* *Suppose a sequence of $k \times 1$ random vectors $Z_n \overset{d}{\to} Z$ as $n \to \infty$ and $g : \mathbb{R}^k \to \mathbb{R}^l$ is a continuous vector-valued function. Then $g(Z_n) \overset{d}{\to} g(Z)$.*

Theorem 7.5 implies that once we know the limiting distribution of Z_n, we can find the limiting distribution of many interesting functions of Z_n. This is particularly useful for deriving the limiting distributions of statistic $T(Z_n)$ once the limiting distribution of Z_n is known.

7.7 Central Limit Theorems

With the concept of convergence in distribution, we now discuss a fundamental theorem in probability and statistics which is called the Central Limit Theorem (CLT).

Theorem 7.6. *[Lindeberg-Levy's Central Limit Theorem (CLT)]:* *Let $\mathbf{X}^n = (X_1, \cdots, X_n)'$ be an IID random sample from a population with mean μ and variance $0 < \sigma^2 < \infty$. Define $\bar{X}_n = n^{-1} \sum_{i=1}^{n} X_i$. Then the standardized sample mean*

$$
\begin{aligned}
Z_n &= \frac{\bar{X}_n - E(\bar{X}_n)}{\sqrt{var(\bar{X}_n)}} \\
&= \frac{\bar{X}_n - \mu}{\sigma/\sqrt{n}} \\
&= \frac{\sqrt{n}(\bar{X}_n - \mu)}{\sigma} \\
&\overset{d}{\to} N(0,1) \; as \; n \to \infty.
\end{aligned}
$$

Let $Z \sim N(0,1)$. Usually, the $N(0,1)$ CDF is denoted as

$$
\Phi(z) = \int_{-\infty}^{z} \frac{1}{\sqrt{2\pi}} e^{-x^2/2} dx.
$$

CLT says that when $n \to \infty$, $Z_n \overset{d}{\to} Z$, i.e., $F_n(z) \equiv P(Z_n \leq z) \to \Phi(z)$ for all $z \in (-\infty, \infty)$.

Proof: Define a standardized random variable

$$Y_i = \frac{X_i - \mu}{\sigma}, \qquad i = 1, \cdots, n,$$

with characteristic function $\varphi_Y(t) = E(e^{\mathbf{i}tY_i})$, where $\mathbf{i} = \sqrt{-1}$. Then Y_i has zero mean and unit variance. It follows that

$$\varphi_Y'(0) = \mathbf{i} \cdot 0 = 0,$$
$$\varphi_Y''(0) = \mathbf{i}^2 \cdot \sigma_Y^2 = -1.$$

We now write the standardized sample mean

$$
\begin{aligned}
Z_n &= \frac{\bar{X}_n - \mu}{\sigma/\sqrt{n}} \\
&= \sqrt{n}\frac{\bar{X}_n - \mu}{\sigma} \\
&= \sqrt{n}\left(n^{-1}\sum_{i=1}^{n}\frac{X_i - \mu}{\sigma}\right) \\
&= \sqrt{n}\bar{Y}_n \\
&= \frac{1}{\sqrt{n}}\sum_{i=1}^{n} Y_i.
\end{aligned}
$$

Since X_i may not have a well-defined MGF, we take a characteristic function approach, that is, we shall show that $\varphi_n(t) \to e^{-\frac{1}{2}t^2}$ as $n \to \infty$, where $\varphi_n(t) = E(e^{\mathbf{i}tZ_n})$ is the characteristic function of Z_n and $e^{-\frac{1}{2}t^2}$ is the characteristic function of $N(0,1)$.

Given the IID assumption, we have

$$
\begin{aligned}
\varphi_n(t) &= E(e^{\mathbf{i}tZ_n}) \\
&= E\left(e^{\mathbf{i}t\sqrt{n}\bar{Y}_n}\right) \\
&= E\left(e^{\frac{\mathbf{i}t}{\sqrt{n}}\Sigma_{i=1}^{n}Y_i}\right) \\
&= E\left(e^{\frac{\mathbf{i}t}{\sqrt{n}}Y_1}e^{\frac{\mathbf{i}t}{\sqrt{n}}Y_2}\cdots e^{\frac{\mathbf{i}t}{\sqrt{n}}Y_n}\right) \\
&= E\left(e^{\frac{\mathbf{i}t}{\sqrt{n}}Y_1}\right)E\left(e^{\frac{\mathbf{i}t}{\sqrt{n}}Y_2}\right)\cdots E\left(e^{\frac{\mathbf{i}t}{\sqrt{n}}Y_n}\right) \\
&= \left[E\left(e^{\frac{\mathbf{i}t}{\sqrt{n}}Y_1}\right)\right]^n \\
&= \left[\varphi_Y\left(\frac{t}{\sqrt{n}}\right)\right]^n,
\end{aligned}
$$

where $\varphi_Y(t) = E(e^{itY_i})$. Now, write

$$\ln\left\{[\varphi_Y(t/\sqrt{n})]^n\right\} = n\ln[\varphi_Y(t/\sqrt{n})]$$
$$= \frac{\ln[\varphi_Y(t/\sqrt{n})]}{1/n}.$$

Noting $\ln[\varphi_Y(t/\sqrt{n})] \to 0$ given $\varphi_Y(0) = 1$, and $1/n \to 0$, we have for any given $t \in (-\infty, \infty)$,

$$\lim_{n\to\infty} \frac{\ln[\varphi_Y(t/\sqrt{n})]}{1/n} = \lim_{n\to\infty} \frac{\frac{\varphi_Y'(t/\sqrt{n})}{\varphi_Y(t/\sqrt{n})}\left(-\frac{t}{2n\sqrt{n}}\right)}{-1/n^2}$$
$$= \frac{t}{2}\lim_{n\to\infty} \frac{\frac{\varphi_Y'(t/\sqrt{n})}{\varphi_Y(t/\sqrt{n})}}{1/\sqrt{n}}$$

by L'Hospital's rule. Since

$$\frac{\varphi_Y'(t/\sqrt{n})}{\varphi_Y(t/\sqrt{n})} \to 0$$

given $\varphi_Y'(0) = 0$ and $1/\sqrt{n} \to 0$, we use L'Hospital's rule again and obtain

$$\lim_{n\to\infty} \frac{\frac{\varphi_Y'(t/\sqrt{n})}{\varphi_Y(t/\sqrt{n})}}{1/\sqrt{n}} = \lim_{n\to\infty} \frac{\frac{\varphi_Y''(t/\sqrt{n})\varphi_Y(t/\sqrt{n})-[\varphi_Y'(t/\sqrt{n})]^2}{[\varphi_Y(t/\sqrt{n})]^2}\cdot\left(-\frac{t}{2n\sqrt{n}}\right)}{-\frac{1}{2n\sqrt{n}}}$$
$$= t\lim_{n\to\infty} \frac{\varphi_Y''(t/\sqrt{n})\varphi_Y(t/\sqrt{n}) - [\varphi_Y'(t/\sqrt{n})]^2}{[\varphi_Y(t/\sqrt{n})]^2}$$
$$= -t,$$

where we have used the fact that $\varphi_Y(0) = 1$, $\varphi_Y'(0) = 0$ and $\varphi_Y''(0) = -1$.

It follows that

$$\lim_{n\to\infty} \ln\varphi_n(t) = \lim_{n\to\infty} \frac{\ln[\varphi_Y(t/\sqrt{n})]}{1/n}$$
$$= -\frac{1}{2}t^2.$$

Because the limit of a continuous function (here, the exponential function) is equal to the function of the limit, we then have

$$\lim_{n\to\infty} \varphi_n(t) = e^{-\frac{t^2}{2}}.$$

Therefore, $Z_n \overset{d}{\to} N(0,1)$ as $n \to \infty$. This completes the proof.

There is an alternative heuristic proof. The characteristic function of Z_n

$$\varphi_n(t) = E\left(e^{it\sqrt{n}\bar{Y}_n}\right)$$
$$= [E(e^{itY_1/\sqrt{n}})]^n$$
$$= [\varphi_Y(t/\sqrt{n})]^n$$
$$= \left[\varphi_Y(0) + \varphi_Y'(0)\frac{t}{\sqrt{n}} + \frac{1}{2}\varphi_Y''(0)\left(\frac{t}{\sqrt{n}}\right)^2 + r\left(\frac{t}{\sqrt{n}}\right)\right]^n$$
$$= \left(1 - \frac{t^2}{2n} + o(n^{-1})\right)^n$$
$$\to e^{-t^2/2},$$

where $r(t/\sqrt{n})$ denotes a reminder term, and we have used the formula $(1 + \frac{a}{n})^n \to e^a$ as $n \to \infty$.

In the proof of CLT, we have used the characteristic function rather than the more familiar moment generating function, because the former covers the population distributions (e.g., the lognormal and Student's t-distributions) whose moment generating functions do not exist. Of course, a price is that the proof is more delicate, since functions of complex variables must be dealt with. Historically, CLT was first established for a random sample from a Bernoulli distribution by A. de Moivre in the early eighteenth century. The proof for a random sample from an arbitrary distribution was given independently by J.W. Lindeberg and P. Levy in the early 1920s.

CLT says that if a large random sample is taken from any population distribution with finite variance, regardless of whether this population distribution is discrete or continuous, then the distribution of the standardized sample mean

$$Z_n = \frac{\sqrt{n}(\bar{X}_n - \mu)}{\sigma}$$

will approximately follow a $N(0,1)$ distribution when n is large. Therefore, for each finite n, the distribution of $\bar{X}_n$ will be approximately a $N(\mu, \sigma^2/n)$ or equivalently, the distribution of the sum $\sum_{i=1}^{n} X_i$ will be approximately a $N(n\mu, n\sigma^2)$ distribution. Figure 7.2 plots the sampling distribution of Z_n for various sample sizes ($n = 1, 2, 5, 10, 30, 50$) under following population distributions: (1) $U[0,1]$; (2) t_4-distribution; (3) EXP(1); and (4) Bernoulli(0.5). The normal approximations work well for $n \geq 30$.

It is important to note that CLT does not say that a large population is approximately normally distributed. It says nothing about the distribution of the population; it is only a statement about the approximate distribution of a standardized sample mean Z_n. Moreover, CLT does not assume that the number of possible values of the population distribution is large (instead, the population can follow a binary distribution); it is the size n of the random sample $\mathbf{X}^n = (X_1, \cdots, X_n)$ that is required to be large.

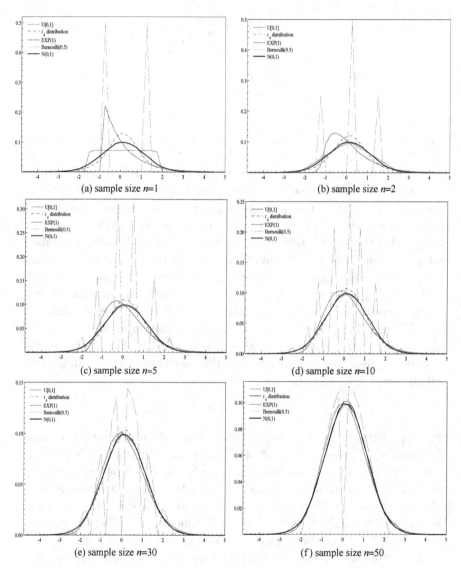

Figure 7.2: Sampling distribution of Z_n for various sample sizes and data generating processes

Sometimes CLT is interpreted incorrectly as implying that the distribution of $\bar{X}_n$ approaches a normal distribution as $n \to \infty$. This is incorrect because $\text{var}(\bar{X}_n) \to 0$ and $\bar{X}_n$ converges to a degenerate distribution $F(\cdot)$ such that $F(x) = 0$ if $x < \mu$ and $F(x) = 1$ if $x \geq \mu$.

CLT provides a plausible explanation for the fact that the distribution of some random variables in economic systems and physical experiments are approximately normal. In economics, many aggregate economic variables are the sums of individual counterparts. In physics, the observed values for important physical variables be the averages of the measurements in a large number of repeated experiments. In these cases, the variables of interest may be approximately normally distributed if the independence assumption is approximately true.

CLT occupies a central position in statistical inference. Although CLT provides a simple and useful general approximation, there is no automatic way of knowing how good the approximation is in general. In fact, the goodness of the approximation is a function of the sample size n and the original population distribution, and differs case by case; see Figure 7.2.

With rapid advance in computing technology, the importance of CLT is somewhat lessened. One can for example use the bootstrap, a computer-based resampling method, to accurately approximate the finite sample distribution of $\bar{X}_n$ for any finite n. For bootstrap, see Hall (1992).

CLT has wide applications in practice. Below is an example.

Example 7.27. [Life Insurance]: Suppose 10,000 people buy insurance from a particular insurance company. The insurance premium is 12 dollars every person per year. The probability of a man dying in one year is 0.006. If an insurance applicant is dead this year, his family will receive 1000 dollar. What is the probability of the insurance company losses?

Solution: Denote:

$$X_i = \begin{cases} 1, & \text{if person } i \text{ dies within a year,} \\ 0, & \text{otherwise.} \end{cases}$$

That is $X_i \sim$ Bernoulli(0.006) distribution. By CLT,

$$100 \times \frac{\frac{1}{10000} \sum_{i=1}^{10000} X_i - 0.006}{\sqrt{0.006 \times (1 - 0.006)}} \sim N(0, 1)$$

That is,

$$\frac{\sum_{i=1}^{10000} X_i - 10000 \times 0.006}{\sqrt{10000 \times 0.006 \times (1 - 0.006)}} \sim N(0, 1)$$

Define the number of the deaths among the insured persons within one year as $Z = \sum_{i=1}^{10000} X_i$. If

$$12 \times 10000 - 1000Z < 0 \text{ or } Z > 120,$$

the company will lose money. So, the probability of the insurance company losses:

$P(Z > 120)$

$$= P\left(\frac{Z - 10000 \times 0.006}{\sqrt{10000 \times 0.006 \times (1 - 0.006)}} > \frac{120 - 10000 \times 0.006}{\sqrt{10000 \times 0.006 \times (1 - 0.006)}}\right)$$

$$\approx 1 - \Phi\left(\frac{120 - 10000 \times 0.006}{\sqrt{10000 \times 0.006 \times (1 - 0.006)}}\right)$$

$$= 1 - \Phi(7.769)$$

$$\approx 0.$$

With CLT, it is easy to understand normal approximation to many well-known distributions. Below we provide some examples.

Example 7.28. [Normal Approximation for the Binomial Distribution $B(n,p)$ When n is Large]: For a binomial $B(n,p)$ random variable Z_n, we can write $Z_n = \sum_{i=1}^{n} X_i$, where $\mathbf{X}^n = (X_1, \cdots, X_n)$ is an IID random sample from a Bernoulli(p) distribution with $P(X_i = 1) = p \in (0,1)$. By CLT, we have that as $n \to \infty$, the standardized random variable

$$\frac{Z_n - E(Z_n)}{\sqrt{\text{var}(Z_n)}} = \frac{Z_n - np}{\sqrt{np(1-p)}}$$

$$\xrightarrow{d} N(0,1).$$

Although this result applies only when $n \to \infty$, the normal distribution is often used to approximate binomial probabilities in practice even when n is fairly small. Figure 7.3 plots the case of $p = 0.4$ for which the normal approximation works well for $n \geq 50$.

In Chapter 4, we approximate the Binomial(n,p) distribution by a Poisson distribution, which is called the law of small numbers. Here we use a normal approximation. A Poisson approximation is better when p is small, while a normal approximation works better when both np and $n(1-p)$ are both larger than 5.

Example 7.29. [Normal Approximation of χ_n^2 When n is Large]: Suppose $\mathbf{X}^n$ is an IID $N(0,1)$ random sample. Put $Y = X_i^2$, $i = 1, \cdots, n$.

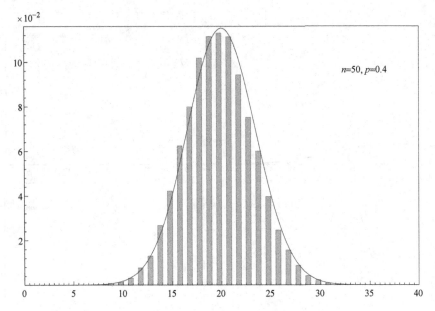

Figure 7.3: Normal approximation for Binomial distribution

Then as $n \to \infty$, the standardized random variable

$$\frac{\sum_{i=1}^{n} Y_i - n\mu_Y}{\sqrt{n\sigma_Y^2}} = \frac{\sum_{i=1}^{n} X_i^2 - n}{\sqrt{2n}}$$

$$\xrightarrow{d} N(0,1).$$

Solution: Put $\bar{Y}_n = n^{-1} \sum_{i=1}^{n} Y_i$. Since $E(Y_i) = 1$ and $\mathrm{var}(Y_i) = 2$, we have from CLT that as $n \to \infty$, the standardized sample mean

$$\frac{\bar{Y}_n - \mu_Y}{\sigma_Y/\sqrt{n}} = \frac{\bar{Y}_n - 1}{\sqrt{2}/\sqrt{n}} \xrightarrow{d} N(0,1),$$

where

$$\frac{\bar{Y}_n - 1}{\sqrt{2}/\sqrt{n}} = \frac{\sqrt{n}(\bar{Y}_n - 1)}{\sqrt{2}}$$

$$= \frac{\frac{1}{\sqrt{n}} \sum_{i=1}^{n}(Y_i - 1)}{\sqrt{2}}$$

$$= \frac{\sum_{i=1}^{n} Y_i - n}{\sqrt{2n}}$$

$$= \frac{\sum_{i=1}^{n} X_i^2 - n}{\sqrt{2n}}.$$

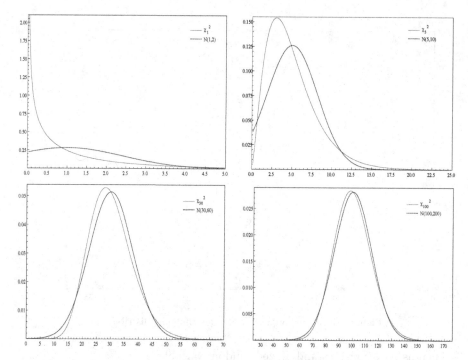

Figure 7.4: $N(n, 2n)$ approximation for χ_n^2 distribution for various n

Note that $\sum_{i=1}^n X_i^2 \sim \chi_n^2$ by Lemma 6.6. It follows that we can approximate a χ_n^2 distribution by a $N(n, 2n)$ distribution when the degree of freedom n is large. As an example, when $\mathbf{X}^n$ is an IID $N(\mu, \sigma^2)$ random sample, we have $(n-1)S_n^2/\sigma^2 \sim \chi_{n-1}^2$ for all $n > 1$. It follows that as $n \to \infty$,

$$\left[\frac{(n-1)S_n^2}{\sigma^2} - (n-1) \right] / \sqrt{2(n-1)} \xrightarrow{d} N(0,1).$$

Figure 7.4 plots the $N(n, 2n)$ approximation for the χ_n^2 distribution for various n.

The normal approximation for the Chi-square distribution is not very accurate when $n \leq 30$. For small n, some continuity refinement may help improve the asymptotic approximation.

Question: How important is the assumption of $\mathrm{var}(X_i) = \sigma^2 < \infty$ in CLT?

We now provide an example to illustrate the importance of the finite variance assumption for CLT.

Example 7.30. [Sum of Independent Cauchy Random Variables]:
Suppose $\mathbf{X}^n$ is an IID random sample from the Cauchy$(0,1)$ distribution.
Then it can be shown that $\bar{X}_n \sim$ Cauchy$(0,1)$ for all $n \geq 1$.

Solution: We use the characteristic function to prove this result. By the
IID property and the fact that the characteristic function of a Cauchy$(0,1)$
random variable is $\varphi(t) = \exp(-|t|)$, as given in Chapter 4, we have

$$\varphi_n(t) = E\left(e^{it\bar{X}_n}\right)$$

$$= \left[\varphi\left(\frac{t}{n}\right)\right]^n$$

$$= e^{-|t|}$$

$$= \varphi(t).$$

Therefore, $\bar{X}_n \sim$ Cauchy$(0,1)$ for all $n \geq 1$. Thus, the sample mean of an
IID Cauchy random sample will not converge to $N(0,1)$ when $n \to \infty$.

Recall from Lemma 6.11 in Chapter 6 that a Cauchy$(0,1)$ random variable follows a Student t_1-distribution. The k-th moment $E(X_i^k)$ of a Cauchy
distribution do not exist for any $k \geq 1$.

In fact, the assumption of a finite variance is essentially necessary for
CLT. It implies that we always obtain an approximate normality from the
sum of "small" (finite variance) independent disturbances. Although the
finite variance assumption can be relaxed somewhat, it cannot be eliminated
(see Casella and Berger, 2000, p. 267).

On the other hand, the identical distribution assumption can be relaxed.
In other words, CLT continues to hold when there exist certain degrees of
heterogeneity in observations.

**Theorem 7.7. [Liapounov's (1901) CLT for Independent Random
Variables]:** *Suppose the random variables $X_1, \cdots, X_n$ are jointly independent and $E|X_i - \mu_i|^3 < \infty$ for $i = 1, \cdots, n$, where $E(X_i) = \mu_i$. Also,
suppose*

$$\lim_{n\to\infty} \frac{\sum_{i=1}^n E|X_i - \mu_i|^3}{\left(\sum_{i=1}^n \sigma_i^2\right)^{3/2}} = 0.$$

Then as $n \to \infty$, we have the standardized random variable

$$Z_n = \frac{\sum_{i=1}^n X_i - \sum_{i=1}^n \mu_i}{\left(\sum_{i=1}^n \sigma_i^2\right)^{1/2}} \xrightarrow{d} N(0,1).$$

CLT also holds when there exists certain degree of dependence among
the random variables $X_1, \cdots, X_n$. Although the dependence allowed cannot

be too strong (see what happens if $X_1 = X_2 = \cdots = X_n$ in an extreme case), one can relax the independence assumption to some extent. See, for example, Billingsley (1995, Section 27). Also see White (1999) for CLT for dependent random samples. This allows application of CLT to some time series data.

We now state an important asymptotic analytic tool that can be used in conjunction with CLT.

Theorem 7.8. *[Slutsky's Theorem]: Suppose $X_n \xrightarrow{d} X$ and $C_n \xrightarrow{p} c$, a constant. Then*

(1) $X_n + C_n \xrightarrow{d} X + c$;

(2) $X_n - C_n \xrightarrow{d} X - c$;

(3) $X_n C_n \xrightarrow{d} cX$;

(4) $\frac{X_n}{C_n} \xrightarrow{d} \frac{X}{c}$, for $c \neq 0$.

Proof: This is left as an exercise by applying the asymptotic equivalence lemma (Lemma 7.10).

Example 7.31. Suppose the standardized sample mean

$$\frac{\sqrt{n}(\bar{X}_n - \mu)}{\sigma} \xrightarrow{d} N(0, 1)$$

and $S_n^2 \xrightarrow{p} \sigma^2$ as $n \to \infty$ (so $S_n \xrightarrow{p} \sigma$ by the continuity of the square root function). Then by the Slutsky theorem

$$\frac{\sqrt{n}(\bar{X}_n - \mu)}{S_n} = \frac{\sigma}{S_n} \frac{\sqrt{n}(\bar{X}_n - \mu)}{\sigma}$$

$$\xrightarrow{d} N(0, 1).$$

This example shows that the replacement of σ by S_n does not change the asymptotic distribution of the t-test statistic, although it changes the finite sample distribution.

Example 7.32. Suppose $X_n \xrightarrow{d} X$ and $Y_n \xrightarrow{d} Y$ as $n \to \infty$. Do we have the following results? Give your reasoning:

(1) $X_n \pm Y_n \xrightarrow{d} X \pm Y$ as $n \to \infty$;

(2) $X_n Y_n \xrightarrow{d} XY$ as $n \to \infty$.

Solution: The answer is generally no, because the dependence between X_n and Y_n is not taken into account. In other words, convergence in marginal distribution does not imply convergence in joint distribution. This is different from other convergence concepts, such as convergence in quadratic mean, convergence in probability and almost sure convergence. For these convergences, element-by-element convergences are equivalent to joint convergence.

Most econometric estimators and statistics are nonlinear functions of random sample $\mathbf{X}^n$. Thus, CLT cannot be applied to them directly. Then, how can one obtain the limiting distribution of a nonlinear statistic $Y_n = g(\bar{X}_n)$?

The asymptotic distribution of $g(\bar{X}_n)$ can be obtained by using the so-called Delta method.

Lemma 7.11. *[Delta Method]: Suppose* $\sqrt{n}(\bar{X}_n - \mu)/\sigma \xrightarrow{d} N(0,1)$ *as* $n \to \infty$, *and function* $g(\cdot)$ *is continuously differentiable with* $g'(\mu) \neq 0$. *Then as* $n \to \infty$,

$$\sqrt{n}\left[g(\bar{X}_n) - g(\mu)\right] \xrightarrow{d} N(0, \sigma^2[g'(\mu)]^2)$$

and

$$\frac{\sqrt{n}\left[g(\bar{X}_n) - g(\mu)\right]}{\sigma g'(\mu)} \xrightarrow{d} N(0,1).$$

Proof: First, by Lemma 7.8, $\sqrt{n}(\bar{X}_n - \mu)/\sigma \xrightarrow{d} N(0,1)$ implies $\sqrt{n}(\bar{X}_n - \mu)/\sigma = O_P(1)$. Therefore, we have $\bar{X}_n - \mu = O_P(n^{-1/2}) = o_P(1)$.

Next, by the mean value theorem, we have

$$Y_n = g(\bar{X}_n) = g(\mu) + g'(\bar{\mu}_n)(\bar{X}_n - \mu),$$

where $\bar{\mu}_n = \lambda\mu + (1 - \lambda)\bar{X}_n$ for some $\lambda \in [0,1]$. Note that $|\bar{\mu}_n - \mu| = |(1 - \lambda)(\bar{X}_n - \mu)| \leq |\bar{X}_n - \mu| = o_P(1)$. It follows by the Slutsky theorem that

$$\sqrt{n}\left[\frac{g(\bar{X}_n) - g(\mu)}{\sigma}\right] = g'(\bar{\mu}_n)\sqrt{n}\frac{\bar{X}_n - \mu}{\sigma}$$

$$\xrightarrow{d} N\left[0, g'(\mu)^2\right],$$

where $g'(\bar{\mu}_n) \xrightarrow{p} g'(\mu)$ by Lemma 7.5 given $\bar{\mu}_n \xrightarrow{p} \mu$ and the continuity of the first derivative $g'(\cdot)$.

By the Slutsky theorem again, we have

$$\frac{\sqrt{n}[g(\bar{X}_n) - g(\mu)]}{\sigma g'(\bar{X}_n)} \xrightarrow{d} N(0,1).$$

This completes the proof.

The Delta method can be viewed as a Taylor series approximation in a statistical context. It linearizes a smooth (i.e., continuously differentiable) nonlinear statistic so that CLT can be applied to the linearized statistic. Therefore, it can be viewed as a generalization of CLT. This method is very useful when more than one parameter makes up the function to be estimated and more than one random variable is used in the estimator.

Example 7.33. Suppose $\sqrt{n}(\bar{X}_n - \mu)/\sigma \xrightarrow{d} N(0,1)$ as $n \to \infty$ and $\mu \neq 0$ and $0 < \sigma < \infty$. Find the limiting distribution of $\sqrt{n}(\bar{X}_n^{-1} - \mu^{-1})$.

Solution: We apply the Delta method with $g(z) = z^{-1}$. Because $\mu \neq 0$, $g(z) = z^{-1}$ is continuously differentiable at $z = \mu$, and its derivative

$$g'(\mu) = \frac{1}{\mu^2}.$$

It follows from the Delta method that

$$\frac{\sqrt{n}(\bar{X}_n^{-1} - \mu^{-1})}{\sigma} \xrightarrow{d} N(0, \mu^{-4}) \text{ as } n \to \infty.$$

By the Slutsky theorem, we have

$$\frac{\bar{X}_n^2 \sqrt{n}(\bar{X}_n^{-1} - \mu^{-1})}{\sigma} \xrightarrow{d} N(0,1) \text{ as } n \to \infty.$$

To apply the Delta method, it is required that $g'(\mu) \neq 0$. What happens to the Delta method if $g'(\mu) = 0$?

The condition of $g'(\mu) \neq 0$ allows the application of the first order Taylor series expansion, which is the basis for the Delta method. When $g'(\mu) = 0$, one has to extend the Delta method by using a second order Taylor series expansion. However, the quadratic form is no longer normally distributed.

Lemma 7.12. *[Second Order Delta Method]: Suppose a sequence of random variables $\sqrt{n}(\bar{X}_n - \mu)/\sigma \xrightarrow{d} N(0,1)$ as $n \to \infty$, and function $g(\cdot)$ is twice continuously differentiable such that $g'(\mu) = 0$ and $g''(\mu) \neq 0$. Then as $n \to \infty$,*

$$\frac{n\left[g(\bar{X}_n) - g(\mu)\right]}{\sigma^2} \xrightarrow{d} \frac{g''(\mu)}{2}\chi_1^2.$$

Proof: This is left as an exercise.

Lemma 7.12 implies that $g(\bar{X}_n) - g(\mu) = O_P(n^{-1})$; that is, $g(\bar{X}_n)$ converges to $g(\mu)$ at a rate of n^{-1}. This is different from the case where $g'(\mu) \neq 0$ where $g(\bar{X}_n) - g(\mu) = O_P(n^{-1/2})$. The statistic $g(\bar{X}_n)$ is called a degenerate statistic because its convergence rate is faster than under the regular case where $g'(\mu) \neq 0$. In the degenerate case where $g'(\mu) = 0$, the asymptotic distribution of $n[g(\bar{X}_n) - g(\mu)]$ is determined by the dominant quadratic from in $(\bar{X}_n - \mu)^2$ from the second order Taylor series expansion, which follows an asymptotic χ_1^2 distribution after suitable standardization.

For the CLT in Theorem 7.6, we have considered the sequence of scalar random variables $\{\bar{X}_n, n = 1, 2, \cdots\}$. How can one derive the asymptotic distribution for a random vector Z_n?

The following Cramer-Wold device will allow us to derive the asymptotic distribution of a sequence of random vectors.

Lemma 7.13. *[Cramer-Wold Device]: Let d be a fixed positive integer. A sequence of random vectors $Z_n = (Z_{1n}, \cdots, Z_{dn})'$ converges in distribution to a random vector Z if $\lim_{n \to \infty} F_n(z) = F(z)$ at every point z where $F(z)$ is continuous, where $F_n(z)$ is the CDF of Z_n and $F(z)$ is the CDF of Z. Then a sequence of random vectors Z_n converges in distribution to a random vector Z if and only if $a'Z_n \xrightarrow{d} a'Z$ for every constant vector $a \neq 0$.*

Example 7.34. Suppose $Z_n \xrightarrow{d} Z \sim N(0, \Sigma)$, where Z is an $m \times 1$ random vector and Σ is an $m \times m$ nonsingular matrix, where the dimension m is fixed. If $\hat{\Sigma}_n \xrightarrow{p} \Sigma$ as $n \to \infty$, then the quadratic form

$$Z_n' \hat{\Sigma}_n^{-1} Z_n \xrightarrow{d} Z' \Sigma^{-1} Z \sim \chi_m^2.$$

Proof: First, by the Cramer-Wold device and the Slutsky theorem, we can show that

$$\hat{\Sigma}^{-\frac{1}{2}} Z_n \xrightarrow{d} \Sigma^{-1/2} Z \sim N(0, I_m) \text{ as } n \to \infty,$$

where I_m is an $m \times m$ identity matrix. It follows from the continuous mapping theorem that

$$\left(\hat{\Sigma}^{-\frac{1}{2}} Z_n \right)' \left(\hat{\Sigma}^{-\frac{1}{2}} Z_n \right) = Z_n' \hat{\Sigma}^{-1} Z_n$$
$$\xrightarrow{d} Z' \Sigma^{-1} Z \sim \chi_m^2.$$

This completes the proof.

7.8 Conclusion

An important empirical stylized fact in economics is that most economic variables follow a probability distribution with heavy tails. The finite sample distributions of most statistics in econometrics are generally unknown or rather complicated, when the random sample are not generated from a normal population. One may like to know the limiting behaviors of econometric statistics when the sample size grows to infinity. This is usually called asymptotic analysis or asymptotic theory. In this chapter we have introduced basic concepts and analytic tools for asymptotic theory. First, we have introduced four convergence concepts — convergence in quadratic mean, convergence in probability, almost sure convergence, and convergence in distribution. The first three convergences characterize, in different manners, the closeness between a sequence of random variables and a random variable, while the last convergence concept characterizes the closeness between the CDFs of a sequence of random variables and the CDF of a random variable, rather than the closeness of random variables themselves. Relationships among these convergence concepts are discussed. We also introduce and show two limit theorems — the law of large numbers and the central limit theorem. These asymptotic tools and methods are rather useful in econometrics and statistics. For more discussion on asymptotic theory, see White (2001).

EXERCISE 7

7.1. Suppose $X_1, X_2, \cdots$ is an uncorrelated sequence with $E(X_i) = \mu$, $\text{var}(X_i) = \sigma_i^2$, and $\Sigma_{i=1}^\infty \sigma_i^2/i^2 < \infty$. Show $\bar{X}_n$ converges to μ in quadratic mean.

7.2. Let $X_1, X_2, \cdots$ be a sequence of random variables that converges in probability to a constant a. Assume that $P(X_i > 0) = 1$ for all i.

(1) Verify that the sequences defined by $Y_i = \sqrt{X_i}$ and $Y_i = a/X_i$ converge in probability;

(2) Use the results in Part (1) to prove that σ/S_n convergences in probability to 1.

7.3. Suppose $\{S, \mathbb{B}, P\}$ is a probability space, where the sample space $S = [0, 1]$, $\mathbb{B}$ is a σ-field, and P is a uniform probability measure on S. Define a sequence of random variables as follow:

$$Z_1(s) = 1 \text{ for } 0 \leq s \leq 1,$$

$$Z_2(s) = \begin{cases} 1, 0 \leq s \leq \frac{1}{2}, \\ 0, \text{ otherwise}, \end{cases}$$

$$Z_3(s) = \begin{cases} 1, 0 < s \leq \frac{1}{2}, \\ 0, \text{ otherwise}, \end{cases}$$

$$Z_4(s) = \begin{cases} 1, 0 \leq s \leq \frac{1}{3}, \\ 0, \frac{1}{3} < s \leq 1, \end{cases}$$

$$Z_5(s) = \begin{cases} 1, 0 < s < \frac{1}{3}, \\ 0, \text{ otherwise}, \end{cases}$$

$$Z_6(s) = \begin{cases} 1, 0 < s \leq \frac{1}{3}, \\ 0, \text{ otherwise}. \end{cases}$$

(1) Does $\{Z_n, n = 1, 2, \cdots\}$ converge in probability to 0? Give your reasoning.

(2) Does $\{Z_n, n = 1, 2, \cdots\}$ converge to 0 almost surely? Give your reasoning.

7.4. Suppose $\mathbf{X}^n$ is an IID random sample from a population with mean μ and variance $\sigma^2 < \infty$. Use the strong law of large numbers to show that the sample variance S_n^2 converges to σ^2 almost surely.

7.5. Let $\mathbf{X}^n$ be an IID random sample from a population with mean μ and variance σ^2. Show that

$$E\left[\frac{\sqrt{n}(\bar{X}_n - \mu)}{\sigma}\right] = 0 \text{ and var}\left[\frac{\sqrt{n}(\bar{X}_n - \mu)}{\sigma}\right] = 1.$$

Thus, the normalization of $\bar{X}_n$ in CLT gives random variables that have the same mean and variance as the limiting $N(0,1)$ distribution.

7.6. Suppose $\mathbf{X}^n = (X_1, \ldots, X_n)$ is an IID random sample from a $N(0, \sigma^2)$ population, where $0 < \sigma^2 < \infty$. Define the sample mean $Z_n = n^{-1} \sum_{i=1}^{n} X_i$.

(1) Find the sampling distribution $F_n(z)$ of Z_n for each $n \geq 1$.
(2) Find the limiting distribution of Z_n as $n \to \infty$.
(3) Is the limiting distribution of Z_n the same as $\lim_{n\to\infty} F_n(z)$? Explain.
(4) Find the limiting distribution of $\sqrt{n} Z_n$ as $n \to \infty$.

7.7. Suppose $\mathbf{X}^n = (X_1, \ldots, X_n)$ is an IID random sample from the uniform distribution $U[\theta, 1]$, where $\theta < 1$. Define an estimator for θ as $Z_n = \min_{1 \leq i \leq n} X_i$.

(1) Show that Z_n is consistent for θ as $n \to \infty$.
(2) Find the limiting distribution of $n(Z_n - \theta)$ as $n \to \infty$.

7.8. Define $X_n = Y_n + Z_n$, where $\{Y_n\}$ is an IID sequence from a $N(0,1)$ population, $\{Z_n\}$ is a sequence of binary random variables with $P(Z_n = \frac{1}{n}) = 1 - \frac{1}{n}$ and $P(Z_n = n) = \frac{1}{n}$, and X_n and Y_n are mutually independent.

(1) Find the limiting distribution of X_n.

(2) The limiting distribution of X_n is also called the asymptotic distribution of X_n, and the mean and variance of the asymptotic distribution are called the asymptotic mean and asymptotic variance respectively. Find $\lim_{n\to\infty} E(X_n)$ and $\lim_{n\to\infty} \text{var}(X_n)$. Are they the same as the asymptotic mean and asymptotic variance of X_n respectively? Show your reasoning.

7.9. The MGF of the Gamma(α, β) distribution is $M_X(t) = (1 - \beta t)^{-\alpha}$. Suppose $\mathbf{X}^n$ is an IID random sample from Gamma(α, β). What is the MGF of $\sqrt{n}(\bar{X}_n - \alpha\beta)$? Drive the limit of $M_{\sqrt{n}(\bar{X}_n - \alpha\beta)}(t)$. From this limit, deduce the limiting distribution of $\sqrt{n}(\bar{X}_n - \alpha\beta)$.

7.10. Prove Lemma 7.12 for the case where Z_n and Z are continuous random variables.

(1) Given t and ϵ, show that $P(Z \leq t - \epsilon) \leq P(Z_n \leq t) + P(|Z_n - Z| \geq \epsilon)$. This gives a lower bound on $P(Z_n \leq t)$.

(2) Use a similar strategy to obtain an upper bound on $P(Z_n \leq t)$.

(3) Show $P(Z_n \leq t) \to P(Z \leq t)$.

7.11. Comment on the validity of the following statement: "the asymptotic distribution function of Z_n is the limit of the distribution function of Z_n as $n \to \infty$." Give your reasoning.

7.12. Let $\{X_i, Y_i\}_{i=1}^n$ be an IID random sample where X_i, Y_i are finite moments up to the fourth order. Define

$$(\hat{\alpha}, \hat{\beta}) = \arg\min_{\alpha, \beta} \sum_{i=1}^n (Y_i - \alpha - \beta X_i)^2.$$

(1) Derive the probability limits (denoted as α^*, β^*) of $\hat{\alpha}, \hat{\beta}$ respectively. Give your reasoning.

(2) Derive the asymptotic distributions of $\sqrt{n}(\hat{\alpha} - \alpha^*)$ and $\sqrt{n}(\hat{\beta} - \beta^*)$ respectively. Give your reasoning.

7.13. Show that if $\sqrt{n}(\bar{X}_n - \mu)/\sigma \xrightarrow{d} N(0,1)$, then $\bar{X}_n \xrightarrow{p} \mu$. Give your reasoning.

7.14. Suppose $\{Z_1, \cdots, Z_n\}$ is an IID $N(0,1)$ random sample. What is the limiting distribution of $(\sum_{i=1}^n Z_i^2 - n)/\sqrt{n}$? Give your reasoning.

7.15. Suppose $\mathbf{X}^n$ is an IID random sample from a population with $E(X_i) = \mu$, $\mathrm{var}(X_i) = \sigma^2$, $E[(X_i - \mu)^4] = \mu_4$. Define $S_n^2 = (n-1)^{-1} \sum_{i=1}^n (X_i - \bar{X}_n)^2$.

(1) Show that $S_n^2 \xrightarrow{p} \sigma^2$ as $n \to \infty$.

(2) Derive the limiting distribution of $\sqrt{n}(S_n^2 - \sigma^2)$ as $n \to \infty$.

Give your reasoning.

7.16. Suppose $\sqrt{n}(\bar{X}_n - \mu)/\sigma \xrightarrow{d} N(0,1)$, where $-\infty < \mu < \infty$ and $0 < \sigma < \infty$. Find a nondegenerate limiting distribution of a suitably normalized version of the following statistics:

(1) $Y_n = e^{-\bar{X}_n}$;

(2) $Y_n = \bar{X}_n^2$, where $\mu = 0$ in this case.

Give your reasoning.

7.17. Let $\mathbf{X}^n$ be an IID Bernoulli(p) random sample and define $\bar{X}_n = n^{-1} \sum_{i=1}^n X_i$. Show:

(1) $\sqrt{n}(\bar{X}_n - p) \xrightarrow{d} N[0, p(1-p)]$ as $n \to \infty$;

(2) for $p \neq 1/2$, the estimator $\bar{X}_n(1 - \bar{X}_n)$ of the population variance satisfies $\sqrt{n}[\bar{X}_n(1 - \bar{X}_n) - p(1-p)] \xrightarrow{d} N(0, (1-2p)^2 p(1-p))$ as $n \to \infty$;

(3) for $p = 1/2$, $n[\bar{X}_n(1 - \bar{X}_n) - \frac{1}{4}] \xrightarrow{d} -\frac{1}{4}\chi_1^2$ as $n \to \infty$.

7.18. Suppose $\mathbf{X}^n = (X_1, \cdots, X_n)$ is an IID $N(0,1)$ random sample. What is the asymptotic distribution of

$$\frac{\sum_{i=1}^n X_i^2 - n}{\sqrt{n}}?$$

7.19. A pharmaceutical factory produces a new drug, and claims that the cure rate of this drug for a disease is 80%. To test the cure rate, we select 100 this disease patients for clinical trials randomly. If at least 75 patients heal, the new drug passes the test. Please calculate the possibility that this drug passes the test for the following two cases:

(1) the actual cure rate of this drug is 80%;

(2) the actual cure rate of this drug is 70%.

7.20. Suppose the death rate of a certain life insured is 0.005. Now 1000 people buy this insurance. Find:

(1) the probability of 40 people died within a year;

(2) the probability of less than 70 people died within a year.

7.21. Prove the Slutsky theorem (Theorem 7.18).

7.22. Prove the second order Delta method (Theorem 7.20).

Chapter 8

Parameter Estimation and Evaluation

Abstract: One of the most important objectives of statistical inference is to estimate unknown model parameters based on an observed data. In this chapter, we will introduce some fundamental estimation methods for distributional model parameters. In particular, the maximum likelihood estimation and the method of moments estimation/generalized method of moments estimation are discussed, and their asymptotic properties investigated. Then we will discuss the methods for evaluating parameter estimators using the mean squared error criterion. The Lagrange multiplier method and the Cramer-Rao lower bound are used to derive the best unbiased estimators.

Key words: Asymptotic normality, Best linear unbiased estimator (BLUE), Bias, Consistency, Cramer-Rao lower bound, Distribution model, Generalized method of moments, Maximum likelihood estimation, Mean squared error, Method of moments.

8.1 Population and Distribution Model

Consider a random sample $\mathbf{X}^n = (X_1, \cdots, X_n)$ from a population distribution $f_X(x)$. A realization $\mathbf{x}^n$ of the random sample $\mathbf{X}^n$ constitutes a data set with sample size n. The primary purpose of statistical inference is to make inference of the population distribution $f_X(x)$ using the observed data $\mathbf{x}^n$.

For this purpose, one often considers a class of parametric candidate distributions

$$\mathbb{F} = \{f(\cdot, \theta) : \theta \in \Theta\},$$

where $f : \Omega \times \Theta \to \mathbb{R}^+$ is a known PMF/PDF function, Ω is the support of random variable X_i, Θ is a parameter space that contains all plausible values for a $p \times 1$ parameter vector θ, where p is a finite and fixed integer. Each value of $\theta \in \Theta$ gives a distribution model for $f_X(x)$.

We assume that the class of functions $\mathbb{F}$ contains the population distribution $f_X(x)$ that generates the observed data. This implies that there exists some parameter value $\theta_0 \in \Theta$ such that

$$f_X(x) = f(x, \theta_0) \text{ for all } x \in \Omega.$$

If $\mathbb{F}$ contains $f_X(\cdot)$, we call that the class of models $\mathbb{F}$ is correctly specified for the population distribution $f_X(\cdot)$, and θ_0 is called the true value of parameter θ. In contrast, $\mathbb{F}$ is said to be misspecified for the population distribution $f_X(\cdot)$ if there exists no value for $\theta \in \Theta$ such that $f_X(x) = f(x, \theta)$ for all $x \in \Omega$. This can occur when, for example, we specify a class of normal distribution models but the true population is a Gamma distribution. In this case, there exists no true parameter value θ_0 such that $f_X(x) = f(x, \theta_0)$ for all $x \in \Omega$. Figure 8.1 below provides simplistic plots for (a) correct model specification where the true population distribution is contained as a member in the family of parametric distribution models, and (b) model misspecification where the true population is outside the family of parametric distribution models:

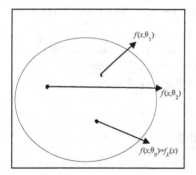 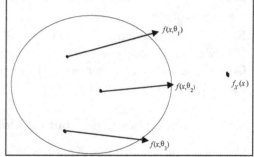

Figure 8.1(a): Correct model
specification

Figure 8.1(b): Model
misspecification

Example 8.1. [Discrete Choice Probit and Logit Models]: The Probit and Logit models are popularly used when a dependent variable has binary outcomes, i.e., there are two possible outcomes 0 and 1. Examples include whether or not an individual is employed, whether or not a consumer makes a purchase, and whether or not a financial crisis (e.g., default risk) occurs.

A probit model assumes

$$P(Y_i = 1|X_i) = \Phi(\theta_1 + \theta_2 X_i), \qquad i = 1, \cdots, n,$$

where $\Phi(\cdot)$ is the N(0,1) CDF, and X_i is an explanatory variable.

On the other hand, a logit model assumes

$$P(Y_i = 1|X_i) = \frac{1}{1 + e^{-(\theta_1 + \theta_2 X_i)}}.$$

Example 8.2. [Duration Analysis in Economics and Finance]: Suppose we are interested in the time it takes for an unemployed person to find a job, the time that elapses between two trades or two price changes, the length of a strike, the length before a cancer patient dies, and the length before a financial crisis (e.g., credit default risk) comes out. Such analysis is called duration analysis or survival analysis.

In practice, the main interest often lies in the question of how long a duration will continue, given that it has not finished yet. The hazard rate measures the chance that the duration will end now, given that it has not ended before. This hazard rate therefore can be interpreted as the chance to find a job, to trade, to end a strike, etc.

Suppose T_i is the duration from a population with the probability density function $f(t)$ and probability distribution function $F(t)$. Then the survival function is

$$S(t) = P(T_i > t) = 1 - F(t),$$

and the hazard rate

$$\lambda(t) = \lim_{\delta \to 0^+} \frac{P(t < T_i \le t + \delta | T_i > t)}{\delta}$$

$$= \frac{f(t)}{S(t)}.$$

Intuitively, the hazard rate $\lambda(t)$ is the instantaneous probability that an event of interest will end at time t given that it has lasted for period t. Note that the specification of $\lambda(t)$ is equivalent to a specification of the probability density $f(t)$, but $\lambda(t)$ is more interpretable from an economic point of view.

The hazard rate may not be the same for all individuals. To control heterogeneity across individuals, we assume that the individual-specific hazard rate depends on some individual characteristics X_i (e.g., age, gender, race, education, work experience) via the form

$$\lambda_i(t) = e^{X_i'\theta} \lambda_0(t),$$

where $\lambda_0(t)$ is called a baseline hazard function. This model, proposed by Cox (1972), is called the proportional hazard model. When the model is correctly specified, the true parameter value

$$\theta_0 = \frac{\partial \ln \lambda_i(t)}{\partial X_i} = \frac{1}{\lambda_i(t)} \frac{\partial \lambda_i(t)}{\partial X_i}$$

can be interpreted as the relative marginal effect of X_i on the hazard rate of individual i. Inference of θ_0 will allow one to examine how individual characteristics affect the duration of interest. For example, suppose T_i is the unemployment duration for individual i, then the inference of θ_0 will allow us to examine how individual characteristics, such as age, education, gender, job training, etc, can affect the unemployment duration. This will provide important policy implication on labor markets.

Because the conditional probability density function of Y_i given X_i is given by

$$f_i(t) = \lambda_i(t)S_i(t),$$

where the survival function

$$S_i(t) = e^{-\int_0^t \lambda_i(s)ds},$$

we can estimate θ_0 by the so-called maximum likelihood method to be discussed below.

For an excellent survey on duration analysis in labor economics, see Kiefer (1988), and for a complete and detailed account, see Lancaster (1990).

Usually, the true parameter value θ_0 is unknown, and one is interested in making inference of θ_0 using an observed data $\mathbf{x}^n$. Traditionally, problems of statistical inference are divided into problems of estimation and hypothesis testing. In this chapter, we are interested in estimating θ_0. An estimator of θ_0 is a statistic whose value can be viewed as an estimate of θ. Below, we introduce two most commonly used estimation methods, namely the Maximum Likelihood Estimation (MLE) and the Generalized Method of Moments (GMM) estimation. The latter includes the classical Method of Moments Estimation (MME) as a special case.

8.2 Maximum Likelihood Estimation

Question: How to estimate θ_0 based on a data set $\mathbf{x}^n$?

R.A. Fisher proposed a general method of estimation called MLE. He demonstrated the advantage of this method by showing that it yields a sufficient estimator for parameter θ whenever it exists and MLE is asymptotically most efficient in some sensible criterion that we will discuss later.

The essential feature of MLE is to look at the observed data $\mathbf{x}^n$ and then choose as the estimate(s) of θ the value(s) for which the probability or probability density of getting the observed data $\mathbf{x}^n$ is a maximum.

We first introduce the likelihood function of the random sample $\mathbf{X}^n$.

Definition 8.1. [Likelihood Function]: Given $\mathbf{X}^n = \mathbf{x}^n$, the joint PMF/PDF of the random sample $\mathbf{X}^n$ as a function of θ,

$$\hat{L}(\theta|\mathbf{x}^n) = f_{\mathbf{X}^n}(\mathbf{x}^n, \theta),$$

is called the likelihood function of the random sample $\mathbf{X}^n$ when $\mathbf{x}^n$ is observed. Also, $\ln \hat{L}(\theta|\mathbf{x}^n)$ is called the log-likelihood function of the random sample $\mathbf{X}^n$ when $\mathbf{x}^n$ is observed.

Mathematically speaking, the likelihood function $\hat{L}(\theta|\mathbf{x}^n)$ is numerically equal to the joint probability or joint probability density of the random sample $\mathbf{X}^n$ when $\mathbf{X}^n = \mathbf{x}^n$. The conceptual difference between them is that the likelihood $\hat{L}(\theta|\mathbf{x}^n)$ is a function of θ, with $\mathbf{x}^n$ held fixed. Given θ, the likelihood $\hat{L}(\theta|\mathbf{x}^n)$ is a measure of the probability or probability density with which the observed sample $\mathbf{x}^n$ will occur.

Note that the joint distribution of the random sample $\mathbf{X}^n$, $f_{\mathbf{X}^n}(\mathbf{x}^n, \theta)$, is different from the population distribution $f(x, \theta)$. The latter is the distribution of each random variable X_i.

The MLE method consists of maximizing the likelihood function with respect to θ over the parameter space Θ and the maximizer is called the MLE.

Definition 8.2. [Maximum Likelihood Estimator (MLE)]: Suppose the statistic $\hat{\theta} = \hat{\theta}_n(\mathbf{X}^n)$ maximizes $\hat{L}(\theta|\mathbf{X}^n)$ over $\theta \in \Theta$, conditional on $\mathbf{X}^n$, where Θ is a finite-dimensional parameter space. That is,

$$\hat{\theta} \equiv \hat{\theta}_n(\mathbf{X}^n) = \arg\max_{\theta \in \Theta} \hat{L}(\theta|\mathbf{X}^n).$$

Then, when exists, $\hat{\theta} \equiv \hat{\theta}_n(\mathbf{X}^n)$ is called the MLE for parameter θ. Given a sample point (or a data set) $\mathbf{x}^n$ for the random sample $\mathbf{X}^n$, $\hat{\theta}_n(\mathbf{x}^n)$ is called a maximum likelihood estimate for θ.

By the nature of the objective function, MLE is the parameter estimate which makes the observed data $\mathbf{x}^n$ most likely to occur. In other words, by choosing a parameter estimate $\hat{\theta}_n(\mathbf{x}^n)$, MLE maximizes the probability that $\mathbf{X}^n = \mathbf{x}^n$, that is, the probability that the random sample $\mathbf{X}^n$ takes the value of the observed data $\mathbf{x}^n$.

In some scenarios, for certain $\mathbf{x}^n$, the maximum value of $\hat{L}(\theta|\mathbf{x}^n)$ may not actually be attainable, for any point θ in Θ. In this case, MLE does not exist. (**Question**: Can you provide such an example?)

We now provide sufficient conditions to ensure the existence of MLE.

Theorem 8.1. *[Existence of MLE]: Suppose, with probability one, $\hat{L}(\theta|\mathbf{X}^n)$ is a continuous function of $\theta \in \Theta$, and Θ is a compact set. Then there exists a global maximizer $\hat{\theta}$ that solves the problem,*

$$\hat{\theta} \equiv \hat{\theta}_n(\mathbf{X}^n) = \arg\max_{\theta \in \Theta} \hat{L}(\theta|\mathbf{X}^n).$$

Proof: Application of the Weierstrass theorem.

Figure 8.2 is a plot for MLE when θ is a scalar parameter.

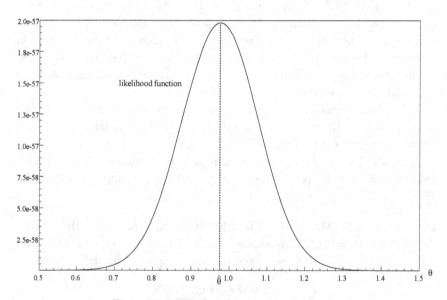

Figure 8.2: MLE when θ is a scalar parameter

It is often convenient to solve $\max_{\theta \in \Theta} \ln \hat{L}(\theta|\mathbf{X}^n)$, where $\ln \hat{L}(\theta|\mathbf{X}^n)$ is called the log-likelihood function, which is a strictly monotonic increasing function of $\hat{L}(\theta|\mathbf{X}^n)$. MLE may not be unique. Given an observed data set

$\mathbf{x}^n$, MLE may be obtained at more than one point in Θ. Thus, multiple solutions for MLE are possible.

The maximum is obtained over parameter space Θ, where Θ may be subject to some restriction. For example, when estimating a Generalized Autoregressive Conditional Heteroskedasticity (GARCH) model (Bollerslev 1986), θ may be subject to some restrictions to ensure that the conditional variance function is always nonnegative.

When $\ln \hat{L}(\theta|\mathbf{X}^n)$ is a smooth function of $\theta \in \Theta$, in particular, when $\ln \hat{L}(\theta|\mathbf{X}^n)$ is twice continuously differentiable with respect to $\theta \in \Theta$, the solution will be easy to find. A necessary condition for MLE is that $\hat{\theta}$ must satisfy the First Order Condition (FOC):

$$\frac{\partial \ln \hat{L}(\theta|\mathbf{X}^n)}{\partial \theta}|_{\theta=\hat{\theta}} = \mathbf{0},$$

which consists of p equations if θ is a $p \times 1$ parameter vector. From this set of first order conditions, we can solve for $\hat{\theta}$. Graphically, the MLE $\hat{\theta}$ responds to a zero slope for the likelihood function; see Figure 8.2.

FOC is only a necessary condition for a maximum, not a sufficient condition. Therefore, it only provides possible candidates for MLE. Points at which the first derivatives are zero may be local maxima, or global minima, or even inflection points. To find a global maximum, we need to check the Second Order Condition (SOC): If the $p \times p$ sample Hessian matrix

$$\hat{H}(\theta) = \frac{\partial^2 \ln \hat{L}(\theta|\mathbf{X}^n)}{\partial \theta \partial \theta'}$$

is negative definite for all $\theta \in \Theta$, then $\hat{\theta}$ is the global maximum. In many cases, it may be difficult to verify that $\hat{H}(\theta)$ is negative definite for all $\theta \in \Theta$. Instead, it is relatively straightforward to verify that $\hat{H}(\hat{\theta})$ is negative definite, which will imply that $\hat{\theta}$ is a local maximizer. When the dimension of θ is high, using the second derivative condition to check for a maximum is a tedious task; other methods might be tried first.

It may be emphasized that the zeros of the first derivatives only locate extreme points in the interior of the domain of a function. If the extrema occur on the boundary, the first derivatives may not be zero. Thus, the boundary must be checked separately for extrema. This can be done by using the Kuhn-Tucker theorem.

It is possible that the first order conditions cannot deliver a closed form solution for $\hat{\theta}$. In this case, numerical solutions for $\hat{\theta}$ will be needed. Most computer software can handle this easily.

There are two inherent drawbacks associated with the general problem of finding the maximum of a function, and hence of the MLE method. The first problem is that of finding the global maximum and verifying that, indeed, a global maximum has been found. The second problem is that of numerical sensitivity. That is, how sensitive is the estimate to small changes in data? It is sometimes the case that a slightly different sample (e.g., adding a few more observations) will produce a vastly different MLE, making its use suspectable. These issues may occur in some economic applications because the likelihood functions may be relatively flat as a function of parameter θ due to the existence of heavy tail probability distributions.

We now summarize the procedure of MLE:

- Find the log-likelihood function, $\ln \hat{L}(\theta|\mathbf{X}^n)$. For an IID random sample with population PMF/PDF $f(x,\theta)$, we have $\ln \hat{L}(\theta|\mathbf{X}^n) = \sum_{i=1}^{n} \ln f(X_i, \theta)$.
- Solve for the FOC and find $\hat{\theta}$.
- Check the SOC to ensure that $\hat{\theta}$ is a global maximizer or at least a local maximizer.

Most computer softwares (e.g., Matlab, GAUSS, SAS, R) have MLE algorithms.

We now consider some numerical examples.

Example 8.3. Let $\mathbf{X}^n$ be an IID $N(\mu, 1)$ random sample. Find the MLE for μ.

Solution: Put $\theta = \mu$. Because $\mathbf{X}^n$ is an IID $N(\mu, 1)$ random sample, the likelihood function of the sample $\mathbf{X}^n$

$$
\begin{aligned}
\hat{L}(\mu|\mathbf{X}^n) &= \prod_{i=1}^{n} f(X_i, \theta) \\
&= \prod_{i=1}^{n} \frac{1}{\sqrt{2\pi}} e^{-\frac{1}{2}(X_i-\mu)^2} \\
&= (2\pi)^{-n/2} e^{-\frac{1}{2}\sum_{i=1}^{n}(X_i-\mu)^2}.
\end{aligned}
$$

It follows that the log-likelihood function

$$
\ln \hat{L}(\mu|\mathbf{X}^n) = -\frac{n}{2}\ln(2\pi) - \frac{1}{2}\sum_{i=1}^{n}(X_i - \mu)^2.
$$

The FOC is given by

$$\frac{d \ln \hat{L}(\hat{\mu}|\mathbf{X}^n)}{d\mu} \equiv \frac{d \ln \hat{L}(\mu|\mathbf{X}^n)}{d\mu}\Bigg|_{\mu=\hat{\mu}}$$

$$= \sum_{i=1}^{n}(X_i - \hat{\mu}) = 0.$$

This gives the sample mean estimator

$$\hat{\mu} = \bar{X}_n.$$

For the SOC, we have

$$\frac{d^2 \ln \hat{L}(\mu, \mathbf{X}^n)}{d\mu^2} = -n < 0 \text{ for all } \mu.$$

It follows that $\hat{\mu} = \bar{X}_n$ is the global maximizer. Note that $\hat{\mu} = \bar{X}_n$ is a sufficient statistic for μ; see Example 6.10 in Chapter 6.

Example 8.4. Suppose $\mathbf{X}^n$ is an IID $N(\mu, \sigma^2)$ random sample. Find the MLE for (μ, σ^2).

Solution: Put $\theta = (\mu, \sigma^2)$. Then the log-likelihood function of $\mathbf{X}^n$

$$\ln \hat{L}(\theta|\mathbf{X}^n) = -\frac{n}{2}\ln(2\pi) - \frac{n}{2}\ln\sigma^2 - \frac{1}{2\sigma^2}\sum_{i=1}^{n}(X_i - \mu)^2.$$

The FOC are:

$$\frac{\partial \ln \hat{L}(\hat{\theta}|\mathbf{X}^n)}{\partial \mu} = \frac{1}{\hat{\sigma}^2}\sum_{i=1}^{n}(X_i - \hat{\mu}) = 0,$$

$$\frac{\partial \ln \hat{L}(\hat{\theta}|\mathbf{X}^n)}{\partial \sigma^2} = -\frac{n}{2\hat{\sigma}^2} + \frac{1}{2\hat{\sigma}^4}\sum_{i=1}^{n}(X_i - \hat{\mu})^2 = 0,$$

where $\hat{\theta} = (\hat{\mu}, \hat{\sigma}^2)$. Note that here, we treat σ^2 instead of σ as a parameter.
It follows that

$$\hat{\mu} = \bar{X}_n,$$

$$\hat{\sigma}^2 = n^{-1}\sum_{i=1}^{n}(X_i - \bar{X}_n)^2.$$

The MLE $\hat{\sigma}^2$ for σ^2 differs slightly from the sample variance $S_n^2 = (n-1)^{-1}\sum_{i=1}^{n}(X_i - \bar{X}_n)^2$.

To check the SOC, we compute the sample Hessian matrix

$$\hat{H}(\theta) = \begin{bmatrix} -\frac{n}{\sigma^2} & -\frac{1}{\sigma^4}\sum_{i=1}^{n}(X_i - \mu) \\ -\frac{1}{\sigma^4}\sum_{i=1}^{n}(X_i - \mu) & \frac{n}{2\sigma^4} - \frac{1}{\sigma^6}\sum_{i=1}^{n}(X_i - \mu)^2 \end{bmatrix}.$$

When $\theta = \hat{\theta}$, we have

$$\hat{H}(\hat{\theta}) = \begin{bmatrix} -\frac{n}{\hat{\sigma}^2} & 0 \\ 0 & -\frac{n}{2\hat{\sigma}^4} \end{bmatrix},$$

which is negative definite. Hence, $\hat{\theta}$ is a local maximizer. Note that the MLE $\hat{\theta} = (\hat{\mu}, \hat{\sigma}^2)$ is a sufficient statistic for $\theta = (\mu, \sigma^2)$; see Example 6.11 in Chapter 6.

Question: What is the value of $\ln \hat{L}(\theta|\mathbf{X}^n)$ when $\theta = \hat{\theta} = (\hat{\mu}, \hat{\sigma}^2)$?

Question: Suppose we treat σ instead of σ^2 as a parameter. Can we obtain the same MLE solution in the above example?

The answer is yes. We will obtain

$$\hat{\sigma} = \sqrt{\hat{\sigma}^2}.$$

This follows from the following invariance property of MLE.

Theorem 8.2. [Invariance Property of MLE]: *Suppose $\hat{\theta}$ is the MLE of $\theta \in \Theta$, and $g(\cdot)$ is a one-to-one function over parameter space Θ. Then $g(\hat{\theta})$ is a MLE of $g(\theta)$.*

Proof: Because $g(\theta)$ is a one-to-one function over Θ, there exists a unique inverse function $h(\cdot)$ such that $h[g(\theta)] = \theta$ for all $\theta \in \Theta$. Define a new parameter $\tau = g(\theta)$ and we are interested in the MLE for τ. Then $\theta = h(\tau)$. It follows that the likelihood function of the random sample $\mathbf{X}^n$

$$\hat{L}(\theta|\mathbf{X}^n) = \hat{L}[h(\tau)|\mathbf{X}^n] = \hat{L}^*(\tau|\mathbf{X}^n),$$

where $\hat{L}^*(\tau|\mathbf{X}^n)$ is the likelihood function of $\mathbf{X}^n$ with respect to the transformed parameter τ.

Now suppose $\hat{\theta}$ is a global MLE of $\theta \in \Theta$. Then we have

$$\hat{L}(\hat{\theta}|\mathbf{X}^n) \geq \hat{L}(\theta|\mathbf{X}^n) \text{ for all } \theta \in \Theta.$$

Put $\hat{\tau} = g(\hat{\theta})$. Then $\hat{\theta} = h(\hat{\tau})$, and for any $\theta \in \Theta$,

$$\begin{aligned} \hat{L}(\hat{\theta}|\mathbf{X}^n) &= \hat{L}\left[h(\hat{\tau})|\mathbf{X}^n\right] \\ &= \hat{L}^*(\hat{\tau}|\mathbf{X}^n) \\ &\geq \hat{L}(\theta|\mathbf{X}^n) = \hat{L}[h(\tau)|\mathbf{X}^n] \end{aligned}$$

where $\tau = g(\theta)$. It follows that

$$\hat{L}^* (\hat{\tau}|\mathbf{X}^n) \geq \hat{L}^* (\tau|\mathbf{X}^n) \text{ for all } \tau \in \Gamma,$$

where $\Gamma = \{\tau : \tau = g(\theta) \text{ for all } \theta \in \Theta\}$ is the parameter space for the transformed parameter τ. Therefore, $\hat{\tau}$ is a MLE for τ. This completes the proof.

The next result shows that MLE $\hat{\theta}$ can also be obtained from maximizing the likelihood of a sufficient statistic $T(\mathbf{X}^n)$ of θ if the latter exists.

Theorem 8.3. *[Sufficiency of MLE]: Suppose $\mathbf{X}^n$ is a random sample with the likelihood function $f_{\mathbf{X}^n}(\mathbf{x}^n, \theta)$, and $T(\mathbf{X}^n)$ is a sufficient statistic for θ, where $\theta \in \Theta$ is a parameter. Then the MLE $\hat{\theta}$ that maximizes the likelihood function $f_{\mathbf{X}^n}(\mathbf{x}^n, \theta)$ of the random sample $\mathbf{X}^n$ is also the MLE that maximizes the likelihood function $f_{T(\mathbf{X}^n)}[T(\mathbf{x}^n), \theta]$ of the sufficient statistic $T(\mathbf{X}^n)$.*

Proof: By definition, we have $\hat{\theta} = \arg\max_{\theta \in \Theta} \ln f_{\mathbf{X}^n}(\mathbf{x}^n, \theta)$. Because $T(\mathbf{X}^n)$ is a sufficient statistic for θ, we have

$$\begin{aligned} f_{\mathbf{X}^n}(\mathbf{x}^n, \theta) &= f_{T(\mathbf{X}^n)}[T(\mathbf{x}^n), \theta] f_{\mathbf{X}^n|T(\mathbf{X}^n)}[\mathbf{x}^n|T(\mathbf{x}^n)] \\ &= f_{T(\mathbf{X}^n)}[T(\mathbf{x}^n), \theta] h(\mathbf{x}^n), \end{aligned}$$

where the conditional probability $f_{\mathbf{X}^n|T(\mathbf{X}^n)}[\mathbf{x}^n|T(\mathbf{x}^n)]$ of $\mathbf{X}^n$ given $T(\mathbf{X}^n) = T(\mathbf{x}^n)$ does not depend on parameter θ and is denoted as function $h(\mathbf{x}^n)$. (See discussion in Section 6.6.) It follows that

$$\ln f_{\mathbf{X}^n}(\mathbf{x}^n, \theta) = \ln f_{T(\mathbf{X}^n)}[T(\mathbf{x}^n), \theta] + \ln h(\mathbf{x}^n)$$

and maximizing $\ln f_{\mathbf{X}^n}(\mathbf{x}^n, \theta)$ by choosing $\theta \in \Theta$ is equivalent to maximizing $\ln f_{T(\mathbf{X}^n)}[T(\mathbf{x}^n), \theta]$ by choosing $\theta \in \Theta$. That is,

$$\begin{aligned} \hat{\theta} &= \arg\max_{\theta \in \Theta} \ln f_{\mathbf{X}^n}(\mathbf{x}^n, \theta) \\ &= \arg\max_{\theta \in \Theta} \ln f_{T(\mathbf{X}^n)}[T(\mathbf{x}^n), \theta]. \end{aligned}$$

This completes the proof.

8.3 Asymptotic Properties of MLE

Because MLE $\hat{\theta}$ is generally a highly nonlinear function of the random sample $\mathbf{X}^n$, it is a prohibitive task to compute the mean, variance and sampling distribution of MLE $\hat{\theta}$ for any given sample size n, when the random sample $\mathbf{X}^n$ is not generated from a normal population. Below, we use the asymptotic theory developed in Chapter 7 to investigate the asymptotic properties (i.e., when $n \to \infty$) of MLE $\hat{\theta}$. In particular, we will show that MLE $\hat{\theta}$ is consistent for the true parameter value $\theta_0 \in \Theta$ and will converge to a normal distribution, after proper standardization.

We first provide a set of regularity conditions. For simplicity, we assume that θ is a scalar here.

Assumption M.1: $\mathbf{X}^n = (X_1, \cdots, X_n)$ is an IID random sample from some unknown population distribution $f_X(x)$.

Assumption M.2: (1) For each $\theta \in \Theta$, $f(x, \theta)$ is a probability PMF/PDF model with $f(x, \theta) > 0$ for all x, where Θ is a finite-dimensional parameter space; (2) there exists a parameter value $\theta_0 \in \Theta$ such that $f(x, \theta_0)$ coincides with the population distribution $f_X(x)$; (3) the function $\ln f(x, \theta)$ is continuous in (x, θ) and its absolute value is bounded by a nonnegative function $b(x)$ such that $E[b(X_i)] < \infty$, where and below the expectation $E(\cdot)$ is taken under the population distribution $f_X(x)$.

Assumption M.3: Θ is closed and bounded, or equivalently Θ is compact.

Assumption M.4: The parameter value θ_0 is the unique maximizer of $E[\ln f(X_i, \theta)]$;

Assumption M.5: θ_0 is in the interior of parameter space Θ.

Assumption M.6: For each interior point $\theta \in \Theta$, $f(x, \theta)$ is twice continuously differentiable with respect to θ such that (1) the functions $\frac{\partial}{\partial \theta} \ln f(x, \theta), \frac{\partial^2}{\partial \theta^2} \ln f(x, \theta)$ are continuous in (x, θ), and their absolute values are bounded by a nonnegative function $b(x)$ such that $E[b(X_i)] < \infty$ and $E[b^2(X_i)] < \infty$; (2) the absolute value of the function $H(\theta) = E\left[\frac{\partial^2}{\partial \theta^2} \ln f(X_i, \theta)\right]$ is bounded by some constant and is nonzero.

The scalar parameter assumption is made for simplicity. The results obtained below can be extended to the case of a parameter vector in a straightforward manner, but it does not offer much additional insight into the asymptotic properties of MLE.

Assumption M.2 is a correct model specification for the probability distribution model $f(x, \theta)$ in the sense that there exists a parameter value θ_0 such that the probability distribution model $f(x, \theta_0)$ coincides with the population distribution $f_X(x)$. The parameter value θ_0 is usually called the true parameter value of θ. The compactness assumption in Assumption M.3 helps ensure the existence of MLE (see Theorem 8.1). Assumption M.4 is called an identification condition. It ensures that θ_0 is the well-defined probability limit of MLE $\hat{\theta}$. It is important to note that

$$E[\ln f(X_i, \theta)] = \int_{-\infty}^{\infty} \ln f(x, \theta) f_X(x) dx$$

$$\neq \int_{-\infty}^{\infty} \ln f(x, \theta) f(x, \theta) dx$$

unless $\theta = \theta_0$.

Assumptions M.5 and M.6 facilitate the application of the Taylor series expansion in order to derive the asymptotic distribution of MLE $\hat{\theta}$. In statistics, the function $\frac{\partial}{\partial \theta} \ln f(x, \theta)$ is called the score function (it is a vector-valued function when θ is a parameter vector), and the function $H(\theta)$ is called the Hessian matrix (it is a square matrix when θ is a parameter vector).

To show the consistency theorem for MLE $\hat{\theta}$, we first state a useful lemma.

Lemma 8.1. *[Extrema Estimator Lemma; White (1994, Theorem 3.4)]: Suppose (1) $Q(\theta)$ is a nonstochastic function continuous in $\theta \in \Theta$, and $\theta_0 \in \Theta$ is the unique maximizer of $Q(\theta)$ over Θ, where Θ is a compact set; (2) with probability one, $\hat{Q}_n(\theta)$ is a sequence of random functions continuous in $\theta \in \Theta$; (3) $\lim_{n \to \infty} \sup_{\theta \in \Theta} |\hat{Q}_n(\theta) - Q(\theta)| = 0$ almost surely. Then $\hat{\theta} = \arg\max_{\theta \in \Theta} \hat{Q}_n(\theta)$ exists and $\hat{\theta} \to \theta_0$ as $n \to \infty$ almost surely.*

Proof: See White (1994, Proof of Theorem 3.4).

Note that condition (3) is a uniform convergence condition which can be ensured by the ULLN given in Lemma 7.10, Chapter 7. It implies that the largest difference between $\hat{Q}_n(\theta)$ and $Q(\theta)$ over the parameter space Θ vanishes to zero as $n \to \infty$ almost surely.

Theorem 8.4. *[Consistency of MLE]: Suppose Assumptions M.1–M.4 hold, and $\hat{\theta} = \arg\max_{\theta \in \Theta} \sum_{i=1}^{n} \ln f(X_i, \theta)$. Then as $n \to \infty$,*

$$\hat{\theta} \to \theta_0 \ a.s.$$

Proof: We apply the above extrema estimator lemma. Given Assumption M.2, we have $Q(\theta) = E[\ln f(X_i, \theta)]$ is a continuous function of $\theta \in \Theta$, and θ_0 is the unique maximizer of $Q(\theta)$ over the compact set Θ by Assumptions M.3 and M.4. Now, put $\hat{Q}_n(\theta) = n^{-1} \sum_{i=1}^{n} \ln f(X_i, \theta)$. Then, given Assumptions M.1–M.3 and the USLLN in Lemma 7.10, Chapter 7, we have $\sup_{\theta \in \Theta} |\hat{Q}_n(\theta) - Q(\theta)| \to 0$ as $n \to \infty$ almost surely. It follows from the extrema estimator lemma that MLE $\hat{\theta} \to \theta_0$ almost surely as $n \to \infty$. This completes the proof.

We note that we do not require θ_0 to be an interior point of parameter space Θ in establishing the consistency of MLE $\hat{\theta}$. In other words, the consistency theorem allows that θ_0 is a corner solution (i.e., θ_0 can be on the boundary of Θ). Also, there is no need to assume differentiability of the log-likelihood function $\ln f(x, \theta)$ with respect to θ. In fact, the FOC condition may fail when there exists a corner solution even if $\ln f(x, \theta)$ is differentiable with respect to θ.

Next, we derive the asymptotic distribution of MLE $\hat{\theta}$ (with suitable normalization). For this purpose, we first state a few useful lemmas.

Lemma 8.2. *Suppose $f(x, \theta)$ is a PDF model and $f(x, \theta)$ is continuously differentiable with respect to $\theta \in \Theta$, where θ is an interior point in parameter space Θ. Then for all θ in the interior of Θ,*

$$\int_{-\infty}^{\infty} \frac{\partial \ln f(x, \theta)}{\partial \theta} f(x, \theta) dx = 0.$$

A similar result holds for a PMF model.

Proof: Given $f(x, \theta)$ is a PDF model, $f(x, \theta)$ is a PDF for any $\theta \in \Theta$. It follows that for any θ in the interior of Θ,

$$\int_{-\infty}^{\infty} f(x, \theta) dx = 1.$$

Differentiating this equation and exchanging the order of integration and differentiation, we have

$$\frac{d}{d\theta} \int_{-\infty}^{\infty} f(x, \theta) dx = \frac{d}{d\theta}(1) = 0,$$

$$\int_{-\infty}^{\infty} \frac{\partial f(x, \theta)}{\partial \theta} dx = 0,$$

$$\int_{-\infty}^{\infty} \left[\frac{\partial \ln f(x, \theta)}{\partial \theta} \right] f(x, \theta) dx = 0.$$

This completes the proof.

It should be noted that

$$\int_{-\infty}^{\infty} \left[\frac{\partial \ln f(x, \theta)}{\partial \theta} \right] f(x, \theta) dx \neq E \left[\frac{\partial \ln f(X_i, \theta)}{\partial \theta} \right]$$

unless $\theta = \theta_0$, where $E(\cdot)$ is the expectation with respect to the population distribution $f_X(x)$.

Lemma 8.3. *[Information Matrix Equality]: Suppose a PDF model $f(x, \theta)$ is twice continuously differentiable with respect to $\theta \in \Theta$, where θ is an interior point in parameter space Θ. Define*

$$I(\theta) = \int_{-\infty}^{\infty} \left[\frac{\partial \ln f(x, \theta)}{\partial \theta}\right]^2 f(x, \theta)dx,$$

$$H(\theta) = \int_{-\infty}^{\infty} \left[\frac{\partial^2 \ln f(x, \theta)}{\partial \theta^2}\right] f(x, \theta)dx.$$

Then for all θ in the interior of Θ,

$$I(\theta) + H(\theta) = 0.$$

A similar result holds for a PMF model.

Proof: By differentiating the identity $\int_{-\infty}^{\infty} f(x, \theta)dx = 1$ with respect to θ, we then obtain

$$\int_{-\infty}^{\infty} \frac{\partial}{\partial \theta} f(x, \theta)dx = 0,$$

This can be rewritten as

$$\int_{-\infty}^{\infty} \frac{\partial \ln f(x, \theta)}{\partial \theta} f(x, \theta)dx = 0.$$

If we further differentiate this equation with respect to θ, we obtain

$$\int_{-\infty}^{\infty} \left\{ \left[\frac{\partial^2 \ln f(x, \theta)}{\partial \theta^2}\right] f(x, \theta) + \left[\frac{\partial \ln f(x, \theta)}{\partial \theta}\right] \frac{\partial f(x, \theta)}{\partial \theta} \right\} dx = 0$$

or equivalently

$$\int_{-\infty}^{\infty} \left[\frac{\partial^2 \ln f(x, \theta)}{\partial \theta^2}\right] f(x, \theta)dx + \int_{-\infty}^{\infty} \left[\frac{\partial \ln f(x, \theta)}{\partial \theta}\right]^2 f(x, \theta)dx = 0.$$

This completes the proof.

In econometrics, there is a well-known model specification test called the information matrix test (White 1982). This test checks correct specification of a parametric likelihood model $f(x, \theta)$ by testing whether the following equality holds:

$$E\left[\frac{\partial \ln f(X_i, \theta)}{\partial \theta}\right]^2 + E\left[\frac{\partial^2 \ln f(X_i, \theta)}{\partial \theta^2}\right] = 0,$$

where the expectation $E(\cdot)$ is taken over the population distribution $f_X(x)$. This equality is not the same as the information matrix equality in Lemma 8.3 (why?). It holds when the model $f(x, \theta)$ is correctly specified for the population distribution $f_X(x)$ and $\theta = \theta_0$ but it generally does not hold when $f(x,\theta)$ is misspecified for $f_X(x)$. (Please verify!)

We now show that MLE $\hat{\theta}$ is asymptotically normally distributed after proper standardization.

Theorem 8.5. *[Asymptotic Normality of MLE]: Suppose Assumptions M.1–M.6 hold. Then as $n \to \infty$,*

$$\sqrt{n}(\hat{\theta} - \theta_0) \xrightarrow{d} N[0, -H(\theta_0)^{-1}].$$

Proof: Because $\hat{\theta} \to \theta_0$ as $n \to \infty$ almost surely, and θ_0 is an interior point of Θ, $\hat{\theta}$ will be also an interior point of Θ for n sufficiently large with probability 1. Thus, we have the FOC:

$$\frac{d \ln \hat{L}(\theta|\mathbf{X}^n)}{d\theta}\Big|_{\theta=\hat{\theta}} = 0,$$

or equivalently,

$$\frac{d}{d\theta} \sum_{i=1}^{n} \ln f(X_i, \hat{\theta}) = 0.$$

By exchanging the differentiation and summation, we obtain

$$\frac{1}{n} \sum_{i=1}^{n} \frac{\partial \ln f(X_i, \hat{\theta})}{\partial \theta} = 0.$$

By the mean value theorem, we have

$$\frac{1}{n} \sum_{i=1}^{n} \frac{\partial \ln f(X_i, \theta_0)}{\partial \theta} + \left[\frac{1}{n} \sum_{i=1}^{n} \frac{\partial^2 \ln f(X_i, \bar{\theta})}{\partial \theta^2} \right] (\hat{\theta} - \theta_0) = 0,$$

where $\bar{\theta}$ lies on the segment between $\hat{\theta}$ and θ_0, that is, $\bar{\theta} = \lambda\hat{\theta} + (1-\lambda)\theta_0$ for some $\lambda \in (0,1)$. Note that $|\bar{\theta} - \theta_0| = |\lambda(\hat{\theta} - \theta_0)| \leq |\hat{\theta} - \theta_0| \to 0$ as $n \to \infty$ almost surely.

Now, define the sample Hessian matrix

$$\hat{H}(\theta) = \frac{1}{n} \sum_{i=1}^{n} \frac{\partial^2 \ln f(X_i, \theta)}{\partial \theta^2}.$$

Then we have

$$\sqrt{n}(\hat{\theta} - \theta_0) = \left[-\hat{H}(\bar{\theta}) \right]^{-1} \frac{1}{\sqrt{n}} \sum_{i=1}^{n} \frac{\partial \ln f(X_i, \theta_0)}{\partial \theta}.$$

We first show that the second term

$$\frac{1}{\sqrt{n}} \sum_{i=1}^{n} \frac{\partial \ln f(X_i, \theta_0)}{\partial \theta} \xrightarrow{d} N(0, I(\theta_0)) \text{ as } n \to \infty.$$

Define the score function

$$S_i(\theta) = \frac{\partial \ln f(X_i, \theta)}{\partial \theta}, \qquad i = 1, \cdots, n.$$

Then $\{S_i(\theta_0)\}_{i=1}^{n}$ is an IID sequence given the IID assumption of the random sample $\mathbf{X}^n$. Given correct model specification in Assumption M.2 (so the population distribution $f_X(x) = f(x, \theta_0)$), we have

$$E\left[S_i(\theta_0)\right] = \int_{-\infty}^{\infty} \frac{\partial \ln f(x, \theta_0)}{\partial \theta} f_X(x) dx$$

$$= \int_{-\infty}^{\infty} \frac{\partial \ln f(x, \theta_0)}{\partial \theta} f(x, \theta_0) dx$$

$$= 0,$$

where the last equality follows by Lemma 8.2. This is actually the FOC of $\max_{\theta \in \Theta} E[\ln f(X_i, \theta)]$, when θ_0 is an interior point of parameter space Θ.

Furthermore, given $E\left[S_i(\theta_0)\right] = 0$, the variance

$$\text{var}\left[S_i(\theta_0)\right] = E\left[S_i(\theta_0)^2\right]$$

$$= E\left[\frac{\partial \ln f(X_i, \theta_0)}{\partial \theta}\right]^2$$

$$= \int_{-\infty}^{\infty} \left[\frac{\partial \ln f(x, \theta_0)}{\partial \theta}\right]^2 f(x, \theta_0) dx$$

$$= I(\theta_0) < \infty$$

by Assumption M.6, where the third equality follows from correct model specification. It follows from the CLT for an IID random sequence (see Theorem 7.16) that as $n \to \infty$,

$$\frac{1}{\sqrt{n}} \sum_{i=1}^{n} \frac{\partial \ln f(X_i, \theta_0)}{\partial \theta} = \frac{1}{\sqrt{n}} \sum_{i=1}^{n} S(X_i, \theta_0)$$

$$\xrightarrow{d} N(0, I(\theta_0)).$$

Next we show $\hat{H}(\bar{\theta}) \to H(\theta_0)$ as $n \to \infty$ almost surely, where $H(\theta)$ is defined in Lemma 8.3.

Put

$$\bar{H}(\theta) = E\left[\frac{\partial^2 \ln f(X_i, \theta)}{\partial \theta^2}\right]$$

$$= \int_{-\infty}^{\infty} \frac{\partial^2 \ln f(x, \theta)}{\partial \theta^2} f_X(x) dx.$$

Note that $\bar{H}(\theta) \neq H(\theta)$ unless $\theta = \theta_0$, where $H(\theta)$ is defined as in Lemma 8.3. We write

$$\hat{H}(\bar{\theta}) - H(\theta_0) = [\hat{H}(\bar{\theta}) - \bar{H}(\bar{\theta})] + [\bar{H}(\bar{\theta}) - H(\theta_0)].$$

For the second term here, we have

$$\bar{H}(\bar{\theta}) - H(\theta_0) = \bar{H}(\bar{\theta}) - \bar{H}(\theta_0) \to 0 \text{ a.s.}$$

by Lemma 7.8, the continuity of the function $\bar{H}(\theta) = E[\frac{\partial^2}{\partial \theta^2} \ln f(X_i, \theta)]$ given Assumption M.6, and $\bar{\theta} - \theta_0 \to 0$ as $n \to \infty$ almost surely.

For the first term here, we have

$$\left|\hat{H}(\bar{\theta}) - \bar{H}(\bar{\theta})\right| = \left|\frac{1}{n}\sum_{i=1}^{n} \frac{\partial^2 \ln f(X_i, \bar{\theta})}{\partial \theta^2} - \left\{E\left[\frac{\partial^2 \ln f(X_i, \theta)}{\partial \theta^2}\right]\right\}_{\theta=\bar{\theta}}\right|$$

$$\leq \sup_{\theta \in \Theta}\left|\frac{1}{n}\sum_{i=1}^{n} \frac{\partial^2 \ln f(X_i, \theta)}{\partial \theta^2} - E\left[\frac{\partial^2 \ln f(X_i, \theta)}{\partial \theta^2}\right]\right|$$

$$\to 0 \text{ a.s.}$$

as $n \to \infty$ by the USLLN in Lemma 7.10, Chapter 7. It follows that as $n \to \infty$, we have $\hat{H}(\bar{\theta}) - H(\theta_0) \to 0$ as $n \to \infty$ almost surely, and so

$$\hat{H}(\bar{\theta})^{-1} \to H(\theta_0)^{-1} \text{ a.s.}$$

given that $H(\theta_0)$ is not zero.

It follows from the Slutsky theorem (see Theorem 7.18) that as $n \to \infty$,

$$\sqrt{n}(\hat{\theta} - \theta_0) = \left[-\hat{H}(\bar{\theta})\right]^{-1} \frac{1}{\sqrt{n}}\sum_{i=1}^{n} S(X_i, \theta_0)$$

$$\xrightarrow{d} N(0, H(\theta_0)^{-1}I(\theta_0)H(\theta_0)^{-1}).$$

Since $I(\theta_0) = -H(\theta_0)$ by Lemma 8.3 (the information matrix equality), we have

$$\sqrt{n}(\hat{\theta} - \theta_0) \xrightarrow{d} N(0, -H(\theta_0)^{-1}).$$

Note that $-H_0(\theta_0)^{-1}$ is positive because $H_0(\theta_0)$ is negative. This completes the proof.

The asymptotic normality $\sqrt{n}(\hat{\theta} - \theta_0) \xrightarrow{d} N(0, -H(\theta_0)^{-1})$ implies that the asymptotic mean of $\sqrt{n}(\hat{\theta} - \theta_0)$ is zero, and the asymptotic variance of $\sqrt{n}(\hat{\theta} - \theta_0)$ is equal to $-H(\theta_0)^{-1}$.

The function

$$H(\theta) \equiv E_\theta \left[\frac{\partial^2 \ln f(X_i, \theta)}{\partial \theta^2} \right]$$

$$= \int_{-\infty}^{\infty} \frac{\partial^2 \ln f(x, \theta)}{\partial \theta^2} f(x, \theta) dx$$

is called the Hessian matrix of the PMF/PDF model $f(x, \theta)$, where the expectation $E_\theta(\cdot)$ is taken under the PDF model $f(x, \theta)$. This function is negative definite and its absolute value magnitude measures the degree of the curvature of the likelihood function at θ. Thus, the efficiency of MLE $\hat{\theta}$ depends on the curvature of the likelihood function at the true parameter value θ_0. If the degree of the curvature of the log-likelihood function is large so that the likelihood function has a sharp peak, it will be easy to estimate θ_0 precisely. On the other hand, if the degree of the curvature is small so that the likelihood function is flat, it will be difficult to estimate θ_0 precisely. Figure 8.3 plots the population likelihood function $E[\ln f(X_i, \theta)]$ for the cases where there is a sharp peak and there is a flat peak for $E[\ln f(X_i, \theta)]$ at the true parameter value θ_0 respectively.

Why is the asymptotic normality result for MLE useful in practice? It can be used to construct confidence interval estimators and hypothesis tests, among other things. For example, an asymptotic $100(1 - \alpha)\%$ confidence interval estimator for θ_0 is given by a random interval $[\hat{\theta}_L, \hat{\theta}_U]$, where $\hat{\theta}_L = \theta_L(\mathbf{X}^n)$ and $\hat{\theta}_U = \theta_U(\mathbf{X}^n)$, such that

$$\lim_{n \to \infty} P\left(\hat{\theta}_L \leq \theta_0 \leq \hat{\theta}_U \right) = 1 - \alpha.$$

That is, the probability that the true parameter value θ_0 lies between $\hat{\theta}_L$ and $\hat{\theta}_U$ will approach $1 - \alpha$ as $n \to \infty$.

Given $\sqrt{n}(\hat{\theta} - \theta_0) \xrightarrow{d} N(0, -H(\theta_0)^{-1})$ and $\hat{H}(\hat{\theta}) \to H(\theta_0)$ as $n \to \infty$ almost surely, we have

$$\sqrt{-n\hat{H}(\hat{\theta})}(\hat{\theta} - \theta_0) \xrightarrow{d} N(0, 1)$$

by the Slutsky theorem (see Theorem 7.18). Therefore, as $n \to \infty$,

$$P\left[-z_{\alpha/2} \leq \sqrt{-n\hat{H}(\hat{\theta})}(\hat{\theta} - \theta_0) \leq z_{\alpha/2} \right] \to 1 - \alpha,$$

where $z_{\alpha/2}$ is the upper-tailed $N(0,1)$ critical value at level $\alpha/2$, namely

$$P(Z \geq z_{\alpha/2}) = \frac{\alpha}{2},$$

where $Z \sim N(0,1)$. For example, $z_{\alpha/2} = 1.65, 1.96, 2.33$ for $\alpha = 0.10, 0.05$ and 0.01 respectively.

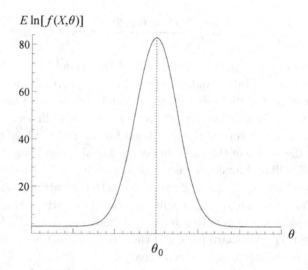

Figure 8.3(a): Sharp peak for $E\ln[f(x,\theta)]$ at θ_0

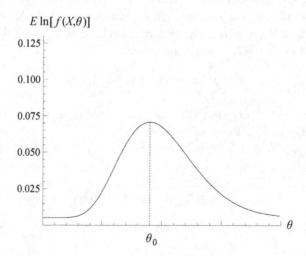

Figure 8.3(b): Flat peak for $E\ln[f(x,\theta)]$ at θ_0

This can be equivalently written as

$$P\left[\hat{\theta} - \frac{z_{\alpha/2}}{\sqrt{n}}\sqrt{\frac{1}{-\hat{H}(\hat{\theta})}} \leq \theta_0 \leq \hat{\theta} + \frac{z_{\alpha/2}}{\sqrt{n}}\sqrt{\frac{1}{-\hat{H}(\hat{\theta})}}\right] \to 1 - \alpha$$

as $n \to \infty$. We then obtain the asymptotic $(1 - \alpha)100\%$ confidence interval estimator as follows:

$$\hat{\theta} - \sqrt{-\frac{1}{n\hat{H}(\hat{\theta})}}z_{\alpha/2} \leq \theta_0 \leq \hat{\theta} + \sqrt{-\frac{1}{n\hat{H}(\hat{\theta})}}z_{\alpha/2},$$

where $z_{\alpha/2}$ is the upper-tailed critical value of $N(0,1)$ at level $\alpha/2$.

Obviously, the larger the sample size n, or the larger the curvature of the likelihood function at the true parameter value θ_0, the sharper confidence bound (i.e., the better interval estimate) for θ_0 will be obtained.

8.4 Method of Moments and Generalized Method of Moments

8.4.1 *Method of Moments Estimation*

Historically MME is one of the oldest methods of estimation in statistics. It consists of equating the first few moments of a population to their sample counterparts, thus getting as many equations as needed to solve for the unknown parameters of the population.

Specifically, suppose $f(x, \theta)$, where $\theta \in \Theta$, is the PMF/PDF model for the unknown population distribution $f_X(x)$ and $f_X(x) = f(x, \theta_0)$ for some parameter value $\theta_0 \in \Theta$. This implies that the parametric probability model $f(x, \theta)$ is correctly specified for the population distribution $f_X(x)$. Suppose $\mathbf{X}^n$ is an IID random sample from the population distribution $f_X(x)$.

We first define a $p \times 1$ sample moment vector

$$\hat{M} = m(\mathbf{X}^n),$$

which will converge in probability to $E[m(\mathbf{X}^n)]$ where $E(\cdot)$ is taken under the unknown true joint distribution of $\mathbf{X}^n$.

Next, we compute its mathematical expectation or population moment

$$M(\theta) = E_\theta[m(\mathbf{X}^n)]$$
$$= \int_{\mathbb{R}^n} m(\mathbf{x}^n)f_{\mathbf{X}^n}(\mathbf{x}^n, \theta)d\mathbf{x}^n,$$

where the mathematical expectation $E_\theta(\cdot)$ is taken under the joint distribution $f_{\mathbf{X}^n}(\mathbf{x}^n, \theta)$ of $\mathbf{X}^n$, and $f_{\mathbf{X}^n}(\mathbf{x}^n, \theta) = \Pi_{i=1}^n f(x_i, \theta)$ when $\mathbf{X}^n$ is an IID random sample.

Then we solve the system of p equations

$$\hat{M} = M(\hat{\theta}).$$

That is, we choose a parameter value $\hat{\theta} = \hat{\theta}_n(\mathbf{X}^n)$ to match the sample moment $\hat{M}$ with the population moment $M(\theta)$. The solution $\hat{\theta}$ is called the MME for the true parameter value θ_0.

Alternatively, we can define a vector-valued sample moment function

$$\hat{m}(\theta) = \hat{M} - M(\theta), \qquad \theta \in \Theta.$$

The method of moments estimator $\hat{\theta}$ is the solution of the equations

$$\hat{m}(\theta) = 0.$$

Often, when $\mathbf{X}^n$ is an IID random sample from population $f_X(x) = f(x, \theta_0)$ for some parameter value θ_0, we can have the following procedures:

- Compute population moments $E_\theta(X_i^k)$, $k = 1, 2, \cdots$, under the PMF/PDF model $f(x, \theta)$:

$$M_k(\theta) = E_\theta(X_i^k)$$
$$= \begin{cases} \int_{-\infty}^{\infty} x^k f(x, \theta) dx & \text{if } X \text{ is a CRV,} \\ \sum_{x \in \Omega_X} x^k f(x, \theta) & \text{if } X \text{ is a DRV;} \end{cases}$$

  Note that the population moment $M_k(\theta)$ depends on parameter θ.
- Compute the sample moments from random sample $\mathbf{X}^n = (X_1, \cdots, X_n)$:

$$\hat{M}_k = n^{-1} \sum_{i=1}^n X_i^k, \qquad k = 1, 2, \cdots;$$

- Match the sample moments and the population moments by choosing some parameter value $\hat{\theta}$. In general, if θ is a $p \times 1$ parameter vector, we need p equations:

$$\begin{cases} \hat{M}_1 = M_1(\hat{\theta}), \\ \hat{M}_2 = M_2(\hat{\theta}), \\ \quad \cdots \\ \hat{M}_p = M_p(\hat{\theta}). \end{cases}$$

Solving for these p equations will yield a MME $\hat{\theta} = \hat{\theta}_n(\mathbf{X}^n)$.

Question: Why can MME $\hat{\theta}$ consistently estimate the true parameter value θ_0?

Intuitively, by the weak law of large numbers, the sample moment

$$\hat{M}_k \xrightarrow{P} E(X_i^k)$$
$$= \int_{-\infty}^{\infty} x^k f_X(x) dx$$
$$= \int_{-\infty}^{\infty} x^k f(x, \theta_0) dx$$
$$= M_k(\theta_0).$$

Thus, if $\hat{M}_k = M(\hat{\theta})$ for all n, we expect that $\hat{\theta}$ will converge in probability to θ_0 as $n \to \infty$.

We now provide some numerical examples.

Example 8.5. Suppose $\mathbf{X}^n$ is an IID EXP(θ) random sample. Find an estimator for θ using the method of moments and maximum likelihood, respectively.

Solution: (1) The method of moments estimation: Because the exponential PDF

$$f(x, \theta) = \begin{cases} \frac{1}{\theta} e^{-x/\theta}, & \text{if } x > 0, \\ 0, & \text{if } x \leq 0, \end{cases}$$

it can be shown that

$$M_1(\theta) = E_\theta(X_i)$$
$$= \int_{-\infty}^{\infty} x f(x, \theta) dx$$
$$= \int_0^{\infty} x \frac{1}{\theta} e^{-x/\theta} dx$$
$$= \theta.$$

On the other hand, the first sample moment is the sample mean:

$$\hat{M}_1 = \bar{X}_n.$$

Moment matching yields

$$\hat{M}_1 = M_1(\hat{\theta}) = \hat{\theta}.$$

We then obtain the method of moment estimator

$$\hat{\theta} = \hat{M}_1 = \bar{X}_n.$$

(2) The MLE method: Given the IID EXP(θ) assumption, the likelihood function of $\mathbf{X}^n$

$$\hat{L}(\theta|\mathbf{X}^n) = \prod_{i=1}^{n} f(X_i, \theta)$$

$$= \left(\frac{1}{\theta}\right)^n e^{-\frac{1}{\theta}\Sigma_{i=1}^n X_i}.$$

Therefore, the log-likelihood function is

$$\ln \hat{L}(\theta|\mathbf{X}^n) = -n \ln \theta - \frac{1}{\theta}\sum_{i=1}^{n} X_i.$$

The FOC is then given by

$$\frac{\partial \ln \hat{L}(\hat{\theta}|\mathbf{X}^n)}{\partial \theta} = -\frac{n}{\hat{\theta}} + \frac{1}{\hat{\theta}}\sum_{i=1}^{n} X_i = 0.$$

It follows that the MLE

$$\hat{\theta} = \bar{X}_n.$$

Both MME and MLE give the same estimator in this example. Therefore, they are equally efficient for estimating θ_0. The reason that MME here is as efficient as MLE is that the sample mean or the first sample moment $\bar{X}_n$ is a sufficient statistic for θ, which contains all information in the random sample $\mathbf{X}^n$ about parameter θ.

Example 8.6. Suppose $\mathbf{X}^n$ is an IID $N(\mu, \sigma^2)$ random sample. Find the MME for $\theta = (\mu, \sigma^2)$.

Solution: The first two population and sample moments are given respectively:

$$M_1(\theta) = E_\theta(X_i) = \mu,$$
$$M_2(\theta) = E_\theta(X_i^2) = \sigma^2 + \mu^2,$$
$$\hat{M}_1 = \bar{X}_n,$$
$$\hat{M}_2 = n^{-1}\sum_{i=1}^{n} X_i^2.$$

We match the first two sample moments with their population counterparts respectively:

$$\bar{X}_n = \hat{\mu},$$
$$n^{-1}\sum_{i=1}^{n} X_i^2 = \hat{\sigma}^2 + \hat{\mu}^2.$$

It follows that

$$\hat{\mu} = \bar{X}_n,$$

$$\hat{\sigma}^2 = n^{-1} \sum_{i=1}^{n} (X_i - \bar{X}_n)^2.$$

These MME's are the same as the MLE for (μ, σ^2). Again, this is due to the fact that $(\bar{X}_n, S_n^2)$ is a sufficient statistic for $\theta = (\mu, \sigma^2)$ for the normal random sample $\mathbf{X}^n$.

It is important to note that although the population moment function $M_k(\theta)$ is calculated using the population distribution model $f(x, \theta)$, MME only uses a finite number of sample moments, which may not capture all information about θ that is contained in the random sample $\mathbf{X}^n$ if the sample moments used are not sufficient statistics for θ. As a result, it may not be a most efficient estimator of θ, even asymptotically. In contrast, MLE captures all information about θ that is contained in $\mathbf{X}^n$ because it uses the entire joint PMF/PDF of the random sample $\mathbf{X}^n$. Therefore, MME may not be as efficient as MLE, unless the sample moments used are sufficient statistics for the parameters of interest.

8.4.2 *Generalized Method of Moments Estimation*

Often in econometrics, the population moment function $M(\theta) = E_\theta[m(\mathbf{X}^n)]$ is not attainable, due to the fact that the population distribution of an economic process is usually not specified. However, economic and financial theory often implies that certain moment conditions must hold when evaluated at the true model parameter θ_0. In other words, economic theories or hypotheses are often characterized by a set of moment conditions. As a result, we can estimate the true model parameter value using the moment conditions implied by economic theory.

Specifically, suppose θ is a $p \times 1$ parameter vector, and there exists a $p \times 1$ moment function $m(X, \theta)$ such that

$$E[m(X, \theta_0)] = \mathbf{0} \text{ for some } \theta_0 \in \Theta,$$

where the expectation is taken over the unknown true distribution of X. This may follow from economic theory (e.g., the equilibrium conditions in rational expectation models). We now provide some economic examples.

Example 8.7. An investor who maximizes an intertemporal utility function

$$\max_{\{C_t\}} U(C_t, C_{t+1}) = u(C_t) + \beta E[u(C_{t+1})|I_t]$$

subject to an intertemporal budget constraint will choose a sequence of consumptions $\{C_t\}$ that satisfies the FOC

$$P_t = \beta E\left[\frac{u'(C_{t+1})}{u'(C_t)}Y_{t+1}\,|I_t\right],$$

where β is a time discount factor parameter, Y_{t+1} is the random payoff of an asset at time $t+1$, and P_t is the price of the asset at time t, and $E_t(\cdot) = E(\cdot|I_t)$ is the conditional expectation given the information set I_t available at time t. This FOC is also called the Euler equation. It states that in the equilibrium, the current asset price should be equal to the expected future payoff of the asset after risk compensation. Here, the factor $\beta\frac{u'(C_{t+1})}{u'(C_t)}$ is called the stochastic discount factor; it measures the risk attitude of the representative economic agent.

Define the stochastic pricing error

$$\varepsilon_{t+1}(\theta) = \beta\frac{u'(C_{t+1})}{u'(C_t)}Y_{t+1} - P_t,$$

where parameter θ contains time discount factor β and any other structural parameters. The Euler equation can be equivalently characterized by the following conditional moment condition:

$$E\left[\varepsilon_{t+1}(\theta_0)|I_t\right] = 0.$$

This implies that a rational economic agent does not make any systematic pricing error in each time period.

Now define a moment function

$$m(X_{t+1}, \theta) = \left[\beta\frac{u'(C_{t+1})}{u'(C_t)}Y_{t+1} - P_t\right]Z_t,$$

where $X_{t+1} = (C_t, C_{t+1}, P_t, Y_{t+1}, Z_t')'$ and $Z_t \in I_t$ is the vector of so-called instrumental variables available at time t. Then by the law of iterated expectations (see Lemma 5.32), we have

$$E[m(X_{t+1}, \theta_0)] = E\{E\left[m(X_{t+1}, \theta_0)|I_t\right]\}$$
$$= 0,$$

where $E(\cdot)$ is the unconditional expectation under the unknown population distribution of X_{t+1}.

In this example, where does parameter θ come from? Besides the time discount parameter β, some parameter(s) may arise from the utility function to characterize risk aversion of the economic agent. For example, when the economic agent has a constant relative risk aversion utility function

$$u(C_t) = \frac{C_t^\gamma - 1}{\gamma},$$

the parameter

$$\gamma = -C_t \frac{u''(C_t)}{u'(C_t)}$$

measures the degree of risk aversion of the economic agent. In this case, $\theta = (\beta, \gamma)$.

Example 8.8. [Capital Asset Pricing Model (CAPM)]: Define Y_t as an $l \times 1$ vector of excess returns for l assets (or portfolios of assets) in period t. For these l assets, the excess returns can be described using the CAPM:

$$Y_t = \alpha + \beta R_{mt} + \varepsilon_t$$
$$= \theta' X_t + \varepsilon_t,$$

where $W_t = (1, R_{mt})'$ is a bivariate vector, R_{mt} is the excess market portfolio return, θ is a $2 \times l$ parameter matrix, and ε_t is an $l \times 1$ disturbance, with $E(\varepsilon_t | W_t) = 0$.

Put $X_t = (Y_t, W_t')'$. Define the $q \times 1$ moment function

$$m(X_t, \theta) = W_t \otimes (Y_t - \theta' W_t),$$

where $q = 2l$ and $\otimes$ denotes the Kronecker product. When the CAPM holds, we have

$$E[m(X_t, \theta_0)] = 0.$$

These $q \times 1$ moment conditions form a basis to estimate and test the CAPM.

In Examples 8.7 and 8.8, one is given a set of moment conditions rather a probability distribution model. In other words, economic theory only delivers a set of moment conditions and is silent about the entire population distribution. As a result, the MLE and MME described above are not applicable. Below, we introduce a GMM which is based on the moment condition as implied by economic theory and does not require knowledge of the population distribution of the data generating process.

Generally speaking, suppose we are given q population moment conditions:

$$E\left[m(X_i, \theta_0)\right] = \mathbf{0},$$

where $m(X_i, \theta)$ is a $q \times 1$ random vector, θ_0 is a $p \times 1$ true parameter vector, the expectation $E(\cdot)$ is taken over the unknown population distribution of X_i, and $\mathbf{0}$ is a $q \times 1$ zero vector. We could estimate θ_0 by finding an estimator $\hat{\theta}$ which matches the sample moment to the population moment $E[m(X_i, \theta_0)] = \mathbf{0}$:

$$\hat{m}(\hat{\theta}) = n^{-1} \sum_{i=1}^{n} m(X_i, \hat{\theta}) = \mathbf{0}.$$

In practice, we can use q moment conditions, where $q \geq p$, namely the number of moment conditions is larger than or at least equal to the number of unknown parameters. In this case, it is generally impossible to find a solution that satisfies the equation $\hat{m}(\theta) = \mathbf{0}$ exactly because the number of equations is larger than the number of unknown parameters. We can only choose an estimator $\hat{\theta}$ to make $\hat{m}(\theta)$ as close to a zero vector as possible. More specifically, we observe the estimator that solves the following minimization problem

$$\hat{\theta} = \arg\min_{\theta \in \Theta} \hat{m}(\theta)' \hat{W}^{-1} \hat{m}(\theta),$$

where $\hat{W}$ is a $q \times q$ random nonsingular symmetric matrix such that $\hat{W} \overset{p}{\to} W$ as $n \to \infty$, and W is a $q \times q$ nonstochastic nonsingular symmetric matrix. One can choose $\hat{W} = I$, a $q \times q$ identity matrix, which yields a convenient procedure. In this case, the objective function

$$\hat{m}(\theta)' \hat{W}^{-1} \hat{m}(\theta) = \sum_{k=1}^{q} \hat{m}_k^2(\theta),$$

namely the sum of q squared components of the sample moment vector $\hat{m}(\theta)$. Here, each component is equally weighted. Note that the classical MME introduced in Section 8.4.1 is a special case of the GMM estimation with $q = p$ and $\hat{W} = I$.

There may exist some optimal choice of the weighting matrix $\hat{W}$ which can give an asymptotically most efficient estimator within a class of estimators, at least asymptotically. Intuitively, the q components in the sample moment vector $\hat{m}(\theta)$ may have different sampling variabilities and may be correlated with each other as well. If the weighting matrix $\hat{W}$ can down-weight the components with larger variances and eliminate their correlations, the resulting estimator will be efficient. This is similar to the idea of the so-called Generalized Least Squares (GLS) estimator in the classical linear regression model, which will be discussed in Chapter 10.

The resulting estimator $\hat{\theta}$ is called the GMM estimator. It is also called the minimum Chi-square estimator in the statistical literature, because the minimized objective function $n\hat{m}(\theta)'\hat{W}^{-1}\hat{m}(\theta)$ follows an asymptotic χ^2 distribution with a suitable choice of $\hat{W}$. Most often, GMM estimation is used in connection with instrumental variables Z_i, which define the vector-valued moment function $m(X_i, \theta)$.

Theorem 8.6. *[Existence of GMM]: Suppose that with probability one, the quadratic form* $\hat{m}(\theta)'\hat{W}^{-1}\hat{m}(\theta)$ *is continuous in* $\theta \in \Theta$, *and that* Θ *is a compact set. Then there exists a global minimizer* $\hat{\theta}$ *that solves the minimization problem*

$$\hat{\theta} = \arg\min_{\theta \in \Theta} \hat{m}(\theta)'\hat{W}^{-1}\hat{m}(\theta).$$

Proof: By the Weierstrass theorem.

Like MLE and MME, the GMM moment conditions are often highly nonlinear so that the GMM estimator $\hat{\theta}$ may have no closed form solution.

As pointed out earlier, the GMM estimation does not require any knowledge on the functional form of the population distribution model $f(x, \theta)$. Most economic theories can be characterized by a set of moment conditions or conditional moment conditions. Thus, GMM has been rather popularly used in econometrics. However, for the reason analogous to that for MME, GMM may be less efficient than MLE if MLE assumes a correct functional form of the population distribution $f(x, \theta)$.

8.5 Asymptotic Properties of GMM

To investigate the asymptotic properties of the GMM estimator $\hat{\theta}$, we first provide a set of regularity conditions.

Assumption G.1: $\mathbf{X}^n = (X_1, \cdots, X_n)$ is an IID random sample from some unknown population distribution $f_X(x)$.

Assumption G.2: The $q \times 1$ moment function $m(x, \theta)$ is continuous in (x, θ) and the absolute values of its components are bounded by a nonnegative function $b(x)$ such that $E[b(X_i)] < \infty$, where and below the expectation $E(\cdot)$ is taken under the unknown population distribution $f_X(x)$.

Assumption G.3: There exists one and only one $p \times 1$ parameter value θ_0 in Θ such that $E[m(X_i, \theta_0)] = 0$.

Assumption G.4: The p-dimensional parameter space Θ is closed and bounded.

Assumption G.5: The $q \times q$ stochastic weighting matrix $\hat{W} \to W$ as $n \to \infty$ almost surely, where W is symmetric, bounded and nonsingular.

Assumption G.6: The parameter value θ_0 is an interior point of the parameter space Θ.

Assumption G.7: (1) The functions $\frac{\partial}{\partial \theta} m(x, \theta)$ and $\frac{\partial^2}{\partial \theta \partial \theta'} m(x, \theta)$ are continuous in (x, θ) and the absolute values of their component functions are bounded by a nonnegative function $b(x)$ such that $E[b(X_i)] < \infty$;

(2) the $q \times q$ symmetric matrix $V = E[m(X_i, \theta_0)m(X_i, \theta_0)']$ is bounded and nonsingular;

(3) the $q \times p$ gradient matrix $G(\theta_0) = E[\frac{\partial}{\partial \theta} m(X_i, \theta_0)]$ is of full rank (which is equal to p given $p \le q$).

Assumption G.3 is an identification condition for θ_0, which is usually called the true parameter value of the model characterized by the population moment condition $E[m(X_i, \theta_0)] = 0$. The identification condition ensures that θ_0 is the unique probability limit of the GMM estimator $\hat{\theta}$. The interior solution assumption for θ_0 is not needed when we establish the consistency of the GMM estimator $\hat{\theta}$, but is needed when we derive the asymptotic normality of the GMM estimator $\hat{\theta}$, where we will use a Taylor series expansion of the FOC of the GMM estimation.

We first establish the consistency of the GMM estimator.

Theorem 8.7. *[Consistency of GMM]: Suppose Assumptions G.1–G.5 hold. Then as $n \to \infty$,*

$$\hat{\theta} \to \theta_0 \ a.s.$$

Proof: The proof is similar to the consistency proof of MLE. Note that here θ_0 could be a corner solution so that the first order conditions may not hold.

Next, we derive the asymptotic distribution of the GMM estimator $\hat{\theta}$ after proper standardization.

Theorem 8.8. *[Asymptotic Normality]: Suppose Assumptions G.1–G.7 hold. Then*

(1) as $n \to \infty$,

$$\sqrt{n}(\hat{\theta} - \theta_0) \xrightarrow{d} N(0, \Omega),$$

where

$$\Omega = \Psi V \Psi',$$

$V = E[m(X_1, \theta_0)m(X_1, \theta_0)']$, and $\Psi = [G(\theta_0)W^{-1}G(\theta_0)']^{-1}G(\theta_0)W^{-1}$.
(2) Moreover, if $W = V$, then as $n \to \infty$,

$$\sqrt{n}(\hat{\theta} - \theta_0) \xrightarrow{d} N(0, [G(\theta_0)V^{-1}G(\theta_0)']^{-1}).$$

Proof: (1) Define the objective function

$$\hat{Q}_n(\theta) = \hat{m}(\theta)'\hat{W}^{-1}\hat{m}(\theta).$$

Noting that the prespecified weighting matrix $\hat{W}$ is not a function of θ, we can obtain the FOC as follows:

$$\frac{d\hat{Q}_n(\hat{\theta})}{d\theta} = 2\hat{G}(\hat{\theta})\hat{W}^{-1}\hat{m}(\hat{\theta}) = \mathbf{0},$$

where the $p \times q$ sample matrix

$$\hat{G}(\theta) = \frac{d\hat{m}(\theta)}{d\theta'}$$

$$= \frac{1}{n}\sum_{i=1}^{n} \frac{\partial m(X_i, \theta)}{\partial \theta'}.$$

By the mean value theorem, we have

$$\hat{m}(\hat{\theta}) = \hat{m}(\theta_0) + \hat{G}(\bar{\theta})'(\hat{\theta} - \theta_0),$$

where $\bar{\theta}$ lies on the segment between $\hat{\theta}$ and θ_0, i.e., $\bar{\theta} = \lambda\hat{\theta} + (1 - \lambda)\theta_0$ for some $\lambda \in [0, 1]$. Substituting this expression into the FOC above, we obtain

$$\hat{G}(\hat{\theta})\hat{W}^{-1}\hat{m}(\theta_0) + \hat{G}(\hat{\theta})\hat{W}^{-1}\hat{G}(\bar{\theta})'(\hat{\theta} - \theta_0) = 0.$$

Following the reasoning similar to that for the sample Hessian matrix $\hat{H}(\bar{\theta})$ in the proof for the asymptotic normality of MLE (see Theorem 8.5), we can show that as $n \to \infty$.

$$\hat{G}(\hat{\theta}) \to G(\theta_0) \text{ a.s.,}$$
$$\hat{G}(\bar{\theta}) \to G(\theta_0) \text{ a.s.,}$$

using the USLLN in Lemma 7.10, the continuity of the gradient function $G(\theta) = E[\frac{\partial}{\partial\theta}m(X_i, \theta)]$, and $||\bar{\theta} - \theta_0|| \le ||\hat{\theta} - \theta_0|| \to 0$ as $n \to \infty$ almost surely. Also, we have $\hat{W} \to W$ as $n \to \infty$ almost surely by Assumption G.5. It follows that as $n \to \infty$,

$$\hat{G}(\hat{\theta})\hat{W}^{-1}G(\tilde{\theta}) \to G(\theta_0)W^{-1}G(\theta_0) \text{ a.s.,}$$

where $G(\theta_0)W^{-1}G(\theta_0)$ is a nonsingular matrix given Assumptions G.5 and G.7(3). Therefore, we have that the stochastic inverse matrix $[\hat{G}(\hat{\theta})\hat{W}^{-1}G(\tilde{\theta})]^{-1}$ exists for n sufficiently large because

$$[\hat{G}(\hat{\theta})\hat{W}^{-1}G(\tilde{\theta})]^{-1} \to [G(\theta_0)W^{-1}G(\theta_0)]^{-1} \text{ a.s.}$$

as $n \to \infty$. It follows from the FOC above that

$$\sqrt{n}(\hat{\theta} - \theta_0) = -\left[\hat{G}(\hat{\theta})\hat{W}^{-1}\hat{G}(\tilde{\theta})'\right]^{-1}\hat{G}(\hat{\theta})\hat{W}^{-1}\sqrt{n}\hat{m}(\theta_0)$$

$$= -\hat{\Psi}\sqrt{n}\hat{m}(\theta_0), \text{ say.}$$

By the CLT for IID random sequences (Theorem 7.16) and the Cramer-Wold device (Lemma 7.21), we have

$$\sqrt{n}\hat{m}(\theta_0) = \frac{1}{\sqrt{n}}\sum_{i=1}^{n} m(X_i, \theta_0) \xrightarrow{d} N(0, V),$$

where $V = E[m(X_i, \theta_0)m(X_i, \theta_0)']$. Furthermore,

$$\left[\hat{G}(\hat{\theta})\hat{W}^{-1}\hat{G}(\tilde{\theta})'\right]^{-1}\hat{G}(\hat{\theta})\hat{W}^{-1} \xrightarrow{a.s.} [G(\theta_0)W^{-1}G(\theta_0)]^{-1}G(\theta_0)W^{-1} \equiv \Psi$$

as $n \to \infty$. It follows from the Slutsky theorem (Theorem 7.18) that as $n \to \infty$,

$$\sqrt{n}(\hat{\theta} - \theta_0) \xrightarrow{d} N(0, \Psi V \Psi').$$

(2) Suppose $W = V$. Then we have

$$\Psi V \Psi' = \{[G(\theta_0)V^{-1}G(\theta_0)']^{-1}G(\theta_0)V^{-1}\}V\{[G(\theta_0)V^{-1}G(\theta_0)']^{-1}G(\theta_0)V^{-1}\}'$$

$$= \{[G(\theta_0)V^{-1}G(\theta_0)']^{-1}G(\theta_0)V^{-1}\}V\{V^{-1}G(\theta_0)'[G(\theta_0)V^{-1}G(\theta_0)']^{-1}\}$$

$$= [G(\theta_0)V^{-1}G(\theta_0)']^{-1}.$$

It follows that when $W = V$, we have as $n \to \infty$,

$$\sqrt{n}(\hat{\theta} - \theta_0) \xrightarrow{d} N(0, [G(\theta_0)V^{-1}G(\theta_0)']^{-1}).$$

This completes the proof.

It can be further shown that the GMM estimator $\hat{\theta}$ with the choice of weighting matrix $\hat{W} \to V$ almost surely as $n \to \infty$ is an asymptotically optimal GMM estimator within a class of GMM estimators in the sense that it has a smaller asymptotic variance than any other GMM estimator with other choice of the weighting matrix $\hat{W}$. This is stated below.

Theorem 8.9. *[Asymptotic Efficiency of GMM]:* Define $\Omega_0 = [G(\theta_0)V^{-1}G(\theta_0)']^{-1}$. Then

$$\Omega - \Omega_0 \text{ is positive semi-definite (PSD)}$$

for all finite and nonsingular matrix W, where Ω is given in Theorem 8.8.

Proof: Observe that $\Omega - \Omega_0$ is PSD if and only if $\Omega_0^{-1} - \Omega^{-1}$ is PSD. For notational simplicity, put $G_0 = G(\theta_0)$ and decompose $V = V^{1/2}V^{1/2}$ where V is a $q \times q$ symmetric and nonsingular matrix. Noting $(ABC)^{-1} = C^{-1}B^{-1}A^{-1}$ for nonsingular matrices A, B, and C, we consider

$$
\begin{aligned}
\Omega_0^{-1} - \Omega^{-1} &= G_0'V^{-1}G_0 - G_0'\Sigma^{-1}G_0[G_0'\Sigma^{-1}V\Sigma^{-1}G_0]^{-1}G_0'\Sigma^{-1}G_0 \\
&= G_0'V^{-\frac{1}{2}}[I - V^{\frac{1}{2}}\Sigma^{-1}G_0[G_0'\Sigma^{-1}V\Sigma^{-1}G_0]^{-1}G_0'\Sigma^{-1}V^{\frac{1}{2}}]V^{-\frac{1}{2}}G_0 \\
&= G_0'V^{-\frac{1}{2}}\Pi V^{-\frac{1}{2}}G_0,
\end{aligned}
$$

where the $q \times q$ matrix

$$\Pi \equiv I - V^{\frac{1}{2}}\Sigma^{-1}G_0[G_0'\Sigma^{-1}V\Sigma^{-1}G_0]^{-1}G_0'\Sigma^{-1}V^{\frac{1}{2}}$$

is an idempotent matrix (i.e., $\Pi = \Pi'$, $\Pi^2 = \Pi$). It follows that we have

$$
\begin{aligned}
\Omega_0^{-1} - \Omega^{-1} &= (G_0'V^{-\frac{1}{2}}\Pi)(\Pi V^{-\frac{1}{2}}G_0) \\
&= [\Pi V^{-\frac{1}{2}}G_0]'[\Pi V^{-\frac{1}{2}}G_0],
\end{aligned}
$$

which is always positive semi-definite (why?). This completes the proof.

In practice, if the GMM estimator $\hat{\theta}$ is not a function of a sufficient statistic, they can always be improved upon by conditioning on a sufficient statistic. In contrast, MLE is always a function of a sufficient statistic when the latter exists, so there is no room for further improvement. Thus, for efficiency consideration, MLE is generally preferred over GMM. However, MLE requires knowledge of the correct specification of the likelihood function, which is often not implied or specified by economic theory.

8.6 Mean Squared Error Criterion

In general, different methods give different estimators for the same parameter θ. A natural question arises here: which is the best estimator for θ? For example, we have obtained two different estimators for population variance σ^2: one is the sample variance $S_n^2 = (n-1)^{-1}\sum_{i=1}^{n}(X_i - \bar{X}_n)^2$ and the other is the MLE estimator $\hat{\sigma}^2 = n^{-1}\sum_{i=1}^{n}(X_i - \bar{X}_n)^2$. Which estimator is better?

Intuitively, a best estimator is the one that is closest to the unknown true parameter θ. To compare different estimators, we need to use some appropriate criterion (a distance or divergence) to measure how close the estimator $\hat{\theta}$ is to θ. There are many divergence measures for $\hat{\theta}$ and θ. In general, any increasing function of the absolute distance $|\hat{\theta} - \theta|$ would serve to measure the goodness of an estimator. However, the mean squared error criterion, defined below, has some advantages over other measures, particularly due to the facts that it is quite tractable analytically and has a nice interpretation of being decomposed as the sum of the variance and so-called squared bias.

Definition 8.3. [Mean Squared Error (MSE)]: Let θ be a population parameter. The MSE of an estimator $\hat{\theta} = \hat{\theta}_n(\mathbf{X}^n)$ of parameter θ is defined as

$$MSE_\theta(\hat{\theta}) = E_\theta(\hat{\theta} - \theta)^2,$$

where $E_\theta(\cdot)$ denotes the expectation which is taken under the joint distribution $f_{\mathbf{X}^n}(\mathbf{x}^n, \theta)$ of the random sample $\mathbf{X}^n$, or equivalently under the sampling distribution of $\mathbf{X}^n$.

$MSE_\theta(\hat{\theta})$ measures the average of variations or deviations of the estimator $\hat{\theta}$ from the parameter θ. The difference $\hat{\theta} - \theta$ is usually called the estimation error, so $MSE_\theta(\hat{\theta})$ is a measure of the magnitude of the estimation error. The smaller is $MSE_\theta(\hat{\theta})$, the better is the estimator $\hat{\theta}$. This is because a smaller MSE means that there is a smaller deviation between $\hat{\theta}$ and θ. A best estimator $\hat{\theta}$ for θ is the one that minimizes $MSE_\theta(\theta)$ over the class of the estimators under investigation.

We should emphasize that MSE is not the only criterion, but it is intuitive and analytically simple, and therefore is most commonly used in practice.

Next, we introduce an important concept called bias.

Definition 8.4. [Bias]: The bias of a point estimator $\hat{\theta}$ of parameter θ is defined as

$$Bias_\theta(\hat{\theta}) = E_\theta(\hat{\theta}) - \theta.$$

An estimator $\hat{\theta}$ for θ is called an unbiased estimator for θ if $Bias_\theta(\hat{\theta}) = 0$.

An unbiased estimator gives the right answer on average in a large number of repeated estimates. That is, there is no systematic upward or downward estimation for the parameter θ.

Example 8.9. Suppose $\mathbf{X}^n$ is an IID random sample from some population with mean μ and variance σ^2. Find an unbiased estimator for $\text{var}_\theta(\bar{X}_n)$.

Solution: Put $\theta = (\mu, \sigma^2)$ and $\tau = \frac{\sigma^2}{n}$. Because $\text{var}_\theta(\bar{X}_n) = \frac{\sigma^2}{n}$, an unbiased estimator of τ can be given as follows:

$$\hat{\tau} = \frac{S_n^2}{n}$$

because $E_\theta(\hat{\tau}) = n^{-1}E_\theta(S_n^2) = \text{var}_\theta(\bar{X}_n) = \tau$. It follows that $\text{Bias}_\theta(\hat{\tau}) = E_\theta(\hat{\tau}) - \tau = 0$.

Example 8.10. Suppose $\mathbf{X}^n$ is an IID random sample from some population with mean μ and variance σ^2. Find an unbiased estimator for μ^2.

Solution: Put $\theta = (\mu, \sigma^2)$. For parameter $\tau = \mu^2$, an unbiased estimator is

$$\hat{\tau} = \bar{X}_n^2 - \frac{S_n^2}{n}.$$

This follows because

$$E(\hat{\tau}) = E_\theta(\bar{X}_n^2) - \frac{E_\theta(S_n^2)}{n}$$

$$= \text{var}_\theta(\bar{X}_n) + [E_\theta(\bar{X}_n)]^2 - \frac{E_\theta(S_n^2)}{n}$$

$$= \frac{\sigma^2}{n} + \mu^2 - \frac{\sigma^2}{n}$$

$$= \mu^2 = \tau.$$

Intuitively, since the sample mean $\bar{X}_n$ is a good estimator for μ, so we expect that $\bar{X}_n^2$ is a good estimator for μ^2. However, $\bar{X}_n^2$ is a nonlinear function of $\bar{X}_n$ which introduces a bias $\frac{\sigma^2}{n}$. This bias can be corrected by subtracting the unbiased estimator $\frac{S_n^2}{n}$ for $\frac{\sigma^2}{n}$.

We now state a very useful decomposition for $\text{MSE}_\theta(\hat{\theta})$.

Theorem 8.10. *[MSE Decomposition]:*

$$E_\theta(\hat{\theta} - \theta)^2 = \text{var}_\theta(\hat{\theta}) + [\text{Bias}_\theta(\hat{\theta})]^2.$$

Proof: Using the formula $(a+b)^2 = a^2 + b^2 + 2ab$, we expand

$$E_\theta(\hat{\theta} - \theta)^2 = E_\theta\left[\hat{\theta} - E_\theta(\hat{\theta}) + E_\theta(\hat{\theta}) - \theta\right]^2$$

$$= E_\theta\left[\hat{\theta} - E_\theta(\hat{\theta})\right]^2 + \left[E_\theta(\hat{\theta}) - \theta\right]^2 + 2E_\theta\left\{\left[\hat{\theta} - E_\theta(\hat{\theta})\right]\left[E_\theta(\hat{\theta}) - \theta\right]\right\}$$

$$= E_\theta\left[\hat{\theta} - E_\theta(\hat{\theta})\right]^2 + \left[E_\theta(\hat{\theta}) - \theta\right]^2,$$

where the cross-product term

$$E_\theta \left\{ \left[\hat{\theta} - E_\theta(\hat{\theta}) \right] \left[E_\theta(\hat{\theta}) - \theta \right] \right\} = E_\theta \left\{ \left[\hat{\theta} - E_\theta(\hat{\theta}) \right] \right\} \left[E_\theta(\hat{\theta}) - \theta \right]$$
$$= 0 \cdot \left[E_\theta(\hat{\theta}) - \theta \right]$$
$$= 0.$$

Therefore, $\text{MSE}_\theta(\hat{\theta})$ can be decomposed into two components: $\text{var}_\theta(\hat{\theta})$ and $\text{Bias}_\theta(\hat{\theta})^2$. Here, $\text{var}_\theta(\hat{\theta})$ measures the variability (precision) of the estimator $\hat{\theta}$ due to the sampling variation, and $\text{Bias}_\theta(\hat{\theta})$ measures the accuracy of the estimator due to the procedure. For any unbiased estimator $\hat{\theta}$, we have $\text{MSE}_\theta(\hat{\theta}) = E_\theta(\hat{\theta} - \theta)^2 = \text{var}_\theta(\hat{\theta})$. Thus, a best unbiased estimator is the one with the smallest variance. Of course, a best unbiased estimator may be dominated by a biased estimator if the latter has a substantially smaller variance that can compensate the bias.

The MSE criterion can be understood intuitively by the example of shooting a target with a large number of repeated shots. A high cumulative score will be obtained if most shots are close to the target. This corresponds to a small MSE. If the shots center at the target (so the bias is small) but they spread over widely around the target (so the variance is large), or if the shots center at a point which is far away from the target (so the bias is large) although they do not spread widely, a low cumulative score will be obtained.

Definition 8.5. [Relative Efficiency]: An estimator $\hat{\theta}$ for parameter θ is said to be more efficient than another estimator $\tilde{\theta}$ for the same parameter θ in terms of MSE if

$$MSE_\theta(\hat{\theta}) \leq MSE_\theta(\tilde{\theta}).$$

It may be emphasized that there exist different criteria to evaluate relative efficiency of estimators, and a different criterion will generally lead to a different ranking between or among different estimators. In this book, we use the MSE criterion, which is most commonly used in practice.

We now consider a few examples to illustrate some fundamental statistical ideas in estimation.

Example 8.11. Suppose $\{X_i\}_{i=1}^{2n}$ is an IID random sample from a population with mean μ and variance σ^2. Define $\hat{\mu}_1 = n^{-1} \sum_{i=1}^{n} X_i$ and $\hat{\mu}_2 = (2n)^{-1} \sum_{i=1}^{2n} X_i$. Which estimator is better in terms of MSE?

Solution: Using the results (Theorems 6.1 and 6.2) in Chapter 6, we have $E_\theta(\hat\mu_1) = \mu$, $\text{var}_\theta(\hat\mu_1) = \frac{\sigma^2}{n}$, $E_\theta(\hat\mu_2) = \mu$ and $\text{var}_\theta(\hat\mu_2) = \frac{\sigma^2}{2n}$. It follows that $\text{MSE}_\theta(\hat\mu_1) = 2\text{MSE}_\theta(\hat\mu_2)$. Therefore, $\hat\mu_2$ is more efficient.

Intuitively, it is always better to use more sample information in estimation. In light of this, sample splitting or sample truncation is generally not a good idea from a statistical point of view because it does not fully utilize the whole sample information contained in $\{X_i\}_{i=1}^{2n}$.

Example 8.12. Let (X_1, X_2) be an IID random sample from a population with mean μ and variance σ^2. Two estimators for μ are

$$\hat\mu_1 = \bar X_n = \frac{1}{2}(X_1 + X_2),$$

$$\hat\mu_2 = \frac{1}{3}(X_1 + 2X_2).$$

Which estimator is better?

Solution: Put $\theta = (\mu, \sigma^2)$. We can check that both estimators are unbiased for μ. Also,

$$\text{var}_\theta(\hat\mu_1) = \frac{1}{2}\sigma^2.$$

$$\text{var}_\theta(\hat\mu_2) = \frac{1}{9}\sigma^2 + \frac{4}{9}\sigma^2$$

$$= \frac{5}{9}\sigma^2$$

$$> \frac{1}{2}\sigma^2.$$

Therefore, $\hat\mu_1$ is more efficient than $\hat\mu_2$ in terms of MSE.

Intuitively, since two random variables X_1 and X_2 are identical distributed, there is no reason to discriminate X_1 and X_2 (i.e. putting different weights on observations which have identical distributions). Equal weighting for each observation will be most efficient for estimation of μ. In Section 8.7, we will consider a best unbiased estimator that generalizes this example.

8.7 Best Unbiased Estimators

We now address the following question: What is the best estimator if a class of estimators of parameter θ is available?

Certainly, we can define the best estimator as the one that has the smallest MSE. Unfortunately, such a best estimator is very difficult to obtain, because the class of estimators we have to compare is very huge. For simplicity, we focus on a class of unbiased estimators and find the best estimator within this class.

First, we consider a generalization of the concept of an unbiased estimator for the transformed parameter $\tau = \tau(\theta)$. An example is $\tau = \mu^2$, which has been considered in Example 8.10 of Section 8.6.

Definition 8.6. [Generalized Unbiased Estimator]: $\hat{\tau} = \hat{\tau}_n(\mathbf{X}^n)$ is an unbiased estimator for the parameter $\tau(\theta)$ if

$$E_\theta(\hat{\tau}) = \tau(\theta) \text{ for all } \theta \in \Theta.$$

When $\tau(\theta) = \theta$, we return to Definition 8.4 of the unbiased estimator for parameter θ.

Example 8.13. Suppose $\mathbf{X}^n$ is an IID (μ, σ^2) random sample. Put $\theta = (\mu, \sigma^2)$ and $\tau(\theta) = (\mu - 2)^2$. Find an unbiased estimator for $\tau(\theta)$.

Solution: We first try an estimator $\tilde{\tau} = (\bar{X}_n - 2)^2$. Then

$$\begin{aligned}
E_\theta(\tilde{\tau}) &= E_\theta(\bar{X}_n - 2)^2 \\
&= E_\theta[(\bar{X}_n - \mu + \mu - 2)^2] \\
&= E_\theta(\bar{X}_n - \mu)^2 + (\mu - 2)^2 \\
&= \frac{\sigma^2}{n} + \tau(\theta).
\end{aligned}$$

It follows that

$$E_\theta(\tilde{\tau}) - \tau(\theta) = \frac{\sigma^2}{n} \neq 0.$$

We then make a necessary bias correction:

$$\hat{\tau} = (\bar{X}_n - 2)^2 - \frac{1}{n}S_n^2.$$

Then this bias-corrected estimator $\hat{\tau}$ is unbiased for the transformed parameter $\tau(\theta) = (\mu - 2)^2$.

Definition 8.7. [Uniform Best Unbiased Estimator]: Let Γ be a class of unbiased estimators of parameter $\tau(\theta)$, where $\theta \in \Theta$ and Θ is a parameter space. An estimator $\hat{\tau}^* \in \Gamma$ is a uniformly best unbiased estimator for $\tau(\theta)$ over parameter space Θ within the class Γ of estimators if
  (1) $E_\theta(\hat{\tau}^*) = \tau(\theta)$ for all $\theta \in \Theta$;
  (2) $\text{var}_\theta(\hat{\tau}^*) \leq \text{var}_\theta(\hat{\tau})$ for any estimator $\hat{\tau}$ of $\tau(\theta)$ in Γ and for all $\theta \in \Theta$.

The estimator $\hat{\tau}^*$ is a Uniform Minimum Variance Unbiased Estimator (UMVUE) of $\tau(\theta)$ over parameter space Θ within the class Γ of estimators for $\tau(\theta)$. Here, uniformity means that $\hat{\tau}^*$ is always a best unbiased estimator for $\tau(\theta)$ no matter what value the parameter θ will take in Θ.

Example 8.14. Let $\mathbf{X}^n$ be an IID (μ, σ^2) random sample. Define a class of linear unbiased estimators of μ as follows:

$$\Gamma = \left\{ \hat{\mu} : \mathbb{R}^n \to \mathbb{R} \mid \hat{\mu} = \sum_{i=1}^{n} c_i X_i \text{ for } (c_1, \cdots, c_n)' \in \mathbb{R}^n \right\}.$$

(1) Show that $\hat{\mu}$ is an unbiased estimator of μ for all $n \geq 1$ if and only if $\sum_{i=1}^{n} c_i = 1$.
(2) Find the uniformly most efficient unbiased estimator of μ within the class Γ of estimators for μ.

Solution: Note that we have $\hat{\tau} = \hat{\mu}$, and $\tau(\theta) = \mu$ in this application.

(1) Given $\hat{\mu} = \sum_{i=1}^{n} c_i X_i$ and $\theta = (\mu, \sigma^2)$, we have $E_\theta(\hat{\mu}) = \mu \sum_{i=1}^{n} c_i$. Therefore, if $\hat{\mu}$ is an unbiased estimator for μ, i.e., if

$$E_\theta(\hat{\mu}) = \mu \text{ for all } \mu \in \mathbb{R},$$

we must have

$$\sum_{i=1}^{n} c_i = 1.$$

On the other hand, if $\sum_{i=1}^{n} c_i = 1$, then from the fact that $E_\theta(\hat{\mu}) = \mu \sum_{i=1}^{n} c_i$, we have

$$E_\theta(\hat{\mu}) = \mu \cdot 1 = \mu \text{ for all } \mu.$$

That is, $\hat{\mu}$ is unbiased for μ. Therefore, $\hat{\mu}$ is unbiased for μ for all possible values of μ if and only if $\sum_{i=1}^{n} c_i = 1$.

(2) To find the uniformly most efficient unbiased estimator for μ, we only need to find the one with the smallest variance within Γ subject to the constraint that $\sum_{i=1}^{n} c_i = 1$. The variance of $\hat{\mu} \in \Gamma$ is

$$\text{var}_\theta(\hat{\mu}) = \text{var}_\theta \left(\sum_{i=1}^{n} c_i X_i \right)$$

$$= \sum_{i=1}^{n} c_i^2 \text{var}_\theta(X_i)$$

$$= \sigma^2 \sum_{i=1}^{n} c_i^2.$$

Because $\hat{\mu}$ is unbiased, we have $\sum_{i=1}^{n} c_i = 1$. Therefore, we can solve for the variance minimization problem

$$\min_{\{c_i\}_{i=1}^{n}} \sigma^2 \sum_{i=1}^{n} c_i^2$$

subject to the constraint that

$$\sum_{i=1}^{n} c_i = 1.$$

Define the Lagrangian function

$$L(c, \lambda) = \sigma^2 \sum_{i=1}^{n} c_i^2 + \lambda \left(1 - \sum_{i=1}^{n} c_i \right),$$

where $c = (c_1, \cdots, c_n)'$ and λ is the Lagrange multiplier.

The $n + 1$ first order conditions are given below:

$$\frac{\partial L(c, \lambda)}{\partial c_i} = 2\sigma^2 c_i - \lambda = 0 \text{ for } i = 1, \cdots, n,$$

$$\frac{\partial L(c, \lambda)}{\partial \lambda} = 1 - \sum_{i=1}^{n} c_i = 0.$$

Solving for these equations, we obtain

$$c_i^* = \frac{1}{n} \text{ for } i = 1, \cdots, n.$$

Thus, the uniformly most efficient unbiased estimator

$$\hat{\mu}^* = \sum_{i=1}^{n} \frac{1}{n} X_i = \bar{X}_n.$$

That is, the most efficient unbiased estimator for μ is the sample mean $\bar{X}_n$. For completeness, we need to verify the second order conditions to ensure that $\hat{\mu}^*$ is a global minimizer. This is indeed the case because it can be shown that the Hessian matrix of $L(c, \lambda)$ is positive definite for all μ (please verify it!).

Intuitively, because the random variables $\{X_i\}_{i=1}^{n}$ are identically distributed, there is no ground to discriminate one against another. The optimal weighting is of course equal weighting for all observations. This is exactly the same idea for the so-called Gauss-Markov theorem in the classical linear regression model, which says that the OLS estimator for a classical linear regression model with IID disturbances

$$Y_i = X_i'\theta_0 + \varepsilon_i, \qquad i = 1, \cdots, n$$

is the Best Linear Unbiased Estimator (BLUE), under the assumption that $\{\varepsilon_i\}_{i=1}^n$ is IID $(0, \sigma_\varepsilon^2)$. The OLS estimator $\hat{\theta}$ is defined as the optimal solution to the problem of minimizing the sum of squared residuals, namely,

$$\hat{\theta} = \arg\min_{\theta} \sum_{i=1}^n (Y_i - X_i'\theta)^2.$$

For comparison, we recall from Theorem 6.1 of Chapter 6 that the sample mean $\bar{X}_n$ is a minimizer to a sum of squared residuals:

$$\bar{X}_n = \arg\min_{a} \sum_{i=1}^n (X_i - a)^2.$$

This is a special case of the linear regression model.

Example 8.15. Suppose $\mathbf{X}^n = (X_1, \cdots, X_n)$ is an independent but not identically distributed random sample, with $E(X_i) = \mu$ and $\text{var}(X_i) = \sigma_i^2 < \infty, i = 1, \cdots, n$. Find a uniformly best linear unbiased estimator of μ within the class of estimators

$$\Gamma = \left\{ \hat{\mu} : \mathbb{R}^n \to \mathbb{R} \mid \hat{\mu} = \sum_{i=1}^n c_i X_i, \ (c_1, \cdots, c_n)' \in \mathbf{R}^n \right\},$$

where $\sum_{i=1}^n c_i = 1$.

Solution: Again, $\hat{\mu}$ is an unbiased estimator for μ if and only if $\sum_{i=1}^n c_i = 1$. Now, using the Lagrange multiplier method, we can find that the optimal estimator within the class Γ is

$$\hat{\mu}^* = \sum_{i=1}^n c_i^* X_i$$

$$= \frac{1}{\sum_{i=1}^n \frac{1}{\sigma_i^2}} \sum_{i=1}^n \frac{1}{\sigma_i^2} X_i,$$

where

$$c_i^* = \frac{\frac{1}{\sigma_i^2}}{\sum_{i=1}^n \frac{1}{\sigma_i^2}} \propto \frac{1}{\sigma_i^2}, \qquad i = 1, \cdots, n.$$

This result suggests that to obtain the most efficient estimator for μ in an independent but not identically distributed random sample (the random variables of which are assumed to have a common mean μ but different variances), one should discount noisy observations (i.e., the observations with large variances should receive small weights, and vice versa). The optimal weight c_i^* is proportional to σ_i^{-2}, the inverse of the variance of the random variable X_i.

This is similar to the idea behind the popular GLS estimator in econometrics. Consider a linear regression model

$$Y_i = X_i'\theta + \varepsilon_i, \qquad i = 1, \cdots, n,$$

where $\{\varepsilon_i\}_{i=1}^n$ is an independent but not identically distributed sequence with $E(\varepsilon_i) = 0$ and $\text{var}(\varepsilon_i^2) = \sigma_i^2$. Here, there may exist unconditional heteroskedasticity because σ_i^2 may be different across i. We consider the transformed regression model

$$\frac{Y_i}{\sigma_i} = \left(\frac{X_i}{\sigma_i}\right)'\theta + \frac{\varepsilon_i}{\sigma_i}$$

or equivalently

$$Y_i^* = X_i^{*\prime}\theta + \varepsilon_i^*,$$

where $\{\varepsilon_i^*\}_{i=1}^n$ is an IID sequence with zero mean and unit variance. Then the OLS estimator of the transformed linear regression model

$$\hat{\theta}^* = \arg\min_{\theta} \sum_{i=1}^n (Y_i^* - X_i^{*\prime}\theta)^2$$

is called the GLS estimator. In other words, GLS is the OLS estimator after correcting heteroskedasticity. This estimator $\hat{\theta}^*$ discounts noisy observations by dividing them by their standard deviations. It could be shown that the GLS estimator is BLUE. For more discussion, see Chapter 10.

The Lagrange multiplier method is an important mathematical method with applications in economics and finance. For example, the problems in the above two examples are directly applicable to the optimal portfolio selection problem, in which an investor will choose portfolio weights to minimize the risk as measured by the variance of the portfolio return subject to the constraint that the expected portfolio return remains constant. The optimal coefficients $\{c_i^*\}$ in either Example 8.14 or Example 8.15 are the optimal portfolio weights. In Example 8.15, for example, we should invest less on asset i if it is more risky (i.e., if it has a larger variance σ_i^2) when all risky assets have the same expected returns. The Lagrange method can also be seen in utility maximization subject to a budget constraint and cost minimization subject to a technology constraint.

We have discussed how to find a uniformly best unbiased estimator within a class of estimators. A question naturally arise here: is an unbiased estimator always better than a biased estimator?

The answer is no. We now illustrate this by an example.

Example 8.16. Let $\mathbf{X}^n$ be an IID random sample from a $N(\mu, \sigma^2)$ distribution. The sample variance $S_n^2 = (n-1)^{-1} \sum_{i=1}^n (X_i - \bar{X}_n)^2$ and the MLE estimator $\hat{\sigma}^2 = n^{-1} \sum_{i=1}^n (X_i - \bar{X}_n)^2$ are two estimators for σ^2. Which is more efficient in terms of MSE?

Solution: Because $\frac{(n-1)S_n^2}{\sigma^2} \sim \chi_{n-1}^2$ by Theorem 6.7 of Chapter 6, we have $E_\theta(S_n^2) = \sigma^2$ and $\text{var}_\theta(S_n^2) = \frac{2\sigma^4}{n-1}$. It follows that

$$\text{MSE}_\theta(S_n^2) = E_\theta \left(S_n^2 - \sigma^2 \right)^2$$
$$= \text{var}_\theta(S_n^2) + \left[\text{Bias}_\theta(S_n^2) \right]^2$$
$$= \frac{2\sigma^4}{n-1}.$$

Next, observing

$$\hat{\sigma}^2 = \frac{n-1}{n} S_n^2,$$

we have

$$\text{Bias}_\theta(\hat{\sigma}^2) = E_\theta(\hat{\sigma}^2) - \sigma^2$$
$$= \frac{n-1}{n} \sigma^2 - \sigma^2$$
$$= -\frac{\sigma^2}{n},$$

$$\text{var}_\theta(\hat{\sigma}^2) = \left(\frac{n-1}{n} \right)^2 \text{var}_\theta(S_n^2)$$
$$= \left(\frac{n-1}{n} \right)^2 \frac{2\sigma^4}{n-1}.$$

It follows that

$$\text{MSE}_\theta(\hat{\sigma}^2) = \left(1 - \frac{1}{n} \right)^2 \frac{2\sigma^4}{n-1} + \frac{\sigma^4}{n^2}$$
$$= \left[\left(1 - \frac{1}{n} \right)^2 + \frac{n-1}{2n^2} \right] \frac{2\sigma^4}{n-1}$$
$$= \frac{n-1}{n} \frac{2n-1}{2n} \frac{2\sigma^4}{n-1}$$
$$< \frac{2\sigma^4}{n-1} = \text{MSE}_\theta(S_n^2)$$

for all $n > 1$. Therefore, the biased estimator $\hat{\sigma}^2$ is better than the unbiased estimator S_n^2.

Example 8.16 shows that unbiased estimators may not be better estimators. It is sometimes the case, as in the above example, that a trade-off occurs between variance and bias in such a way that a small increase in bias can be traded for a larger decrease in variance, resulting an improvement in MSE.

The conclusion in Example 8.16 does not mean that S_n^2 should be abandoned as an estimator for σ^2. It is a result based on the MSE criterion. Whether MSE is most suitable for measuring goodness of variance estimators is unknown. Furthermore, when n is large, there is virtually no difference between S_n^2 and $\hat{\sigma}^2$ even in terms of MSE.

8.8   Cramer-Rao Lower Bound — An Alternative Method

In general, it is still quite difficult to find a best efficient estimator within an unbiased estimator class. There is an alternative method of evaluating parameter estimators when the population distribution model $f(x, \theta)$ is available. For simplicity, we assume that parameter θ is a scalar here.

Theorem 8.11. *[Cramer-Rao Lower Bound; Cramer-Rao Inequality; Information Inequality]: Let $\mathbf{X}^n$ be a random sample with joint PMF/PDF $f_{\mathbf{X}^n}(\mathbf{x}^n, \theta)$, and let $\hat{\tau} = \hat{\tau}_n(\mathbf{X}^n)$ be any estimator of parameter $\tau(\theta)$ where $E_\theta(\hat{\tau})$ is a differentiable function of θ. Suppose the joint PMF/PDF $f_{\mathbf{X}^n}(\mathbf{x}^n, \theta)$ of the random sample $\mathbf{X}^n$ satisfies the condition that*

$$\frac{d}{d\theta} \int_{\mathbb{R}^n} h(\mathbf{x}^n) f_{\mathbf{X}^n}(\mathbf{x}^n, \theta) dx^n = \int_{\mathbb{R}^n} h(\mathbf{x}^n) \frac{\partial f_{\mathbf{X}^n}(\mathbf{x}^n, \theta)}{\partial \theta} dx^n,$$

for any function $h : \mathbb{R}^n \to \mathbb{R}$ with $E_\theta|h(\mathbf{X}^n)| < \infty$, where $E_\theta(\cdot)$ is taken over the joint PMF/PDF $f_{\mathbf{X}^n}(\mathbf{x}^n, \theta)$ of $\mathbf{X}^n$. Then for all $n > 0$ and all $\theta \in \Theta$,

$$var_\theta(\hat{\tau}) \geq B_n(\theta) \equiv \frac{\left[\frac{dE_\theta(\hat{\tau})}{d\theta}\right]^2}{E_\theta \left[\frac{\partial \ln f_{\mathbf{X}^n}(\mathbf{X}^n, \theta)}{\partial \theta}\right]^2},$$

where $B_n(\theta)$ is called the Cramer-Rao lower bound. In particular, when $\hat{\tau}$ is unbiased for parameter $\tau(\theta)$, we have

$$B_n(\theta) = \frac{[\tau'(\theta)]^2}{E_\theta \left[\frac{\partial \ln f_{\mathbf{X}^n}(\mathbf{X}^n, \theta)}{\partial \theta}\right]^2}.$$

Proof: We first consider the case of continuous distributions; the proof for the case of discrete distributions is similar. Suppose we have: (1)

$$E_\theta \left[\frac{\partial \ln f_{\mathbf{X}^n}(\mathbf{X}^n, \theta)}{\partial \theta} \right]^2 = \text{var}_\theta \left[\frac{\partial \ln f_{\mathbf{X}^n}(\mathbf{X}^n, \theta)}{\partial \theta} \right]$$

and (2)

$$\frac{dE_\theta(\hat{\tau})}{d\theta} = \text{cov}_\theta \left[\hat{\tau}_n(\mathbf{X}^n), \frac{\partial \ln f_{\mathbf{X}^n}(\mathbf{X}^n, \theta)}{\partial \theta} \right].$$

Then by the Cauchy-Schwarz inequality, we have

$$\left\{ \text{cov}_\theta \left[\hat{\tau}_n(\mathbf{X}^n), \frac{\partial \ln f_{\mathbf{X}^n}(\mathbf{X}^n, \theta)}{\partial \theta} \right] \right\}^2 \leq \text{var}_\theta(\hat{\tau}) \cdot \text{var}_\theta \left[\frac{\partial \ln f_{\mathbf{X}^n}(\mathbf{X}^n, \theta)}{\partial \theta} \right].$$

This implies that

$$\text{var}_\theta(\hat{\tau}) \geq \frac{\left\{ \text{cov}_\theta \left[\hat{\tau}, \frac{\partial \ln f_{\mathbf{X}^n}(\mathbf{X}^n, \theta)}{\partial \theta} \right] \right\}^2}{\text{var}_\theta \left[\frac{\partial \ln f_{\mathbf{X}^n}(\mathbf{X}^n, \theta)}{\partial \theta} \right]}$$

$$= \frac{\left[\frac{dE_\theta(\hat{\tau})}{d\theta} \right]^2}{E_\theta \left[\frac{\partial \ln f_{\mathbf{X}^n}(\mathbf{X}^n, \theta)}{\partial \theta} \right]^2} = B_n(\theta)$$

given results (1) and (2) above. Therefore, it suffices to show results (1) and (2).

We first prove result (1). We shall show that the mean of the score function $\frac{\partial \ln f_{\mathbf{X}^n}(\mathbf{x}^n, \theta)}{\partial \theta}$ of the random sample $\mathbf{X}^n$ under the joint distribution $f_{\mathbf{X}^n}(\mathbf{X}^n, \theta)$ is zero. This follows because

$$E_\theta \left[\frac{\partial \ln f_{\mathbf{X}^n}(\mathbf{X}^n, \theta)}{\partial \theta} \right] = \int_{\mathbb{R}^n} \left[\frac{\partial \ln f_{\mathbf{X}^n}(\mathbf{x}^n, \theta)}{\partial \theta} \right] f_{\mathbf{X}^n}(\mathbf{x}^n, \theta) d\mathbf{x}^n$$

$$= \int_{\mathbb{R}^n} \frac{\partial f_{\mathbf{X}^n}(\mathbf{x}^n, \theta)}{\partial \theta} d\mathbf{x}^n$$

$$= \frac{d}{d\theta} \int_{\mathbb{R}^n} f_{\mathbf{X}^n}(\mathbf{x}^n, \theta) d\mathbf{x}^n$$

$$= \frac{d(1)}{d\theta}$$

$$= 0,$$

where the third equality follows by exchanging the orders of integration and differentiation. Therefore, using the formula $\text{var}(Y) = E(Y^2) - \mu_Y^2$, we obtain the variance of the score function

$$\text{var}_\theta \left[\frac{\partial f_{\mathbf{X}^n}(\mathbf{X}^n, \theta)}{\partial \theta} \right] = E_\theta \left[\frac{\partial \ln f_{\mathbf{X}^n}(\mathbf{X}^n, \theta)}{\partial \theta} \right]^2 - \left\{ E_\theta \left[\frac{\partial \ln f_{\mathbf{X}^n}(\mathbf{X}^n, \theta)}{\partial \theta} \right] \right\}^2$$

$$= E_\theta \left[\frac{\partial \ln f_{\mathbf{X}^n}(\mathbf{X}^n, \theta)}{\partial \theta} \right]^2.$$

This has proved result (1).

Next, we prove result (2). Given $E_\theta \left[\frac{\partial f_{\mathbf{X}^n}(\mathbf{X}^n, \theta)}{\partial \theta} \right] = 0$ and $\hat{\tau} = \hat{\tau}(\mathbf{X}^n)$, and using the formula $\text{cov}(Y, Z) = E(YZ) - \mu_Y \mu_Z$, we have

$$\text{cov}_\theta \left[\hat{\tau}, \frac{\partial \ln f_{\mathbf{X}^n}(\mathbf{X}^n, \theta)}{\partial \theta} \right]$$

$$= E_\theta \left[\hat{\tau} \frac{\partial \ln f_{\mathbf{X}^n}(\mathbf{X}^n, \theta)}{\partial \theta} \right] - E_\theta(\hat{\tau}) E_\theta \left[\frac{\partial \ln f_{\mathbf{X}^n}(\mathbf{X}^n, \theta)}{\partial \theta} \right]$$

$$= E_\theta \left[\hat{\tau} \frac{\partial \ln f_{\mathbf{X}^n}(\mathbf{X}^n, \theta)}{\partial \theta} \right]$$

$$= \int_{\mathbb{R}^n} \hat{\tau}_n(\mathbf{x}^n) \left[\frac{\partial \ln f_{\mathbf{X}^n}(\mathbf{x}^n, \theta)}{\partial \theta} \right] f_{\mathbf{X}^n}(\mathbf{x}^n, \theta) d\mathbf{x}^n$$

$$= \int_{\mathbb{R}^n} \hat{\tau}_n(\mathbf{x}^n) \frac{\partial f_{\mathbf{X}^n}(\mathbf{x}^n, \theta)}{\partial \theta} d\mathbf{x}^n$$

$$= \frac{d}{d\theta} \int_{\mathbb{R}^n} \hat{\tau}_n(\mathbf{x}^n) f_{\mathbf{X}^n}(\mathbf{x}^n, \theta) d\mathbf{x}$$

$$= \frac{dE_\theta(\hat{\tau})}{d\theta},$$

where the last to second equality follows from the exchange condition between integration and differentiation. We have proved result (2).

The Cramer-Rao lower bound applies to discrete distributions as well. The only change is to replace the integrals involved by summations, with $f_{\mathbf{X}^n}(\mathbf{x}^n, \theta)$ being a joint PMF rather than a joint PDF of the random sample $\mathbf{X}^n$. Thus, the proof of the theorem is completed.

It is important to emphasize that a key assumption in the Cramer-Rao theorem is the ability to differentiate under the integral sign, which, of course, is somewhat restrictive. The condition on the exchange between differentiation and integration is crucial. This condition called a regularity condition which usually holds under "regular cases." It can be written as follows:

$$\frac{d}{d\theta} \int_{\mathbb{R}^n} h(\mathbf{x}^n) f_{\mathbf{X}^n}(\mathbf{x}^n, \theta) d\mathbf{x}^n = \int_{\mathbb{R}^n} h(\mathbf{x}^n) \frac{\partial \ln f_{\mathbf{X}^n}(\mathbf{x}^n, \theta)}{\partial \theta} f_{\mathbf{X}^n}(\mathbf{x}^n, \theta) d\mathbf{x}^n,$$

or equivalently,

$$\frac{dE_\theta[h(\mathbf{X}^n)]}{d\theta} = E_\theta \left[h(\mathbf{X}^n) \frac{\partial \ln f_{\mathbf{X}^n}(\mathbf{X}^n, \theta)}{\partial \theta} \right].$$

As we have seen, densities in the exponential class will satisfy the assumption but in general, such an assumption needs to be checked; otherwise contradictions may arise. Newey and McFadden (1994, Lemma 3.6) provide sufficient conditions for such an assumption to hold.

Theorem 8.11 implies that $\text{var}_\theta(\hat{\tau})$ is greater than or at least equal to the Cramer-Rao lower bound

$$B_n(\theta) \equiv \frac{\left[\frac{dE_\theta(\hat{\tau})}{d\theta} \right]^2}{E_\theta \left[\frac{\partial \ln f_{\mathbf{X}^n}(\mathbf{X}^n, \theta)}{\partial \theta} \right]^2}.$$

Suppose the variance of an unbiased estimator $\hat{\tau}$ of $\tau(\theta)$ achieves the Cramer-Rao lower bound, that is,

$$\text{var}_\theta(\hat{\tau}) = \frac{[\tau'(\theta)]^2}{E_\theta \left[\frac{\partial \ln f_{\mathbf{X}^n}(\mathbf{X}^n, \theta)}{\partial \theta} \right]^2}.$$

Then the unbiased estimator $\hat{\tau}$ must be a most efficient estimator for $\tau(\theta)$.

Question: How to show the Cramer-Rao lower bound when θ is a parameter vector rather than a scalar parameter?

The Cramer-Rao lower bound in Theorem 8.11 is a general result, because it is based on the joint likelihood function $f_{\mathbf{X}^n}(\mathbf{x}^n, \theta)$ of the random sample $\mathbf{X}^n$, which may not be IID. In other words, it applies to the random samples which are not IID, because the IID assumption is not imposed. We now consider a special case where $\mathbf{X}^n$ is IID with population distribution $f(x, \theta)$. In this case, the Cramer-Rao lower bound can be simplified as a function of the parametric population model $f(x, \theta)$ and the sample size n.

Corollary 8.1. *[Cramer-Rao Lower Bound Under an IID Random Sample]: Let $\mathbf{X}^n$ be an IID random sample from population PMF/PDF $f(x, \theta)$ and let $\hat{\tau} = \hat{\tau}_n(\mathbf{X}^n)$ be any estimator of $\tau(\theta)$, where $E_\theta[\hat{\tau}_n(\mathbf{X}^n)]$ is* *a differentiable function of $\theta \in \Theta$. Suppose*

$$\frac{d}{d\theta} \int_{-\infty}^{\infty} h(x) f(x, \theta) dx = \int_{-\infty}^{\infty} h(x) \frac{\partial f(x, \theta)}{\partial \theta} dx$$

for all $h(x)$ such that $E_\theta |h(X_i)| < \infty$. Then for all n,

$$var_\theta(\hat{\tau}) \geq B_n(\theta) \equiv \frac{\left[\frac{dE_\theta(\hat{\tau})}{d\theta} \right]^2}{nI(\theta)},$$

where

$$I(\theta) = E_\theta \left[\frac{\partial \ln f(X_i, \theta)}{\partial \theta} \right]^2$$

$$= \int_{-\infty}^{\infty} \left[\frac{\partial \ln f(x, \theta)}{\partial \theta} \right]^2 f(x, \theta) dx$$

is Fisher's information matrix of the population PMF/PDF $f(x, \theta)$.
When $\hat{\tau}$ is unbiased for $\tau(\theta)$, then

$$var_\theta(\hat{\tau}) \geq B_n(\theta) \equiv \frac{[\tau'(\theta)]^2}{nI(\theta)}.$$

Proof: Given the Cramer-Rao lower bound formula in Theorem 8.11, it suffices to show that under the IID assumption for $\mathbf{X}^n$, the denominator in the Cramer-Rao lower bound

$$E_\theta \left[\frac{\partial \ln f_{\mathbf{X}^n}(\mathbf{X}^n, \theta)}{\partial \theta} \right]^2 = nE_\theta \left[\frac{\partial \ln f(X_i, \theta)}{\partial \theta} \right]^2 = nI(\theta).$$

Because $\mathbf{X}^n$ is an IID sample from population $f(x, \theta)$, we have $f_{\mathbf{X}^n}(\mathbf{X}^n, \theta) = \prod_{i=1}^{n} f(X_i, \theta)$. It follows that

$$\ln f_{\mathbf{X}^n}(\mathbf{X}^n, \theta) = \sum_{i=1}^{n} \ln f(X_i, \theta).$$

Hence, the score function of the random sample $\mathbf{X}^n$

$$\frac{\partial \ln f_{\mathbf{X}^n}(\mathbf{X}^n, \theta)}{\partial \theta} = \sum_{i=1}^{n} \frac{\partial \ln f(X_i, \theta)}{\partial \theta}.$$

By Lemma 8.2, the expectation of the score function $\frac{\partial}{\partial \theta} \ln f(X_i, \theta)$ under the population distribution $f(x, \theta)$ is zero:

$$E_\theta \left[\frac{\partial \ln f(X_i, \theta)}{\partial \theta} \right] = \int_{-\infty}^{\infty} \frac{\partial \ln f(x, \theta)}{\partial \theta} f(x, \theta) dx = 0.$$

It follows from the IID assumption that

$$\begin{aligned}
E_\theta \left[\frac{\partial \ln f_{\mathbf{X}^n}(\mathbf{X}^n, \theta)}{\partial \theta} \right]^2 &= \mathrm{var}_\theta \left[\frac{\partial \ln f_{\mathbf{X}^n}(\mathbf{X}^n, \theta)}{\partial \theta} \right] \\
&= \mathrm{var}_\theta \left[\sum_{i=1}^{n} \frac{\partial \ln f(X_i, \theta)}{\partial \theta} \right] \\
&= \sum_{i=1}^{n} \mathrm{var}_\theta \left[\frac{\partial \ln f(X_i, \theta)}{\partial \theta} \right] \\
&= n \cdot \mathrm{var}_\theta \left[\frac{\partial \ln f(X, \theta)}{\partial \theta} \right] \\
&= n \cdot E_\theta \left[\frac{\partial \ln f(X, \theta)}{\partial \theta} \right]^2 \\
&= nI(\theta),
\end{aligned}$$

where the third and fourth equalities follows from the IID assumption, and the fifth equality follows from the fact that $E_\theta \left[\frac{\partial}{\partial \theta} \ln f(X_i, \theta) \right] = 0$. This completes the proof.

In this corollary, $f(x, \theta)$ is the PMF/PDF of each random variable X_i, while $f_{\mathbf{X}^n}(\mathbf{x}^n, \theta)$ in the previous theorem is the joint PMF/PDF of the random sample $\mathbf{X}^n$. Under the IID assumption, it is clear that the Cramer-Rao lower bound $B_n(\theta)$ vanishes to zero at a rate n.

In addition to the scale factor n^{-1}, the Cramer-Rao lower bound is proportional to the inverse of Fisher's information matrix $I(\theta)$. Fisher's information matrix $I(\theta)$ measures the degree of information about θ that is contained in the population PMF/PDF $f(x, \theta)$ or equivalently in each random variable X_i. The larger $I(\theta)$, the more information about θ is contained in the random variable X_i. As a consequence, we have a smaller $B_n(\theta)$.

The calculation of $I(\theta)$ sometimes is tedious. By using the information matrix equality $I(\theta) + H(\theta) = 0$ in Lemma 8.3, where

$$\begin{aligned}
H(\theta) &\equiv E_\theta \left[\frac{\partial^2 \ln f(X_i, \theta)}{\partial \theta^2} \right] \\
&= \int_{-\infty}^{\infty} \frac{\partial^2 \ln f(x, \theta)}{\partial \theta^2} f(x, \theta) dx
\end{aligned}$$

is called the Hessian matrix, we can write the Cramer-Rao lower bound

$$B_n(\theta) = \frac{[\tau'(\theta)]^2}{-nH(\theta)}.$$

Thus, the Cramer-Rao lower bound $B_n(\theta)$ depends on the inverse of the curvature of the log-likelihood function $\ln f(x, \theta)$. The larger the curvature in absolute value, the smaller the Cramer-Rao lower bound $B_n(\theta)$.

We have shown in Theorem 8.5 that the MLE

$$\sqrt{n}(\hat{\theta} - \theta) \xrightarrow{d} N(0, -H(\theta)^{-1}) \text{ as } n \to \infty,$$

where θ denotes the true parameter value in Θ. This implies $\hat{\theta} - \theta$ approximately follows a $N(0, [-nH(\theta)]^{-1})$ for n sufficiently large. Thus, the variance of MLE $\hat{\theta}$ approximately achieves the Cramer-Rao lower bound when n is large. This is true for any population distribution $f(x, \theta)$. Given the asymptotic normality of $\sqrt{n}(\hat{\theta} - \theta_0)$ in Theorem 8.5, the asymptotic bias of MLE $\hat{\theta}$ is zero. It follows that MLE $\hat{\theta}$ is asymptotically most efficient.

Example 8.17. Let $\mathbf{X}^n$ be an IID Poisson(λ) random sample. Use the Cramer-Rao lower bound method to show that the sample mean $\bar{X}_n$ is a best unbiased estimator of λ.

Solution: In this application we have $\tau(\theta) = \theta = \lambda$, and $\hat{\tau} = \bar{X}_n$, $E_\lambda(\hat{\tau}) = \lambda$. Therefore, $\hat{\tau}$ is an unbiased estimator for λ. Because

$$\text{var}_\lambda(\bar{X}_n) = \frac{\sigma^2}{n} = \frac{\lambda}{n}$$

given $\sigma^2 = \lambda$ for the Poisson(λ) distribution, it suffices to show that the Cramer-Rao lower bound $B_n(\lambda) = \frac{\lambda}{n}$.

Because the log-likelihood function

$$\ln f(X_i, \lambda) = \ln \left(\frac{e^{-\lambda} \lambda^{X_i}}{X_i!} \right)$$
$$= -\lambda + X_i \ln \lambda - \ln X_i!,$$

we have

$$\frac{\partial^2 \ln f(X_i, \lambda)}{\partial \lambda^2} = \frac{\partial^2 (-\lambda + X_i \ln \lambda - \ln X_i!)}{\partial \lambda^2}$$
$$= -\frac{X_i}{\lambda^2}.$$

It follows that

$$H_\lambda(\lambda) = E_\lambda \left[\frac{\partial^2 \ln f(X_i, \lambda)}{\partial \lambda^2} \right]$$

$$= E_\lambda \left(-\frac{X_i}{\lambda^2} \right)$$

$$= -\frac{E_\lambda(X_i)}{\lambda^2}$$

$$= -\frac{1}{\lambda}.$$

Also, because $\hat\tau = \bar{X}_n$, we have $E_\lambda(\bar{X}_n) = \lambda$, so

$$\frac{dE_\lambda(\hat\tau)}{d\lambda} = \frac{d\lambda}{d\lambda} = 1.$$

It follows that

$$B_n(\lambda) = \frac{\left[\frac{d}{d\lambda} E_\lambda(\hat\tau) \right]^2}{-nH_\lambda(\lambda)}$$

$$= \frac{1^2}{-n(-\frac{1}{\lambda})}$$

$$= \frac{\lambda}{n}.$$

Because $\mathrm{var}_\lambda(\bar{X}_n) = B_n(\lambda)$, i.e., the sample mean estimator $\bar{X}_n$ achieves the Cramer-Rao lower bound, it is a best unbiased estimator of λ.

The Cramer-Rao lower bound approach may not reach a decisive conclusion when $\mathrm{var}_\theta(\hat\tau)$ does not attain the Cramer-Rao lower bound $B_n(\theta)$. This drawback is illustrated by the following example.

Example 8.18. Let $\mathbf{X}^n$ be an IID $N(\mu, \sigma^2)$ random sample, where μ and σ^2 are unknown. Show that S_n^2 does not attain the Cramer-Rao lower bound.

Solution: In this application, we have $\theta = (\mu, \sigma^2)$, $\hat\tau = S_n^2$ and $\tau(\theta) = \sigma^2$. The log-likelihood function of a $N(\mu, \sigma^2)$ distribution is

$$\ln f(X_i, \theta) = -\ln \sqrt{2\pi} - \frac{1}{2} \ln \sigma^2 - \frac{(X_i - \mu)^2}{2\sigma^2}.$$

Hence, we have

$$\frac{\partial^2 \ln f(X_i, \theta)}{\partial(\sigma^2)^2} = \frac{1}{2\sigma^4} - \frac{(X_i - \mu)^2}{\sigma^6}.$$

It follows that the Hessian matrix

$$E_\theta \left[\frac{\partial^2 \ln f(X_i, \theta)}{\partial (\sigma^2)^2} \right] = \frac{1}{2\sigma^4} - \frac{E_\theta (X_i - \mu)^2}{\sigma^6}$$

$$= \frac{1}{2\sigma^4} - \frac{1}{\sigma^4}$$

$$= -\frac{1}{2\sigma^4}.$$

In this example, it is easy to understand why the information matrix $I(\theta)$ or the Hessian matrix $H(\theta)$ is a measure of the information contained in the random variable X_i. The smaller the variance σ^2, the less noisy the random variable X_i, and so the more precise the estimation of the parameter σ^2.

For any unbiased estimator $\hat{\tau} = \hat{\tau}_n(\mathbf{X}^n)$ of σ^2, we have $E_\theta[\hat{\tau}_n(\mathbf{X}^n)] = \sigma^2$,

$$\frac{dE_\theta[\hat{\tau}_n(\mathbf{X}^n)]}{d\sigma^2} = 1.$$

It follows that the Cramer-Rao lower bound

$$B_n(\theta) = \frac{1}{n} \frac{1^2}{\frac{1}{2\sigma^4}}$$

$$= \frac{2\sigma^4}{n}$$

$$< \frac{2\sigma^4}{n-1} = \text{var}_\theta(S_n^2).$$

Hence, S_n^2 does not attain the Cramer-Rao lower bound $2\sigma^4/n$. Here, what we can say is that S_n^2 does not attain the Cramer-Rao lower bound. Given the information, we cannot say that S_n^2 is not a best unbiased estimator, because there exist two possibilities: (a) there may be some unbiased estimator that attains $B_n(\theta)$; or (b) $B_n(\theta)$ is unattainable by any unbiased estimator for σ^2.

In general, the Cramer-Rao lower bound approach has a shortcoming in finding best unbiased estimators: if the Cramer-Rao lower bound is not attainable by an unbiased estimator, there is no guarantee that it is a most efficient estimator. That is to say, the value of the Cramer-Rao lower bound may be strictly smaller than the variance of any unbiased estimator. In fact, in the usually favorable case of the population $f(x, \theta)$ being a one-parameter exponential family, the most that we can say is that there exists a parameter $\tau(\theta)$ with an unbiased estimator that achieves the Cramer-Rao lower bound. However, in other typical situations, for other parameters, the bound may not be attainable. These situations cause concern because, if we cannot find an estimator that attains the Cramer-Rao lower bound, we have to decide whether no estimator can attain it or whether we must look at more estimators.

We now provide a characterization under which an estimator can achieve the Cramer-Rao lower bound.

Theorem 8.12. *Suppose* $f_{\mathbf{X}^n}(\mathbf{x}^n, \theta)$ *is the joint PMF/PDF of the random sample* $\mathbf{X}^n$ *and* $\hat{\tau} = \hat{\tau}_n(\mathbf{X}^n)$ *is an unbiased estimator for parameter* $\tau(\theta)$, *where* $f_{\mathbf{X}^n}(\mathbf{x}^n, \theta)$ *and* $\hat{\tau}$ *satisfy the conditions in the Cramer-Rao lower bound theorem (Theorem 8.11). Then the estimator* $\hat{\tau}$ *attains the Cramer-Rao lower bound if and only if*

$$\hat{\tau} - \tau(\theta) = a(\theta) \frac{\partial \ln \hat{L}(\theta|\mathbf{X}^n)}{\partial \theta}$$

for some function $a : \Theta \to \mathbb{R}$.

Proof: By the Cauchy-Schwarz inequality, we have

$$\left\{ \text{cov}_\theta \left[\hat{\tau}, \frac{\partial \ln \hat{L}(\theta|\mathbf{X}^n)}{\partial \theta} \right] \right\}^2 \le \text{var}_\theta(\hat{\tau}) \text{var}_\theta \left[\frac{\partial \ln \hat{L}(\theta|\mathbf{X}^n)}{\partial \theta} \right].$$

The equality holds if and only if the centered parameter estimator $\hat{\tau} - \tau(\theta)$ is proportional to the score function $\frac{\partial}{\partial \theta} \ln \hat{L}(\theta|X^n)$ of the random sample $\mathbf{X}^n$. This follows because, when $\hat{\tau} - \tau(\theta)$ is a linear function of the score function $\frac{\partial}{\partial \theta} \ln \hat{L}(\theta|X^n)$, their correlation will be equal to unity in absolute value. This completes the proof.

Example 8.19. [Continuation of Example 8.18]: The likelihood function of an IID $N(\mu, \sigma^2)$ random sample $\mathbf{X}^n$ is

$$\hat{L}(\theta|\mathbf{X}^n) = \frac{1}{(2\pi\sigma^2)^{n/2}} e^{-\frac{1}{2\sigma^2} \sum_{i=1}^n (X_i - \mu)^2},$$

where $\theta = (\mu, \sigma^2)$. Hence, the score function

$$\frac{\partial \ln \hat{L}(\mu, \sigma^2|x)}{\partial \sigma^2} = \frac{n}{2\sigma^4} \left[n^{-1} \sum_{i=1}^n (X_i - \mu)^2 - \sigma^2 \right].$$

Thus, taking $a(\theta) = n/(2\sigma^4)$ shows that the best unbiased estimator of σ^2 is $n^{-1} \sum_{i=1}^n (x_i - \mu)^2$ if μ is known. If μ is unknown, the Cramer-Rao lower bound is not attainable.

8.9   Conclusion

Estimation is one of the most important objectives in statistical inference. In this chapter, we first introduce two important estimation methods: MLE and MME, the latter having been extended by econometricians to GMM. MLE is based on a correctly specified parametric PMF/PDF model for the population distribution, while MME and GMM are based on a set of moment conditions on the population. Because it utilizes the information on the joint distribution of the random sample, MLE is expected to be more efficient than MME and GMM, unless for the latter, the sample moments used are sufficient statistics for the parameters of interest. On the other hand, GMM does not require the knowledge of the population distribution of the data generating process. To gain insight into the properties of MLE and MME/GMM, we develop the asymptotic theory for the proposed two estimators.

In order to evaluate different estimators for the same population parameter, we introduce the mean squared error criterion to measure the closeness between an estimator and the population parameter. We then develop two important methods to evaluate unbiased parameter estimators. One is the Lagrange multiplier approach, which does not require the knowledge of the population PMF/PDF; the other is the Cramer-Rao lower bound approach which requires the knowledgement of the joint likelihood function of the random sample.

EXERCISE 8

8.1. One observation is taken on a discrete random variable X with PMF $f(x, \theta)$ given below

x	$f(x, 1)$	$f(x, 2)$	$f(x, 3)$
0	$\frac{1}{3}$	$\frac{1}{4}$	0
1	$\frac{1}{3}$	$\frac{1}{4}$	0
2	0	$\frac{1}{4}$	$\frac{1}{4}$
3	$\frac{1}{6}$	$\frac{1}{4}$	$\frac{1}{2}$
4	$\frac{1}{6}$	0	$\frac{1}{4}$

where $\theta \in \Theta = \{1, 2, 3\}$. Find the MLE of θ.

8.2. Let $\mathbf{X}^n$ be an IID random sample with one of two PDF's. If $\theta = 0$, then

$$f(x, \theta) = \begin{cases} 1, & 0 < x < 1, \\ 0, & \text{otherwise,} \end{cases}$$

while if $\theta = 1$, then

$$f(x, \theta) = \begin{cases} 1/(2\sqrt{x}), & 0 < x < 1, \\ 0, & \text{otherwise.} \end{cases}$$

Find the MLE of θ.

8.3. One observation, X, is taken from a $N(0, \sigma^2)$ population.
   (1) Find an unbiased estimation of σ^2.
   (2) Find the MLE of σ.
   (3) Discuss how the method of moments estimator of σ might be found.

8.4. Suppose the random variables $\{Y_1, \cdots, Y_n\}$ satisfy

$$Y_i = \beta x_i + \varepsilon_i, \qquad i = 1, \cdots, n,$$

where $x_1, \cdots, x_n$ are fixed constants, and $\{\varepsilon_i\}$ is an IID sequence from a $N(0, \sigma^2)$ distribution, with σ^2 unknown. Find:
   (1) a two-dimensional sufficient statistic for (β, σ^2);
   (2) the MLE of β, and show that it is an unbiased estimator of β;
   (3) the distribution of the MLE of β.

8.5. Let $\mathbf{X}^n$ be an IID Bernoulli(p) random sample. Show:
   (1) $\bar{X}_n$ is the MLE for unknown parameter p;
   (2) the variance of $\bar{X}_n$ attains the Cramer-Rao Lower Bound, and hence $\bar{X}_n$ is the best unbiased estimator of p.

8.6. Let $\hat{\theta}_1, \cdots, \hat{\theta}_k$ be unbiased estimators of a parameter θ with $\text{var}_\theta(\hat{\theta}_i) = \sigma_i^2$ and $\text{Cov}(\hat{\theta}_i, \hat{\theta}_j) = 0$ if $i \neq j$, where $i, j = 1, \cdots, k$. Show:

(1) of all estimators of the form $\sum_{i=1}^{k} a_i \hat{\theta}_i$, where the a_i's are constants and $E_\theta(\sum_{i=1}^{k} a_i \hat{\theta}_i) = \theta$, the estimator $\hat{\theta}^* = \frac{\sum_{i=1}^{k} \hat{\theta}_i / \sigma_i^2}{\sum_{i=1}^{k} 1/\sigma_i^2}$ has minimum variance;

(2) $\text{var}(\hat{\theta}^*) = \frac{1}{\sum_{i=1}^{k} 1/\sigma_i^2}$.

8.7. Let $\mathbf{X}^n$ be an IID random sample from the $N(\mu, \sigma^2)$ population, where σ^2 is known. Show:

(1) $\bar{X}_n$ is the MLE for μ;

(2) $\bar{X}_n$ achieves the Cramer-Rao lower bound, and so is best unbiased estimator of μ.

8.8. Let $\mathbf{X}^n$ be an IID random sample from the $N(\theta, 1)$ population. The estimator $\hat{\theta} = \bar{X}_n^2 - (1/n)$ is the best unbiased estimator of θ^2. Calculate its variance and show that it is greater than the Cramer-Rao lower bound. [Hint: You can calculate the variance of $\bar{X}_n^2 - \frac{1}{n}$ using the property of χ^2.]

8.9. Suppose $\mathbf{X}^n$ be an IID random sample from a $N(\mu, \sigma^2)$ population. Let S_n^2 be the sample variance. Determine the value of constant c such that cS_n is an unbiased estimator for σ.

8.10. Let $\mathbf{X}^n$ be an IID random sample from a $N(\theta, \theta^2)$ population, where $\theta > 0$. For this model both $\bar{X}_n$ and cS are unbiased estimators of θ, where $c = \frac{\sqrt{n-1}\,\Gamma[(n-1)/2]}{\sqrt{2}\,\Gamma(n/2)}$.

(1) Prove that for any number a the estimator $a\bar{X}_n + (1-a)cS_n$ is an unbiased estimator of θ.

(2) Find the value of a that produces the estimator with minimum variance.

8.11. Suppose $\hat{\theta}_1, \hat{\theta}_2$ and $\hat{\theta}_3$ are estimators of θ, and we know that $E(\hat{\theta}_1) = E(\hat{\theta}_2) = \theta$, $E(\hat{\theta}_3) \neq \theta$, $\text{var}(\hat{\theta}_1) = 12$, $\text{var}(\hat{\theta}_2) = 10$, and $E(\hat{\theta}_3 - \theta)^2 = 6$. Which estimator is the best in terms of the MSE criterion?

8.12. Suppose $\mathbf{X}^n$ is an IID random sample from some population with unknown mean μ and variance σ^2. Define parameter $\theta = (\mu - 2)^2$.

(1) Suppose $\hat{\theta} = (\bar{X}_n - 2)^2$ is an estimator for θ, where $\bar{X}_n$ is the sample mean. Show that $\hat{\theta}$ is not unbiased for θ. [Hint: $\bar{X}_n - 2 = \bar{X}_n - \mu + \mu - 2$.]

(2) Find an unbiased estimator for θ.

8.13. Let $\mathbf{X}^n$ be an IID $U[0, \theta]$ random sample, where θ is unknown. Define two estimators of θ:

$$\hat{\theta}_1 = \frac{n+1}{n} \max_{1 \leq i \leq n} X_i,$$

$$\hat{\theta}_2 = \frac{2}{n} \sum_{i=1}^{n} X_i.$$

(1) Show $P(\max_{1 \leq i \leq n} X_i \leq t) = [F_X(t)]^n$, where $F_X(\cdot)$ is the CDF of the population distribution $U[0, \theta]$.

(2) Compute $E_\theta(\hat{\theta}_1)$ and $\text{var}_\theta(\hat{\theta}_1)$.

(3) Show that $\hat{\theta}_1$ converges to θ in probability.

(4) Compute $E_\theta(\hat{\theta}_2)$ and $\text{var}_\theta(\hat{\theta}_2)$.

(5) Show that $\hat{\theta}_2$ converges to θ in probability;

(6) Which estimator, $\hat{\theta}_1$ or $\hat{\theta}_2$, is more efficient? Explain.

8.14. An IID random sample $\mathbf{X}^n$ is taken from a population with mean μ and variance σ^2. Consider the following estimator of μ:

$$\hat{\mu} = \frac{2}{n(n+1)} \sum_{i=1}^{n} i \cdot X_i.$$

(1) Show that $\hat{\mu}$ is unbiased for μ.

(2) Which estimator, $\hat{\mu}$ or $\bar{X}_n$, is more efficient? Explain. [Hint: $\sum_{i=1}^{n} i = \frac{n(n+1)}{2}$ and $\sum_{i=1}^{n} i^2 = \frac{n(n+1)(2n+1)}{6}$.]

8.15. Suppose $X_1, \cdots, X_n$ are independent random variables with $E(X_i) = \alpha i$ and $\text{var}(X_i) = \sigma^2$ for $i = 1, \cdots, n$. Consider the following class of estimators for α

$$\hat{\alpha} = \sum_{i=1}^{n} c_i X_i,$$

where the c_i are some constants.

(1) Show that $\hat{\alpha}$ is unbiased for μ if and only if $\sum_{i=1}^{n} i c_i = 1$.

(2) Determine the values of the c_i to obtain the best unbiased estimator of α.

8.16. Suppose $\mathbf{X}^n$ is an IID $N(0, \sigma^2)$ random sample. Define

$$S_n^2 = (n-1)^{-1} \sum_{i=1}^{n} (X_i - \bar{X}_n)^2,$$

where $\bar{X}_n = n^{-1} \sum_{i=1}^{n} X_i$, and

$$\hat{\sigma}^2 = n^{-1} \sum_{i=1}^{n} X_i^2.$$

Which estimator is more efficient? Give your reasoning.

8.17. Suppose we have an IID random sample $\mathbf{X}^{2n} = (X_1, \cdots, X_n, X_{n+1}, \cdots, X_{2n})$ from a $N(\mu, \sigma^2)$ population. Let S_1^2, S_2^2 and S^2 be the sample variances based on the first half sample $(X_1, \cdots, X_n)$, the second half sample $(X_{n+1}, \cdots, X_{2n})$, and the whole sample $\mathbf{X}^{2n}$, respectively.

(1) Compare the relative efficiency among S_1^2, S_2^2 and S^2.

(2) Define $\bar{S}^2 = \frac{1}{2}(S_1^2 + S_2^2)$. Which estimator, $\bar{S}^2$ or S^2, is more efficient?

8.18. Suppose that S_1^2, S_2^2 and S_3^2 are the sample variances based on three IID random samples $\{X_1, \cdots, X_{n_1}\}$, $\{Y_1, \cdots, Y_{n_2}\}$ and $\{Z_1, \cdots, Z_{n_3}\}$ respectively, where these three random samples are mutually independent and come from the same $N(\mu, \sigma^2)$ distribution. The sample sizes n_1, n_2, n_3 are given and may not be necessarily the same.

Define a class of estimators for σ^2 as

$$S^2 = c_1 S_1^2 + c_1 S_2^2 + c_3 S_3^2,$$

where c_1, c_2, c_3 are some constants to be chosen.

(1) Under what conditions will S^2 be an unbiased estimator for σ^2? Explain.

(2) Find the best unbiased estimator for σ^2 within the class of estimators S^2. Show your reasoning in detail.

8.19. Let $X_1, \cdots, X_n$ be an IID random sample from the following distribution:

$$P(X = -1) = \frac{1 - \theta}{2}, \quad P(X = 0) = \frac{1}{2}, \quad P(X = 1) = \frac{\theta}{2}.$$

(1) Find the MLE of θ and check whether it is unbiased estimator.

(2) Find the method of moments estimator of θ.

(3) Calculate the Cramer-Rao lower bound for the variance of an unbiased estimator of θ.

8.20. Let $X_1, \cdots, X_n$ be an IID random sample from the population with PMF

$$f(x, \theta) = \begin{cases} \theta, & \text{if } x = 1, \\ 1 - \theta, & \text{if } x = 0, \end{cases}$$

where $0 < \theta < 1$.

(1) Find the MLE $\hat{\theta}$ of θ.

(2) Is $\hat{\theta}$ the best unbiased estimator of θ?

8.21. A random variable X with PDF

$$f(x, \mu, \sigma^2) = \frac{1}{\sqrt{2\pi}\sigma x} e^{-\frac{(\ln x - \mu)^2}{2\sigma^2}}, \quad 0 < x < \infty,$$

is called a Lognormal (μ, σ^2) random variable because its logarithm $\ln X$ is $N(\mu, \sigma^2)$.

Suppose $\{X_1, \cdots, X_n\}$ is an IID random sample from a Lognormal $(0, \sigma^2)$ population. There are two estimators for σ^2: $\hat{\sigma}^2 = n^{-1} \sum_{i=1}^n (\ln X_i)^2$, and $S^2 = (n-1)^{-1} \sum_{i=1}^n (\ln X_i - \hat{\mu})^2$, where $\hat{\mu} = n^{-1} \sum_{i=1}^n \ln X_i$.

Which estimator is more efficient in terms of the MSE criterion? Give your reasoning clearly.

8.22. Put $\theta = (\mu, \sigma^2)$. A random variable X with PDF

$$f(x, \theta) = \frac{1}{\sqrt{2\pi}\sigma x} e^{-\frac{(\ln x - \mu)^2}{2\sigma^2}}, \ 0 < x < \infty,$$

is called a Lognormal(μ, σ^2) random variable because its logarithm $\ln X$ is $N(\mu, \sigma^2)$, namely,

$$\ln X \sim N(\mu, \sigma^2).$$

Suppose $(X_1, \cdots, X_n)$ is an IID random sample from a Lognormal (μ, σ^2) population.

(1) Find the MLE for (μ, σ^2).

(2) Denote the MLE estimator for μ by $\hat{\mu}$. Is $\hat{\mu}$ the best unbiased estimator of μ?

8.23. Suppose $X^n = (X_1, \cdots, X_n)$ is an IID random sample from a Poisson(α) distribution with probability mass function

$$f_X(x) = e^{-\alpha} \frac{\alpha^x}{x!} \text{ for } x = 0, 1, 2, \cdots,$$

where α is unknown.

(1) Find a sufficient statistic for α.

(2) Find the MLE for α.

(3) Is the MLE for α the best unbiased estimator for α? Give your reasoning.

8.24. Suppose $(X_1, \cdots, X_n)$ is an independent and identically distributed random sample from a Weibull distribution whose probability density function is given by

$$f_X(x) = \begin{cases} \frac{\alpha}{\beta} x^{\alpha-1} \exp(-x^\alpha/\beta), & \text{if } x > 0, \\ 0, & \text{otherwise,} \end{cases}$$

where $\alpha > 0, \beta > 0$. Suppose α is known, so β is the only unknown parameter.

(1) Find a sufficient statistic for β.

(2) Find the MLE for β.

(3) Is the MLE obtained in Part (b) an unbiased estimator for β?

(4) Does the MLE obtained in Part (b) achieve the Cramer-Rao lower bound?

For each part, give your reasoning clearly.

8.25. Let $(X_1, \cdots, X_{n+1})$ be an IID Bernoulli(p) random sample, and define the function $h(p)$ by

$$h(p) = P\left(\sum_{i=1}^{n} X_i > X_{n+1}\right),$$

i.e., the probability that the sum of the first n observations exceed the $(n+1)th$.

(1) Show that

$$T(X_1, \cdots, X_{n+1}) = \begin{cases} 1, & \text{if } \sum_{i=1}^{n} X_i > X_{n+1}, \\ 0, & \text{otherwise,} \end{cases}$$

is an unbiased estimator of $h(p)$.

(2) Find the best unbiased estimator of $h(p)$.

8.26. Let X be an observation from the PDF

$$f(x|\theta) = \left(\frac{\theta}{2}\right)^{|x|} (1-\theta)^{1-|x|}, \quad x = -1, 0, 1; \quad 0 \le \theta \le 1.$$

(1) Find the MLE of θ.

(2) Define the estimator $T(X)$ by

$$T(X) = \begin{cases} 2, & \text{if } X = 1, \\ 0, & \text{otherwise.} \end{cases}$$

Show that $T(X)$ is an unbiased estimator of θ.

(3) Find a better estimator than $T(X)$ and prove that it is better.

8.27. Suppose $\mathbf{X}^n = (X_1, X_2, \ldots, X_n)$ is an observed random sample of size n. Consider the model

$$X_i = \theta X_{i-1} + \varepsilon_i, \quad i = 1, \cdots, n,$$

where $\{\varepsilon_i\}_{i=1}^{n} \sim$ IID $N(0, \sigma_\varepsilon^2)$, $X_0 \sim f_0(x)$, and θ is an unknown scalar parameter. The density function $f_0(x)$ of X_0 is known.

In statistics, the so-called Bayesian school of statistics develops an important method to estimate the unknown parameter θ. The first step is to assume that parameter θ is random and follows a prior distribution. Suppose the prior distribution of θ is a $N(0, \sigma_\theta^2)$ distribution, and σ_ε^2 and σ_θ^2 are known constants.

(1) Derive the joint probability density $f(\theta, \mathbf{x}^n)$ of random vector $(\theta, \mathbf{X}^n)$, where $\mathbf{x}^n = (x_1, x_2, \ldots, x_n)$.

(2) Derive the conditional probability density $f(\theta|\mathbf{x}^n)$ of θ given the sample $\mathbf{X}^n$. This is called the posterior probability density.

(3) The Bayesian estimator $\hat{\theta} = \hat{\theta}_n(\mathbf{x}^n)$ minimizes the following average mean squared error

$$\hat{\theta} = \arg\min_a \int (a - \theta)^2 f(\theta, \mathbf{x}^n)d\theta.$$

Find the Bayesian estimator of θ.

In each step, state your reasoning clearly.

Chapter 9

Hypothesis Testing

Abstract: Hypothesis testing is one of the two most important objectives in statistical inference. In this chapter, we will introduce basic concepts in hypothesis testing, and discuss the three fundamental principles of hypothesis testing — the Wald test, the Lagrange multiplier test, and the likelihood ratio test.

Key words: Hypothesis testing, Lagrange multiplier test, Level, Likelihood ratio test, Neyman-Pearson lemma, Power, Type I error, Type II error, Size, Uniformly most powerful test, Wald test.

9.1 Introduction to Hypothesis Testing

In Chapter 8, we introduced some important methods to estimate unknown parameters in a population distribution model $f(x, \theta)$. Beside that, given a realization $\mathbf{x}^n$ of a random sample $\mathbf{X}^n$ from the population distribution $f(x, \theta)$, we would also like to know whether the parameter value θ_0 belongs to some specific subset Θ_0 of the parameter space Θ.

Example 9.1. [Return to Education]: Let θ measure the change in hourly wage given another year of education, holding all other factors fixed. Labor economists are interested in testing whether the return to education, controlling other factors, is zero. That is, whether or not θ is equal to zero.

Definition 9.1. [Hypothesis]: A hypothesis is a statement about the population or some attributes of the population distribution. The two complementary hypotheses in a hypothesis testing problem are called the null hypothesis and the alternative hypothesis, denoted by $\mathbb{H}_0$ and $\mathbb{H}_A$ respectively.

A null hypothesis is a statement about the population or some attributes of the population. The alternative hypothesis is the statement that the null hypothesis is false. The goal of hypothesis testing is to decide, based on an observed data set $\mathbf{x}^n$ generated from a population, which of two complementary hypotheses is true. Suppose a random sample $\mathbf{X}^n$ is generated from a population distribution $f(x, \theta)$ with some unknown value of parameter $\theta \in \Theta$, where Θ is a finite-dimensional parameter space. In hypothesis testing, the parameter space Θ is divided into two mutually exclusive and collectively exhaustive subsets Θ_0 and Θ_A, namely $\Theta_0 \cap \Theta_A = \varnothing$ and $\Theta_0 \cup \Theta_A = \Theta$. The problem is to determine to which of these two subsets the true value of θ belongs. That is, based upon an observed data set $\mathbf{x}^n$, one is trying to choose between the two hypotheses

$$\mathbb{H}_0 : \theta \in \Theta_0$$

versus

$$\mathbb{H}_A : \theta \in \Theta_A.$$

The first hypothesis $\mathbb{H}_0$ is called the null hypothesis, and the second, $\mathbb{H}_A$, is called the alternative hypothesis. They have labels because, as we shall see, they are treated asymmetrically in the theory of hypothesis testing. In particular, $\mathbb{H}_0$ is called the "null" hypothesis because it is often stated as "no effects" or "no relationship". One example, is $\mathbb{H}_0 : \theta = 0$ versus $\mathbb{H}_A : \theta \neq 0$, as is the case of Example 9.1. Nowadays the term "null hypothesis" is applied to any hypothesis that one may like to test.

We now consider some other examples.

Example 9.2. If θ denotes the proportion of defective items for some manufactured product. We might be interested in testing $\mathbb{H}_0 : \theta \leq \theta_0$ versus $\mathbb{H}_A : \theta > \theta_0$, where θ_0 is the maximum acceptable proportion of defective items. The null hypothesis states that the proportion of defective items is below an unacceptable threshold. This hypothesis testing is the basic idea for statistical quality control.

Example 9.3. [Constant Return to Scale Hypothesis]: A production function

$$Y = F(L, K)$$

tells how much output Y to produce using inputs of labor L and capital K. A production technology is said to display a constant return to scale if the output increases by the same proportion as the inputs increase; that is, for all $\lambda > 0$,

$$\lambda F(L, K) = F(\lambda L, \lambda K).$$

Suppose a production function is given by the Cobb-Douglas function

$$Y = AK^{\alpha}L^{\beta},$$

where Y is the output, K and L are the capital and labor inputs, A is a constant, and $\theta = (\alpha, \beta)$ is a parameter vector. Then the constant return to scale hypothesis can be stated as

$$\mathbb{H}_0 : \alpha + \beta = 1.$$

The alternative hypothesis $\mathbb{H}_0 : \alpha + \beta \neq 1$ consists of two cases: $\alpha + \beta > 1$ and $\alpha + \beta < 1$, which imply an increasing return to scale and a decreasing return to scale respectively.

The hypotheses can be divided into two basic categories — simple hypotheses and composite hypotheses.

Definition 9.2. [Simple Hypothesis Versus Composite Hypothesis]: A hypothesis is *simple* if and only if it contains exactly one population. If the hypothesis contains more than one population, it is called a *composite hypothesis*.

For a parametric distribution model $f(x, \theta)$, one parameter value of θ yields a population distribution, and different parameter values of θ yield different population distributions. In Example 9.1, the null hypothesis $\mathbb{H}_0$ contains only one parameter value, so $\mathbb{H}_0$ is a simple hypothesis. In contrast, the null hypotheses in Examples 9.2 and 9.3 contain more than one parameter values, so they are composite hypotheses.

Definition 9.3. [Hypothesis Testing]: A hypothesis testing procedure or a hypothesis test is a decision rule that specifies:

- for what sample values $\mathbf{x}^n$ the decision is made to accept $\mathbb{H}_0$ as true, and
- for what sample values $\mathbf{x}^n$ $\mathbb{H}_0$ is rejected and $\mathbb{H}_A$ is accepted as true.

The key to the construction of the decision rule for a hypothesis testing procedure is to determine the rules of rejection or acceptance of the null hypothesis.

Definition 9.4. [Critical Region or Rejection Region]: The set $\mathbb{C}$ of the sample points of the random sample $\mathbf{X}^n$ for which $\mathbb{H}_0$ will be rejected is called the rejection region or critical region. The complement of the rejection region is called the acceptance region.

A standard approach to hypothesis testing is to choose a statistic $T(\mathbf{X}^n)$ and use it to divide the sample space of $\mathbf{X}^n$ into two mutually exclusive and exhaustive regions

$$\mathbb{A}_n(c) = \{\mathbf{x}^n : T(\mathbf{x}^n) \leq c\}$$

and

$$\mathbb{C}_n(c) = \{\mathbf{x}^n : T(\mathbf{x}^n) > c\}$$

for some prespecified constant c. The first region, $\mathbb{A}_n(c)$, is the acceptance region, and the second, $\mathbb{C}_n(c)$, is the rejection region. The cutoff point c is called the critical value and $T(\mathbf{X}^n)$ is called a test statistic. An important issue in hypothesis testing is to determine a suitable value of c given a data set $\mathbf{x}^n$. In general, we need to know the sampling distribution of $T(\mathbf{X}^n)$ under $\mathbb{H}_0$ in order to determine the threshold value c. We now consider a simple example.

Example 9.4. Suppose $\mathbf{X}^n$ is an IID $N(\mu, \sigma^2)$ random sample, where μ is unknown but σ^2 is known. We are interested in testing for $\mathbb{H}_0 : \mu = \mu_0$ versus $\mathbb{H}_A : \mu \neq \mu_0$, where μ_0 is a known number. Here, $\Theta_0 = \{\mu_0\}$ contains one parameter value μ_0 and Θ_A contains all parameter values on the real line $\mathbb{R}$ except μ_0.

To test the null hypothesis $\mathbb{H}_0 : \mu = \mu_0$, we consider the following test statistic

$$T(\mathbf{X}^n) = \frac{\bar{X}_n - \mu_0}{\sigma/\sqrt{n}},$$

where σ is a known number by assumption. Under the null hypothesis $\mathbb{H}_0 : \mu = \mu_0$, we have

$$T(\mathbf{X}^n) = \frac{\bar{X}_n - \mu_0}{\sigma/\sqrt{n}} \sim N(0,1).$$

Under $\mathbb{H}_A : \mu \neq \mu_0$,

$$T(\mathbf{X}^n) \sim N\left(\frac{\sqrt{n}(\mu - \mu_0)}{\sigma}, 1\right),$$

which diverges to infinity with probability approaching one as $n \to \infty$. Thus, one can accept $\mathbb{H}_A$ if $|T(\mathbf{X}^n)|$ is large, and reject $\mathbb{H}_A$ if $|T(\mathbf{X}^n)|$ is small. How large $T(\mathbf{X}^n)$ should be in order to be considered as "large" is determined by the sampling distribution ($N(0,1)$) of $T(\mathbf{X}^n)$ under $\mathbb{H}_0$. Specifically, by setting $c = z_{\alpha/2}$, where $z_{\alpha/2}$ is the upper-tailed critical value of $N(0,1)$ at level $\frac{\alpha}{2} \in (0,1)$, i.e., $P(Z > z_{\alpha/2}) = \frac{\alpha}{2}$ where $Z \sim N(0,1)$, we can define the acceptance and rejection regions as follows:

$$\mathbb{A}_n(c) = \left\{ \mathbf{x}^n : \left| \frac{\bar{X}_n - \mu_0}{\sigma/\sqrt{n}} \right| \le z_{\frac{\alpha}{2}} \right\},$$

$$\mathbb{C}_n(c) = \left\{ \mathbf{x}^n : \left| \frac{\bar{X}_n - \mu_0}{\sigma/\sqrt{n}} \right| > z_{\frac{\alpha}{2}} \right\}.$$

Then we obtain the following decision rule:

- Accept $\mathbb{H}_0 : \mu = \mu_0$ at the significance level α if

$$\mathbf{x}^n \in \mathbb{A}_n(c).$$

- Reject $\mathbb{H}_0 : \mu = \mu_0$ at the significance level α if

$$\mathbf{x}^n \in \mathbb{C}_n(c).$$

Figure 9.1 plots the rejection and acceptance regions at the 5% significance level in terms of the value of the normal test statistic $T(\mathbf{X}^n)$.

As we will explain below, the significance level α is the probability that the above decision rule will wrongly reject a correct null hypothesis. This error is called Type I error and is unavoidable given any finite sample size n. This is because that the test statistic $T(\mathbf{X}^n)$ follows a $N(0,1)$ distribution under $\mathbb{H}_0$, and so there is still a small probability of taking a large value.

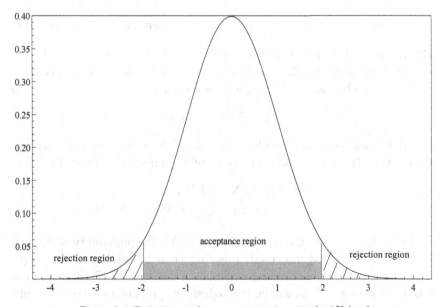

Figure 9.1: Rejection and acceptance regions at the 5% level

When the random sample $\mathbf{X}^n$ is not normally distributed or the test statistic $T(\mathbf{X}^n)$ is a nonlinear function of $\mathbf{X}^n$, the sampling distribution of $T(\mathbf{X}^n)$ is generally unknown or tedious to obtain for a finite sample size n. In this chapter, we will use the asymptotic theory to obtain the asymptotic critical value for c. This provides an approximate but convenient testing decision rule in practice. There exist other methods to choose c. One example is the bootstrap procedure, which can approximate the finite sample distribution of $T(\mathbf{X}^n)$ accurately although it may be computationally intensive (see Hall 1992).

We now introduce the power function of a test procedure.

Definition 9.5. [**Power of Test**]: If $\mathbb{C}$ is the rejection region of a test of the null hypothesis $\mathbb{H}_0 : \theta \in \Theta_0$, then the function $\pi(\theta) = P_\theta(\mathbf{X}^n \in \mathbb{C})$ is called the power of the test with the rejection region $\mathbb{C}$, where $P_\theta(\cdot)$ is the probability measure when the random sample follows the distribution $f_{\mathbf{X}^n}(\mathbf{x}^n, \theta)$.

The power function $\pi(\theta)$ is the probability of rejecting $\mathbb{H}_0$. In Example 9.4, the power of the test statistic $T(\mathbf{X}^n) = \sqrt{n}(\bar{X}_n - \mu_0)/\sigma$ is given by

$$\pi(\mu) = P\left(\left|\frac{\bar{X}_n - \mu_0}{\sigma/\sqrt{n}}\right| > z_{\alpha/2}\right),$$

which is plotted in Figure 9.2 below for $\alpha = 0.05$ and $\alpha = 0.10$ respectively.

Definition 9.6. [**Type I and Type II Errors**]: If $\mathbb{H}_0 : \theta \in \Theta_0$ holds and the observed data $\mathbf{x}^n$ falls into the critical region $\mathbb{C}$, then a Type I error is made. The probability of making a Type I error is

$$\alpha(\theta) \equiv P_\theta(\mathbf{X}^n \in \mathbb{C}|\mathbb{H}_0).$$

If $\mathbb{H}_A : \theta \in \Theta_0^c$ holds and the observed data $\mathbf{x}^n$ is in the acceptance region, then a Type II error is made. The probability of making a Type II error is

$$\beta(\theta) \equiv P_\theta(\mathbf{X}^n \in \mathbb{A}|\mathbb{H}_A)$$
$$= 1 - P_\theta(\mathbf{X}^n \in \mathbb{C}|\mathbb{H}_A).$$

Under $\mathbb{H}_0 : \theta \in \Theta_0$, the power function $\pi(\theta)$ is the probability of making a Type I error, namely incorrectly rejecting a correct null hypothesis. Type I errors are unavoidable because under $\mathbb{H}_0$, a test statistic $T(\mathbf{X}^n)$ may still take large values with nontrivial (though small) probabilities. One example is the test statistic $T(\mathbf{X}^n) = \sqrt{n}(\bar{X}_n - \mu_0)/\sigma$ in Example 9.4, which follows a $N(0,1)$ distribution under $\mathbb{H}_0$ and thus may still take a large value with a small but nontrivial probability. Under $\mathbb{H}_A : \theta \in \Theta_0^c$, the power function $\pi(\theta)$ is the probability of rejecting an false null hypothesis, and it is equal to $1 - \beta(\theta)$, where $\beta(\theta)$ is the probability of making a Type II error, namely accepting a false null hypothesis. Type II errors may occur for various reasons. For example, the sample size n may be small so that a test statistic $T(\mathbf{X}^n)$ may still have a nontrivial probability of taking small values. A test is called unbiased if $P[T(\mathbf{X}^n) > c|\mathbb{H}_A] > P[T(\mathbf{X}^n) > c|\mathbb{H}_0]$. That is, the probability to reject the null hypothesis $\mathbb{H}_0$ when $\mathbb{H}_0$ is false is strictly larger than when it is true.

Ideally, one would like to have a test with power function $\pi(\theta)$ being 0 for all $\theta \in \Theta_0$ and 1 for all $\theta \in \Theta_A$. However, for any test procedure, there exists a trade-off between Type I errors and Type II errors given any sample size n. For any given n, if the critical region $\mathbb{C}$ shrinks, the probability of making a Type I error decreases, but the probability of making a Type II error increases. Similarly, if the critical region $\mathbb{C}$ increases, the Type II error $\beta(\theta)$ decreases, but the Type I error $\alpha(\theta)$ increases. The only way to reduce both types of errors is to increase the sample size n and to find a most powerful test.

Usually, hypothesis tests are evaluated and compared through their probabilities of making mistakes. In many cases, since the probabilities of making these two type of errors are inversely related, the classical approach to hypothesis testing is to bound the probability of a Type I error by some value $\alpha \in (0,1)$ over all values of θ in Θ_0 and to try to find a test that minimizes the probability of a Type II error over all values of θ in Θ_A. Specifically, one is trying to find a test statistic $T(\mathbf{X}^n)$ that satisfies $P[T(\mathbf{X}^n) > c|\mathbb{H}_0] \le \alpha$ for all $\theta \in \Theta_0$ and has the property that $P[T(\mathbf{X}^n) > c|\mathbb{H}_A] \ge P[G(\mathbf{X}^n) > c|\mathbb{H}_A]$ for any other test statistic $G(\mathbf{X}^n)$ for all $\theta \in \Theta_A$. In other words, one is trying to find a test statistic $T(\mathbf{X}^n)$ that has the best power while its Type I error is under control. A test $T(\mathbf{X}^n)$ that has these properties is called uniformly most powerful, where by uniformity it is meant that the test is most powerful for all $\theta \in \Theta_A$.

We now provide a formal definition.

Definition 9.7. [Uniformly Most Powerful Test]: Let $\mathbb{T}$ be a class of tests for testing $\mathbb{H}_0 : \theta \in \Theta_0$ versus $\mathbb{H}_A : \theta \in \Theta_A$. A test $T(\mathbf{X}^n)$ in class $\mathbb{T}$, with power function $\pi(\theta)$, is a uniformly most powerful test over $\mathbb{T}$ if $\pi(\theta) \ge \tilde{\pi}(\theta)$ for all $\theta \in \Theta_A$, where $\tilde{\pi}(\theta)$ is the power function of any other test $G(\mathbf{X}^n)$ in class $\mathbb{T}$.

Suppose for a test statistic $T(\mathbf{X}^n)$, $P[T(\mathbf{X}^n) > c|\mathbb{H}_0] \leq \alpha$. Then the value of α gives the maximum Type I error for the test statistic $T(\mathbf{X}^n)$, and is called the level of the test. If $T(\mathbf{X}^n)$ has level α and $P[T(\mathbf{X}^n) > c|\mathbb{H}_0] = \alpha$, then the test is called a size α test. Obviously, the class of level α tests contains the set of size α tests.

A size α test gives a precise Type I error for the test. Usually, we will choose $\mathbb{T}$ to be a class of tests with the same level or same size α. In some complicated testing situations, however, it may be computationally impossible to construct a size α test. In such situations, a researcher must be satisfied with a level α test, realizing that some compromises may be made. An example is the so-called Bonferroni's method, which is used to address the problem of multiple comparisons when we have k individual tests for the null hypothesis $\mathbb{H}_0$. It can be described as follows: suppose for each $i \in \{1, \cdots, k\}$, $T_i(\mathbf{X}^n)$ is a level α/k test for $\mathbb{H}_0$, namely

$$P\left[T_i(\mathbf{X}^n) > c_i|\mathbb{H}_0\right] \leq \frac{\alpha}{k}, \qquad i = 1, \cdots, k,$$

where c_i is a critical value for $T_i(\mathbf{X}^n)$. Then the maximum test statistic

$$T(\mathbf{X}^n) = \max_{1 \leq i \leq k} T_i(\mathbf{X}^n)$$

is a level α test, because

$$P[T(\mathbf{X}^n) > c|\mathbb{H}_0] = P\left[\max_{1 \leq i \leq k} T_i(\mathbf{X}^n) > c|\mathbb{H}_0\right]$$

$$= P\left[\bigcup_{i=1}^{k} \{T_i(\mathbf{X}^n) > c\}\,|\mathbb{H}_0\right]$$

$$\leq \sum_{i=1}^{k} P\left[T_i(\mathbf{X}^n) > c|\mathbb{H}_0\right] = \alpha,$$

where the inequality follows by Boole's inequality in Theorem 2.8, Chapter 2. For applications of this method in econometrics, see Lee, White and Granger (1993), Hong and White (1995), or Campbell and Yogo (2006).

In practice, a researcher commonly specifies the level of the test he wishes to use, with typical choices being $\alpha = 0.01$, 0.05 and 0.10. In fixing the level of the test, the researcher is controlling only the Type I error probability.

All hypotheses we discuss here are statistical hypotheses, i.e., hypotheses or restrictions on unknown parameters in a population distribution model.

In econometric and economic analysis, we are primarily interested in economic hypotheses. To test a economic hypothesis, we have to transform it into a statistical hypothesis, and then test the statistical hypothesis using an observed economic data. In transforming an economic hypothesis into a statistical hypothesis, some auxiliary assumptions are often imposed. This induces a gap between the original economic hypothesis and the resulting statistical hypothesis. This gap may cause some problem in economic interpretation of the empirical results on testing the statistical hypothesis. We illustrate this point by considering the following example on testing the efficient market hypothesis.

Example 9.5. [Efficient Market Hypothesis (EMH)]: Suppose R_t is the return on some asset or portfolio in time period t, and $I_{t-1} = (R_{t-1}, R_{t-2}, \cdots)$ denotes the historical asset return information available at time $t - 1$. The asset market is called informationally weakly efficient if

$$E(R_t | I_{t-1}) = E(R_t).$$

That is, the historical asset return information has no predictive power for future asset return.

To test this economic hypothesis, one can consider a linear autoregressive model

$$R_t = \alpha_0 + \sum_{j=1}^{k} \alpha_j R_{t-j} + \varepsilon_t,$$

where ε_t is a stochastic disturbance. Under EMH, we have

$$\mathbb{H}_0 : \alpha_1 = \alpha_2 = \cdots = \alpha_k = 0.$$

If one has evidence that at least one α_j, $j \in \{1, \cdots, k\}$, is not zero, it will imply that the efficient market hypothesis must be rejected.

However, when one does not reject the statistical hypothesis $\mathbb{H}_0$, this does not necessarily imply that the original economic hypothesis — EMH holds. The reason is that the linear autoregressive model is just one of many (possibly infinite) ways to test predictability of the historical asset returns for future asset returns. In other words, the predictability may arise in a nonlinear manner. Therefore, there exists a gap between the efficient market hypothesis and the statistical hypothesis $\mathbb{H}_0$. Because of this gap, when one does not reject $\mathbb{H}_0$, one can only say that no evidence against EMH is found rather than concludes that EMH holds.

9.2   Neyman-Pearson Lemma

We now provide a characterization for a class of uniformly most powerful tests when $\mathbb{H}_0$ and $\mathbb{H}_A$ are both simple hypothesis. This is the well-known Neyman-Pearson lemma. The basic idea of this lemma is that a likelihood ratio-based test is uniformly most powerful.

Theorem 9.1. *[Neyman-Pearson Lemma]: Consider testing a simple null hypothesis $\mathbb{H}_0 : \theta = \theta_0$ versus a simple alternative hypothesis $\mathbb{H}_A : \theta = \theta_1$, where the PMF/PDF of the random sample $\mathbf{X}^n$ corresponding to $\theta_i, i \in \{0,1\}$, is $f_{\mathbf{X}^n}(\mathbf{x}^n, \theta_i)$. Suppose a test with rejection and acceptance regions $\mathbb{C}_n(c)$ and $\mathbb{A}_n(c)$ respectively is defined as follows:*

(a)

$$\mathbb{C}_n(c) = \left\{ \mathbf{x}^n : \frac{f_{\mathbf{X}^n}(\mathbf{x}^n, \theta_1)}{f_{\mathbf{X}^n}(\mathbf{x}^n, \theta_0)} > c \right\}$$

and

$$\mathbb{A}_n(c) = \left\{ \mathbf{x}^n : \frac{f_{\mathbf{X}^n}(\mathbf{x}^n, \theta_1)}{f_{\mathbf{X}^n}(\mathbf{x}^n, \theta_0)} \leq c \right\}$$

for some constant $c \geq 0$, and

(b)

$$P[\mathbf{X}^n \in \mathbb{C}_n(c)|\mathbb{H}_0] = \alpha.$$

Then

(1) [Sufficiency] Any test that satisfies Conditions (a) and (b) is a uniformly most powerful level α test.

(2) [Necessity] If there exists a test satisfying Conditions (a) and (b) with $c > 0$, then every uniformly most powerful level α test is a size α test (i.e., satisfying Condition (b)), and every uniformly most powerful level α test satisfies Condition (a) except perhaps on a set A in the sample space of $\mathbf{X}^n$ satisfying $P(\mathbf{X}^n \in A|\mathbb{H}_0) = P(\mathbf{X}^n \in A|\mathbb{H}_A) = 0$.

Proof: Notice that

$$P[\mathbf{X}^n \in \mathbb{C}_n(c)|\mathbb{H}_0] = E\{\mathbf{1}[\mathbf{X}^n \in \mathbb{C}_n(c)]|\mathbb{H}_0\}$$
$$= \int \mathbf{1}[\mathbf{x}^n \in \mathbb{C}_n(c)]f(\mathbf{x}^n, \theta_0)dx^n,$$

where $\mathbf{1}(\cdot)$ is the indicator function which takes value 1 if the condition inside holds and takes value 0 otherwise.

(1) We first show that a test (denoted as $T(\mathbf{X}^n)$) that satisfies Conditions (a) and (b) is uniformly most powerful. Suppose there is any other test (denoted $T_1(\mathbf{X}^n)$) with $E\{\mathbf{1}[\mathbf{X}^n \in \mathbb{C}_{1n}(c)]|\mathbb{H}_0\} \leq \alpha$. (The test $T_1(\mathbf{X}^n)$ need not be a likelihood ratio test.) We shall show that $T_1(\mathbf{X}^n)$ is not more

powerful than $T(\mathbf{X}^n)$. Observe that if $\mathbf{1}[\mathbf{x}^n \in \mathbb{C}_n(c)] > \mathbf{1}[\mathbf{x}^n \in \mathbb{C}_{1n}]$, then the sample point $\mathbf{x}^n$ is in the critical region $\mathbb{C}_n(c)$ of the test $T(\mathbf{X}^n)$, and so $f_{\mathbf{X}^n}(\mathbf{x}^n, \theta_1) > cf_{\mathbf{X}^n}(\mathbf{x}^n, \theta_0)$; if $\mathbf{1}[\mathbf{x}^n \in \mathbb{C}_n(c)] < \mathbf{1}[\mathbf{x}^n \in \mathbb{C}_{1n}]$, then the sample point $\mathbf{x}^n$ is in the acceptance region $\mathbb{A}_n(c)$ of the test $T(\mathbf{X}^n)$, and so $f_{\mathbf{X}^n}(\mathbf{x}^n, \theta_1) \le cf_{\mathbf{X}^n}(\mathbf{x}^n, \theta_0)$. In either case, we have

$$\{\mathbf{1}[\mathbf{x}^n \in \mathbb{C}_n(c)] - \mathbf{1}[\mathbf{x}^n \in \mathbb{C}_{1n}]\}\,[f_{\mathbf{X}^n}(\mathbf{x}^n, \theta_1) - cf_{\mathbf{X}^n}(\mathbf{x}^n, \theta_0)] \ge 0.$$

Thus,

$$\int_{\mathbb{R}^n} \{\mathbf{1}[\mathbf{x}^n \in \mathbb{C}_n(c)] - \mathbf{1}[\mathbf{x}^n \in \mathbb{C}_{1n}]\}\,[f_{\mathbf{X}^n}(\mathbf{x}^n, \theta_1) - cf_{\mathbf{X}^n}(\mathbf{x}^n, \theta_0)]\,d\mathbf{x}^n \ge 0.$$

This implies

$$\int_{\mathbb{R}^n} \{\mathbf{1}[\mathbf{x}^n \in \mathbb{C}_n(c)] - \mathbf{1}[\mathbf{x}^n \in \mathbb{C}_{1n}]\}\,f_{\mathbf{X}^n}(\mathbf{x}^n, \theta_1)d\mathbf{x}^n$$

$$\ge c\int_{\mathbb{R}^n} \{\mathbf{1}[\mathbf{x}^n \in \mathbb{C}_n(c)] - \mathbf{1}[\mathbf{x}^n \in \mathbb{C}_{1n}]\}\,f_{\mathbf{X}^n}(\mathbf{x}^n, \theta_0)d\mathbf{x}^n.$$

Because $\int_{\mathbb{R}^n} \mathbf{1}[\mathbf{x}^n \in \mathbb{C}_{1n}]f_{\mathbf{X}^n}(\mathbf{x}^n, \theta_0)d\mathbf{x}^n \le \alpha$, $\int_{\mathbb{R}^n} \mathbf{1}[\mathbf{x}^n \in \mathbb{C}_n(c)]f_{\mathbf{X}^n}(\mathbf{x}^n,$ $\theta_0)\mathbf{x}^n = \alpha$, and $c \ge 0$, we have

$$c\int_{\mathbb{R}^n} \{\mathbf{1}[\mathbf{x}^n \in \mathbb{C}_n(c)] - \mathbf{1}[\mathbf{x}^n \in \mathbb{C}_{1n}]\}\,f_{\mathbf{X}^n}(\mathbf{x}^n, \theta_0)d\mathbf{x}^n \ge 0.$$

It follows that

$$\int_{\mathbb{R}^n} \{\mathbf{1}[\mathbf{x}^n \in \mathbb{C}_n(c)] - \mathbf{1}[\mathbf{x}^n \in \mathbb{C}_{1n}]\}\,f_{\mathbf{X}^n}(\mathbf{x}^n, \theta_1)d\mathbf{x}^n \ge 0,$$

which implies

$$P[\mathbf{X}^n \in \mathbb{C}_n(c)|\mathbb{H}_A] \ge P[\mathbf{X}^n \in \mathbb{C}_{1n}|\mathbb{H}_A].$$

That is, the test $T_1(\mathbf{X}^n)$ is not more powerful than $T(\mathbf{X}^n)$.

(2) Assume that $T(\mathbf{X}^n)$ is a test that satisfies Conditions (a) and (b) with $c > 0$.

(i) We first show that a uniformly most powerful level α test (denoted as $T_2(\mathbf{X}^n)$) is a size α test (i.e., satisfies Condition (b)). Suppose it is not a size α test. Then $\int_{\mathbb{R}^n} \mathbf{1}[\mathbf{x}^n \in \mathbb{C}_{2n}]f_{\mathbf{X}^n}(\mathbf{x}^n, \theta_0)d\mathbf{x}^n < \alpha$. Given Condition (b) for the test $T(\mathbf{X}^n)$, we have $\int_{\mathbb{R}^n} \mathbf{1}[\mathbf{x}^n \in \mathbb{C}_n(c)]f_{\mathbf{X}^n}(\mathbf{x}^n, \theta_0)d\mathbf{x}^n = \alpha$. It follows that

$$\int_{\mathbb{R}^n} \{\mathbf{1}[\mathbf{x}^n \in \mathbb{C}_n(c)] - \mathbf{1}[\mathbf{x}^n \in \mathbb{C}_{2n}]\}\,f_{\mathbf{X}^n}(\mathbf{x}^n, \theta_0)d\mathbf{x}^n > 0.$$

Observe that, if $1[\mathbf{x}^n \in \mathbb{C}_n(c)] - 1[\mathbf{x}^n \in \mathbb{C}_{2n}] > 0$, then $\mathbf{x}^n$ is in the rejection region $\mathbb{C}_n(c)$ of the test $T(\mathbf{X}^n)$, and so $f_{\mathbf{X}^n}(\mathbf{x}^n, \theta_1) > cf_{\mathbf{X}^n}(\mathbf{x}^n, \theta_0)$; if $1[\mathbf{x}^n \in \mathbb{C}_n(c)] - 1[\mathbf{x}^n \in \mathbb{C}_{2n}] < 0$, then $\mathbf{x}^n$ is in the acceptance region $\mathbb{A}_n(c)$ of the test $T(\mathbf{X}^n)$, and so $f_{\mathbf{X}^n}(\mathbf{x}^n, \theta_1) \leq cf_{\mathbf{X}^n}(\mathbf{x}^n, \theta_0)$. It follows that

$$\int_{\mathbb{R}^n} \{1[\mathbf{x}^n \in \mathbb{C}_n(c)] - 1[\mathbf{x}^n \in \mathbb{C}_{2n}]\} [f_{\mathbf{X}^n}(\mathbf{x}^n, \theta_1) - cf_{\mathbf{X}^n}(\mathbf{x}^n, \theta_0)] d\mathbf{x}^n \geq 0.$$

Therefore,

$$\int_{\mathbb{R}^n} \{1[\mathbf{x}^n \in \mathbb{C}_n(c)] - 1[\mathbf{x}^n \in \mathbb{C}_{2n}]\} f_{\mathbf{X}^n}(\mathbf{x}^n, \theta_1) d\mathbf{x}^n$$

$$\geq c \int_{\mathbb{R}^n} \{1[\mathbf{x}^n \in \mathbb{C}_n(c)] - 1[\mathbf{x}^n \in \mathbb{C}_{2n}]\} f_{\mathbf{X}^n}(\mathbf{x}^n, \theta_0) d\mathbf{x}^n$$

$$> 0$$

given $c > 0$. This implies $P[\mathbf{X}^n \in \mathbb{C}_n(c)|\mathbb{H}_A] > P[\mathbf{X}^n \in \mathbb{C}_{2n}||\mathbb{H}_A]$, suggesting that the test $T_2(\mathbf{X}^n)$ is not uniformly most powerful, a contradiction.

(ii) Next, we show that every uniformly most powerful level α test must satisfy Conditions (a) and (b) except perhaps on a set A in the sample space of $\mathbf{X}^n$ satisfying $P(\mathbf{X}^n \in A|\mathbb{H}_0) = P(\mathbf{X}^n \in A|\mathbb{H}_A) = 0$. Suppose $T^*(\mathbf{X}^n)$ is a most powerful level α test with a rejection region $\mathbb{C}_n^*$. Because every most powerful test with level α is a size α test, we have

$$\int_{\mathbb{R}^n} 1[\mathbf{x}^n \in \mathbb{C}_n^*] f_{\mathbf{X}^n}(\mathbf{x}^n, \theta_0) d\mathbf{x}^n = \alpha = \int_{\mathbb{R}^n} 1[\mathbf{x}^n \in \mathbb{C}_n(c)] f_{\mathbf{X}^n}(\mathbf{x}^n, \theta_0) d\mathbf{x}^n.$$

Also, both $T(\mathbf{X}^n)$ and $T^*(\mathbf{X}^n)$ are most powerful tests, so they are equally powerful under $\mathbb{H}_A$, namely

$$\int_{\mathbb{R}^n} \{1[\mathbf{x}^n \in \mathbb{C}_n(c)] - 1[\mathbf{x}^n \in \mathbb{C}_n^*]\} f_{\mathbf{X}^n}(\mathbf{x}^n, \theta_1) d\mathbf{x}^n = 0.$$

Thus

$$\int_{\mathbb{R}^n} \{1[\mathbf{x}^n \in \mathbb{C}_n(c)] - 1[\mathbf{x}^n \in \mathbb{C}_n^*]\} [f_{\mathbf{X}^n}(\mathbf{x}^n, \theta_1) - cf_{\mathbf{X}^n}(\mathbf{x}^n, \theta_0)] d\mathbf{x}^n = 0.$$

Now, observe that if $1[\mathbf{x}^n \in \mathbb{C}_n(c)] - 1[\mathbf{x}^n \in \mathbb{C}_n^*] > 0$, then $\mathbf{x}^n$ must be in the rejection region $\mathbb{C}_n(c)$ of the test $T(\mathbf{X}^n)$, and so $f_{\mathbf{X}^n}(\mathbf{x}^n, \theta_1) - cf_{\mathbf{X}^n}(\mathbf{x}^n, \theta_0) > 0$; if $1[\mathbf{x}^n \in \mathbb{C}_n(c)] - 1[\mathbf{x}^n \in \mathbb{C}_n^*] < 0$, then $\mathbf{x}^n$ must be in the acceptance region of the test $T(\mathbf{X}^n)$, and so $f_{\mathbf{X}^n}(\mathbf{x}^n, \theta_1) - cf_{\mathbf{X}^n}(\mathbf{x}^n, \theta_0) \leq 0$. In both cases, the product $\{1[\mathbf{x}^n \in \mathbb{C}_n(c)] - 1[\mathbf{x}^n \in \mathbb{C}_n^*]\} [f_{\mathbf{X}^n}(\mathbf{x}^n, \theta_1) - cf_{\mathbf{X}^n}(\mathbf{x}^n, \theta_0)] \geq 0$. Given that the integral of this nonnegative product is 0, we have that $1[\mathbf{x}^n \in \mathbb{C}_n(c)] - 1[\mathbf{x}^n \in \mathbb{C}_n^*] = 0$ for all $\mathbf{x}^n$ in the sample space of $\mathbf{X}^n$ except for a set with zero probability. Therefore, the sets $\mathbb{C}_n^*$ and $\mathbb{C}_n(c)$ are identical except on the set of probability 0. The proof is completed.

The Neyman-Pearson lemma provides a method of finding a uniformly most powerful test when both the null and the alternative are simple hypotheses. That is, the likelihood ratio-based test is the uniformly most powerful test. However, when a hypothesis is composite, i.e., the hypothesis contains more than one parameter value, the lemma may not hold. See Hong and Lee (2013) for an example.

We note that for a given size $\alpha \in (0,1)$, the constant c is determined by the equation $P[\mathbf{X}^n \in \mathbb{C}_n(c)|\mathbb{H}_0] = \alpha$.

The following corollary links the Neyman-Pearson lemma to a sufficient statistic for parameter θ.

Corollary 9.1. [Likelihood Ratio Test and Sufficient Statistic]: *Suppose $T(\mathbf{X}^n)$ is a sufficient statistic for θ and $g(t, \theta_i)$ is the PMF/PDF of $T(\mathbf{X}^n)$ corresponding to $\theta_i, i \in \{0,1\}$. Then any test based on $T(\mathbf{X}^n)$ is a uniformly most powerful level α test for $\mathbb{H}_0 : \theta = \theta_0$ against $\mathbb{H}_A : \theta = \theta_1$ if the test has the rejection and acceptance regions*

$$\mathbb{C}_n(c) = \left\{ t : \frac{g(t, \theta_1)}{g(t, \theta_0)} > c \right\}$$

and

$$\mathbb{A}_n(c) = \left\{ t : \frac{g(t, \theta_1)}{g(t, \theta_0)} \leq c \right\}$$

for some $c \geq 0$, where $P[T(\mathbf{X}^n) \in \mathbb{C}_n(c)|\mathbb{H}_0] = \alpha$.

Proof: This is left as an exercise.

Thus, the likelihood ratio test based on the random sample X^n can be reduced to a likelihood ratio test based on the sufficient statistic $T(\mathbf{X}^n)$ of θ which remains to be a uniformly most powerful test.

Example 9.6. Suppose $\mathbf{X}^n$ is an IID random sample from an Exponential(θ) PDF $f(x, \theta) = \frac{1}{\theta} e^{-x/\theta}$ for $x \geq 0$. Find a uniformly most powerful level α test for $\mathbb{H}_0 : \theta = 1$ versus $\mathbb{H}_A : \theta = 2$.

Solution: The likelihood function of the random sample $\mathbf{X}^n$ is

$$f_{\mathbf{X}^n}(\mathbf{x}^n, \theta) = \prod_{i=1}^{n} f(x_i, \theta)$$

$$= \frac{1}{\theta^n} e^{-n\bar{x}_n/\theta},$$

where $\bar{x}_n = n^{-1}\sum_{i=1}^{n} x_i$. By the factorization theorem in Theorem 6.14, $\bar{X}_n$ is a sufficient statistic for θ. Since the sum of IID Exponential(θ) random variables follows a Gamma(n, θ) distribution (see Example 5.34), and so the sample mean $\bar{X}_n \sim$ Gamma($n, \frac{\theta}{n}$). Thus, its PDF

$$g(\bar{x}_n, \theta) = \frac{n^n}{(n-1)!\theta^n}\bar{x}_n^{n-1}e^{-n\bar{x}_n/\theta}, \text{ for } \bar{x}_n > 0.$$

It follows that the likelihood ratio

$$\frac{f_{\mathbf{X}^n}(\mathbf{x}^n, \theta_1)}{f_{\mathbf{X}^n}(\mathbf{x}^n, \theta_0)} = \frac{g(\bar{x}_n, \theta_1)}{g(\bar{x}_n, \theta_0)}$$

$$= \frac{\frac{1}{2^n}e^{-\frac{n}{2}\bar{x}_n}}{e^{-n\bar{x}_n}}$$

$$= \frac{1}{2^n}e^{\frac{n}{2}\bar{x}_n}.$$

Define a test with the following one-dimensional rejection region

$$\bar{x}_n \in \mathbb{C}_n(c) \text{ if } \frac{1}{2^n}e^{\frac{n}{2}\bar{x}_n} > c,$$

or equivalently,

$$\bar{x}_n \in \mathbb{C}_n(c) \text{ if } \bar{x}_n > 2\ln 2 + \frac{2\ln c}{n}.$$

Then we determine the value of c so that the test has size α. This requires

$$\alpha = \int_{2\ln 2 + 2n^{-1}\ln c}^{\infty} g(\bar{x}_n, \theta_0)d\bar{x}_n$$

$$= \int_{2\ln 2 + 2n^{-1}\ln c}^{\infty} \frac{n^n}{(n-1)!}\bar{x}_n^{n-1}e^{-n\bar{x}_n}d\bar{x}_n.$$

Solving this nonlinear equation, we can obtain $c = c(\alpha, n)$ as a function of α and n but it has no closed form solution. By the Neyman-Pearson lemma and Corollary 9.1, the above test is a uniformly most powerful size α test.

9.3 Wald Test

In the rest of this chapter, we will focus on the following hypotheses of interest

$$\mathbb{H}_0 : g(\theta) = \mathbf{0}$$

versus

$$\mathbb{H}_A : g(\theta) \neq \mathbf{0}$$

where $g : \mathbb{R}^p \rightarrow \mathbb{R}^q$ is a continuously differentiable q-dimensional vector-valued function of a p-dimensional parameter vector θ. The integer q is the number of restrictions on parameter vector θ. We assume $q \leq p$, that is, the number of restrictions is not larger than the number of parameters.

One important example is a linear vector-valued function

$$g(\theta) = R\theta - r,$$

where R is a $q \times p$ known constant matrix, r is a $q \times 1$ known constant vector. The null hypothesis

$$\mathbb{H}_0 : R\theta = r$$

imposes q linear restrictions on the p-dimensional parameter vector θ. The constant matrix R can be viewed as a selection matrix. For example, if one choose R to be a $q \times p$ identity matrix and r is a $q \times 1$ zero vector. Then $R\theta = r$ implies that all p components of parameter vector θ are jointly equal to zero.

There are three most commonly used classical testing principles in statistical inference, namely the Wald test, the Lagrange Multiplier (LM) test, and the Likelihood Ratio (LR) test. In subsequent sections, we discuss each of these important tests and derive the asymptotic distributions of the test statistics under the null hypothesis $\mathbb{H}_0$. For hypothesis testing in this book, we maintain an assumption that X^n is generated from $f_X(x) = f(x, \theta_0)$ for some unknown $\theta_0 \in \Theta$, so the hypotheses we are testing are $\mathbb{H}_0 : g(\theta_0) = \mathbf{0}$ and $\mathbb{H}_A : g(\theta_0) \neq \mathbf{0}$.

We first discuss the Wald test. We provide a set of regularity conditions.

Assumption W.1: $\sqrt{n}(\hat{\theta} - \theta_0) \xrightarrow{d} N(0, V)$, where V is a $p \times p$ symmetric bounded and nonsingular matrix, θ_0 is the true parameter value which is an interior point in Θ, and Θ is a compact parameter space.

Assumption W.2: $\hat{V} \xrightarrow{p} V$ as $n \rightarrow \infty$.

Assumption W.3: $g : \mathbb{R}^p \rightarrow \mathbb{R}^q$ is a continuously differentiable function of $\theta \in \Theta$, and $G(\theta_0) = \frac{\partial}{\partial \theta} g(\theta_0)$ has rank q, where $q \leq p$.

Assumption W.1 allows the estimator $\hat{\theta}$ to be any root-n consistent asymptotically normal estimator. For example, it can be the MLE or MME. Assumption W.2 assumes that there exists a consistent estimator $\hat{V}$ for the asymptotic variance V of $\sqrt{n}(\hat{\theta} - \theta_0)$. Suppose $\hat{\theta}$ admits the following asymptotic expansion

$$\sqrt{n}\left(\hat{\theta} - \theta_0\right) = n^{-\frac{1}{2}} \sum_{i=1}^{n} \psi(X_i, \theta_0) + o_P(1),$$

for some function $\psi(X_i, \theta_0)$, where $\mathbf{X}^n = (X_1, \cdots, X_n)$ is an IID sequence and $E[\psi(X_i, \theta_0)] = 0$, and the expectation $E(\cdot)$ is taken under the population distribution $f_X(x) = f(x, \theta_0)$. Then we have $V = E[\psi(X_i, \theta_0)\psi(X_i, \theta_0)']$. Therefore, a consistent estimator

$$\hat{V} = \frac{1}{n} \sum_{i=1}^{n} \psi(X_i, \hat{\theta})\psi(X_i, \hat{\theta})'.$$

By the uniform law of large numbers for an IID random sample in Theorem 7.10, primitive regularity conditions can be provided to ensure $\hat{V} \to V$ as $n \to \infty$ almost surely. As an example, if $\hat{\theta}$ is an MLE, then from the proof of the asymptotic normality of MLE (see Theorem 8.8), we have

$$\psi_i(X_i, \theta_0) = -H^{-1}(\theta_0)\frac{\partial \ln f(X_i, \theta_0)}{\partial \theta}.$$

It follows that

$$V = E\left[\psi(X_i, \theta_0)\psi(X_i, \theta)'\right]$$
$$= H^{-1}(\theta_0)I(\theta_0)H^{-1}(\theta_0)$$
$$= -H^{-1}(\theta_0)$$

where the last equality follows from the information matrix equality $I(\theta_0) + H(\theta_0) = \mathbf{0}$ in Lemma 8.7, where both the information matrix $I(\theta)$ and the Hessian matrix $H(\theta)$ are defined in Section 8.3.

Assumption W.3 is a regularity condition on the restriction function $g(\cdot)$. The full rank condition for the $q \times p$ matrix $G(\theta_0)$ and $q \leq p$ ensure that the $q \times q$ symmetric matrix $G(\theta_0)VG(\theta_0)'$ is nonsingular.

To test $\mathbb{H}_0 : g(\theta_0) = 0$, a natural approach is to base a test on statistic $g(\hat{\theta})$, where $\hat{\theta}$ is a consistent estimator of θ_0. Because $g(\cdot)$ is continuous, we always have $g(\hat{\theta}) \xrightarrow{p} g(\theta_0)$ whenever $\hat{\theta} \xrightarrow{p} \theta_0$ as $n \to \infty$ by Lemma 7.6 in Chapter 7. It follows that $g(\hat{\theta})$ will be close to zero under $\mathbb{H}_0$ and will converge to a nonzero limit under $\mathbb{H}_A$. Thus, we can test $\mathbb{H}_0$ by checking whether $g(\hat{\theta})$ is close to zero. If $g(\hat{\theta})$ is close to zero, then $\mathbb{H}_0$ holds; otherwise, if $g(\hat{\theta})$ is significantly different from zero, then $\mathbb{H}_A$ holds.

How large the value of $g(\hat{\theta})$ should be in order to be considered as significantly different from zero will be determined by the sampling distribution of $g(\hat{\theta})$ under $\mathbb{H}_0$. The sampling distribution of $g(\hat{\theta})$ precisely describes the distance of $g(\hat{\theta})$ from $g(\theta_0)$. However, the sampling distribution of $g(\hat{\theta})$ is often difficult to obtain, particularly when $g(\cdot)$ is a nonlinear function. We now use the asymptotic distribution theory from Chapter 7 to derive the asymptotic distribution of a test statistic based on $g(\hat{\theta})$.

By the mean value theorem (Bartle 1976, p. 206), we have

$$g(\hat{\theta}) = g(\theta_0) + G(\bar{\theta})\left(\hat{\theta} - \theta_0\right),$$

where $\bar{\theta} = \lambda\hat{\theta} + (1 - \lambda)\theta_0$ for some $\lambda \in [0, 1]$, and the gradient function

$$G(\theta) = \frac{dg(\theta)}{d\theta}$$

is a $q \times p$ matrix, where its i-th row is the derivative of the i-th component of $g(\theta)$ with respect to each component of parameter vector θ.

Given that $\|\bar{\theta} - \theta_0\| = \|\lambda(\hat{\theta} - \theta_0)\| \leq \|\hat{\theta} - \theta_0\| \xrightarrow{p} 0$ as $n \to \infty$ and the continuity of $G(\cdot)$, we have $G(\bar{\theta}) \xrightarrow{p} G(\theta_0)$ as $n \to \infty$ by Lemma 7.6. Then by the asymptotic normality that $\sqrt{n}(\hat{\theta} - \theta_0) \xrightarrow{d} N(0, V)$, and the Slutsky theorem (see Theorem 7.18), we have

$$\sqrt{n}\left[g(\hat{\theta}) - g(\theta_0)\right] \xrightarrow{d} N\left(0, G(\theta_0)VG(\theta_0)'\right).$$

Under $\mathbb{H}_0 : g(\theta_0) = 0$, we have

$$\sqrt{n}g(\hat{\theta}) \xrightarrow{d} N\left(0, G(\theta_0)VG(\theta_0)'\right).$$

Since the $q \times q$ matrix $G(\theta_0)VG(\theta_0)'$ is nonsingular given the full rank condition on $G(\theta_0)$ and the nonsingularity condition on V, the quadratic form

$$\sqrt{n}g(\hat{\theta})' \left[G(\theta_0)VG(\theta_0)'\right]^{-1} \sqrt{n}g(\hat{\theta}) \xrightarrow{d} \chi_q^2.$$

Because $G(\hat{\theta}) \xrightarrow{p} G(\theta_0)$ as $n \to \infty$ by continuity of $G(\cdot)$ and $\hat{\theta} \xrightarrow{p} \theta_0$, and $\hat{V} \xrightarrow{p} V$ by Assumption W.2, we have

$$G(\hat{\theta})\hat{V}G(\hat{\theta})' \xrightarrow{p} G(\theta_0)VG(\theta_0)'.$$

Therefore, the stochastic matrix $G(\hat{\theta})\hat{V}G(\hat{\theta})'$ is nonsingular for n sufficiently large, and by the Slutsky theorem (see Theorem 7.18), the Wald test statistic

$$W = n \cdot g(\hat{\theta})' \left[G(\hat{\theta})\hat{V}G(\hat{\theta})'\right]^{-1} g(\hat{\theta}) \xrightarrow{d} \chi_q^2$$

under $\mathbb{H}_0$.

Theorem 9.2. [Wald Test]: *Suppose Assumptions W.1–W.3 and* $\mathbb{H}_0$ *hold. Then as* $n \to \infty$,

$$W = n \cdot g(\hat{\theta})' \left[G(\hat{\theta})\hat{V}G(\hat{\theta})'\right]^{-1} g(\hat{\theta}) \xrightarrow{d} \chi_q^2.$$

By construction, the Wald test statistic W is a quadratic form in the difference between $\sqrt{n}g(\hat{\theta})$ and $\sqrt{n}g(\theta_0) = \mathbf{0}$, weighted by the asymptotic variance estimator $G(\hat{\theta})\hat{V}G(\hat{\theta})'$ of $\sqrt{n}[g(\hat{\theta}) - g(\theta_0)]$. The Wald test is an asymptotically size α test that rejects $\mathbb{H}_0 : g(\theta_0) = \mathbf{0}$ when W exceeds the $(1 - \alpha)$th quantile of the χ_q^2 distribution.

On the other hand, under $\mathbb{H}_A : g(\theta_0) \neq \mathbf{0}$, we have $g(\hat{\theta}) \overset{p}{\to} g(\theta_0) \neq \mathbf{0}$, $G(\hat{\theta}) \overset{p}{\to} G(\theta_0)$, and $\hat{V} \overset{p}{\to} V$. It follows that

$$\frac{W}{n} \to g(\theta_0)' \left[G(\theta_0)VG(\theta_0)' \right]^{-1} g(\theta_0) > 0$$

under $\mathbb{H}_A$. In other words, the Wald statistic W diverges to positive infinity at the rate of n, thus ensuring asymptotic power one of the test under $\mathbb{H}_A$ at any given significance level $\alpha \in (0,1)$. This suggests that the Wald test is a consistent test in the sense that it can detect all alternatives to $\mathbb{H}_0$.

The Wald test statistic W in Theorem 9.2 is applicable for many root-n consistent estimators $\hat{\theta}$ (i.e., $\sqrt{n}(\hat{\theta}-\theta_0) = O_P(1)$). We now consider a special but important case: that is, when $\hat{\theta}$ is an MLE. In this case, $V = -H^{-1}(\theta_0)$, and we can use the asymptotic variance estimator $\hat{V} = [-\hat{H}(\hat{\theta})]^{-1}$, where the sample Hessian matrix

$$\hat{H}(\theta) = \frac{1}{n} \sum_{i=1}^{n} \frac{\partial^2 \ln f(X_i, \theta)}{\partial\theta\partial\theta'}.$$

The resulting Wald test statistic can be constructed as follows:

$$W = ng(\hat{\theta})' \left[-G(\hat{\theta})\hat{H}^{-1}(\hat{\theta})G(\hat{\theta})' \right]^{-1} g(\hat{\theta}).$$

If in addition the regularity conditions of Section 8.3 hold, we can show $\hat{H}(\hat{\theta}) \to H(\theta_0)$ as $n \to \infty$ almost surely, and therefore $W \to \chi_q^2$ under $\mathbb{H}_0$.

Note that the Wald test statistic W only involves estimation under the alternative $\mathbb{H}_A$.

9.4 Lagrange Multiplier (LM) Test

Next, we introduce the Lagrange Multiplier (LM) test. It is also called Rao's (1959) efficient score test in statistics.

Suppose we have an IID random sample $\mathbf{X}^n$ from the population $f(x, \theta_0)$, where θ_0 is an unknown parameter value in Θ. Consider the normalized log-likelihood

$$\hat{l}(\theta) = \frac{1}{n} \sum_{i=1}^{n} \ln f(X_i, \theta),$$

and the constrained maximum likelihood estimator that solves the constrained maximization problem

$$\tilde{\theta} = \arg\max_{\theta \in \Theta} \hat{l}(\theta)$$

subject to the constraint that $g(\theta) = \mathbf{0}$.

Define the Lagrangian function

$$L(\theta, \lambda) = \hat{l}(\theta) + \lambda' g(\theta),$$

where λ is the Lagrange multiplier. Let $\tilde{\lambda}$ be the corresponding maximizing value of λ. Then the FOC are

$$\frac{\partial L(\tilde{\theta}, \tilde{\lambda})}{\partial \theta} = \frac{\partial \hat{l}(\tilde{\theta})}{\partial \theta} + G(\tilde{\theta})' \tilde{\lambda} = \mathbf{0},$$

$$\frac{\partial L(\tilde{\theta}, \tilde{\lambda})}{\partial \lambda} = g(\tilde{\theta}) = \mathbf{0}.$$

By the mean value theorem, we have

$$G(\tilde{\theta})' \tilde{\lambda} = -\frac{d\hat{l}(\tilde{\theta})}{d\theta} = -\frac{d\hat{l}(\theta_0)}{d\theta} - \frac{d^2\hat{l}(\bar{\theta}_a)}{d\theta d\theta'}(\tilde{\theta} - \theta_0),$$

where $\bar{\theta}_a = a\tilde{\theta} + (1-a)\theta_0$ for some $a \in [0,1]$, which lies on the segment between $\tilde{\theta}$ and θ_0. Note

$$\frac{d^2\hat{l}(\theta)}{d\theta d\theta'} = \hat{H}(\theta)$$

is the sample Hessian matrix. Given the regularity conditions in Section 8.3, we have shown there that $\hat{H}(\hat{\theta}) \to H(\theta_0)$ as $n \to \infty$ almost surely for any consistent estimator $\hat{\theta}$ of θ_0. Because $H(\theta_0)$ is nonsingular, $\hat{H}^{-1}(\hat{\theta}) \to H^{-1}(\theta_0)$ as $n \to \infty$ almost surely, and $\hat{H}^{-1}(\hat{\theta})$ exists for n sufficiently large. It follows that

$$\hat{H}(\bar{\theta}_a)^{-1} G(\tilde{\theta})' \tilde{\lambda} = -\hat{H}(\bar{\theta}_a)^{-1} \frac{d\hat{l}(\theta_0)}{d\theta} - (\tilde{\theta} - \theta_0). \tag{9.1}$$

Next, by the mean value theorem again, we have

$$\mathbf{0} = g(\tilde{\theta}) = g(\theta_0) + G(\bar{\theta}_b)(\tilde{\theta} - \theta_0),$$

where $\bar{\theta}_b = b\tilde{\theta} + (1-b)\theta_0$ for some $b \in [0,1]$, which lies on the segment between $\tilde{\theta}$ and θ_0. It follows that under $\mathbb{H}_0 : g(\theta_0) = 0$, we have

$$G(\bar{\theta}_b)(\tilde{\theta} - \theta_0) = \mathbf{0}. \tag{9.2}$$

Hence, multiplying Eq. (9.1) by $G(\bar{\theta}_b)$ and using Eq. (9.2), we obtain

$$G(\bar{\theta}_b)\hat{H}(\bar{\theta}_a)^{-1}G(\tilde{\theta})'\tilde{\lambda} = -G(\bar{\theta}_b)\hat{H}(\bar{\theta}_a)^{-1}\frac{d\hat{l}(\theta_0)}{d\theta} - G(\bar{\theta}_b)(\tilde{\theta} - \theta_0)$$

$$= -G(\bar{\theta}_b)\hat{H}(\bar{\theta}_a)^{-1}\frac{d\hat{l}(\theta_0)}{d\theta}.$$

By the continuity of function $G(\cdot)$, $||\bar{\theta}_a - \theta_0|| \leq ||\tilde{\theta} - \theta_0|| \to 0$, $||\bar{\theta}_b - \theta_0||$
$\leq ||\tilde{\theta} - \theta_0|| \to 0$ and $\hat{H}(\hat{\theta}) \to H(\theta_0)$ as $n \to \infty$ almost surely, we have

$$G(\bar{\theta}_b)\hat{H}(\bar{\theta}_a)^{-1}G(\tilde{\theta})' \to G(\theta_0)H^{-1}(\theta_0)G(\theta_0)' \text{ a.s.}$$

and the latter is nonsingular. Therefore, for n sufficiently large, $G(\bar{\theta}_b)\hat{H}(\bar{\theta}_a)^{-1}G(\tilde{\theta})'$ is nonsingular as well. Hence,

$$\sqrt{n}\,\tilde{\lambda} = -\left[G(\bar{\theta}_b)\hat{H}(\bar{\theta}_a)^{-1}G(\tilde{\theta})'\right]^{-1}G(\bar{\theta}_b)\hat{H}(\bar{\theta}_a)^{-1}\sqrt{n}\frac{d\hat{l}(\theta_0)}{d\theta}$$

$$= -\hat{A}\sqrt{n}\frac{d\hat{l}(\theta_0)}{d\theta}, \text{ say.}$$

By the CLT for an IID random sequence in Theorem 7.16, we have

$$\sqrt{n}\frac{d\hat{l}(\theta_0)}{d\theta} = \frac{1}{\sqrt{n}}\sum_{i=1}^{n}\frac{\partial \ln f(X_i, \theta_0)}{\partial \theta} \xrightarrow{d} N(0, I(\theta_0)),$$

where $I(\theta_0)$ is Fisher's information matrix evaluated at $\theta = \theta_0$. On the other hand,

$$\hat{A} \xrightarrow{p} \left[G(\theta_0)H(\theta_0)^{-1}G(\theta_0)'\right]^{-1}G(\theta_0)H(\theta_0) = A_0, \text{ say.}$$

It follows from the Slutsky theorem (see Theorem 7.18) that

$$\sqrt{n}\,\tilde{\lambda} \xrightarrow{d} N(0, A_0 I(\theta_0)A_0') \sim N\left(0, -[G(\theta_0)H(\theta_0)^{-1}G(\theta_0)']^{-1}\right),$$

where we have used the fact that

$$A_0 I(\theta_0)A_0' = \left[G(\theta_0)H(\theta_0)^{-1}G(\theta_0)'\right]^{-1}G(\theta_0)H(\theta_0)^{-1}I(\theta_0)H(\theta_0)^{-1}G(\theta_0)'$$
$$\left[G(\theta_0)H(\theta_0)^{-1}G(\theta_0)'\right]^{-1}$$

$$= -\left[G(\theta_0)H(\theta_0)^{-1}G(\theta_0)'\right]^{-1}$$

given the information matrix equality $I(\theta_0) + H(\theta_0) = \mathbf{0}$ in Lemma 8.7. It follows that the quadratic form

$$-n\tilde{\lambda}'G(\theta_0)H(\theta_0)^{-1}G(\theta_0)'\tilde{\lambda} \xrightarrow{d} \chi_q^2$$

under $\mathbb{H}_0$. We can now construct a LM test statistic by forming a quadratic form in $\sqrt{n}\,\tilde{\lambda}$.

Theorem 9.3. *[LM Test]: Suppose Assumptions M.1–M.6, Assumption W.3, and $\mathbb{H}_0$ hold. Define*

$$LM = n\tilde{\lambda}'G(\tilde{\theta})\left[-\hat{H}(\tilde{\theta})\right]^{-1}G(\tilde{\theta})'\tilde{\lambda}.$$

Then under $\mathbb{H}_0$,

$$LM \xrightarrow{d} \chi_q^2 \text{ as } n \to \infty.$$

Therefore, an asymptotically size α LM test will reject the null hypothesis $\mathbb{H}_0 : g(\theta_0) = 0$ when the LM test statistic LM exceeds the $(1 - \alpha)$th quantile of the χ_q^2 distribution.

Question: What is the interpretation for the Lagrange multiplier $\tilde{\lambda}$?

When the parametric restriction $g(\theta_0) = 0$ is valid, the restricted estimator $\tilde{\theta}$ should be near the point that maximizes the log-likelihood. Therefore, the slope of the log-likelihood function should be close to zero at the restricted estimator $\tilde{\theta}$. As can be seen from the FOC that $G(\tilde{\theta})'\tilde{\lambda} = -\frac{d\hat{l}(\tilde{\theta})}{d\theta}$, the LM test is based on the slope of the log-likelihood at the point where the function is maximized subject to the constraint. If the constraint $g(\theta_0) = 0$ holds, $\tilde{\lambda}$ will be close to 0. On the other hand, if the constraint $g(\theta_0) \neq 0$, then $\tilde{\lambda}$ will be large, giving the LM test power to reject $\mathbb{H}_0$. How large $\sqrt{n}\,\tilde{\lambda}$ is considered as "large" is determined by its sampling distribution.

The LM test is convenient to use in practice, because only the null model (or called restricted model), which is usually simpler, has to be estimated. Breusch and Pagan (1980) provided examples of applying the LM principle in econometrics. See also Engle (1984).

9.5 Likelihood Ratio Test

We now discuss the Likelihood Ratio (LR) test. In Section 9.2, the LR test has been shown to be uniformly most powerful in testing a simple null hypothesis versus a simple alternative hypothesis. This is the well-known Neyman-Pearson Lemma. We now consider a LR test for $\mathbb{H}_0 : g(\theta_0) = \mathbf{0}$ versus $\mathbb{H}_A : g(\theta_0) \neq \mathbf{0}$.

Since $\mathbf{X}^n$ is an IID random sample from the population $f_X(x) = f(x, \theta_0)$, where θ_0 is unknown, we have the likelihood function of $\mathbf{X}^n$

$$f_{\mathbf{X}^n}(\mathbf{X}^n, \theta) = \prod_{i=1}^{n} f(X_i, \theta).$$

We define the likelihood ratio statistic

$$\hat{\Lambda} = \frac{\max_{\theta \in \Theta} f_{\mathbf{X}^n}(\mathbf{X}^n, \theta)}{\max_{\theta \in \Theta_0} f_{\mathbf{X}^n}(\mathbf{X}^n, \theta)} = \frac{\prod_{i=1}^{n} f(X_i, \hat{\theta})}{\prod_{i=1}^{n} f(X_i, \tilde{\theta})},$$

where $\hat{\theta}$ and $\tilde{\theta}$ are the unconstrained and constrained MLE estimators respectively, namely,

$$\hat{\theta} = \arg \max_{\theta \in \Theta} \hat{l}(\theta),$$

$$\tilde{\theta} = \arg \max_{\theta \in \Theta_0} \hat{l}(\theta),$$

with

$$\hat{l}(\theta) = \frac{1}{n} \sum_{i=1}^{n} \ln f(X_i, \theta)$$

is the sample average of log-likelihood functions, and Θ_0 is the parameter space Θ subject to the constraint $g(\theta) = \mathbf{0}$, i.e., $\Theta_0 = \{\theta \in \Theta : g(\theta) = \mathbf{0}\}$.

Suppose the null hypothesis $\mathbb{H}_0 : g(\theta_0) = 0$ holds. Then both unconstrained and constrained MLEs $\hat{\theta}$ and $\tilde{\theta}$ are consistent for θ_0, and imposing the restriction should not lead to a large reduction in the log-likelihood function. Therefore, we expect that the likelihood ratio $\hat{\Lambda}$ will be close to unity. On the other hand, if $\mathbb{H}_0$ is false, then the unconstrained MLE $\hat{\theta}$ is consistent for θ_0 but the constrained MLE $\tilde{\theta}$ is not. As a consequence, we expect that the likelihood ratio $\hat{\Lambda}$ is larger than unity. Hence, we can test $\mathbb{H}_0$ by comparing whether $\hat{\Lambda}$ is significantly larger than unity or whether $\ln \hat{\Lambda}$ is significantly greater than 0. How large $\hat{\Lambda}$ or $\ln \hat{\Lambda}$ must be in order to be considered as significantly large will be determined by the sampling distribution of $\hat{\Lambda}$.

Formally, we define the LR test statistic as follows:

$$LR = 2 \ln \hat{\Lambda} = 2n \left[\hat{l}(\hat{\theta}) - \hat{l}(\tilde{\theta}) \right].$$

Using a second order Taylor's series expansion of $\hat{l}(\tilde{\theta})$ around the unconstrained MLE $\hat{\theta}$, we have

$$LR = 2n \left\{ \hat{l}(\hat{\theta}) - \left[\hat{l}(\hat{\theta}) + \frac{d\hat{l}(\hat{\theta})}{d\theta}(\tilde{\theta} - \hat{\theta}) + \frac{1}{2}(\tilde{\theta} - \hat{\theta})' \frac{d^2\hat{l}(\bar{\theta}_a)}{d\theta d\theta'}(\tilde{\theta} - \hat{\theta}) \right] \right\}$$

$$= \sqrt{n}(\tilde{\theta} - \hat{\theta})'[-\hat{H}(\bar{\theta}_a)]\sqrt{n}(\tilde{\theta} - \hat{\theta}),$$

where $\bar{\theta}_a = a\tilde{\theta} + (1-a)\hat{\theta}$ for some $a \in [0,1]$, lies on the segment between $\tilde{\theta}$ and $\hat{\theta}$, and $\frac{d}{d\theta}\hat{l}(\hat{\theta}) = 0$, which is the FOC of the unconstrained MLE $\hat{\theta}$, and again,

$$\hat{H}(\theta) = \frac{d^2\hat{l}(\theta)}{d\theta d\theta'}$$

is the sample Hessian matrix.

Next, applying the mean value theorem for $\frac{d}{d\theta}\hat{l}(\hat{\theta})$ around the constrained MLE $\tilde{\theta}$, we obtain

$$\mathbf{0} = \frac{d\hat{l}(\hat{\theta})}{d\theta} = \frac{d\hat{l}(\tilde{\theta})}{d\theta} + \hat{H}(\bar{\theta}_b)(\hat{\theta} - \tilde{\theta}),$$

where $\bar{\theta}_b = b\tilde{\theta} + (1-b)\hat{\theta}$ for some $b \in [0,1]$ lies on the segment between $\tilde{\theta}$ and $\hat{\theta}$. This and the FOC of the constrained MLE that $G(\tilde{\theta})'\tilde{\lambda} = -\frac{d}{d\theta}\hat{l}(\tilde{\theta})$ imply

$$\sqrt{n}(\hat{\theta} - \tilde{\theta}) = -\hat{H}(\bar{\theta}_b)^{-1}\sqrt{n}\frac{d\hat{l}(\tilde{\theta})}{d\theta}$$

$$= \hat{H}(\bar{\theta}_b)^{-1}G(\tilde{\theta})'\sqrt{n}\,\tilde{\lambda}.$$

This relationship provides an alternative interpretation for the multiplier $\tilde{\lambda}$, namely, it measures the difference between the unconstrained and constrained MLEs $\hat{\theta}$ and $\tilde{\theta}$.

It follows that

$$LR = -\sqrt{n}\,\tilde{\lambda}G(\tilde{\theta})\hat{H}(\bar{\theta}_b)^{-1}\hat{H}(\bar{\theta}_a)\hat{H}(\bar{\theta}_b)^{-1}G(\tilde{\theta})'\sqrt{n}\,\tilde{\lambda}.$$

Because

$$LR - LM$$
$$= -\sqrt{n}\,\tilde{\lambda}\left[G(\tilde{\theta})\hat{H}(\bar{\theta}_b)^{-1}\hat{H}(\bar{\theta}_a)\hat{H}(\bar{\theta}_b)^{-1}G(\tilde{\theta})' - G(\tilde{\theta})\hat{H}(\tilde{\theta})^{-1}G(\tilde{\theta})'\right]\sqrt{n}\,\tilde{\lambda}$$
$$= O_P(1)o_P(1)O_P(1)$$
$$= o_P(1),$$

where $\sqrt{n}\,\tilde{\lambda} = O_P(1)$ by Lemma 7.11 given $\sqrt{n}\tilde{\lambda} \xrightarrow{d} N(0, -[G(\theta_0)H(\theta_0)^{-1}G(\theta_0)']^{-1})$ as shown in Section 9.4, and

$$G(\tilde{\theta})\hat{H}(\bar{\theta}_b)^{-1}\hat{H}(\bar{\theta}_a)\hat{H}(\bar{\theta}_b)^{-1}G(\tilde{\theta})' - G(\tilde{\theta})\hat{H}(\tilde{\theta})^{-1}G(\tilde{\theta})'$$
$$\xrightarrow{p} G(\theta_0)H(\theta_0)^{-1}H(\theta_0)H(\theta_0)^{-1}G(\theta_0)' - G(\theta_0)H(\theta_0)^{-1}G(\theta_0)' = \mathbf{0}.$$

It follows that LR and LM are asymptotically equivalent under $\mathbb{H}_0$. As a result, by the asymptotic equivalence lemma (see Lemma 7.13) and $LM \xrightarrow{d} \chi_q^2$ as $n \to \infty$ under $\mathbb{H}_0$ (see Theorem 9.3), we have the following result:

Theorem 9.4. *[LR Test]: Suppose Assumptions M.1–M.6, Assumption W.3, and $\mathbb{H}_0$ hold. Then under $\mathbb{H}_0$,*

$$LR \xrightarrow{d} \chi_q^2 \text{ as } n \to \infty.$$

In fact, it can also be shown that the Wald test statistic W based on MLE $\hat\theta$ and the LM test statistic LM are asymptotically equivalent under $\mathbb{H}_0$. This implies that all three tests are asymptotically equivalent under $\mathbb{H}_0$.

The LR test statistic LR involves both constrained and unconstrained MLE estimators. However, it is very convenient to compute, because the sample log-likelihood value is the objective function and is usually reported by statistical software when a probability distribution model $f(x, \theta)$ is estimated by the MLE.

We now link the LR test statistic with a sufficient statistic when the latter exists. Suppose $T(\mathbf{X}^n)$ is a sufficient statistic for θ with PMF/PDF $g[T(\mathbf{X}^n), \theta]$. Then we may consider constructing a LR test based on $T(\mathbf{X}^n)$ and its likelihood function $\hat{L}^*[\theta|T_n(\mathbf{X}^n)] = g[T(\mathbf{X}^n), \theta]$, rather than on the sample $\mathbf{X}^n$ and its likelihood function $\hat{L}(\theta|\mathbf{X}^n)$. Given that all the information about θ in $\mathbf{X}^n$ is contained in $T(\mathbf{X}^n)$, the test based on $T(\mathbf{X}^n)$ should be as good as the test based on the original sample $\mathbf{X}^n$. In fact, they are equivalent, as is shown below.

Theorem 9.5. *[LR Test Based on Sufficient Statistic]: If $T(\mathbf{X}^n)$ is a sufficient statistic for θ, and $LR(\mathbf{X}^n)$ and $LR[T(\mathbf{X}^n)]$ are the likelihood ratio tests based on $\mathbf{X}^n$ and $T(\mathbf{X}^n)$ respectively, then*

$$LR(\mathbf{X}^n) = LR[T(\mathbf{X}^n)].$$

Proof: By the factorization theorem (see Theorem 6.14), the PMF/PDF of $\mathbf{X}^n$ can be written as

$$f_{\mathbf{X}^n}(\mathbf{x}^n, \theta) = g[T(\mathbf{x}^n), \theta]h(\mathbf{x}^n), \text{ for all } \theta \in \Theta,$$

where $g(t, \theta)$ is the PMF/PDF of $T(\mathbf{X}^n)$ and $h(\mathbf{x}^n)$ does not depend on θ. It follows that

$$LR(\mathbf{X}^n) = 2n \ln \hat{\Lambda}$$

$$= 2n \ln \left[\frac{f_{\mathbf{X}^n}(\mathbf{X}^n, \hat{\theta})}{f_{\mathbf{X}^n}(\mathbf{X}^n, \tilde{\theta})} \right]$$

$$= 2n \ln \left\{ \frac{g[T(\mathbf{X}^n), \hat{\theta}] h(\mathbf{X}^n)}{g[T(\mathbf{X}^n), \tilde{\theta}] h(\mathbf{X}^n)} \right\}$$

$$= 2n \ln \left\{ \frac{g[T(\mathbf{X}^n), \hat{\theta}]}{g[T(\mathbf{X}^n), \tilde{\theta}]} \right\}$$

$$= LR[T(\mathbf{X}^n)].$$

This result implies that the LR test statistic depends on $\mathbf{X}^n$ only through the statistic $T(\mathbf{X}^n)$ when $T(\mathbf{X}^n)$ is a sufficient statistic for θ.

9.6 Illustrative Examples

We now provide two simple examples to illustrate how to compute the Wald test statistic W, the LM test statistic LM, and the LR test statistic LR. The first example is an IID random sample from a Bernoulli(θ) distribution, and the second example is an IID random sample from a $N(\mu, \sigma^2)$ distribution.

9.6.1 *Hypothesis Testing under the Bernoulli Distribution*

Suppose $\mathbf{X}^n$ is an IID random sample from a Bernoulli(θ) distribution, where a Bernoulli random variable takes two possible values:

$$X_i = \begin{cases} 1, & \text{with probability } \theta, \\ 0, & \text{with probability } 1 - \theta. \end{cases}$$

The parameter $\theta \in (0, 1)$ is unknown. Suppose we are interested in testing

$$\mathbb{H}_0 : \theta = \theta_0$$

versus

$$\mathbb{H}_1 : \theta \neq \theta_0.$$

Hence, we have $g(\theta) = \theta - \theta_0$, and the gradient

$$G(\theta) = \frac{dg(\theta)}{d\theta} = 1.$$

Since the population PMF $f(x,\theta) = \theta^x(1-\theta)^{1-x}$ for $x - 0,1$, the log-likelihood function of the IID random sample $\mathbf{X}^n$ is given by:

$$\ln \hat{L}(\theta|\mathbf{X}^n) = \sum_{i=1}^{n} \ln f(X_i,\theta)$$
$$= n\bar{X}_n \ln\theta + n\left(1 - \bar{X}_n\right)\ln(1-\theta),$$

where the sample mean $\bar{X}_n$ is a sufficient statistic for θ. The FOC of MLE is

$$\frac{\partial \ln \hat{L}(\theta|\mathbf{X}^n)}{\partial\theta} = \frac{n\bar{X}_n}{\hat{\theta}} - \frac{n - n\bar{X}_n}{1-\hat{\theta}} = 0.$$

Thus we have MLE $\hat{\theta} = \bar{X}_n$.

We first consider the Wald Test. Recall the Hessian matrix

$$H(\theta) = E_\theta\left[\frac{\partial^2 \ln f(X_i,\theta)}{\partial\theta^2}\right].$$

Because

$$\frac{\partial^2 \ln f(X_i,\theta)}{\partial\theta^2} = -\frac{X_i}{\theta^2} - \frac{1 - X_i}{(1-\theta)^2},$$

the sample Hessian matrix

$$\hat{H}(\theta) = n^{-1}\sum_{i=1}^{n}\frac{\partial^2 \ln f(X_i,\theta)}{\partial\theta}$$
$$= -\frac{\sum_{i=1}^{n} X_i}{n\theta^2} - \frac{\sum_{i=1}^{n}(1 - X_i)}{n(1-\theta)^2}$$
$$= -\frac{\bar{X}_n}{\theta^2} - \frac{1 - \bar{X}_n}{(1-\theta)^2}.$$

Hence, we have

$$\hat{H}(\hat{\theta}) = -\frac{1}{\bar{X}_n\left(1 - \bar{X}_n\right)}.$$

The Wald test statistic

$$W = ng(\hat{\theta})'\left[-G(\hat{\theta})\hat{H}^{-1}(\hat{\theta})G(\hat{\theta})'\right]^{-1}g(\hat{\theta})$$
$$= \frac{n\left(\hat{\theta} - \theta_0\right)^2}{\bar{X}_n\left(1 - \bar{X}_n\right)}$$
$$= \frac{n\left(\bar{X}_n - \theta_0\right)^2}{\bar{X}_n\left(1 - \bar{X}_n\right)} \xrightarrow{d} \chi_1^2 \text{ as } n \to \infty$$

under $\mathbb{H}_0$. Hence, $\sqrt{n}(\bar{X}_n - \theta_0) \xrightarrow{d} N(0,\sigma^2)$.

Next, we consider the LM test. Define the Lagrangian function

$$L(\theta, \lambda) = \hat{l}(\theta) + \lambda' g(\theta) = \hat{l}(\theta) + \lambda (\theta - \theta_0)$$

where the normalized log-likelihood

$$\hat{l}(\theta) = \bar{X}_n \ln \theta + \left(1 - \bar{X}_n\right) \ln \left(1 - \theta\right).$$

The first order conditions for the constrained MLE are

$$\frac{\partial L(\tilde{\theta}, \tilde{\lambda})}{\partial \theta} = \frac{\partial \hat{l}(\tilde{\theta})}{\partial \theta} + \tilde{\lambda} = 0,$$

$$\frac{\partial L(\tilde{\theta}, \tilde{\lambda})}{\partial \lambda} = g(\tilde{\theta}) = \tilde{\theta} - \theta_0 = 0.$$

It follows that $\tilde{\theta} = \theta_0$, and

$$
\begin{aligned}
\tilde{\lambda} &= -\frac{\partial \hat{l}(\tilde{\theta})}{\partial \theta} \\
&= -\frac{\bar{X}_n}{\tilde{\theta}} + \frac{1 - \bar{X}_n}{1 - \tilde{\theta}} \\
&= -\frac{\bar{X}_n - \tilde{\theta}}{\tilde{\theta}\left(1 - \tilde{\theta}\right)} \\
&= -\frac{\bar{X}_n - \theta_0}{\theta_0(1 - \theta_0)}.
\end{aligned}
$$

This indicates that $\tilde{\lambda}$ measures the difference between the unconstrained MLE $\hat{\theta}$ and the constrained MLE $\tilde{\theta} = \theta_0$.

Also, the sample Hessian matrix

$$
\begin{aligned}
\hat{H}(\tilde{\theta}) &= -\frac{\bar{X}_n}{\theta_0^2} - \frac{1 - \bar{X}_n}{(1 - \theta_0)^2} \\
&= -\frac{\bar{X}_n(1 - \theta_0)^2 + (1 - \bar{X}_n)\theta_0^2}{\theta_0^2(1 - \theta_0)^2}.
\end{aligned}
$$

Therefore, we have

$$
\begin{aligned}
LM &= -n\tilde{\lambda}_n' G(\tilde{\theta})\hat{H}(\tilde{\theta})^{-1}G(\tilde{\theta})'\tilde{\lambda} \\
&= n\left[-\frac{\bar{X}_n - \theta_0}{\theta_0(1 - \theta_0)}\right]^2 \left[\frac{\bar{X}_n(1 - \theta_0)^2 + (1 - \bar{X}_n)\theta_0^2}{\theta_0^2(1 - \theta_0)^2}\right]^{-1} \\
&= \frac{n(\bar{X}_n - \theta_0)^2}{\bar{X}_n(1 - \theta_0)^2 + (1 - \bar{X}_n)\theta_0^2}.
\end{aligned}
$$

Finally, we calculate the LR test statistic:

$$LR = 2n \left[\hat{l}(\hat{\theta}) - \hat{l}(\tilde{\theta}) \right]$$

$$= 2n \left[\bar{X}_n \ln \left(\frac{\bar{X}_n}{\theta_0} \right) + (1 - \bar{X}_n) \ln \left(\frac{1 - \bar{X}_n}{1 - \theta_0} \right) \right].$$

9.6.2 *Hypothesis Testing under the Normal Distribution*

Suppose now X^n is an IID random sample from a $N(\mu, \sigma^2)$ population, where $\theta = (\mu, \sigma^2)$ is unknown. We are interested in testing the hypotheses

$$\mathbb{H}_0 : \mu = \mu_0$$

versus

$$\mathbb{H}_A : \mu \neq \mu_0,$$

where μ_0 is a known number. This is equivalent to choosing the test function

$$g(\theta) = \mu - \mu_0.$$

It follows that

$$G(\theta) = \frac{dg(\theta)}{d\theta} = (1, 0)$$

is a two-dimensional row vector.

Since the PDF of a $N(\mu, \sigma^2)$ population is

$$f(x, \theta) = \frac{1}{\sqrt{2\pi\sigma^2}} e^{-\frac{(x-\mu)^2}{2\sigma^2}}.$$

The normalized log-likelihood function of the random sample $\mathbf{X}^n$ is

$$\hat{l}(\theta|\mathbf{X}^n) = -\frac{1}{2} \ln(2\pi) - \frac{1}{2} \ln(\sigma^2) - \frac{1}{2\sigma^2} \frac{1}{n} \sum_{i=1}^{n} (X_i - \mu)^2.$$

For the unconstrained MLE, we have obtained $\hat{\theta} = (\hat{\mu}, \hat{\sigma}^2)$ in Example 8.4, Chapter 8, where

$$\hat{\mu} = \bar{X}_n,$$

$$\hat{\sigma}^2 = \frac{1}{n} \sum_{i=1}^{n} (X_i - \bar{X}_n)^2.$$

Also, the sample Hessian matrix

$$\hat{H}(\theta) = \begin{bmatrix} -\frac{1}{\sigma^2} & -\frac{1}{\sigma^4} \frac{1}{n} \sum_{i=1}^{n} (X_i - \mu) \\ -\frac{1}{\sigma^4} \frac{1}{n} \sum_{i=1}^{n} (X_i - \mu) & \frac{1}{2\sigma^4} - \frac{1}{\sigma^6} \frac{1}{n} \sum_{i=1}^{n} (X_i - \mu)^2 \end{bmatrix}.$$

When $\theta = \hat{\theta}$, we have

$$\hat{H}(\hat{\theta}) = \begin{bmatrix} -\frac{1}{\hat{\sigma}^2} & 0 \\ 0 & -\frac{1}{2\hat{\sigma}^4} \end{bmatrix}.$$

It follows that the Wald test statistic

$$W = -ng(\hat{\theta})' \left[G(\hat{\theta})\hat{H}^{-1}(\hat{\theta})G(\hat{\theta})' \right]^{-1} g(\hat{\theta})$$

$$= \frac{n(\bar{X}_n - \mu_0)^2}{\hat{\sigma}^2}$$

$$\xrightarrow{d} \chi_1^2$$

under $\mathbb{H}_0$.

Next, we construct the LM test statistic. Consider the constrained MLE

$$\tilde{\theta} = \max_{\theta \in \Theta} \hat{l}(\theta|\mathbf{X}^n)$$

subject to the constrain that $\mu = \mu_0$. Define the Lagrangian function

$$L(\theta, \lambda) = \hat{l}(\theta|\mathbf{X}^n) + \lambda(\mu - \mu_0).$$

The FOCs are:

$$\frac{\partial L(\tilde{\theta}, \tilde{\lambda})}{\partial \mu} = \frac{1}{\tilde{\sigma}^2} \frac{1}{n} \sum_{i=1}^{n} (X_i - \tilde{\mu}) + \tilde{\lambda} = 0,$$

$$\frac{\partial L(\tilde{\theta}, \tilde{\lambda})}{\partial \sigma^2} = -\frac{1}{2\tilde{\sigma}^2} + \frac{1}{2\tilde{\sigma}^4} \frac{1}{n} \sum_{i=1}^{n} (X_i - \tilde{\mu})^2,$$

$$\frac{\partial L(\tilde{\theta}, \tilde{\lambda})}{\partial \lambda} = \tilde{\mu} - \mu_0 = 0.$$

Solving for the FOC's, we obtain

$$\tilde{\mu} = \mu_0,$$

$$\tilde{\sigma}^2 = \frac{1}{n} \sum_{i=1}^{n} (X_i - \mu_0)^2,$$

$$\tilde{\lambda} = -\frac{1}{\tilde{\sigma}^2} (\bar{X}_n - \mu_0),$$

and the sample Hessian matrix

$$\hat{H}(\tilde{\theta}) = \begin{bmatrix} -\frac{1}{\tilde{\sigma}^2} & -\frac{1}{\tilde{\sigma}^4}(\bar{X}_n - \mu_0) \\ -\frac{1}{\tilde{\sigma}^4}(\bar{X}_n - \mu_0) & -\frac{1}{2\tilde{\sigma}^4} \end{bmatrix}.$$

It follows that the LM test statistic

$$LM = -n\tilde{\lambda}'G(\tilde{\theta})\hat{H}^{-1}(\tilde{\theta})G(\tilde{\theta})'\tilde{\lambda}$$
$$= \frac{n(\bar{X}_n - \mu_0)^2}{\tilde{\sigma}^2 - 2(\bar{X}_n - \mu_0)^2}.$$

Finally, we construct the LR test statistic *LR*. Since

$$\hat{l}(\hat{\theta}) = -\frac{1}{2}\ln(2\pi) - \frac{1}{2}\ln(\hat{\sigma}^2) - \frac{1}{2},$$

$$\hat{l}(\tilde{\theta}) = -\frac{1}{2}\ln(2\pi) - \frac{1}{2}\ln(\tilde{\sigma}^2) - \frac{1}{2}.$$

It follows that

$$LR = 2n\left[\hat{l}(\hat{\theta}|\mathbf{X}^n) - \hat{l}(\tilde{\theta}|\mathbf{X}^n)\right]$$

$$= n\ln\left(\frac{\tilde{\sigma}^2}{\hat{\sigma}^2}\right).$$

It is interesting to observe that the LR principle tests the hypothesis on the population mean by comparing the variance estimators under the null and alternative hypotheses. Here, the likelihood ratio is a log-function of the sample variance ratio. Intuitively, when $\mu \neq \mu_0$, $\tilde{\sigma}^2 = n^{-1}\sum_{i=1}^{n}(X_i - \mu_0)^2$ is not a consistent estimator for σ^2. It will be larger than $\hat{\sigma}^2$ when n is large, giving the LR test its power.

9.7 Conclusion

Hypothesis testing is one of the most important tasks in statistical inference. In this chapter we have introduced basic ideas of hypothesis testing in statistical inference. We introduce the well-known Neyman-Pearson lemma that the likelihood ratio based test will be uniformly most powerful test for simple hypotheses. We then discuss the three important testing methods, namely the Wald test, the LM test, and the LR test, and show they are asymptotically equivalent to each other under the null hypothesis.

It is important to note that all hypothesis tests considered in this chapter assume that the population distribution model is correctly specified. That is, we are conducting hypothesis testing on model parameters when the population distribution model is correctly specified in the sense that the population distribution $f_X(x) = f(x, \theta_0)$ for some unknown parameter value θ_0. When the population distribution model is misspecified, the Wald test statistic and the LM test statistics have to be modified by using a consistent asymptotic variance estimator which is robust to model misspecification. However, it is impossible to modify the LR test statistic, which is based on comparison between the likelihood values under the null and alternative hypotheses.

Finally, we note that when testing economic hypotheses, we usually need to transform an economic hypothesis into a statistical hypothesis on model parameters. Since some auxiliary conditions are often imposed in such a transformation, there usually exists a gap between the original economic hypothesis and the resulting statistical hypothesis. Caution is needed to interpret the testing results of the statistical hypotheses.

EXERCISE 9

9.1. Suppose $\mathbf{X}^n$ is an IID random sample from a $N(\mu, \sigma^2)$ population, where μ is unknown but σ^2 is known. Consider a test statistic $T = \sqrt{n}(\bar{X}_n - \mu_0)/\sigma$ at the significance level α for $\mathbb{H}_0 : \mu = \mu_0$ versus $\mathbb{H}_A : \mu \neq \mu_0$.

(1) Find the Type I error of this test.

(2) Find the Type II error of this test.

(3) Derive the power function of this test under the alternative $\mathbb{H}_A : \mu = \mu_0 + \delta$, where $\delta \neq 0$. What happens to the power of the test if $|\delta|$ increases? And what happens to the power of the test if the sample size n increases?

9.2. Suppose $\mathbf{X}^n$ is an IID random sample from a $N(\mu, \sigma^2)$ population, where both μ and σ^2 are unknown. Find the acceptance and rejection regions at the significance level α of the t-test statistic for $\mathbb{H}_0 : \mu = \mu_0$ versus $\mathbb{H}_A : \mu \neq \mu_0$.

9.3. Suppose $(X_i, Y_i)_{i=1}^n$ is an IID random sample from the joint population PMF $f(x, y | \beta, \rho) = \frac{(\beta + x)^{-\rho}}{\Gamma(\rho)} y^{\rho-1} e^{-y/(\beta+x)}$. We are interested in testing the null hypothesis $\mathbb{H}_0 : \rho = 1$, which implies that the population distribution is given by the joint PMF $f(x, y | \beta) = \frac{1}{\beta+x} e^{-y/(\beta+x)}$. Derive:

(1) the LR test statistic for $\mathbb{H}_0$;

(2) the LM test statistic for $\mathbb{H}_0$;

(3) the Wald test statistic for $\mathbb{H}_0$ that is based on the MLE.

9.4. Suppose $\mathbf{X}^n$ is an IID random sample from a $N(\mu, \sigma^2)$ population, where $\sigma^2 = \sigma_0^2$ is known.

(1) Derive the LR test statistic for the null hypothesis $\mathbb{H}_0 : \mu = \mu_0$ versus the alternative hypothesis $\mathbb{H}_A : \mu \neq \mu_0$.

(2) What is the distribution of the LR test statistic under $\mathbb{H}_0$?

(3) Show that the LR test is equivalent to the test based on the test statistic $Z_n = \sqrt{n}(\bar{X}_n - \mu_0)/\sigma_0$, which follows a $N(0, 1)$ distribution under $\mathbb{H}_0$.

9.5. Suppose $\mathbf{X}^n$ is an IID random sample from a $N(\mu, \sigma^2)$ population, where both μ and σ^2 are unknown.

(1) Derive the LR test statistic for the null hypothesis $\mathbb{H}_0 : \mu = \mu_0$ versus the alternative hypothesis $\mathbb{H}_A : \mu \neq \mu_0$.

(2) What is the distribution of the LR test statistic under $\mathbb{H}_0$?

(3) Show that the LR test is equivalent to the test based on the test statistic $Z_n = \sqrt{n}(\bar{X}_n - \mu_0)/S_n$, which follows a Student t_{n-1} distribution under $\mathbb{H}_0$. Here, S_n is the sample standard deviation.

9.6. Suppose $\mathbf{X}^n$ is an IID random sample from a $N(\mu, \sigma^2)$ population, where μ is unknown.

(1) Derive the LR test statistic for the null hypothesis $\mathbb{H}_0 : \sigma^2 = \sigma_0^2$ versus the alternative hypothesis $\mathbb{H}_A : \sigma^2 \neq \sigma_0^2$.

(2) What is the distribution of the LR test statistic under $\mathbb{H}_0$?

9.7. Suppose $\mathbf{X}^{n_1}$ is an IID random sample from a $N(\mu_1, \sigma_1^2)$ population, $\mathbf{Y}^{n_2}$ is an IID random sample from a $N(\mu_2, \sigma_2^2)$ population, and $\mathbf{X}^{n_1}$ and $\mathbf{Y}^{n_2}$ are independent, where $\mu_1, \mu_2, \sigma_1^2, \sigma_2^2$ are unknown.

(1) Derive the LR test statistic for the null hypothesis $\mathbb{H}_0 : \sigma_1^2 = \sigma_2^2$ versus the alternative hypothesis $\mathbb{H}_A : \sigma_1^2 \neq \sigma_2^2$.

(2) What is the distribution of the LR test statistic under $\mathbb{H}_0$?

9.8. Suppose $\mathbf{X}^{n_1}$ is an IID random sample from a $N(\mu_1, \sigma^2)$ population, $\mathbf{Y}^{n_2}$ is an IID random sample from a $N(\mu_2, \sigma^2)$ population, and $\mathbf{X}^{n_1}$ and $\mathbf{Y}^{n_2}$ are independent, where μ_1, μ_2, σ^2 are unknown.

(1) Derive the LR test statistic for the null hypothesis $\mathbb{H}_0 : \mu_1 = \mu_2$ versus the alternative hypothesis $\mathbb{H}_A : \mu_1 \neq \mu_2$.

(2) What is the distribution of the LR test statistic under $\mathbb{H}_0$?

9.9. Show the LM test statistic LM can be written as follows:

$$LM = n \left[\frac{d\hat{l}(\tilde{\theta})}{d\theta} \right]' \left[-G(\tilde{\theta})\hat{H}(\tilde{\theta})G(\tilde{\theta})' \right]^{-1} \left[\frac{d\hat{l}(\tilde{\theta})}{d\theta} \right].$$

9.10. Suppose X^n be an IID random sample from the population $f(x, \theta_0)$, where θ_0 is an unknown parameter. Let $\hat{S}(\theta) = n^{-1}\Sigma_{i=1}^n \frac{\partial}{\partial \theta} \ln f(X_i, \theta)$ and $\hat{\theta}$ be a MLE. Define

$$W = ng(\hat{\theta})'G(\hat{\theta})\hat{S}(\hat{\theta})\hat{S}(\hat{\theta})'G(\hat{\theta})'g(\hat{\theta}).$$

Show $W \overset{d}{\to} \chi_q^2$ under $\mathbb{H}_0$.

9.11. Let $\hat{S}(\theta)$ be defined as in 9.2. Define

$$LM = n\tilde{\lambda}'\hat{G}(\tilde{\theta})\hat{S}(\tilde{\theta})\hat{S}(\tilde{\theta})'\hat{G}(\tilde{\theta})'\tilde{\lambda}.$$

Show $LM \overset{d}{\to} \chi_q^2$ under $\mathbb{H}_0$.

9.12. Show that the Wald test statistic W based on MLE $\hat{\theta}$ is asymptotically equivalent to the Lagrange multiplier test statistic LM under $\mathbb{H}_0$.

9.13. What is the role of the information matrix equality in deriving the asymptotic distribution of the LR test statistic? Suppose the information matrix equality does not hold. Can one surely obtain $LR \xrightarrow{d} \chi_q^2$ under $\mathbb{H}_0$. Give your reasoning.

9.14. Find a most uniformly powerful test for $\mathbb{H}_0 : p = p_0$ versus $\mathbb{H}_A : p \neq p_0$ for an IID sample from the Bernoulli(p) distribution. Give your reasoning.

9.15. Construct the Wald test statistic W, the Lagrange multiplier test statistic LM and the likelihood ratio test statistic LR given an IID random sample from the Poisson(λ) distribution in testing $\mathbb{H}_0 : \lambda = \lambda_0$ versus $\mathbb{H}_A : \lambda \neq \lambda_0$.

9.16. Under the regularity conditions of Section 8.3, show that LM and LR are asymptotically equivalent under the null hypothesis $\mathbb{H}_0 : g(\theta) = \mathbf{0}$.

Chapter 10

Classical Linear Regression

Abstract: In this chapter, we introduce the classical linear regression theory, including the classical model assumptions, the statistical properties of the Ordinary Least Squares (OLS) estimator, the t-test and the F-test, as well as the Generalized Least Squares (GLS) estimator and related procedures. Various applications in economics and finance are also used to illustrate the applications of the statistical procedures.

Key words: Autocorrelation, Classical linear regression, Conditional heteroskedasticity, Conditional homoskedasticity, F-test, GLS, Hypothesis testing, Model selection criterion, OLS, R^2, t-test.

10.1 Classical Linear Regression Model

Suppose we have an observable random sample $\{Z_i\}_{i=1}^n$ of size n, where $Z_i = (Y_i, X_i')'$, Y_i is a scalar, $X_i = (1, X_{1i}, X_{2i}, \ldots, X_{ki})'$ is a $(k+1) \times 1$ vector, the index i may denote an individual unit (e.g., a firm, a household, or a country) for cross-sectional observations, or a time period (e.g., day, week, month, or year) for time series observations, and n is the sample size. We are interested in making inferences of the conditional mean $E(Y_i|X_i)$ based on an observed realization (i.e., a data set) of the random sample $\{Y_i, X_i'\}_{i=1}^n$. Throughout this chapter, we set $p \equiv k+1$, where k is the number of explanatory variables except the intercept, and p is the number of unknown parameters.

We consider the following linear regression model

$$Y_i = \alpha + \sum_{j=1}^k \beta_j X_{ji} + \varepsilon_i$$
$$= X_i'\theta + \varepsilon_i, \qquad i = 1, \ldots, n,$$

497

where $\theta = (\alpha, \beta_1, \ldots, \beta_k)'$ is a $p \times 1$ parameter vector, and ε_i is an unobservable disturbance. Here, Y_i is called the dependent variable (or regressand), the $p \times 1$ vector X_i is called the set of regressors (or independent variables, or explanatory variables).

Suppose there exists a unique parameter value θ_0 such that

$$E(Y_i|X_i) = X_i'\theta_0.$$

Then the linear regression model $Y_i = X_i'\theta + \varepsilon_i$ is said to be correctly specified for the conditional mean $E(Y_i|X_i)$, and θ_0 is usually called the true parameter value. Correct model specification implies

$$E(\varepsilon_i|X_i) = 0$$

and vice versa.

In contrast, if for all parameter values θ, we have

$$E(Y_i|X_i) \neq X_i'\theta,$$

then the linear regression model $Y_i = X_i'\theta + \varepsilon_i$ is said to be misspecified for the conditional mean $E(Y_i|X_i)$. In this chapter, we assume that the linear regression model is correctly specified for $E(Y_i|X_i)$.

When the linear model is correctly specified for the conditional mean $E(Y_i|X_i)$, the true model parameter value

$$\theta_0 = \frac{d}{dX_i} E(Y_i|X_i)$$

can be interpreted as the expected marginal effect of X_i on Y_i. For example, if X_i is income and Y_i is consumption, then θ_0 is the expected marginal propensity to consume.

The key notion of *linearity* in the classical linear regression model is that the regression model is linear in θ rather than in X_i. In other words, the class of linear regression models covers some models for which Y_i has a nonlinear relationship with X_i. For example, when $X_i = (1, X_{1i}, X_{1i}^2, \ldots, X_{1i}^k)'$, Y_i is a k-th order polynomial in X_{1i}, but it is still a linear regression model.

The linear regression model, even if correctly specified, does not necessarily imply a causal relationship from X_i to Y_i, namely, a change in X_i causing a change in Y_i. As Kendall and Stuart (1961, V.2, Ch. 26, p. 279) point out, "a statistical relationship, however strong and however suggestive, can never establish causal connection. Our ideas of causation must come from outside statistics ultimately, from some theory or other." A linear regression model only describes a predictive linear relationship: given X_i, can we predict Y_i linearly?

Denote

$$\mathbf{Y} = (Y_1, \ldots, Y_n)', \qquad n \times 1,$$
$$\varepsilon = (\varepsilon_1, \ldots, \varepsilon_n)', \qquad n \times 1,$$
$$\mathbf{X} = (X_1, \ldots, X_n)', \qquad n \times K,$$

where the i-th row of $\mathbf{X}$ is $X_i' = (1, X_{1i}, \ldots, X_{ki})$. With these matrix notations, we can compactly write the linear regression model as

$$\mathbf{Y} = \mathbf{X}\theta_0 + \varepsilon,$$
$$n \times 1 = (n \times p)(p \times 1) + n \times 1.$$

Throughout this chapter, we assume the following condition holds:

$$E(\varepsilon_i | \mathbf{X}) = E(\varepsilon_i | X_1, \ldots, X_i, \ldots, X_n) = 0, \qquad i = 1, \ldots, n.$$

This is called a *strict exogeneity* condition. Among other things, it implies correct model specification for $E(Y_i | X_i)$, because it implies $E(\varepsilon_i | X_i) = 0$ by the law of iterated expectations. It also implies $E(\varepsilon_i) = 0$.

Under the strict exogeneity condition, we have $E(X_j \varepsilon_i) = 0$ for any (i, j), where $i, j \in \{1, \ldots, n\}$. This follows because

$$E(X_j \varepsilon_i) = E[E(X_j \varepsilon_i | \mathbf{X})]$$
$$= E[X_j E(\varepsilon_i | \mathbf{X})]$$
$$= E(X_j \cdot 0)$$
$$= 0.$$

Given $E(\varepsilon_i) = 0$, we have $\text{cov}(X_j, \varepsilon_i) = 0$ for all $i, j \in \{1, \ldots, n\}$.

Because $\mathbf{X}$ contains regressors $\{X_j\}$ for both $j \leq i$ and $j > i$, the strict exogeneity condition essentially requires that the disturbance ε_i do not depend on the past and future values of regressors if i is a time index. This rules out dynamic time series models for which the regressors X_i may contain lagged dependent variables (e.g., Y_{i-1}, Y_{i-2}). In a dynamic regression model, ε_i may be correlated with the future values of regressors. For example, we consider a first order AutoRegressive (AR(1)) model

$$Y_i = \alpha_0 + \beta_0 Y_{i-1} + \varepsilon_i,$$
$$= X_i' \theta_0 + \varepsilon_i, \qquad i = 1, \ldots, n,$$
$$\{\varepsilon_i\} \sim \text{IID}(0, \sigma^2),$$

where $X_i = (1, Y_{i-1})'$ contains a lagged dependent variable Y_{i-1}. This is a dynamic regression model because the term $\beta_0 Y_{i-1}$ represents the "memory" or "feedback" of the past into the present value of the process, which induces a correlation between Y_i and the past. The term autoregression refers to the regression of Y_i on its own past values. The parameter β_0 determines the strength of feedback, with a large absolute value of β_0 resulting in stronger feedback. The disturbance ε_i can be viewed as representing the effect of "new information" that arrives at time i. The new shock cannot be anticipated so that the effect of today's news should be unrelated to the effect of yesterday's news in the sense that $E(\varepsilon_i|X_i) = 0$. Here, we make a stronger assumption that the new information $\{\varepsilon_i\}$ follows an IID $(0, \sigma^2)$ sequence.

Obviously, $E(X_i\varepsilon_i) = E(X_i)E(\varepsilon_i) = 0$ but $E(X_{i+1}\varepsilon_i) \neq 0$. Thus, we have $E(\varepsilon_i|\mathbf{X}) \neq 0$, and so strict exogeneity does not hold. Here, the lagged dependent variable Y_{i-1} contained in X_i is called a predetermined variable, since it is orthogonal to ε_i but depends on the past history of $\{\varepsilon_i\}$.

We would like to emphasize that the main reason of imposing the strict exogeneity condition is to obtain a finite sample distribution theory. For a large sample theory (i.e., an asymptotic theory), the strict exogeneity condition will not be needed.

In econometrics, there are alternative definitions of exogeneity. For example, one can assume that ε_i and $\mathbf{X}$ are independent. Another example is that $\mathbf{X}$ is nonstochastic. These examples rule out conditional heteroskedasticity (i.e., var$(\varepsilon_i|\mathbf{X})$ depends on $\mathbf{X}$). Under the strict exogeneity condition, we still allow for conditional heteroskedasticity, because we do not assume that ε_i and $\mathbf{X}$ are independent. We only assume that the conditional mean $E(\varepsilon_i|\mathbf{X})$ does not depend on $\mathbf{X}$. It is possible that var$(\varepsilon_i|\mathbf{X})$ depends on $\mathbf{X}$, as in the case of GLS estimation. See Section 10.9 for detailed discussion.

We now study two special cases. The first case is that $\mathbf{X}$ is nonstochastic. In this case, we have

$$E(\varepsilon_i|\mathbf{X}) = E(\varepsilon_i) = 0,$$

namely, the strict exogeneity condition is equivalent to the condition of $E(\varepsilon_i) = 0$. An example of nonstochastic $\mathbf{X}$ is $X_i = (1, i, \ldots, i^k)'$, where i denotes time. This corresponds to a time-trend regression model

$$Y_i = X_i'\theta_0 + \varepsilon_i$$
$$= \alpha_0 + \sum_{j=1}^{k} \beta_{j0}i^j + \varepsilon_i.$$

Next, we consider another case where $Z_i = (Y_i, X'_i)'$ is an independent random sample (i.e., Z_i and Z_j are independent whenever $i \neq j$, although Y_i and X_i may not be independent). In this case, we have

$$E(\varepsilon_i | \mathbf{X}) = E(\varepsilon_i | X_1, \ldots X_i, \ldots, X_n)$$
$$= E(\varepsilon_i | X_i)$$
$$= 0.$$

Thus, when $\{Z_i\}$ is IID, $E(\varepsilon_i | \mathbf{X}) = 0$ is equivalent to $E(\varepsilon_i | X_i) = 0$, the correct model specification condition.

In addition to strict exogeneity, we also impose a so-called *spherical error variance* assumption:

- *(a) [conditional homoskedasticity]:*

$$E(\varepsilon_i^2 | \mathbf{X}) = \sigma^2 > 0, \qquad i = 1, \ldots, n;$$

- *(b) [conditional non-autocorrelation]:*

$$E(\varepsilon_i \varepsilon_j | \mathbf{X}) = 0, \qquad i \neq j, i, j \in \{1, \ldots, n\}.$$

Condition (a) implies that there exists conditional homoskedasticity in $\{\varepsilon_i\}$, since

$$\mathrm{var}(\varepsilon_i | \mathbf{X}) = E(\varepsilon_i^2 | \mathbf{X}) - [E(\varepsilon_i | \mathbf{X})]^2$$
$$= \sigma^2.$$

Condition (b) implies that there exists no autocorrelation in $\{\varepsilon_i\}$, since for all $i \neq j$,

$$\mathrm{cov}(\varepsilon_i, \varepsilon_j | \mathbf{X}) = E(\varepsilon_i \varepsilon_j | \mathbf{X}) - E(\varepsilon_i | \mathbf{X}) E(\varepsilon_j | \mathbf{X})$$
$$= 0.$$

Thus, there exists no serial correlation between ε_i and its lagged values when i is an index for time, or there exists no spatial correlation between the disturbances associated with different cross-sectional units when i denotes a cross-sectional unit (e.g., consumer, firm, or household). In both cases, we say that there exists no autocorrelation in $\{\varepsilon_i\}$.

The strict exogeneity and spherical error variance assumptions can be compactly written as

$$E(\varepsilon | \mathbf{X}) = 0,$$

$$E(\varepsilon \varepsilon' | \mathbf{X}) = \sigma^2 \mathbf{I},$$

where $\mathbf{I}$ is a $n \times n$ identity matrix.

The strict exogeneity and spherical error variance assumptions together do not necessarily imply that ε_i and $\mathbf{X}$ are independent. They only impose restrictions on the conditional first two moments of ε_i and allow conditional higher order moments (e.g., skewness and kurtosis) of ε_i to depend on $\mathbf{X}$.

To identify the true parameter value θ_0, we shall assume that the $p \times p$ square matrix $\mathbf{X'X}$ is not singular throughout this chapter. This rules out multicollinearity among the p regressors in X_i. We say that there exists multicollinearity (sometimes called exact or perfect multicollinearity in the literature) among the components in X_i if for all $i \in \{1, \ldots, n\}$ and for some $j \in \{0, 1, \ldots, k\}$, the variable X_{ji} is a linear combination of all other variables $\{X_{li}, l \neq j\}$. In this case, the matrix $\mathbf{X'X}$ is singular, and as a result, the true parameter value θ_0 is not identifiable.

The nonsingularity of $\mathbf{X'X}$ implies that $\mathbf{X}$ must be of full rank of p. Thus, we need $p \leq n$. That is, the number of regressors cannot be larger than the sample size. This is a necessary condition for identification of parameter θ_0. Intuitively, if there are no or little variations in the values of X_i, it will be difficult to determine the relationship between Y_i and X_i. Indeed, the purpose of classical linear regression analysis is to investigate how a change in X_i can predict a change in Y_i. In certain sense, one may call $\mathbf{X'X}$ the "information matrix" of the random sample $\mathbf{X}$ because it is a measure of the information contained in $\mathbf{X}$. The magnitude of $\mathbf{X'X}$ will affect the preciseness of parameter estimation of θ_0. If there exists an approximate linear relationship among the sample values of explanatory variables in X_i such that although $\mathbf{X'X}$ is nonsingular, its minimum eigenvalue does not grow as the sample size n increases. This is called near-multicollinearity. In this case, the OLS estimator to be introduced below is well-defined and has a well-behaved finite sample distribution, but its variance never vanishes to zero as $n \to \infty$. In other words, the OLS estimator will never converge in probability to the true parameter value θ_0, although it will still have a well-defined finite sample distribution.

10.2 OLS Estimation

Question: How to estimate θ_0 using an observed data set generated from the random sample $\{Z_i\}_{i=1}^n$, where $Z_i = (Y_i, X_i')'$?

We first introduce the method of OLS estimation.

Definition 10.1. [OLS Estimator]: Define the sum of squared residuals (SSR) of the linear regression model $Y_i = X_i'\theta + \varepsilon_i$ as

$$SSR(\theta) \equiv (\mathbf{Y} - \mathbf{X}\theta)'(\mathbf{Y} - \mathbf{X}\theta)$$

$$= \sum_{i=1}^{n}(Y_i - X_i'\theta)^2.$$

Then the OLS estimator $\hat{\theta}$ is the solution to

$$\hat{\theta} = \arg\min_{\theta \in \mathbb{R}^p} SSR(\theta).$$

Theorem 10.1. *[Existence of OLS]:* *Suppose the $p \times p$ matrix $\mathbf{X}'\mathbf{X}$ is nonsingular. Then the OLS estimator $\hat{\theta}$ exists and is given by*

$$\hat{\theta} = (\mathbf{X}'\mathbf{X})^{-1}\mathbf{X}'\mathbf{Y}$$

$$= \left(\sum_{i=1}^{n} X_i X_i'\right)^{-1} \sum_{i=1}^{n} X_i Y_i.$$

Proof: Using the formula that for $p \times 1$ vectors A and θ, the derivative

$$\frac{\partial(A'\theta)}{\partial\theta} = A,$$

we obtain

$$\frac{dSSR(\theta)}{d\theta} = \frac{d}{d\theta}\sum_{i=1}^{n}(Y_i - X_i'\theta)^2$$

$$= \sum_{i=1}^{n}\frac{\partial}{\partial\theta}(Y_i - X_i'\theta)^2$$

$$= \sum_{i=1}^{n}2(Y_i - X_i'\theta)\frac{\partial}{\partial\theta}(Y_i - X_i'\theta)$$

$$= -2\sum_{i=1}^{n}X_i(Y_i - X_i'\theta)$$

$$= -2\mathbf{X}'(Y - \mathbf{X}\theta).$$

The OLS estimator must satisfy the FOC:

$$-2\mathbf{X}'(\mathbf{Y} - \mathbf{X}\hat{\theta}) = 0.$$

It follows that

$$\mathbf{X}'\mathbf{X}\hat{\theta} = \mathbf{X}'\mathbf{Y}.$$

Since $\mathbf{X'X}$ is nonsingular, we have

$$\hat{\theta} = (\mathbf{X'X})^{-1}\mathbf{X'Y}.$$

Checking the SOC, the $p \times p$ Hessian matrix

$$\frac{\partial^2 SSR(\theta)}{\partial\theta\partial\theta'} = -2\sum_{i=1}^{n} \frac{\partial}{\partial\theta'}\left[(Y_i - X_i'\theta)X_i\right]$$
$$= 2\mathbf{X'X}$$

is positive definite (why?). Thus, $\hat{\theta}$ is the global minimizer. This completes the proof.

Put

$$\hat{Y}_i \equiv X_i'\hat{\theta}.$$

This is called the fitted value or predicted value for Y_i, and

$$e_i \equiv Y_i - \hat{Y}_i$$

is called the estimated residual or prediction error for Y_i. Note that

$$\begin{aligned}
e_i &= Y_i - \hat{Y}_i \\
&= (X_i'\theta_0 + \varepsilon_i) - X_i'\hat{\theta} \\
&= \varepsilon_i - X_i'(\hat{\theta} - \theta_0),
\end{aligned}$$

where ε_i is the unavoidable true disturbance ε_i, and $X_i'(\hat{\theta} - \theta_0)$ is an estimation error, which becomes smaller when a larger data set is available (so $\hat{\theta}$ becomes closer to θ_0 as n increases).

The FOC implies that the $n \times 1$ estimated residual vector

$$e = \mathbf{Y} - \mathbf{X}\hat{\theta}$$

is orthogonal to the $p \times n$ regressor matrix $\mathbf{X}$ in the sense that

$$\mathbf{X}'e = \sum_{i=1}^{n} X_i e_i = 0.$$

This is the result from the very nature of OLS, as implied by the FOC of $\min_{\theta\in\mathbb{R}^p} SSR(\theta)$. It always holds no matter whether the strict exogeneity condition $(E(\varepsilon_i|\mathbf{X}) = 0)$ holds, which implies correct model specification. Note that if X_i contains an intercept, then $\mathbf{X}'e = 0$ implies

$$\mathbf{X}'\mathbf{i} = \sum_{i=1}^{n} e_i = 0,$$

where $\mathbf{i} = (1,\ldots,1)'$ is a $n \times 1$ vector of ones.

10.3 Goodness of Fit and Model Selection Criteria

Question: How well does a linear regression model fit the data? That is, how well do the predicted values $\{\hat{Y}_i\}_{i=1}^{n}$ explain the variations of the observed data of $\{Y_i\}_{i=1}^{n}$?

We need some criteria or measures to characterize goodness of fit. We first introduce two measures for goodness of fit. The first is called the uncentered squared multi-correlation coefficient.

Definition 10.2. [Uncentered R^2]: The uncentered squared multi-correlation coefficient, denoted as R_{uc}^2, is defined as

$$R_{uc}^2 = \frac{\hat{\mathbf{Y}}'\hat{\mathbf{Y}}}{\mathbf{Y}'\mathbf{Y}} = 1 - \frac{e'e}{\mathbf{Y}'\mathbf{Y}},$$

where $\hat{\mathbf{Y}}$ is the $n \times 1$ vector of fitted values, and the second equality follows from the first order condition of the OLS estimation.

The measure R_{uc}^2 has a nice interpretation: the proportion of the uncentered sample quadratic variations in the dependent variable Y_i that can be attributed to the uncentered sample quadratic variations of the predictor $\hat{Y}_i$. By definition, we always have $0 \le R_{uc}^2 \le 1$.

Next, we define a closely related measure called the centered squared multi-correlation coefficient, denoted as R^2.

Definition 10.3. [Centered R^2 — Coefficient of Determination]: The coefficient of determination

$$R^2 \equiv 1 - \frac{\sum_{i=1}^{n} e_i^2}{\sum_{i=1}^{n}(Y_i - \bar{Y})^2},$$

where $\bar{Y} = n^{-1} \sum_{i=1}^{n} Y_i$ is the sample mean.

When X_i contains an intercept, we have the following orthogonal decomposition:

$$
\begin{aligned}
\sum_{i=1}^{n}(Y_i - \bar{Y})^2 &= \sum_{i=1}^{n}(\hat{Y}_i - \bar{Y} + Y_i - \hat{Y}_i)^2 \\
&= \sum_{i=1}^{n}(\hat{Y}_i - \bar{Y})^2 + \sum_{i=1}^{n} e_i^2 \\
&\quad + 2\sum_{i=1}^{n}(\hat{Y}_i - \bar{Y})e_i \\
&= \sum_{i=1}^{n}(\hat{Y}_i - \bar{Y})^2 + \sum_{i=1}^{n} e_i^2,
\end{aligned}
$$

where the cross-product term

$$\sum_{i=1}^{n}(\hat{Y}_i - \bar{Y})e_i = \sum_{i=1}^{n}\hat{Y}_i e_i - \bar{Y}\sum_{i=1}^{n}e_i$$

$$= \hat{\theta}'\sum_{i=1}^{n}X_i e_i - \bar{Y}\sum_{i=1}^{n}e_i$$

$$= \hat{\theta}'(\mathbf{X}'e) - \bar{Y}\mathbf{i}'e$$

$$= \hat{\theta}' \cdot 0 - \bar{Y} \cdot 0$$

$$= 0,$$

where we have made use of the facts that $\mathbf{X}'e = 0$ and $\mathbf{i}'e = 0$ from the FOC of OLS estimation and the fact that X_i contains an intercept. It follows that

$$R^2 \equiv 1 - \frac{e'e}{\sum_{i=1}^{n}(Y_i - \bar{Y})^2}$$

$$= \frac{\sum_{i=1}^{n}(Y_i - \bar{Y})^2 - \sum_{i=1}^{n}e_i^2}{\sum_{i=1}^{n}(Y_i - \bar{Y})^2}$$

$$= \frac{\sum_{i=1}^{n}(\hat{Y}_i - \bar{Y})^2}{\sum_{i=1}^{n}(Y_i - \bar{Y})^2},$$

and consequently we have

$$0 \le R^2 \le 1.$$

On the other hand, if X_i does not contain an intercept, then the orthogonal decomposition identity

$$\sum_{i=1}^{n}(Y_i - \bar{Y})^2 = \sum_{i=1}^{n}(\hat{Y}_i - \bar{Y})^2 + \sum_{i=1}^{n}e_i^2$$

will no longer holds. As a result, R^2 may be negative because the cross-product term $2\sum_{i=1}^{n}(\hat{Y}_i - \bar{Y})e_i$ may be negative.

When X_i contains an intercept, the centered R^2 has a similar interpretation to the uncentered R_{uc}^2. That is, R^2 measures the proportion of the sample variance of $\{Y_i\}_{i=1}^{n}$ that can be explained by the predicted values $\{\hat{Y}_i\}_{i=1}^{n}$.

Example 10.1. [**Capital Asset Pricing Model (CAPM) and Economic Interpretation of** R^2]: The classical CAPM is characterized by the equation

$$r_i - r_{fi} = \alpha + \beta(r_{mi} - r_{fi}) + \varepsilon_i, \qquad i = 1, \ldots, n,$$

where r_i is the return on some portfolio (or asset) in time period i, r_{fi} is the return on a risk-free asset in time period i, and r_{mi} is the return on the market portfolio in time period i. Here, $r_i - r_{fi}$ is the risk premium of the portfolio, $r_{mi} - r_{fi}$ is the risk premium of the market portfolio, which is the only systematic risk factor, and ε_i is the idiosyncratic risk of the portfolio or asset in time period i. In this model, R^2 has an interesting economic interpretation: it is the proportion of the risk of the portfolio (as measured by the sample variance of its risk premium $r_i - r_{fi}$) that is attributed to the market risk factor $(r_{mi} - r_{fi})$. In contrast, $1 - R^2$ is the proportion of the risk of the portfolio that is contributed by the idiosyncratic risk factor ε_i.

The centered R^2 is essentially the squared sample correlation coefficient between $\{Y_i\}_{i=1}^n$ and $\{\hat{Y}_i\}_{i=1}^n$, as is stated below.

Theorem 10.2. $R^2 = \hat{\rho}_{Y\hat{Y}'}^2$, where $\hat{\rho}_{Y\hat{Y}}$ is the sample correlation coefficient between $\{Y_i\}_{i=1}^n$ and $\{\hat{Y}_i\}_{i=1}^n$.

Proof: Left as an exercise.

Because the fitted value $\hat{Y}_i = X_i'\hat{\theta} = \hat{\alpha} + \sum_{j=1}^k \hat{\beta}_j X_{ji}$ is a linear combination of variables $\{X_{ji}\}_{j=0}^k$, where $\hat{\theta} = (\hat{\alpha}, \hat{\beta}_1, \ldots, \hat{\beta}_k)'$, R^2 can be viewed as the square of a weighted average of multi-sample correlation coefficients between Y_i and $\{X_{ji}\}_{j=1}^k$. This is the reason why R^2 is called a squared multi-correlation coefficient.

Theorem 10.3 below shows that for any given random sample $\{Y_i, X_i'\}_{i=1}^n$, R^2 is nondecreasing in the number of explanatory variables $\{X_{ji}\}_{j=0}^k$. In other words, the more explanatory variables are added in the linear regression, the higher R^2 is. This is true no matter whether X_i has any true explanatory power for Y_i.

Theorem 10.3. *Suppose* $\{Y_i, X_{1i}, \ldots, X_{(k+q)i}\}_{i=1}^n$ *is a random sample of size* n. *Let* R_1^2 *be the centered* R^2 *from the linear regression*

$$Y_i = X_i'\theta + \varepsilon_i,$$

where $X_i = (1, X_{1i}, \ldots, X_{ki})'$, *and* θ *is a* $p \times 1$ *parameter vector; also,* R_2^2 *is the centered* R^2 *from the extended linear regression*

$$Y_i = \tilde{X}_i'\gamma + u_i,$$

where $\tilde{X}_i = (1, X_{1i}, \ldots, X_{ki}, X_{(k+1)i}, \ldots, X_{(k+q)i})'$, γ *is a* $(p+q) \times 1$ *parameter vector, and* q *is a positive integer. Then*

$$R_2^2 \geq R_1^2.$$

Proof: By definition of R^2, we have

$$R_1^2 = 1 - \frac{e'e}{\sum_{i=1}^n (Y_i - \bar{Y})^2},$$

$$R_2^2 = 1 - \frac{\tilde{e}'\tilde{e}}{\sum_{i=1}^n (Y_i - \bar{Y})^2},$$

where e is the $n \times 1$ estimated residual vector from the regression of $\mathbf{Y}$ on $\mathbf{X}$, and $\tilde{e}$ is the $n \times 1$ estimated residual vector from the regression of $\mathbf{Y}$ on $\tilde{\mathbf{X}}$. It suffices to show $\tilde{e}'\tilde{e} \leq e'e$. Because the OLS estimator $\hat{\gamma} = (\tilde{\mathbf{X}}'\tilde{\mathbf{X}})^{-1}\tilde{\mathbf{X}}'\mathbf{Y}$ minimizes $SSR(\gamma)$ for the extended regression model, we have

$$\tilde{e}'\tilde{e} = \sum_{i=1}^n (Y_i - \tilde{X}_i'\hat{\gamma})^2 \leq \sum_{i=1}^n (Y_i - \tilde{X}_i'\gamma)^2 \text{ for all } \gamma \in \mathbb{R}^{K+q}.$$

Now we choose

$$\gamma = (\hat{\theta}', 0')',$$

where $\hat{\theta} = (\mathbf{X}'\mathbf{X})^{-1}\mathbf{X}'\mathbf{Y}$ is the OLS from the first regression. It follows that

$$\tilde{e}'\tilde{e} \leq \sum_{i=1}^n \left(Y_i - X_i'\hat{\theta} - \sum_{j=k+1}^{k+q} 0 \cdot X_{ji} \right)^2$$

$$= \sum_{i=1}^n (Y_i - X_i'\hat{\theta})^2$$

$$= e'e.$$

Hence, we have $R_1^2 \leq R_2^2$. This completes the proof.

Theorem 10.3 has important implications. First, R^2 can be used to compare models with the same number of predictors, but it is not a useful criterion for comparing models of different sizes because it is biased in favor of large models. Second, R^2 is not a suitable criterion for correct model specification. It is a measure for sampling variations rather than a measure of population. A high value of R^2 does not necessarily imply correct model specification, and correct model specification also does not necessarily imply a high value of R^2.

Strictly speaking, R^2 is a measure merely of association with nothing to say about causality. High values of R^2 are often easy to obtain when dealing with economic time series data, even when the causal link between two variables is extremely tenuous or perhaps nonexistent. For example, in the so-called spurious regressions where the dependent variable Y_i and the regressors X_i have no causal relationship but they display similar trending behaviors over time, it is often found that R^2 is close to unity.

Finally, R^2 is a measure of the strength of linear association between the dependent variable Y_i and the regressor vector X_i. It is not a suitable measure for goodness of fit of a nonlinear regression model where $E(Y_i|X_i)$ is a nonlinear function of X_i. For example, consider a linear regression model

$$\ln Y_i = \alpha_0 + \beta_{10} \ln L_i + \beta_{20} \ln K_i + \varepsilon_i,$$

where Y_i is output, L_i is labor and K_i is capital. Here, output Y_i is not a linear function of inputs L_i and K_i. In this case, R^2 is the proportion of the total sample variations in $\ln Y_i$ that can be attributed to the sample variations in $\ln L_i$ and $\ln K_i$. It is not the proportion of the sample quadratic variations in Y_i that can be attributed to the sample variations of L_i and K_i.

Question: Then, what will be appropriate model selection criteria for linear regression models?

Often, a large number of potential explanatory variables are available, but we do not necessarily want to include all of them. There are two conflicting factors here: on one hand, a larger model has less systematic bias and it would give the best predictions if all parameters could be estimated without error. On the other hand, when unknown parameters are replaced by their estimates, the prediction becomes less accurate, and this effect is worse when there are more parameters to estimate. An important idea in statistics is to use a simple model to capture essential information contained in a data as much as possible. This is often called the KISS principle, namely *"Keep It Sophisticatedly Simple"*!

Below, we introduce three popular model selection criteria that reflect such an idea.

Criterion 1: Akaike Information Criterion (AIC)

A linear regression model can be selected by minimizing the following AIC criterion with a suitable choice of p:

$$AIC = \ln(s^2) + \frac{2p}{n},$$

where

$$s^2 = \frac{e'e}{n-p}$$

is called the residual variance estimator for $E(\varepsilon_i^2) = \sigma^2$, and $p = k + 1$ is the number of regressors. The first term $\ln(s^2)$ measures goodness of fit of the model, and the second term $2p/n$ measures model complexity. AIC is proposed by Akaike (1973).

Criterion 2: Bayesian Information Criterion (BIC, Schwarz 1978)

A linear regression model can be selected by minimizing the following criterion with a suitable choice of p:

$$BIC = \ln(s^2) + \frac{p\ln(n)}{n}.$$

This is called the Bayesian information criterion (BIC), proposed by Schwarz (1978).

Both AIC and BIC try to trade off the goodness of fit to data measured by $\ln(s^2)$ with the desire to use as few parameters as possible. When $\ln n \geq 2$, which is the case when $n > 7$, BIC gives a heavier penalty for model complexity than AIC. As a result, BIC will tend to choose a more parsimonious linear regression model than AIC.

The difference between AIC and BIC is due to the way in which they are constructed. AIC is designed to select a model that will predict best and is less concerned than BIC with having too many parameters. BIC is designed to select the true value of dimension p exactly. Under certain regularity conditions, BIC is strongly consistent in the sense that it determines the true model asymptotically (i.e., as $n \to \infty$), whereas for AIC, an overparameterized model often emerges no matter how large the sample is. Of course, such properties are not necessarily guaranteed in finite samples. In practice, the best AIC model is usually close to the best BIC model and often they deliver the same model.

Criterion 3: Adjusted R^2

In addition to AIC and BIC, there are other criteria such as $\bar{R}^2$, the so-called adjusted R^2, that can also be used to select a linear regression model. The adjusted R^2 is defined as

$$\bar{R}^2 = 1 - \frac{e'e/(n-p)}{(\mathbf{Y} - \bar{Y}\mathbf{i})'(\mathbf{Y} - \bar{Y}\mathbf{i})/(n-1)},$$

where $\mathbf{i} = (1, \ldots, 1)'$ is a $n \times 1$ vector with each element equal to unity. This differs from

$$R^2 = 1 - \frac{e'e}{(\mathbf{Y} - \bar{Y}\mathbf{i})'(\mathbf{Y} - \bar{Y}\mathbf{i})}.$$

For $\bar{R}^2$, the adjustment is made according to the degrees of freedom, or the number of explanatory variables in X_i. It may be shown that

$$\bar{R}^2 = 1 - \frac{n-1}{n-p}(1 - R^2).$$

We note that $\bar{R}^2$ may take a negative value although there is an intercept in X_i.

All model criteria are structured in terms of the estimated residual variance s^2 plus a penalty adjustment involving the number of estimated parameters, and it is in the extent of this penalty that the criteria differ from each other. For more discussion about these and other selection criteria, see Judge *et al.* (1988, Section 7.5).

Question: Why is it not a good practice to use a complicated model?

A complicated model contains many unknown parameters. Given a fixed amount of data information, parameter estimation will become less precise if more parameters have to be estimated. As a result, the out-of-sample forecast may become less precise than the forecast of a simpler model. The latter may have a larger bias but more precise parameter estimates. Intuitively, a complicated model is too flexible in the sense that it may capture not only systematic components but also some spurious features in the data which may not show up again in the future. Thus, it cannot forecast futures well.

10.4 Consistency and Efficiency of the OLS Estimator

We now investigate the statistical properties of the OLS estimator $\hat{\theta}$. We are interested in addressing the following questions:

- Is $\hat{\theta}$ a good estimator for θ_0 (consistency)?
- Is $\hat{\theta}$ the best estimator (efficiency)?
- What is the sampling distribution of $\hat{\theta}$ (normality)?

The distribution of $\hat{\theta}$ is called the sampling distribution of $\hat{\theta}$, because $\hat{\theta}$ is a function of the random sample $\{Y_i, X_i'\}_{i=1}^n$. The sampling distribution of $\hat{\theta}$ is useful for any statistical inference involving $\hat{\theta}$, such as confidence interval estimation and hypothesis testing.

We first investigate the statistical properties of $\hat{\theta}$.

Theorem 10.4. *Suppose* $\mathbf{X}'\mathbf{X}$ *is nonsingular,* $E(\varepsilon|\mathbf{X}) = 0$ *and* $E(\varepsilon\varepsilon'|\mathbf{X}) = \sigma^2\mathbf{I}$, *where* $\mathbf{I}$ *is a* $n \times n$ *identity matrix. Then for all* $n > p$,
  (a) [Unbiasedness]

$$E(\hat{\theta}|\mathbf{X}) = \theta_0 \text{ and } E(\hat{\theta}) = \theta_0.$$

  (b) [Variance Structure]

$$var(\hat{\theta}|\mathbf{X}) = E\left\{ \left[\hat{\theta} - E(\hat{\theta}|\mathbf{X})\right] \left[\hat{\theta} - E(\hat{\theta}|\mathbf{X})\right]' |\mathbf{X}\right\}$$
$$= \sigma^2(\mathbf{X}'\mathbf{X})^{-1}.$$

(c) [Orthogonality Between e and $\hat{\theta}$]

$$cov(\hat{\theta}, e|\mathbf{X}) = E\{[\hat{\theta} - E(\hat{\theta}|\mathbf{X})]e'|\mathbf{X}\} = 0.$$

(d) [Gauss-Markov Theorem]

$$var(\hat{b}|\mathbf{X}) - var(\hat{\theta}|\mathbf{X}) \text{ is positive semi-definite (PSD)}$$

for any unbiased estimator $\hat{b}$ that is linear in Y with $E(\hat{b}|\mathbf{X}) = \theta_0$.
(e) [Residual Variance Estimator]

$$s^2 = \frac{e'e}{n-p} = \frac{1}{n-p}\sum_{i=1}^{n} e_i^2$$

is unbiased for $\sigma^2 = E(\varepsilon_i^2)$. That is, $E(s^2|\mathbf{X}) = \sigma^2$.

Proof: Given $\hat{\theta} = (\mathbf{X'X})^{-1}\mathbf{X'Y}$ and $\mathbf{Y} = \mathbf{X'}\theta_0 + \varepsilon$, we have

$$\hat{\theta} - \theta_0 = (\mathbf{X'X})^{-1}\mathbf{X'}\varepsilon.$$

It follows that

$$\begin{aligned}
E[(\hat{\theta} - \theta_0)|\mathbf{X}] &= E[(\mathbf{X'X})^{-1}\mathbf{X'}\varepsilon|\mathbf{X}] \\
&= (\mathbf{X'X})^{-1}\mathbf{X'}E(\varepsilon|\mathbf{X}) \\
&= (\mathbf{X'X})^{-1}\mathbf{X'}0 \\
&= 0,
\end{aligned}$$

where we have made use of the strict exogeneity condition that $E(\varepsilon|\mathbf{X}) = 0$.
(b) Given $\hat{\theta} - \theta_0 = (\mathbf{X'X})^{-1}\mathbf{X'}\varepsilon$ and $E(\varepsilon\varepsilon'|\mathbf{X}) = \sigma^2\mathbf{I}$, we have

$$\begin{aligned}
var(\hat{\theta}|\mathbf{X}) &\equiv E\left\{\left[(\hat{\theta} - E(\hat{\theta}|\mathbf{X}))\right]\left[\hat{\theta} - E(\hat{\theta}|\mathbf{X})\right]'|\mathbf{X}\right\} \\
&= E\left[(\hat{\theta} - \theta_0)(\hat{\theta} - \theta_0)'|\mathbf{X}\right] \\
&= E[(\mathbf{X'X})^{-1}\mathbf{X'}\varepsilon\varepsilon'\mathbf{X}(\mathbf{X'X})^{-1}|\mathbf{X}] \\
&= (\mathbf{X'X})^{-1}\mathbf{X'}E(\varepsilon\varepsilon'|\mathbf{X})\mathbf{X}(\mathbf{X'X})^{-1} \\
&= (\mathbf{X'X})^{-1}\mathbf{X'}\sigma^2\mathbf{I}\mathbf{X}(\mathbf{X'X})^{-1} \\
&= \sigma^2(\mathbf{X'X})^{-1}\mathbf{X'X}(\mathbf{X'X})^{-1} \\
&= \sigma^2(\mathbf{X'X})^{-1},
\end{aligned}$$

where we have made use of the spherical error variance assumption that $E(\varepsilon\varepsilon'|\mathbf{X}) = \sigma^2\mathbf{I}$, which is the key to obtain the expression of $\sigma^2(\mathbf{X'X})^{-1}$ for $var(\hat{\theta}|\mathbf{X})$.

(c) Define a $n \times n$ projection matrix

$$\mathbf{P} = \mathbf{X}(\mathbf{X}'\mathbf{X})^{-1}\mathbf{X}'$$

and a $n \times n$ matrix

$$\mathbf{M} = \mathbf{I} - \mathbf{P}.$$

Then both $\mathbf{P}$ and $\mathbf{M}$ are symmetric matrices, with $\mathbf{PX} = \mathbf{X}, \mathbf{MX} = 0, \mathbf{P}^2 = \mathbf{P}$ and $\mathbf{M}^2 = \mathbf{M}$. Given $\hat{\theta} - \theta_0 = (\mathbf{X}'\mathbf{X})^{-1}\mathbf{X}'\varepsilon$ and $\mathbf{MX} = 0$, we have

$$e = \mathbf{Y} - \mathbf{X}\hat{\theta} = \mathbf{MY} = \mathbf{M}\varepsilon.$$

It follows that

$$
\begin{aligned}
\mathrm{cov}(\hat{\theta}, e | \mathbf{X}) &= E\left\{ \left[\hat{\theta} - E(\hat{\theta}|\mathbf{X})\right] [e - E(e|\mathbf{X})]' \,|\mathbf{X}\right\} \\
&= E\left[(\hat{\theta} - \theta_0)e'|\mathbf{X}\right] \\
&= E[(\mathbf{X}'\mathbf{X})^{-1}\mathbf{X}'\varepsilon\varepsilon'\mathbf{M}|\mathbf{X}] \\
&= (\mathbf{X}'\mathbf{X})^{-1}\mathbf{X}'E(\varepsilon\varepsilon'|\mathbf{X})\mathbf{M} \\
&= (\mathbf{X}'\mathbf{X})^{-1}\mathbf{X}'\sigma^2\mathbf{I}\mathbf{M} \\
&= \sigma^2(\mathbf{X}'\mathbf{X})^{-1}\mathbf{X}'\mathbf{M} \\
&= 0
\end{aligned}
$$

where we have made use of the assumptions that $E(\varepsilon|\mathbf{X}) = 0$ and $E(\varepsilon\varepsilon'|\mathbf{X}) = \sigma^2\mathbf{I}$, which are the keys to ensure zero correlation between $\hat{\theta}$ and e.

(d) Consider a linear estimator of θ_0:

$$\hat{b} = \mathbf{C}'\mathbf{Y},$$

where $\mathbf{C} = C(\mathbf{X})$ is a $n \times p$ matrix. This estimator is conditionally unbiased for θ_0 regardless of the value of θ_0 if and only if

$$
\begin{aligned}
E(\hat{b}|\mathbf{X}) &= \mathbf{C}'\mathbf{X}\theta_0 + \mathbf{C}'E(\varepsilon|\mathbf{X}) \\
&= \mathbf{C}'\mathbf{X}\theta_0 \\
&= \theta_0.
\end{aligned}
$$

This follows if and only if

$$\mathbf{C}'\mathbf{X} = \mathbf{I}.$$

Because

$$
\begin{aligned}
\hat{b} &= \mathbf{C}'\mathbf{Y} \\
&= \mathbf{C}'(\mathbf{X}\theta_0 + \varepsilon) \\
&= \mathbf{C}'\mathbf{X}\theta_0 + \mathbf{C}'\varepsilon \\
&= \theta_0 + \mathbf{C}'\varepsilon,
\end{aligned}
$$

the conditional variance of $\hat{b}$

$$
\begin{aligned}
\operatorname{var}(\hat{b}|\mathbf{X}) &= E\left\{ \left[\hat{b} - E(\hat{b}|\mathbf{X})\right]\left[\hat{b} - E(\hat{b}|\mathbf{X})\right]' |\mathbf{X}\right\} \\
&= E\left[(\hat{b} - \theta_0)(\hat{b} - \theta_0)'|\mathbf{X}\right] \\
&= E\left(\mathbf{C}'\varepsilon\varepsilon'\mathbf{C}|\mathbf{X}\right) \\
&= \mathbf{C}'E(\varepsilon\varepsilon'|\mathbf{X})\mathbf{C} \\
&= \mathbf{C}'\sigma^2\mathbf{I}\mathbf{C} \\
&= \sigma^2\mathbf{C}'\mathbf{C}.
\end{aligned}
$$

Using $\mathbf{C}'\mathbf{X} = \mathbf{I}$ and $\mathbf{M}^2 = \mathbf{M}$, we now have

$$
\begin{aligned}
\operatorname{var}(\hat{b}|\mathbf{X}) - \operatorname{var}(\hat{\theta}|\mathbf{X}) &= \sigma^2\mathbf{C}'\mathbf{C} - \sigma^2(\mathbf{X}'\mathbf{X})^{-1} \\
&= \sigma^2[\mathbf{C}'\mathbf{C} - \mathbf{C}'\mathbf{X}(\mathbf{X}'\mathbf{X})^{-1}\mathbf{X}'\mathbf{C}] \\
&= \sigma^2\mathbf{C}'[\mathbf{I} - \mathbf{X}(\mathbf{X}'\mathbf{X})^{-1}\mathbf{X}']\mathbf{C} \\
&= \sigma^2\mathbf{C}'\mathbf{M}\mathbf{C} \\
&= \sigma^2\mathbf{C}'\mathbf{M}\mathbf{M}\mathbf{C} \\
&= \sigma^2\mathbf{C}'\mathbf{M}'\mathbf{M}\mathbf{C} \\
&= \sigma^2(\mathbf{M}\mathbf{C})'(\mathbf{M}\mathbf{C}) \\
&= \sigma^2\mathbf{D}'\mathbf{D} \\
&\sim \text{PSD},
\end{aligned}
$$

where we have used the fact that for any real-valued $p \times n$ matrix $\mathbf{D} = \mathbf{M}\mathbf{C}$, the squared matrix $\mathbf{D}'\mathbf{D}$ is always positive semi-definite (PSD).

(e) Now we show $E(s^2|\mathbf{X}) = \sigma^2$. Because $e'e = \varepsilon'\mathbf{M}^2\varepsilon = \varepsilon'\mathbf{M}\varepsilon$ and $\operatorname{tr}(AB) = \operatorname{tr}(BA)$, we have

$$
\begin{aligned}
E(e'e|\mathbf{X}) &= E(\varepsilon'\mathbf{M}\varepsilon|\mathbf{X}) \\
&= E[\operatorname{tr}(\varepsilon'\mathbf{M}\varepsilon)|\mathbf{X}] \\
&= E[\operatorname{tr}(\varepsilon\varepsilon'\mathbf{M})|\mathbf{X}] \\
&= \operatorname{tr}[E(\varepsilon\varepsilon'|\mathbf{X})\mathbf{M}] \\
&= \operatorname{tr}(\sigma^2\mathbf{I}\mathbf{M}) \\
&= \sigma^2\operatorname{tr}(\mathbf{M}) \\
&= \sigma^2(n - p)
\end{aligned}
$$

where

$$
\begin{aligned}
\operatorname{tr}(\mathbf{M}) &= \operatorname{tr}(\mathbf{I}) - \operatorname{tr}(\mathbf{X}(\mathbf{X}'\mathbf{X})^{-1}\mathbf{X}') \\
&= \operatorname{tr}(\mathbf{I}) - \operatorname{tr}(\mathbf{X}'\mathbf{X}(\mathbf{X}'\mathbf{X})^{-1}) \\
&= n - p
\end{aligned}
$$

using $\text{tr}(AB) = \text{tr}(BA)$ again. It follows that

$$E(s^2|\mathbf{X}) = \frac{E(e'e|\mathbf{X})}{n-p}$$
$$= \frac{\sigma^2(n-p)}{(n-p)}$$
$$= \sigma^2.$$

Note that the residual variance estimator s^2 can be viewed as a generalization of the sample variance S_n^2 investigated in Chapter 6. This completes the proof.

Both Theorem 10.4(a) and (b) imply that the conditional MSE

$$MSE(\hat{\theta}|\mathbf{X}) = E\left[(\hat{\theta} - \theta_0)(\hat{\theta} - \theta_0)'|\mathbf{X}\right]$$
$$= \text{var}(\hat{\theta}|\mathbf{X}) + \text{Bias}(\hat{\theta}|\mathbf{X})\text{Bias}(\hat{\theta}|\mathbf{X})'$$
$$= \text{var}(\hat{\theta}|\mathbf{X}),$$

where we have used the fact that

$$\text{Bias}(\hat{\theta}|\mathbf{X}) \equiv E(\hat{\theta}|\mathbf{X}) - \theta_0 = 0.$$

Recall that MSE measures how close the estimator $\hat{\theta}$ is to the true parameter value θ_0.

Theorem 10.4(d) implies that $\hat{\theta}$ is the Best Linear Unbiased Estimator (BLUE) for θ_0 because $\text{var}(\hat{\theta}|\mathbf{X})$ is the smallest among all unbiased linear estimators for θ_0. This is called the Gauss-Markov theorem.

Formally, we can define a related concept for comparing two unbiased estimators:

Definition 10.4. **[Efficiency]:** An unbiased estimator $\hat{\theta}$ of parameter θ_0 is more efficient than another unbiased estimator $\hat{b}$ of parameter θ_0 if

$$\text{var}(\hat{b}|\mathbf{X}) - \text{var}(\hat{\theta}|\mathbf{X}) \text{ is PSD}.$$

When $\hat{\theta}$ is more efficient than $\hat{b}$, we have that for any $\tau \in \mathbf{R}^p$ such that $\tau'\tau = 1$,

$$\tau'\left[\text{var}(\hat{b}|\mathbf{X}) - \text{var}(\hat{\theta}|\mathbf{X})\right]\tau \geq 0.$$

Choosing $\tau = (0, \ldots, 1, 0, \ldots, 0)'$, for example, where the j-th element is unity and all other elements are zero, we have

$$\text{var}(\hat{b}_j) - \text{var}(\hat{\theta}_j) \geq 0, \text{ for } 1 \leq j \leq p.$$

We note that the OLS estimator $\hat{\theta}$ is still BLUE under the conditions of Theorem 10.4 even when there exists near-multicollinearity, where $\mathbf{X}'\mathbf{X}$ is nonsingular but its minimum eigenvalue does not grow as the sample size n increases. Near-multicollinearity is essentially a data problem which we cannot remedy or improve upon when the objective is to estimate the unknown parameter θ_0.

10.5 Sampling Distribution of the OLS Estimator

To obtain the finite sample sampling distribution of $\hat{\theta}$, we impose a conditional normality assumption on ε, namely,

$$\varepsilon | \mathbf{X} \sim N(0, \sigma^2 \mathbf{I}).$$

This implies both the strict exogeneity assumption $(E(\varepsilon | \mathbf{X}) = 0)$ and the spherical error variance assumption $(E(\varepsilon\varepsilon | \mathbf{X}) = \sigma^2 \mathbf{I})$. Moreover, the conditional PDF of ε given $\mathbf{X}$ is given by

$$f(\varepsilon | \mathbf{X}) = \frac{1}{(\sqrt{2\pi\sigma^2})^n} e^{-\frac{\varepsilon'\varepsilon}{2\sigma^2}} = f(\varepsilon),$$

which does not depend on $\mathbf{X}$. Therefore, the disturbance ε is independent of $\mathbf{X}$. All conditional moments of ε given $\mathbf{X}$ do not depend on $\mathbf{X}$.

Under the conditional normality assumption on ε, we can derive the finite sample distributions of $\hat{\theta}$ and related statistics, i.e., the distributions of $\hat{\theta}$ and related statistics when the sample size n is a finite integer. The conditional normality assumption may be reasonable for observations that are computed as the averages of the outcomes of many repeated experiments, due to the effect of the central limit theorem. This may occur in physics, for example. In economics, however, the normality assumption may not always be reasonable. For example, many high-frequency financial time series usually display heavy tails (with kurtosis larger than 3).

Question: What is the sampling distribution of $\hat{\theta}$?

Define a $p \times 1$ weighting vector

$$C_i = (\mathbf{X}'\mathbf{X})^{-1} X_i,$$

which is called the leverage of observation X_i. Then we can write

$$\hat{\theta} - \theta_0 = (\mathbf{X'X})^{-1}\mathbf{X'}\varepsilon$$
$$= (\mathbf{X'X})^{-1}\sum_{i=1}^{n} X_i\varepsilon_i$$
$$= \sum_{i=1}^{n} C_i\varepsilon_i.$$

Conditional on $\mathbf{X}, \hat{\theta} - \theta_0$ is a linear combination of an IID sequence $\{\varepsilon_i\}$. By the reproductive property of the normal distribution, $\hat{\theta} - \theta_0$ follows a conditional normal distribution, as is stated below.

Theorem 10.5. *[Conditional Normality of $\hat{\theta}$]: Suppose $X'X$ is non-singular and $\varepsilon|\mathbf{X} \sim N(0, \sigma^2\mathbf{I})$. Then for any given $n > p$,*

$$(\hat{\theta} - \theta_0)|\mathbf{X} \sim N(0, \sigma^2(\mathbf{X'X})^{-1}).$$

Proof: Conditional on $\mathbf{X}$, $\hat{\theta} - \theta_0$ is a weighted sum of IID normal random variables $\{\varepsilon_i\}$, and so it is also normally distributed by the reproductive property of the normal distribution.

We note that the OLS estimator $\hat{\theta} - \theta_0$ still has the conditional finite sample normal distribution $N(0, \sigma^2(\mathbf{X'X})^{-1})$ even when there exists near-multicollinearity.

A useful corollary follows from Theorem 10.5 immediately.

Corollary 10.1. *[Conditional Normality of $R(\hat{\theta} - \theta_0)$]: Suppose $\mathbf{X'X}$ is nonsingular and $\varepsilon|\mathbf{X} \sim N(0, \sigma^2\mathbf{I})$. Then for any nonstochastic $J \times p$ matrix R and any given sample size $n > p$, we have*

$$R(\hat{\theta} - \theta_0)|\mathbf{X} \sim N(0, \sigma^2 R(\mathbf{X'X})^{-1}R').$$

Proof: Conditional on $\mathbf{X}$, $\hat{\theta} - \theta_0$ is normally distributed. Therefore, conditional on $\mathbf{X}$, the linear combination $R(\hat{\theta} - \theta_0)$ is also normally distributed, with the conditional mean

$$E[R(\hat{\theta} - \theta_0)|\mathbf{X}] = RE[(\hat{\theta} - \theta_0)|\mathbf{X}] = 0$$

and the variance-covariance matrix

$$\text{var}[R(\hat{\theta} - \theta_0)|\mathbf{X}] = E\left\{R(\hat{\theta} - \theta_0)\left[R(\hat{\theta} - \theta_0)\right]'|\mathbf{X}\right\}$$

$$= E\left[R(\hat{\theta} - \theta_0)(\hat{\theta} - \theta_0)'R'|\mathbf{X}\right]$$

$$= RE\left[(\hat{\theta} - \theta_0)(\hat{\theta} - \theta_0)'|\mathbf{X}\right]R'$$

$$= R\text{var}(\hat{\theta}|\mathbf{X})R'$$

$$= \sigma^2 R(\mathbf{X}'\mathbf{X})^{-1}R'.$$

It follows that

$$R(\hat{\theta} - \theta_0)|\mathbf{X} \sim N(0, \sigma^2 R(\mathbf{X}'\mathbf{X})^{-1}R').$$

The $J \times K$ nonstochastic matrix R may be viewed as a selection matrix. For example, when $R = (0, \ldots, 1, \ldots, 0)$, where the j-th element is unity and all other elements are zero, we have $R(\hat{\theta} - \theta_0) = \hat{\theta}_j - \theta_j$ for $j \in \{1, \ldots, p\}$.

The sampling distribution of $R(\hat{\theta} - \theta_0)$ is important for confidence interval estimation and hypothesis testing, as will be seen below.

10.6 Variance Estimator of the OLS Estimator

Since the error variance $\text{var}(\varepsilon_i) = \sigma^2$ is unknown, $\text{var}[R(\hat{\theta} - \theta_0)|\mathbf{X}] = \sigma^2 R(\mathbf{X}'\mathbf{X})^{-1}R'$ is also unknown. We need to estimate σ^2. We can use the residual variance estimator

$$s^2 = \frac{e'e}{n - p}.$$

To investigate the statistical properties of s^2, we first introduce a lemma.

Lemma 10.1. *[Quadratic Form of Normal Random Variables]: Suppose a $m \times 1$ random vector $v \sim N(0, \mathbf{I})$ and Q is a $m \times m$ nonstochastic symmetric idempotent matrix with rank $q \le m$. Then the quadratic form*

$$v'Qv \sim \chi_q^2.$$

In our application below, we set $v = \varepsilon/\sigma$, and $Q = \mathbf{M}$. Since $\text{rank}(\mathbf{M}) = n - p$, we have

$$\frac{e'e}{\sigma^2}|\mathbf{X} \sim \chi_{n-p}^2.$$

Theorem 10.6. *[Residual Variance Estimator]: Suppose $\mathbf{X}'\mathbf{X}$ is nonsingular and $\varepsilon|\mathbf{X} \sim N(0, \sigma^2 \mathbf{I})$. Then for all $n > p$,*

(a)
$$\frac{(n-K)s^2}{\sigma^2}|\mathbf{X} = \frac{e'e}{\sigma^2}|\mathbf{X} \sim \chi^2_{n-p};$$

(b) *conditional on* $\mathbf{X}$, s^2 *and* $\hat{\theta}$ *are mutually independent.*

Proof: (a) Because $e = \mathbf{M}\varepsilon$, we have

$$\frac{e'e}{\sigma^2} = \frac{\varepsilon'\mathbf{M}\varepsilon}{\sigma^2} = \left(\frac{\varepsilon}{\sigma}\right)' \mathbf{M} \left(\frac{\varepsilon}{\sigma}\right).$$

In addition, because $\varepsilon|\mathbf{X} \sim N(0,\sigma^2\mathbf{I})$, and $\mathbf{M}$ is an idempotent matrix with rank $q = n - p$, we have the quadratic form

$$\frac{e'e}{\sigma^2}|\mathbf{X} = \frac{\varepsilon'\mathbf{M}\varepsilon}{\sigma^2}|\mathbf{X} \sim \chi^2_{n-p}$$

by Lemma 10.1.

(b) Next, we show that conditional on $\mathbf{X}$, s^2 and $\hat{\theta}$ are independent. Because s^2 is a function of e, it suffices to show that e and $\hat{\theta}$ are independent conditional on $\mathbf{X}$. Our strategy is to first prove that conditional on $\mathbf{X}$, both e and $\hat{\theta}$ have a joint normal distribution. Then, given $\text{cov}(\hat{\theta}, e|\mathbf{X}) = 0$ as shown in Theorem 10.4(e), e and $\hat{\theta}$ are conditionally independent, because for a joint normal distribution, zero correlation implies independence (cf. Theorem 5.26).

To show that e and $\hat{\theta}$ have a conditional joint normal distribution, we write

$$\begin{bmatrix} e \\ \hat{\theta} - \theta_0 \end{bmatrix} = \begin{bmatrix} \mathbf{M}\varepsilon \\ (\mathbf{X}'\mathbf{X})^{-1}\mathbf{X}'\varepsilon \end{bmatrix}$$
$$= \begin{bmatrix} \mathbf{M} \\ (\mathbf{X}'\mathbf{X})^{-1}\mathbf{X}' \end{bmatrix} \varepsilon.$$
$$= \mathbf{A}\varepsilon,$$

where the $(n+p) \times 1$ vector $\mathbf{A}$ depends on $\mathbf{X}$. Because $\varepsilon|\mathbf{X} \sim N(0,\sigma^2\mathbf{I})$, the linear combination $\mathbf{A}\varepsilon$ also follows a conditional normal distribution. It follows that e and $\hat{\theta}$ are conditionally independent given $\text{cov}(\hat{\theta}, e|\mathbf{X}) = 0$. This completes the proof.

Theorem 10.6 is a generalization of Theorems 6.2 and 6.8. To discuss the implications of Theorem 10.6, recall that the mean and variance of a χ^2_q distribution are equal to q and $2q$ respectively. Thus, Theorem 10.6(a) implies that

$$E\left[\frac{(n-p)s^2}{\sigma^2}|\mathbf{X}\right] = n - p$$

and

$$\text{var}\left[\frac{(n-p)s^2}{\sigma^2}\Big|\mathbf{X}\right] = 2(n-p).$$

It follows that $E(s^2|\mathbf{X}) = \sigma^2$ and

$$\text{var}(s^2|\mathbf{X}) = \frac{2\sigma^4}{n-p}.$$

Therefore, the conditional MSE of s^2

$$
\begin{aligned}
MSE(s^2|\mathbf{X}) &= E\left[(s^2 - \sigma^2)^2|\mathbf{X}\right] \\
&= \text{var}(s^2|\mathbf{X}) + [E(s^2|\mathbf{X}) - \sigma^2]^2 \\
&= \frac{2\sigma^4}{n-p},
\end{aligned}
$$

which vanishes to zero as $n \to \infty$.

The sample residual variance s^2 is a generalization of the sample variance $S_n^2 = (n-1)^{-1}\sum_{i=1}^{n}(Y_i - \bar{Y})^2$ for the random sample $\{Y_i\}_{i=1}^{n}$. The factor $n - p$ is called the number of degrees of freedom of the estimated residual sample $\{e_i\}_{i=1}^{n}$. To gain intuition why the number of degrees of freedom is equal to $n - p$, note that the original sample $\{Y_i, X_i'\}_{i=1}^{n}$ has n observations, which can be viewed to have n degrees of freedom. Now to estimate σ^2, we have to use the estimated residual sample $\{e_i\}_{i=1}^{n}$. These n estimated residuals are not linearly independent because they have to satisfy the FOC of OLS estimation, namely,

$$\mathbf{X}'e = 0.$$

$$(p \times n) \times (n \times 1) = p \times 1.$$

The FOC imposes p restrictions on $\{e_i\}_{i=1}^{n}$, which are required in order to estimate p unknown parameters in θ_0. Thus, the number of degrees of freedom of the estimated residual sample $\{e_i\}_{i=1}^{n}$ is $n - p$. Note that the sample variance S_n^2 is the residual variance estimator for a simple linear regression model with an intercept only: $Y_i = \theta_0 + \varepsilon_i$.

Question: Why are these sampling distributions of $\hat{\theta}$ and s^2 useful in practice?

The sampling distributions of $\hat{\theta}$ and s^2 are useful in confidence interval estimation and hypothesis testing on the true model parameter θ_0. In this chapter, we will focus on hypothesis testing. Statistically speaking, confidence interval estimation and hypothesis testing are just two sides of the same coin.

10.7 Hypothesis Testing

We now use the sampling distributions of $\hat{\theta}$ and s^2 to develop test procedures for hypotheses of interest. We consider testing the following linear hypothesis:

$$\mathbb{H}_0 \; : \; R\theta_0 = r,$$
$$(J \times p)(p \times 1) = J \times 1,$$

where R is a nonstochastic selection matrix, r is a nonstochastic row vector, and J is the number of restrictions on the p parameters contained in θ_0. We assume $J \leq p$, namely the number of restrictions is not more than the number of unknown parameters. It is important to emphasize that we shall test $\mathbb{H}_0$ under correct model specification for the conditional mean $E(Y_i | X_i)$.

We first provide some motivating examples.

Example 10.2. [Reforms Have No Effect]: Consider the extended production function

$$\ln(Y_i) = \alpha_0 + \beta_{10} \ln(L_i) + \beta_{20} \ln(K_i) + \beta_{30} AU_i + \beta_{40} PS_i + \varepsilon_i,$$

where AU_i is a dummy variable indicating whether firm i is granted autonomy, and PS_i is the profit share of firm i with the state. Suppose we are interested in testing whether autonomy AU_i has an effect on firm productivity. Then with $\theta_0 = (\alpha_0, \beta_{10}, \beta_{20}, \beta_{30}, \beta_{40})'$, we can write the null hypothesis

$$\mathbb{H}_0^a : \beta_{30} = 0.$$

This is equivalent to the choices of $R = (0, 0, 0, 1, 0)$ and $r = 0$.

If we are interested in testing whether profit sharing has an effect on productivity, we can consider the null hypothesis

$$\mathbb{H}_0^b : \beta_{40} = 0.$$

Alternatively, to test whether the production technology exhibits the constant return to scale (CRS), we can write the null hypothesis

$$\mathbb{H}_0^c : \beta_{10} + \beta_{20} = 1.$$

This is equivalent to the choices of $R = (0, 1, 1, 0, 0)$ and $r = 1$.

Finally, if we are interested in examining the joint effect of both autonomy and profit sharing, we can test the hypothesis that neither autonomy nor profit sharing has impact:

$$\mathbb{H}_0^d : \beta_{30} = \beta_{40} = 0.$$

This is equivalent to the choices of

$$R = \begin{bmatrix} 0\,0\,0\,1\,0 \\ 0\,0\,0\,0\,1 \end{bmatrix}, \qquad r = \begin{bmatrix} 0 \\ 0 \end{bmatrix}.$$

Example 10.3. [**Optimal Prediction for Future Spot Exchange Rates**]: Consider

$$S_{i+\tau} = \alpha_0 + \beta_0 F_i(\tau) + \varepsilon_{i+\tau}, \qquad i = 1, \ldots, n,$$

where $S_{i+\tau}$ is the spot exchange rate in period $i+\tau$, and $F_i(\tau)$ is the forward exchange rate with maturity τ, namely the period i's price for the foreign currency to be delivered in period $i + \tau$. The null hypothesis of interest is that the forward exchange rate $F_i(\tau)$ is an optimal predictor for the future spot rate $S_{i+\tau}$ in the sense that

$$E(S_{i+\tau}|I_i) = F_i(\tau),$$

where I_i is the information set available in time period i. This is called the *expectations hypothesis* in economics. Under a linear regression model, this hypothesis can be written as

$$\mathbb{H}_0^e : \alpha_0 = 0, \beta_0 = 1,$$

and $E(\varepsilon_{i+\tau}|I_i) = 0$. This is equivalent to the choices of

$$R = \begin{bmatrix} 1\,0 \\ 0\,1 \end{bmatrix}, \qquad r = \begin{bmatrix} 0 \\ 1 \end{bmatrix}.$$

We now discuss the basic idea of constructing a test for the null hypothesis

$$\mathbb{H}_0 : R\theta_0 = r.$$

We consider the statistic:

$$R\hat{\theta} - r$$

and check if the difference is statistically significantly different from zero.

Under $\mathbb{H}_0 : R\theta_0 = r$, we have

$$R\hat{\theta} - r = R\hat{\theta} - R\theta_0$$
$$= R(\hat{\theta} - \theta_0) \xrightarrow{p} 0$$

as $n \to \infty$ because $\text{MSE}(\hat{\theta}|\mathbf{X}) \to 0$ as $n \to \infty$.

Under the alternative to $\mathbb{H}_0$, we have $R\theta_0 \neq r$ and $\mathrm{MSE}(\hat{\theta}|\mathbf{X}) \to 0$ as $n \to \infty$. It follows that

$$R\hat{\theta} - r = R(\hat{\theta} - \theta_0) + R\theta_0 - r \xrightarrow{p} R\theta_0 - r$$

as $n \to \infty$, where $R\theta_0 - r$ is a nonzero limit.

The fact that the behavior of statistic $R\hat{\theta} - r$ is different under $\mathbb{H}_0$ and under the alternative hypothesis to $\mathbb{H}_0$ provides a basis to construct hypothesis tests. In particular, we can test $\mathbb{H}_0$ by examining whether the difference $R\hat{\theta} - r$ is significantly different from zero.

How large should the magnitude of the difference $R\hat{\theta} - r$ be in order to claim that $R\hat{\theta} - r$ is significantly different from zero? We need a decision rule which specifies a threshold value with which we can compare the value of $R\hat{\theta} - r$. Because $R\hat{\theta} - r$ is random and so it can take many values. Given a data generated from the random sample $\{Y_i, X_i'\}_{i=1}^n$, we only obtain one realization of $R\hat{\theta} - r$. Whether such a realization of $R\hat{\theta} - r$ is close to zero should be judged using the critical value of its sampling distribution, which depends on the sample size n and the prespecified significance level $\alpha \in (0, 1)$.

To determine the critical value at a given significance level α, we now derive the sampling distribution of $R\hat{\theta} - r$ under $\mathbb{H}_0$. By Corollary 10.1,

$$R(\hat{\theta} - \theta_0)|\mathbf{X} \sim N(0, \sigma^2 R(\mathbf{X}'\mathbf{X})^{-1}R').$$

Therefore, conditional on $\mathbf{X}$,

$$R\hat{\theta} - r = R(\hat{\theta} - \theta_0) + R\theta_0 - r$$
$$\sim N(R\theta_0 - r, \sigma^2 R(\mathbf{X}'\mathbf{X})^{-1}R').$$

Corollary 10.2: *Suppose $\mathbf{X}'\mathbf{X}$ is nonsingular and $\varepsilon|\mathbf{X} \sim N(0, \sigma^2 \mathbf{I})$. Then under the null hypothesis $\mathbb{H}_0 : R\theta_0 = r$, we have for each $n > p$,*

$$(R\hat{\theta} - r)|\mathbf{X} \sim N(0, \sigma^2 R(\mathbf{X}'\mathbf{X})^{-1}R').$$

The difference $R\hat{\theta} - r$ cannot be directly used as a test statistic for $\mathbb{H}_0$, because its conditional variance $\mathrm{var}(R\hat{\theta}|\mathbf{X}) = \sigma^2 R(\mathbf{X}'\mathbf{X})^{-1}R'$ involves the unknown error variance σ^2. Therefore, we cannot calculate the critical values of the sampling distribution of $R\hat{\theta} - r$.

Then, how can we construct a feasible test statistic? The form of the test statistic will depend on whether the number of parameter restrictions $J = 1$ or $J > 1$. We first consider the case of $J = 1$.

10.7.1 *Student's t-Test*

Recall that under $\mathbb{H}_0 : R\theta_0 = r$,

$$(R\hat{\theta} - r)|\mathbf{X} \sim N(0, \sigma^2 R(\mathbf{X}'\mathbf{X})^{-1}R').$$

When $J = 1$, the conditional variance

$$\text{var}[(R\hat{\theta} - r)|\mathbf{X}] = \sigma^2 R(\mathbf{X}'\mathbf{X})^{-1}R'$$

is a scalar. It follows that conditional on $\mathbf{X}$, we have

$$\frac{R\hat{\theta} - r}{\sqrt{\text{var}[(R\hat{\theta} - r)|\mathbf{X}]}} = \frac{R\hat{\theta} - r}{\sqrt{\sigma^2 R(\mathbf{X}'\mathbf{X})^{-1}R'}}$$
$$\sim N(0, 1).$$

Because the $N(0, 1)$ distribution does not depend on $\mathbf{X}$, the ratio

$$\frac{R\hat{\theta} - r}{\sqrt{\sigma^2 R(\mathbf{X}'\mathbf{X})^{-1}R'}} \sim N(0, 1).$$

That is, its unconditional distribution is also $N(0, 1)$.

However, σ^2 is unknown, so we cannot use the ratio

$$\frac{R\hat{\theta} - r}{\sqrt{\sigma^2 R(\mathbf{X}'\mathbf{X})^{-1}R'}}$$

as a test statistic. We have to replace σ^2 by the residual variance estimator s^2. This yields a feasible test statistic

$$T = \frac{R\hat{\theta} - r}{\sqrt{s^2 R(\mathbf{X}'\mathbf{X})^{-1}R'}}.$$

However, the test statistic T no longer has a normal distribution. Instead,

$$T = \frac{R\hat{\theta} - r}{\sqrt{s^2 R(\mathbf{X}'\mathbf{X})^{-1}R'}}$$

$$= \frac{\dfrac{R\hat{\theta} - r}{\sqrt{\sigma^2 R(\mathbf{X}'\mathbf{X})^{-1}R'}}}{\sqrt{\dfrac{(n-p)s^2}{\sigma^2}/(n-p)}}$$

$$\sim \frac{N(0, 1)}{\sqrt{\chi^2_{n-p}/(n-p)}}$$

$$\sim t_{n-p},$$

where t_{n-p} denotes a Student's t-distribution with $n-p$ degrees of freedom. In deriving the Student t_{n-p} distribution of T, we have made use of the property that the numerator and denominator of T are independent conditional on $\mathbf{X}$, which follows because $\hat{\theta}$ and s^2 are independent conditional on $\mathbf{X}$. The feasible statistic T is called a t-test statistic because it follows a Student's t_{n-p} distribution under $\mathbb{H}_0 : R\theta_0 = r$.

Recall that $t_q \to N(0,1)$ as $q \to \infty$, we have

$$T = \frac{R\hat{\theta} - r}{\sqrt{s^2 R(\mathbf{X}'\mathbf{X})^{-1}R'}} \xrightarrow{d} N(0,1) \text{ as } n \to \infty,$$

under $\mathbb{H}_0$. This result has an important implication: for a large sample size n, it makes no difference to use either the critical values of t_{n-K} or those of $N(0,1)$.

Having obtained the sampling distribution for the test statistic T, we can now describe a critical value-based decision rule for testing $\mathbb{H}_0$ when $J = 1$:

- Reject $\mathbb{H}_0 : R\theta_0 = r$ at a prespecified significance level $\alpha \in (0,1)$ if

$$|T| > C_{t_{n-p},\frac{\alpha}{2}},$$

  where $C_{t_{n-p},\frac{\alpha}{2}}$ is the upper-tailed critical value of the t_{n-p} distribution at level $\frac{\alpha}{2}$, which is determined by

$$P\left[t_{n-p} > C_{t_{n-p},\frac{\alpha}{2}}\right] = \frac{\alpha}{2}$$

  or equivalently

$$P\left[|t_{n-p}| > C_{t_{n-p},\frac{\alpha}{2}}\right] = \alpha.$$

- Do not reject $\mathbb{H}_0$ at the significance level α if

$$|T| \le C_{t_{n-p},\frac{\alpha}{2}}.$$

In testing $\mathbb{H}_0$, there exist two types of errors, due to the limited information about the population in a random sample $\{Y_i, X_i'\}_{i=1}^n$. One possibility is that $\mathbb{H}_0$ is true but we reject it. This is called a Type I error. The significance level α is the probability of making a Type I error. If

$$P\left[|T| > C_{t_{n-K},\frac{\alpha}{2}} \,\big|\, \mathbb{H}_0\right] = \alpha,$$

we say that the decision rule is a test with size α.

On the other hand, the probability $P[|T| > C_{t_{n-K},\frac{\alpha}{2}} | \mathbb{H}_0$ is false] is called the power function of a size α test. When

$$P\left[|T| > C_{t_{n-K},\frac{\alpha}{2}} \,\big|\, \mathbb{H}_0 \text{ is false}\right] < 1,$$

there exists a possibility that one may fail to reject $\mathbb{H}_0$ when it is false. This is called a Type II error.

Ideally one would like to minimize both the Type I error and Type II error, but this is impossible for any given finite sample. In practice, one usually presets the level for the Type I error, the so-called significance level, and then minimizes the Type II error. Conventional choices for significance level α are 10%, 5% and 1% respectively.

Next, we describe an alternative but equivalent decision rule for testing $\mathbb{H}_0$ when $J = 1$, based on the so-called P-value of test statistic T.

Given a data set $\mathbf{z}^n = \{y_i, x_i'\}_{i=1}^n$, which is a realization of the random sample $\mathbf{Z}^n = \{Y_i, X_i'\}_{i=1}^n$, we can compute a realization (i.e., a real number) for the t-test statistic T, namely

$$T(\mathbf{z}^n) = \frac{R\hat{\theta} - r}{\sqrt{s^2 R(\mathbf{x}'\mathbf{x})^{-1}R'}}.$$

Then the probability

$$p(\mathbf{z}^n) = P\left[|T| > |T(\mathbf{z}^n)| \,|\, \mathbb{H}_0\right]$$
$$= P\left[|t_{n-p}| > |T(\mathbf{z}^n)|\right]$$

is called the P-value (i.e., probability value) of the test statistic T when $\mathbf{Z}^n = \mathbf{z}^n$ is observed, where t_{n-p} is a Student's t random variable with $n - p$ degrees of freedom, and $T(\mathbf{z}^n)$ is a realization for the test statistic $T = T(\mathbf{Z}^n)$ when $\mathbf{Z}^n = \mathbf{z}^n$ is observed. Intuitively, the P-value is the smallest value of significance level at which the null hypothesis can be rejected. It is the tail probability that the absolute value of a Student's t_{n-p} random variable takes values larger than the absolute value of the test statistic $T(\mathbf{z}^n)$. If this probability is very small relative to the significance level α, then it is unlikely that the test statistic $T(\mathbf{Z}^n)$ will follow a Student's t_{n-p} distribution. In this case, the null hypothesis $\mathbb{H}_0$ is likely to be false, and so should be rejected.

The above decision rule can be described equivalently as follows:

- Reject $\mathbb{H}_0$ at the significance level α if $p(\mathbf{z}^n) < \alpha$.
- Do not reject $\mathbb{H}_0$ at the significance level α if $p(\mathbf{z}^n) \geq \alpha$.

Intuitively, a small P-value is evidence against the null hypothesis $\mathbb{H}_0$, and a large P-value shows that the data are consistent with $\mathbb{H}_0$.

P-values are more informative than only rejecting or accepting the null hypothesis at some significance level α. A P-value not only tells us whether the null hypothesis $\mathbb{H}_0$ should be accepted or rejected, but also tells us whether the decision to accept or reject $\mathbb{H}_0$ is a close call.

When we reject the null hypothesis $\mathbb{H}_0$, we say that there is a statistically significant effect. This does not mean that there is an effect of practical importance (i.e., an effect of economic importance). This is because when a large sample is used, small and practically unimportant effects are likely to be statistically significant.

The t-test and associated procedures are valid even when there exists near-multicollinearity. In other words, Type I errors are not affected by near-multicollinearity. However, the degree of near-multicollinearity, as measured by sample correlations between explanatory variables, will affect the precision of the OLS estimator $\hat{\theta}$. Other things being equal, the higher degree of near-multicollinearity, the larger the variance of $\hat{\theta}$. As a result, the t-statistic is often insignificant even when the null hypothesis $\mathbb{H}_0$ is false.

We now discuss some examples of using t-tests.

Example 10.4. [Reforms Have No Effects (continued.)]: In the extended production function of Example 10.2, we first consider testing the null hypothesis

$$\mathbb{H}_0^a : \beta_{30} = 0,$$

where β_{30} is the coefficient of the autonomy AU_i. This is equivalent to the selection of $R = (0, 0, 0, 1, 0)$. In this case, we have

$$s^2 R(\mathbf{X'X})^{-1} R' = \left[s^2(\mathbf{X'X})^{-1}\right]_{(4,4)}$$
$$= S_{\hat{\beta}_3}^2,$$

which is the estimator of $\text{var}(\hat{\beta}_3 | \mathbf{X})$. The squared root of $\text{var}(\hat{\beta}_3 | X)$ is called the standard error of estimator $\hat{\beta}_3$, and $S_{\hat{\beta}_3}$ is called the estimated standard error of $\hat{\beta}_3$. The t-test statistic

$$T = \frac{R\hat{\theta} - r}{\sqrt{s^2 R(\mathbf{X'X})^{-1} R'}}$$
$$= \frac{\hat{\beta}_3}{\sqrt{S_{\hat{\beta}_3}^2}}$$
$$\sim t_{n-5}.$$

Next, we consider testing the CRS hypothesis

$$\mathbb{H}_0^c : \beta_{10} + \beta_{20} = 1,$$

which corresponds to $R = (0, 1, 1, 0, 0)$ and $r = 1$. In this case,

$$
\begin{aligned}
s^2 R(\mathbf{X}'\mathbf{X})^{-1}R' &= S_{\hat{\beta}_1}^2 + S_{\hat{\beta}_2}^2 + 2\hat{c}\mathrm{ov}(\hat{\beta}_1, \hat{\beta}_2) \\
&= \left[s^2(\mathbf{X}'\mathbf{X})^{-1} \right]_{(2,2)} \\
&\quad + \left[s^2(\mathbf{X}'\mathbf{X})^{-1} \right]_{(3,3)} \\
&\quad + 2 \left[s^2(\mathbf{X}'\mathbf{X})^{-1} \right]_{(2,3)} \\
&= S_{\hat{\beta} + \hat{\beta}_2}^2,
\end{aligned}
$$

which is the estimator of $\mathrm{var}(\hat{\beta}_1 + \hat{\beta}_2 | \mathbf{X})$. Here, $\hat{c}\mathrm{ov}(\hat{\beta}_1, \hat{\beta}_2)$ is the estimator for $\mathrm{cov}(\hat{\beta}_1, \hat{\beta}_2 | \mathbf{X})$, the covariance between $\hat{\beta}_1$ and $\hat{\beta}_2$ conditional on $\mathbf{X}$.

The t-test statistic is

$$
\begin{aligned}
T &= \frac{R\hat{\theta} - r}{\sqrt{s^2 R(\mathbf{X}'\mathbf{X})^{-1}R'}} \\
&= \frac{\hat{\beta}_1 + \hat{\beta}_2 - 1}{S_{\hat{\beta}_1 + \hat{\beta}_2}} \\
&\sim t_{n-5}.
\end{aligned}
$$

10.7.2   F-Test

Question: How to construct a test statistic for $\mathbb{H}_0$ if $J > 1$?

We first introduce a useful lemma.

Lemma 10.2. *If $Z \sim N(0, V)$, where $V = \mathrm{var}(Z)$ is a nonsingular $J \times J$ variance-covariance matrix, then*

$$
Z'V^{-1}Z \sim \chi_J^2.
$$

Proof: Because V is symmetric and positive definite, we can find a symmetric and invertible matrix $V^{1/2}$ such that

$$
V^{1/2}V^{1/2} = V,
$$
$$
V^{-1/2}V^{-1/2} = V^{-1}.
$$

Question: What is this decomposition called?

Now, define

$$
Y = V^{-1/2}Z.
$$

Then we have $E(Y) = 0$, and

$$
\begin{aligned}
\text{var}(Y) &= E\left\{[Y - E(Y)][Y - E(Y)]'\right\} \\
&= E(YY') \\
&= E(V^{-1/2}ZZ'V^{-1/2}) \\
&= V^{-1/2}E(ZZ')V^{-1/2} \\
&= V^{-1/2}VV^{-1/2} \\
&= V^{-1/2}V^{1/2}V^{1/2}V^{-1/2} \\
&= I,
\end{aligned}
$$

where I is a $J \times J$ identity matrix. It follows that $Y \sim N(0, I)$. Therefore, we have

$$
Y'Y \sim \chi_J^2.
$$

Applying this lemma and using the result that

$$
(R\hat{\theta} - r)|\mathbf{X} \sim N(0, \sigma^2 R(\mathbf{X}'\mathbf{X})^{-1}R')
$$

under $\mathbb{H}_0$, we have the quadratic form

$$
\frac{(R\hat{\theta} - r)'[R(\mathbf{X}'\mathbf{X})^{-1}R']^{-1}(R\hat{\theta} - r)}{\sigma^2} \sim \chi_J^2
$$

conditional on $\mathbf{X}$. Because χ_J^2 does not depend on $\mathbf{X}$, we also have

$$
\frac{(R\hat{\theta} - r)'[R(\mathbf{X}'\mathbf{X})^{-1}R']^{-1}(R\hat{\theta} - r)}{\sigma^2} \sim \chi_J^2
$$

unconditionally.

Like in constructing a t-test statistic, we should replace σ^2 by s^2. This renders the quadratic form no longer follows a χ^2 distribution. Instead, after proper scaling, the quadratic form will follow an F-distribution with degrees of freedom $(J, n - p)$.

To see this, we observe

$$
\begin{aligned}
&\frac{(R\hat{\theta} - r)'[R(\mathbf{X}'\mathbf{X})^{-1}R']^{-1}(R\hat{\theta} - r)}{s^2} \\
&= J \cdot \frac{\frac{(R\hat{\theta}-r)'[R(\mathbf{X}'\mathbf{X})^{-1}R']^{-1}(R\hat{\theta}-r)}{\sigma^2}/J}{\frac{(n-K)s^2}{\sigma^2}/(n-K)} \\
&\sim J \cdot \frac{\chi_J^2/J}{\chi_{n-p}^2/(n-p)} \\
&\sim F_{J, n-p},
\end{aligned}
$$

where $F_{J,n-p}$ denotes the F distribution with degrees of J and $n-p$. We note that conditional on $\mathbf{X}$, the numerator and denominator of the quadratic form are independent χ^2 random variables, which is crucial for an F-distribution.

We now define the F-test statistic for testing $\mathbb{H}_0$:

$$F \equiv \frac{(R\hat{\theta} - r)'[R(\mathbf{X'X})^{-1}R']^{-1}(R\hat{\theta} - r)/J}{s^2}.$$

Theorem 10.7. *Suppose* $\mathbf{X'X}$ *is nonsingular and* $\varepsilon|\mathbf{X} \sim N(0, \sigma^2\mathbf{I})$. *Then under* $\mathbb{H}_0 : R\theta_0 = r$, *we have for all* $n < p$,

$$F \sim F_{J,n-p}.$$

In fact, the F test is also applicable to testing a single parameter restriction ($J = 1$). Since $F_{1,q} \sim t_q^2$, a t-test is equivalent to an F-test when $J = 1$.

A practical issue is how to compute the F statistic. One can of course compute the F statistic using the definition of the F statistic. However, there is a convenient way to compute the F statistic, as shown below.

Theorem 10.8. *Suppose* $\mathbf{X'X}$ *is nonsingular. Let* $SSR_u = e'e$ *be the sum of squared residuals from the unrestricted model*

$$Y = \mathbf{X}\theta + \varepsilon.$$

Let $SSR_r = \tilde{e}'\tilde{e}$ *be the sum of squared residuals from the restricted model*

$$Y = \mathbf{X}\theta + \varepsilon$$

subject to the constraint that

$$R\theta = r.$$

Then the F-test statistic

$$F = \frac{(\tilde{e}'\tilde{e} - e'e)/J}{e'e/(n - K)}.$$

Proof: Let $\tilde{\theta}$ be the OLS under $\mathbb{H}_0$, namely

$$\tilde{\theta} = \arg\min_{\theta \in \mathbf{R}^p} (Y - \mathbf{X}\theta)'(Y - \mathbf{X}\theta)$$

subject to the constraint that $R\theta = r$. We first form a Lagrangian function

$$L(\theta, \lambda) = (Y - \mathbf{X}\theta)'(Y - \mathbf{X}\theta) + 2\lambda'(r - R\theta),$$

where λ is a $J \times 1$ vector called a Lagrange multiplier vector.

We have the following FOC:

$$\frac{\partial L(\tilde{\theta}, \tilde{\lambda})}{\partial \beta} = -2\mathbf{X}'(Y - \mathbf{X}\tilde{\theta}) - 2R'\tilde{\lambda} = 0,$$

$$\frac{\partial L(\tilde{\theta}, \tilde{\lambda})}{\partial \lambda} = 2(r - R\tilde{\theta}) = 0.$$

With the unconstrained OLS estimator $\hat{\theta} = (\mathbf{X}'\mathbf{X})^{-1}\mathbf{X}'Y$, and from the first equation of FOC, we obtain

$$-(\hat{\theta} - \tilde{\theta}) = (\mathbf{X}'\mathbf{X})^{-1}R'\tilde{\lambda},$$
$$R(\mathbf{X}'\mathbf{X})^{-1}R'\tilde{\lambda} = -R(\hat{\theta} - \tilde{\theta}).$$

Hence, the Lagrangian multiplier

$$\tilde{\lambda} = -[R(\mathbf{X}'\mathbf{X})^{-1}R']^{-1}R(\hat{\theta} - \tilde{\theta}).$$
$$= -[R(\mathbf{X}'\mathbf{X})^{-1}R']^{-1}(R\hat{\theta} - r),$$

where we have made use of the constraint that $R\tilde{\theta} = r$. It follows that

$$\hat{\theta} - \tilde{\theta} = (\mathbf{X}'\mathbf{X})^{-1}R'[R(\mathbf{X}'\mathbf{X})^{-1}R']^{-1}(R\hat{\theta} - r).$$

Now, the restricted estimated residual

$$\tilde{e} = Y - \mathbf{X}\tilde{\theta}$$
$$= Y - \mathbf{X}\hat{\theta} + \mathbf{X}(\hat{\theta} - \tilde{\theta})$$
$$= e + \mathbf{X}(\hat{\theta} - \tilde{\theta}).$$

It follows that

$$\tilde{e}'\tilde{e} = e'e + (\hat{\theta} - \tilde{\theta})'\mathbf{X}'\mathbf{X}(\hat{\theta} - \tilde{\theta})$$
$$= e'e + (R\hat{\theta} - r)'[R(\mathbf{X}'\mathbf{X})^{-1}R']^{-1}(R\hat{\theta} - r).$$

We have

$$(R\hat{\theta} - r)'[R(\mathbf{X}'\mathbf{X})^{-1}R']^{-1}(R\hat{\theta} - r) = \tilde{e}'\tilde{e} - e'e$$

and

$$F \equiv \frac{(R\hat{\theta} - r)'[R(\mathbf{X}'\mathbf{X})^{-1}R']^{-1}(R\hat{\theta} - r)/J}{s^2}$$

$$= \frac{(\tilde{e}'\tilde{e} - e'e)/J}{e'e/(n - K)}.$$

This completes the proof.

Theorem 10.8 shows that the F statistic is rather convenient to compute! One only needs to compute the sums of squared residuals under the restricted and unrestricted models respectively. Intuitively, the sum of squared residuals SSR_u of the restricted model is always larger than that of the unrestricted model. When the null hypothesis $\mathbb{H}_0$ is true, the sum of squared residuals SSR_r of the restricted model is more or less similar to that of the unrestricted model, subject to the difference due to sampling variations. If SSR_r is sufficiently larger than SSR_u, then there exists evidence against $\mathbb{H}_0$. How large the difference between SSR_r and SSR_u is considered as sufficiently large has to be determined by the critical value of the $F_{J,n-p}$ distribution.

Question: What is the interpretation for the Lagrange multiplier $\tilde{\lambda}$?

Recall that we have obtained

$$\tilde{\lambda} = -[R(\mathbf{X'X})^{-1}R']^{-1}R(\hat{\theta} - \tilde{\theta})$$
$$= -[R(\mathbf{X'X})^{-1}R']^{-1}(R\hat{\theta} - r).$$

Thus, $\tilde{\lambda}$ is an indicator of the departure of $R\hat{\theta}$ from r. The value of $\tilde{\lambda}$ indicates whether $R\hat{\theta} - r$ is significantly different from zero.

Question: What happens to the distribution of the F statistic when $n \to \infty$?

Since $J \cdot F_{J,q} \overset{d}{\to} \chi^2_J$ when $q \to \infty$, and our F-statistic follows an $F_{J,n-p}$ distribution under $\mathbb{H}_0$, we have the quadratic form

$$J \cdot F = (R\hat{\theta} - r)' \left[s^2 R(\mathbf{X'X})^{-1}R' \right]^{-1} (R\hat{\theta} - r)$$
$$\overset{d}{\to} \chi^2_J$$

as $n \to \infty$. This is the Wald test statistic. We formally state this result below.

Theorem 10.9. *[Wald Test]: Suppose* $\mathbf{X'X}$ *is nonsingular and* $\varepsilon|\mathbf{X} \sim N(0, \sigma^2\mathbf{I})$. *Then under* $\mathbb{H}_0$ *and as* $n \to \infty$, *the Wald test statistic*

$$W \equiv \frac{(R\hat{\theta} - r)'[R(\mathbf{X'X})^{-1}R']^{-1}(R\hat{\theta} - r)}{s^2}$$
$$\overset{d}{\to} \chi^2_J.$$

Thus, when the sample size n is sufficiently large, using the F-statistic with the exact $F_{J,n-p}$ distribution or using the Wald test statistic W with a simpler χ^2_J approximation will make no essential difference in statistical inference.

10.8 Applications and Important Examples

We now consider some special but important cases often encountered in economics and econometrics.

10.8.1 Testing for Joint Significance of All Explanatory Variables

Consider a linear regression model

$$Y_i = X_i'\theta_0 + \varepsilon_i$$
$$= \alpha_0 + \sum_{j=1}^{k} \beta_{j0}X_{ji} + \varepsilon_i, \qquad i = 1,\ldots,n.$$

We are interested in testing whether the combined effect of all explanatory variables except the intercept is zero. The null hypothesis is

$$\mathbb{H}_0 : \beta_{10} = \cdots = \beta_{k0} = 0,$$

which implies that none of the explanatory variables $\{X_{ji}\}_{i=1}^{k}$ has impact on Y_i.

The alternative hypothesis is

$$\mathbb{H}_A : \beta_{j0} \neq 0 \text{ at least for some } j \in \{1, \cdots, k\}.$$

One can use the F-test statistic $F \sim F_{k,n-(k+1)}$. In fact, the restricted model under $\mathbb{H}_0$ is quite simple:

$$Y_i = \alpha_0 + \varepsilon_i.$$

The restricted OLS estimator $\tilde{\beta} = (\bar{Y}, 0, \cdots, 0)'$. It follows that

$$\tilde{e} = Y - \mathbf{X}\tilde{\beta} = \mathbf{Y} - \bar{Y}\mathbf{i},$$

where $\mathbf{i} = (1,\ldots,1)'$ is a $n \times 1$ vector of ones. Hence, we have

$$\tilde{e}'\tilde{e} = (\mathbf{Y} - \bar{Y}\mathbf{i})'(\mathbf{Y} - \bar{Y}\mathbf{i}).$$

Following the definition of R^2, namely,

$$R^2 = 1 - \frac{e'e}{(\mathbf{Y} - \bar{Y}\mathbf{i})'(\mathbf{Y} - \bar{Y}\mathbf{i})}$$
$$= 1 - \frac{e'e}{\tilde{e}'\tilde{e}},$$

we obtain

$$F = \frac{(\tilde{e}'\tilde{e} - e'e)/k}{e'e/(n-k-1)}$$

$$= \frac{(1 - \frac{e'e}{\tilde{e}'\tilde{e}})/k}{\frac{e'e}{\tilde{e}'\tilde{e}}/(n-k-1)}$$

$$= \frac{R^2/k}{(1-R^2)/(n-k-1)}.$$

Thus, it suffices to run only one regression, namely the unrestricted model. We emphasize that this formula is valid only when one is testing for joint significance of all explanatory variables except the intercept.

We now consider some applications.

Example 10.5. [**Efficient Market Hypothesis (EMH)**]: Suppose Y_i is the exchange rate change in time period i, and I_{i-1} is the information available in time period $i-1$. Then a classical version of EMH can be stated as

$$E(Y_i | I_{i-1}) = E(Y_i).$$

To check whether the exchange rate change is unpredictable using its past history, we specify the following linear regression model

$$Y_i = X_i'\theta_0 + \varepsilon_i,$$

where

$$X_i = (1, Y_{i-1}, \ldots, Y_{i-k})'.$$

This is an AR(k) model. Under EMH, we have

$$\mathbb{H}_0 : \beta_{10} = \cdots = \beta_{k0} = 0.$$

If the alternative

$$\mathbb{H}_A : \beta_{j0} \neq 0 \text{ at least for some } j \in \{1, \ldots, k\}$$

holds, then the exchange rate change is predictable using its past information.

Question: What is the appropriate interpretation if $\mathbb{H}_0$ is not rejected?

Note that there exists a gap between the economic hypothesis EMH and the statistical hypothesis $\mathbb{H}_0$, since a linear regression model is just one of many different ways to check EMH. When $\mathbb{H}_0$ is not rejected, at most can we only say that no evidence against EMH is found. We should not conclude that EMH holds. It is possible that the exchange rate change is not predictable using a linear regression model but it is predictable in a nonlinear manner.

Strictly speaking, the strict exogeneity condition ($E(\varepsilon_i|\mathbf{X}) = 0$) rules out this application, since an AR(k) model is a dynamic regression model. However, it could be shown that as $n \to \infty$,

$$k \cdot F = \frac{R^2}{(1 - R^2)/(n - k - 1)}$$

$$\xrightarrow{d} \chi_k^2$$

under conditional homoskedasticity even for a linear dynamic regression model. See Hong (2017, Ch. 5) for more discussion.

In fact, we can use an even simpler version when n is large:

$$(n - k - 1)R^2 \xrightarrow{d} \chi_k^2 \text{ as } n \to \infty.$$

This follows from the Slutsky theorem (cf. Theorem 7.18) because $R^2 \xrightarrow{p} 0$ under $\mathbb{H}_0$. Although the strict exogeneity condition is not needed for this result, conditional homoskedasticity is required, which rules out AutoRegressive Conditional Heteroskedasticity (ARCH) in the time series context (Engle 1982).

We now examine a concrete numerical application.

Example 10.6. [**Consumption Function and Wealth Effect**]: Let Y_i be consumption, X_{1i} be labor income, X_{2i} be liquidity asset wealth. Based on a data set, a regression estimation delivers the following results:

$$Y_i = 33.88 - 26.00X_{1i} + 6.71X_{2i} + e_i, \qquad R^2 = 0.742, n = 25.$$
$$[1.77] \qquad [-0.74] \qquad [0.77]$$

where the numbers inside [·] are t-statistics.

Suppose we are interested in whether labor income or liquidity asset wealth has impact on consumption. Both t-statistics for labor income and wealth are not significant at the 5% significance level. However, due to possible near-multicorrellinearity, we should also check whether labor income and wealth are jointly insignificant. For this purpose, we compute the F-test statistic

$$F = \frac{R^2/2}{(1 - R^2)/(n - 3)}$$
$$= (0.742/2)/[(1 - 0.742)/(25 - 3)]$$
$$= 31.636.$$

Comparing it with the critical value 4.38 of $F_{J,n-p} = F_{2,22}$ at the 5% significance level, we reject the null hypothesis that neither income nor liquidity asset has impact on consumption.

10.8.2 *Testing for Omitted Variables*

Suppose X_i is a $(k+1) \times 1$ vector and Z_i is a $q \times 1$ vector. We say that the random vector Z_i has no explanatory power for the conditional expectation of Y_i if

$$E(Y_i|X_i, Z_i) = E(Y_i|X_i).$$

Alternatively, Z_i has explanatory power for the conditional expectation of Y_i if

$$E(Y_i|X_i, Z_i) \neq E(Y_i|X_i).$$

When Z_i has explanatory power for Y_i but is not included in the regression, we call that the explanatory variables in Z_i are omitted variables.

Question: How to test whether the variables in Z_i are omitted variables in the linear regression model?

Consider the restricted model

$$Y_i = \alpha_0 + \beta_{10}X_{1i} + \cdots + \beta_{k0}X_{ki} + \varepsilon_i, \qquad i = 1, \ldots, n.$$

Suppose we have additional q explanatory variables $(Z_{1i}, \cdots, Z_{qi})$, and so the unrestricted regression model is

$$Y_i = \alpha_0 + \beta_{10}X_{1i} + \cdots + \beta_{k0}X_{ki} + \gamma_{10}Z_{1i} + \cdots + \gamma_{q0}Z_{qi} + \varepsilon_i.$$

The null hypothesis is that the q additional variables $\{Z_{ji}\}_{j=1}^q$ have no effect on Y_i. If this is the case, we will have

$$\mathbb{H}_0 : \gamma_{10} = \gamma_{20} = \cdots = \gamma_{q0} = 0.$$

The alternative is that at least one of the additional variables $\{Z_{ji}\}_{i=1}^q$ has effect on Y_i.

The F-test statistic is

$$F = \frac{(\tilde{e}'\tilde{e} - e'e)/k_2}{e'e/(n - p - q - 1)}$$
$$\sim F_{q, n-(p+q+1)}.$$

If we reject the null hypothesis, then some explanatory variables are omitted, and they should be included in the regression. On the other hand, if the F-test statistic does not reject the null hypothesis, can we say that there is no omitted variable?

The answer is no. There may exist a nonlinear relationship for additional variables which a linear regression model cannot capture.

Example 10.7. [**Testing for Effect of Reforms**]: Consider the extended production function given in Example 10.2:

$$
\begin{aligned}
Y_i = \alpha_0 &+ \beta_{10} \ln(L_i) + \beta_{20} \ln(K_i) \\
&+ \beta_{30} AU_i + \beta_{40} PS_i + \beta_{50} CM_i + \varepsilon_i, \qquad i = 1, \ldots, n,
\end{aligned}
$$

where AU_i is the autonomy dummy, PS_i is the profit sharing ratio, and CM_i is the dummy for change of manager. The null hypothesis of interest here is that none of these reforms has impact, which implies

$$
\mathbb{H}_0 : \beta_{30} = \beta_{40} = \beta_{50} = 0.
$$

We can use the F-test, and $F \sim F_{3, n-6}$ under $\mathbb{H}_0$.

Suppose rejection occurs. Then there exists evidence against $\mathbb{H}_0$, namely, at least one of the three reforms has impact on firm productivity. However, if no rejection occurs, we can only say that we find no evidence against $\mathbb{H}_0$ and we cannot claim that these reforms have no effect. It is possible that the effect of these reforms is of nonlinear form. In this case, we may obtain zero coefficients for these reforms in a linear regression model because it may not be able to capture nonlinear effects.

Example 10.8. [**Testing for Granger Causality**]: Consider two time series $\{Y_i, X_i\}$, where i is a time index, $I_{i-1}^Y = \{Y_{i-1}, \ldots, Y_1\}$ and $I_{i-1}^X = \{X_{i-1}, \ldots, X_1\}$ are the past histories of $\{Y_i\}$ and $\{X_i\}$ respectively. For example, Y_i is the GDP growth in year i, and X_i is the money supply growth in year i. We say that $\{X_i\}$ does not Granger-cause $\{Y_i\}$ in conditional mean with respect to the information set $I_{i-1} = \{I_{i-1}^{(Y)}, I_{i-1}^{(Z)}\}$ if

$$
E(Y_i | I_{i-1}^{(Y)}, I_{i-1}^{(X)}) = E(Y_i | I_{i-1}^{(Y)}).
$$

In other words, conditional on the past history of $\{Y_i\}$, the lagged variables of X_i have no predictive power for Y_i in conditional mean.

Granger causality is defined in terms of incremental predictability in a time series context rather than the real cause-effect relationship. From an econometric point of view, this can be viewed as an omitted variables problem in a time series context.

How to test Granger causality? Consider a linear regression model

$$Y_i = \alpha_0 + \beta_{10}Y_{i-1} + \cdots + \beta_{p0}Y_{i-p}$$
$$+ \gamma_{10}X_{i-1} + \cdots + \gamma_{q0}X_{i-q} + \varepsilon_i, \qquad i = 1, \ldots, n.$$

Under non-Granger causality, we have

$$\mathbb{H}_0 : \gamma_{10} = \cdots = \gamma_{q0} = 0.$$

The F-test statistic

$$F \sim F_{q, n-(p+q+1)}.$$

This is first introduced by Granger (1969). However, the strict exogeneity assumption rules out this application, because it is a dynamic regression model. Nevertheless, one could justify that when $\mathbb{H}_0$ holds,

$$q \cdot F \xrightarrow{d} \chi_q^2 \text{ as } n \to \infty$$

under conditional homoskedasticity. See Hong (2017, Ch.5) for more discussion.

Example 10.9. [Testing for Structural Changes]: Consider a simple bivariate regression model

$$Y_i = \alpha_0 + \beta_0 X_{1i} + \varepsilon_i, \qquad i = 1, \ldots, n,$$

where i is a time index, and $\{X_i\}$ and $\{\varepsilon_i\}$ are mutually independent. If parameters α_0 and/or β_0 change at time i_0 and afterward, then we say that there exists a structural change at time i_0.

To test whether exists a structural change at a known time i_0, we consider the extended regression model

$$Y_i = (\alpha_0 + \alpha_1 D_i) + (\beta_0 + \beta_1 D_i)X_{1i} + \varepsilon_i$$
$$= \alpha_0 + \beta_0 X_{1i} + \alpha_1 D_i + \beta_1(D_i X_{1i}) + \varepsilon_i,$$

where $D_i = 0$ if $i \leq i_0$ and $D_i = 1$ otherwise. The variable D_i is called a time dummy variable, indicating whether it is a pre- or post-structural break period.

When there exists no structural change, the null hypothesis

$$\mathbb{H}_0 : \alpha_1 = \beta_1 = 0$$

holds. If there exists a structural change, then the alternative hypothesis

$$\mathbb{H}_A : \alpha_1 \neq 0 \text{ or } \beta_1 \neq 0$$

holds. The F-test statistic

$$F \sim F_{2,n-4}.$$

The idea of such an F-test for structural change is first proposed by Chow (1960). It is therefore called Chow's test.

10.8.3 *Testing for Linear Restrictions*

We first consider an example of testing the hypothesis of constant return to scale.

Example 10.10. [Testing for Constant Return to Scale (CRS)]: Consider the extended production function of Example 10.7:

$$\ln(Y_i) = \alpha_0 + \beta_{10}\ln(L_i) + \beta_{20}\ln(K_i) + \beta_{30}AU_i + \beta_{40}PS_i + \beta_{50}CM_i + \varepsilon_i.$$

We will test the null hypothesis of CRS:

$$\mathbb{H}_0 : \beta_{10} + \beta_{20} = 1.$$

We now construct an F-test statistic. Under $\mathbb{H}_0$, the restricted model is given by

$$\ln(Y_i) = \alpha_0 + \beta_{10}\ln(L_i) + (1 - \beta_{10})\ln(K_i) + \beta_{30}AU_i + \beta_{40}PS_i + \beta_{50}CM_i + \varepsilon_i$$

or equivalently

$$\ln(Y_i/K_i) = \alpha_0 + \beta_{10}\ln(L_i/K_i) + \beta_{30}AU_i + \beta_{40}CON_i + \beta_{50}CM_i + \varepsilon_i.$$

The F-test statistic

$$F \sim F_{1,n-6}.$$

Since there is only one restriction, both t- and F- tests are applicable to test CRS and they are equivalent to each other.

Example 10.11. [Wage Determination]: Consider the wage function in a time series context

$$W_i = \alpha_0 + \beta_{10}P_i + \beta_{20}P_{i-1} + \beta_{30}U_i$$
$$+ \beta_{40}V_i + \beta_{50}W_{i-1} + \varepsilon_i, \qquad i = 1, \ldots, n,$$

where i is a time index, W_i is wage, P_i is price, U_i is unemployment, and V_i is number of unfilled vacancies.

We will test the null hypothesis

$$\mathbb{H}_0 : \beta_{10} + \beta_{20} = 0, \ \beta_{30} + \beta_{40} = 0, \text{ and } \beta_{50} = 1.$$

To provide an economic interpretation for $\mathbb{H}_0$, we obtain the following restricted wage equation

$$\Delta W_i = \alpha_0 + \beta_{10}\Delta P_i + \beta_{40}D_i + \varepsilon_i,$$

where $\Delta W_i = W_i - W_{i-1}$ is the wage growth rate, $\Delta P_i = P_i - P_{i-1}$ is the inflation rate, and $D_i = V_i - U_i$ is an index for excess job supply in the job market. The null hypothesis $\mathbb{H}_0$ implies that the wage increase depends on the inflation rate and the excess job supply.

The F-test statistic for $\mathbb{H}_0$ is

$$F \sim F_{3,n-6}.$$

Again, the strict exogeneity condition rules out this application. However, it could be shown that under $\mathbb{H}_0$,

$$3F \rightarrow^d \chi_3^2 \text{ as } n \rightarrow \infty$$

under conditional homoskedasticity. See Hong (2017, Ch. 5) for more discussion.

10.9   Generalized Least Squares (GLS) Estimation

The classical linear regression theory crucially depends on the conditional normality assumption that $\varepsilon|\mathbf{X} \sim N(0,\sigma^2\mathbf{I})$, which implies that $\{\varepsilon_i\} \sim$ IID $N(0,\sigma^2)$, and $\{X_i\}$ and $\{\varepsilon_i\}$ are mutually independent. What may happen if some classical assumptions, such as conditional homoskedasticity and conditional uncorrelatedness, do not hold?

Suppose we impose a weaker assumption:

$$\varepsilon|\mathbf{X} \sim N(0,\sigma^2\mathbf{V}),$$

where $\mathbf{V} = V(\mathbf{X})$ is a known $n \times n$ symmetric, finite and positive definite matrix, and $0 < \sigma^2 < \infty$ is unknown.

Here, ε still follows a conditional normal distribution, and the strict exogeneity condition ($E(\varepsilon|\mathbf{X}) = 0$) holds. However,

$$\text{var}(\varepsilon|\mathbf{X}) = \sigma^2\mathbf{V}$$

is known up to an unknown constant σ^2. Because $\mathbf{V} = V(\mathbf{X})$ depends on $\mathbf{X}$ and may have nonzero off-diagonal elements, it allows for conditional heteroskedasticity and conditional autocorrelation, both being of a known form. If i is a time index, this implies that there exists serial correlation. If

i is an index for a cross-sectional unit, this implies that there exists spatial correlation.

However, the assumption that $\mathbf{V}$ is known is still restrictive from a practical point of view.

We first investigate the statistical property of the OLS estimator $\hat{\theta}$ under the more general conditional normality assumption.

Theorem 10.10. *Suppose $\mathbf{X}'\mathbf{X}$ is nonsingular, and $\varepsilon|\mathbf{X} \sim N(0, \sigma^2\mathbf{V})$, where $\mathbf{V} = V(\mathbf{X})$ is a known $n \times n$ symmetric, finite and positive definite matrix, and $0 < \sigma^2 < \infty$ is unknown. Then for all $n > p$,*
  (a) [Unbiasedness]

$$E(\hat{\theta}|\mathbf{X}) = \theta_0;$$

  (b) [Variance Structure]

$$var(\hat{\theta}|\mathbf{X}) = \sigma^2(\mathbf{X}'\mathbf{X})^{-1}\mathbf{X}'\mathbf{V}\mathbf{X}(\mathbf{X}'\mathbf{X})^{-1}$$
$$\neq \sigma^2(\mathbf{X}'\mathbf{X})^{-1};$$

  (c) [Normality]

$$(\hat{\theta} - \theta_0)|\mathbf{X} \sim N(0, \sigma^2(\mathbf{X}'\mathbf{X})^{-1}\mathbf{X}'\mathbf{V}\mathbf{X}(\mathbf{X}'\mathbf{X})^{-1});$$

  (d) [Correlation Between $\hat{\theta}$ and e]: $\hat{\theta}$ and e are conditionally correlated, namely,

$$cov(\hat{\theta}, e|\mathbf{X}) \neq 0.$$

Proof: (a) Using $\hat{\theta} - \theta_0 = (\mathbf{X}'\mathbf{X})^{-1}\mathbf{X}'\varepsilon$, we have

$$
\begin{aligned}
E[(\hat{\theta} - \theta_0)|\mathbf{X}] &= (\mathbf{X}'\mathbf{X})^{-1}\mathbf{X}'E(\varepsilon|\mathbf{X}) \\
&= (\mathbf{X}'\mathbf{X})^{-1}\mathbf{X}'0 \\
&= 0.
\end{aligned}
$$

(b) Given (a), we have

$$
\begin{aligned}
var(\hat{\theta}|\mathbf{X}) &= E\left\{ \left[\hat{\theta} - E(\hat{\theta}|\mathbf{X})\right]\left[\hat{\theta} - E(\hat{\theta}|\mathbf{X})\right]' |\mathbf{X}\right\} \\
&= E[(\hat{\theta} - \theta_0)(\hat{\theta} - \theta_0)'|\mathbf{X}] \\
&= E[(\mathbf{X}'\mathbf{X})^{-1}\mathbf{X}'\varepsilon\varepsilon'\mathbf{X}(\mathbf{X}'\mathbf{X})^{-1}|\mathbf{X}] \\
&= (\mathbf{X}'\mathbf{X})^{-1}\mathbf{X}'E(\varepsilon\varepsilon'|\mathbf{X})\mathbf{X}(\mathbf{X}'\mathbf{X})^{-1} \\
&= \sigma^2(\mathbf{X}'\mathbf{X})^{-1}\mathbf{X}'\mathbf{V}\mathbf{X}(\mathbf{X}'\mathbf{X})^{-1}.
\end{aligned}
$$

We cannot further simplify this expression because generally $\mathbf{V} \neq \mathbf{I}$.

(c) Because

$$\hat{\theta} - \theta_0 = (\mathbf{X}'\mathbf{X})^{-1}\mathbf{X}'\varepsilon$$

$$= \sum_{i=1}^{n} C_i \varepsilon_i,$$

is a weighed sum of normal random variables, where the $p \times 1$ weighting vector

$$C_i = (\mathbf{X}'\mathbf{X})^{-1}X_i,$$

$\hat{\theta} - \theta_0$ follows a normal distribution given $\mathbf{X}$. As a result,

$$(\hat{\theta} - \theta_0)|\mathbf{X} \sim N(0, \sigma^2(\mathbf{X}'\mathbf{X})^{-1}\mathbf{X}'\mathbf{V}\mathbf{X}(\mathbf{X}'\mathbf{X})^{-1}).$$

(d) Because generally $\mathbf{X}'\mathbf{V}\mathbf{M} \neq 0$, we have

$$
\begin{aligned}
\operatorname{cov}(\hat{\theta}, e|\mathbf{X}) &= E[(\hat{\theta} - \theta_0)e'|\mathbf{X}] \\
&= E[(\mathbf{X}'\mathbf{X})^{-1}\mathbf{X}'\varepsilon\varepsilon'\mathbf{M}|\mathbf{X}] \\
&= (\mathbf{X}'\mathbf{X})^{-1}\mathbf{X}'E(\varepsilon\varepsilon'|\mathbf{X})\mathbf{M} \\
&= \sigma^2(\mathbf{X}'\mathbf{X})^{-1}\mathbf{X}'\mathbf{V}\mathbf{M} \\
&\neq 0.
\end{aligned}
$$

Note that it is conditional heteroskedasticity and/or autocorrelation in $\{\varepsilon_i\}$ that cause $\hat{\theta}$ to be correlated with e. This completes the proof.

Theorem 10.10 implies that the OLS estimator $\hat{\theta}$ is still unbiased and one could show that its variance goes to zero as $n \to \infty$. However, the variance of $\hat{\theta}$ no longer has the simple expression of $\sigma^2(\mathbf{X}'\mathbf{X})^{-1}$ under the condition that $\varepsilon|\mathbf{X} \sim N(0, \sigma^2\mathbf{V})$. As a result, the classical t- and F-test statistics are invalid because they are based on an incorrect variance formula of $\hat{\theta}$. That is, they use an incorrect variance $\sigma^2(\mathbf{X}'\mathbf{X})^{-1}$ rather than the true variance $\sigma^2(\mathbf{X}'\mathbf{X})^{-1}\mathbf{X}'\mathbf{V}\mathbf{X}(\mathbf{X}'\mathbf{X})^{-1}$. Theorem 10.10(d) implies that even if we can obtain a consistent estimator for the correct variance $\sigma^2(\mathbf{X}'\mathbf{X})^{-1}\mathbf{X}'\mathbf{V}\mathbf{X}(\mathbf{X}'\mathbf{X})^{-1}$ and use it to construct test statistics, we still cannot obtain the Student t-distribution and F-distribution, because the numerators and the denominators of the t and F statistics are no longer independent.

To solve the aforementioned difficulty, we consider a new method called GLS estimation. We first state a useful lemma.

Lemma 10.3. *For any $n \times n$ symmetric positive definite matrix $\mathbf{V}$, we can always write*

$$\mathbf{V}^{-1} = \mathbf{C}'\mathbf{C},$$
$$\mathbf{V} = \mathbf{C}^{-1}(\mathbf{C}')^{-1}$$

where $\mathbf{C}$ is a $n \times n$ nonsingular matrix.

This is called the Cholesky factorization, where $\mathbf{C}$ may not be symmetric.

We now consider the original linear regression model

$$Y = \mathbf{X}\theta_0 + \varepsilon.$$

If we multiply this equation by $\mathbf{C}$, we obtain a transformed regression model

$$\mathbf{C}Y = (\mathbf{C}\mathbf{X})\theta_0 + \mathbf{C}\varepsilon$$

or

$$\mathbf{Y}^* = \mathbf{X}^*\theta_0 + \varepsilon^*,$$

where $\mathbf{Y}^* = \mathbf{C}Y, \mathbf{X}^* = C\mathbf{X}$ and $\varepsilon^* = \mathbf{C}\varepsilon$. Then the OLS estimator of the transformed model

$$\hat{\theta}^* = (\mathbf{X}^{*\prime}\mathbf{X}^*)^{-1}\mathbf{X}^{*\prime}\mathbf{Y}^*$$
$$= (\mathbf{X}'\mathbf{C}'\mathbf{C}\mathbf{X})^{-1}(\mathbf{X}'\mathbf{C}'\mathbf{C}Y)$$
$$= (\mathbf{X}'\mathbf{V}^{-1}\mathbf{X})^{-1}\mathbf{X}'\mathbf{V}^{-1}Y$$

is called the GLS estimator.

To investigate the statistical properties of GLS, we observe that

$$E(\varepsilon^*|\mathbf{X}) = E(\mathbf{C}\varepsilon|\mathbf{X})$$
$$= \mathbf{C}E(\varepsilon|\mathbf{X})$$
$$= \mathbf{C} \cdot 0$$
$$= 0.$$

Also, note that

$$\text{var}(\varepsilon^*|\mathbf{X}) = E(\varepsilon^*\varepsilon^{*\prime}|\mathbf{X})$$
$$= E(\mathbf{C}\varepsilon\varepsilon'\mathbf{C}'|\mathbf{X})$$
$$= \mathbf{C}E(\varepsilon\varepsilon'|\mathbf{X})\mathbf{C}'$$
$$= \sigma^2\mathbf{C}\mathbf{V}\mathbf{C}'$$
$$= \sigma^2\mathbf{C}[\mathbf{C}^{-1}(\mathbf{C}')^{-1}]\mathbf{C}'$$
$$= \sigma^2\mathbf{I}.$$

It follows that

$$\varepsilon^*|\mathbf{X} \sim N(0, \sigma^2 \mathbf{I}).$$

Thus, the new disturbance ε^* is conditionally homoskedastic and auto-uncorrelated, while maintaining the conditional normality distribution. Suppose that for some i, the error ε_i has a large variance σ_i^2. Then the transformation will discount ε_i with its conditional standard deviation so that ε_i^* becomes conditionally homoskedastic. Moreover, the transformation also removes possible autocorrelation in $\{\varepsilon_i\}$. As a result, GLS becomes BLUE for θ_0 in term of the Gauss-Markov theorem.

To appreciate how the transformation by matrix C can remove conditional heteroskedasticity and autocorrelation, we examine two examples.

Example 10.12. [**Removing Heteroskedasticity**]: Suppose

$$\mathbf{V} = \begin{bmatrix} \sigma_1^2 & 0 & \cdots & 0 \\ 0 & \sigma_2^2 & \cdots & 0 \\ \multicolumn{4}{c}{\cdots\cdots\cdots\cdots 0} \\ 0 & \cdots\cdots & \sigma_n^2 \end{bmatrix},$$

Then

$$\mathbf{C} = \begin{bmatrix} \sigma_1^{-1} & 0 & \cdots & 0 \\ 0 & \sigma_2^{-1} & \cdots & 0 \\ \multicolumn{4}{c}{\cdots\cdots\cdots\cdots 0} \\ 0 & \cdots\cdots & \sigma_n^{-1} \end{bmatrix},$$

where $\sigma_i^2 = \sigma_i^2(\mathbf{X})$, and

$$\varepsilon^* = \mathbf{C}\varepsilon = \left(\frac{\varepsilon_1}{\sigma_1}, \frac{\varepsilon_2}{\sigma_2}, \cdots, \frac{\varepsilon_n}{\sigma_n} \right)'.$$

The transformed regression model is

$$Y_i^* = X_i^{*\prime}\theta_0 + \varepsilon_i^*, \qquad i = 1, \ldots, n,$$

where

$$Y_i^* = Y_i/\sigma_i,$$
$$X_i^* = X_i/\sigma_i,$$
$$\varepsilon_i^* = \varepsilon_i/\sigma_i.$$

Example 10.13. [Eliminating Serial Correlation]: Let $|\rho| < 1$. Suppose

$$
\mathbf{V} = \begin{bmatrix}
1 & \rho & \rho^2 & \cdots & \rho^{n-2} & \rho^{n-1} \\
\rho & 1 & \rho & \cdots & \rho^{n-3} & \rho^{n-2} \\
\rho^2 & \rho & 1 & \cdots & \rho^{n-4} & \rho^{n-3} \\
\cdots & \cdots & \cdots & \cdots & \cdots & \cdots \\
\rho^{n-2} & \rho^{n-3} & \rho^{n-4} & \cdots & 1 & \rho \\
\rho^{n-1} & \rho^{n-2} & \rho^{n-3} & \cdots & \rho & 1
\end{bmatrix}.
$$

This matrix arises due to the fact that $\{\varepsilon_i\}$ follows an AR(1) process, namely,

$$
\varepsilon_i = \rho\varepsilon_{i-1} + v_i, \qquad i = 1, \ldots, n,
$$

where $\{v_i\} \sim IID\,(0, \sigma^2)$. We then have

$$
\mathbf{V}^{-1} = \begin{bmatrix}
1 & -\rho & 0 & \cdots & 0 & 0 \\
-\rho & 1+\rho^2 & -\rho & \cdots & \rho^{n-3} & 0 \\
0 & -\rho & 1+\rho^2 & \cdots & \rho^{n-4} & 0 \\
\cdots & \cdots & \cdots & \cdots & \cdots & \cdots \\
0 & 0 & 0 & \cdots & 1+\rho^2 & -\rho \\
0 & 0 & 0 & \cdots & -\rho & 1
\end{bmatrix},
$$

and

$$
\mathbf{C} = \begin{bmatrix}
\sqrt{1-\rho^2} & 0 & 0 & \cdots & 0 & 0 \\
-\rho & 1 & 0 & \cdots & 0 & 0 \\
0 & -\rho & 1 & \cdots & 0 & 0 \\
\cdots & \cdots & \cdots & \cdots & \cdots & \cdots \\
0 & 0 & 0 & \cdots & 1 & 0 \\
0 & 0 & 0 & \cdots & -\rho & 1
\end{bmatrix}.
$$

It follows that

$$
\begin{aligned}
\varepsilon^* &= \mathbf{C}\varepsilon \\
&= \left(\sqrt{1-\rho^2}\varepsilon_1,\, \varepsilon_2 - \rho\varepsilon_1,\, \cdots,\, \varepsilon_n - \rho\varepsilon_{n-1}\right)'.
\end{aligned}
$$

The transformed regression model is

$$
Y_i^* = X_i^{*\prime}\theta_0 + \varepsilon_i^*, \qquad i = 1, \ldots, n,
$$

where

$$Y_1^* = \sqrt{1 - \rho^2}\, Y_1, \qquad Y_i^* = Y_i - \rho Y_{i-1}, \qquad i = 2, \ldots, n,$$

$$X_1^* = \sqrt{1 - \rho^2}\, X_1, \qquad X_i^* = X_i - \rho X_{i-1}, \qquad i = 2, \ldots, n,$$

$$\varepsilon_1^* = \sqrt{1 - \rho^2}\, \varepsilon_1, \qquad \varepsilon_i^* = \varepsilon_i - \rho \varepsilon_{i-1}, \qquad i = 2, \ldots, n.$$

The transformation $\sqrt{1 - \rho^2}$ for the first observation ($i = 1$) is called the Prais-Winsten transformation. Note that autocorrelation in ε_i is eliminated by differencing.

Theorem 10.11. *Suppose* $\mathbf{X}'\mathbf{X}$ *is nonsingular, and* $\varepsilon|\mathbf{X} \sim N(0, \sigma^2 \mathbf{V})$, *where* $\mathbf{V} = V(\mathbf{X})$ *is a known* $n \times n$ *symmetric, finite and positive definite matrix, and* $0 < \sigma^2 < \infty$ *is unknown. Then for all* $n > p$,

(a) *[Unbiasedness]*

$$E(\hat{\theta}^* | \mathbf{X}) = \theta_0;$$

(b) *[Variance Structure]*

$$var(\hat{\theta}^* | \mathbf{X}) = \sigma^2 (\mathbf{X}^{*\prime} \mathbf{X}^*)^{-1}$$
$$= \sigma^2 (\mathbf{X}' \mathbf{V}^{-1} \mathbf{X})^{-1};$$

(c) *[Orthogonality Between $\hat{\theta}^*$ and e^*] Put* $e^* = Y^* - \mathbf{X}^* \hat{\theta}^*$. *Then*

$$cov(\hat{\theta}^*, e^* | \mathbf{X}) = 0;$$

(d) *[Gauss-Markov Theorem]* $\hat{\theta}^*$ *is a linear best unbiased estimator* (BLUE) *for* θ_0;

(e) *[Residual Variance Estimator] Put* $s^{*2} = e^{*\prime} e^* / (n - p)$. *Then*

$$E(s^{*2} | \mathbf{X}) = \sigma^2.$$

Proof: The transformed linear regression model satisfies all assumptions for Theorem 10.4, and the GLS estimator $\hat{\theta}^*$ is the OLS of the transformed model. Therefore, all results (a)–(e) of Theorem 10.11 follow immediately from Theorem 10.4. This completes the proof.

Because $\hat{\theta}^*$ is the OLS of the transformed linear regression model with $\varepsilon^* | \mathbf{X} \sim N(0, \sigma^2 \mathbf{I})$, the t-test and F-test are applicable, and these test statistics are defined as

$$T^* = \frac{R\hat{\theta}^* - r}{\sqrt{s^{*2} R (\mathbf{X}^{*\prime} \mathbf{X}^*)^{-1} R'}}$$
$$\sim t_{n-K},$$

$$F^* = \frac{(R\hat{\theta}^* - r)'[R(\mathbf{X}^{*\prime}\mathbf{X}^*)^{-1}R']^{-1}(R\hat{\theta}^* - r)/J}{s^{*2}}$$

$$\sim F_{J,n-K}.$$

Note that we still have to estimate the unknown constant σ^2 in spite of the fact that $\mathbf{V} = V(\mathbf{X})$ is known.

The most important message of GLS is the insight it provides into the impact of conditional heteroskedasticity and serial correlation on estimation and inference of a linear regression model. In practice, GLS is generally not feasible, because the matrix $\mathbf{V}$ is of unknown form. There are adaptive feasible GLS procedures that employ consistent estimators for $\mathbf{V}$ (e.g., White and Stinchcombe 1991), but these are beyond the scope of this book.

10.10 Conclusion

In this chapter, we have developed the statistical theory for classical linear regression models. We first discuss the key assumptions (strict exogeneity and spherical error variance) upon which the classical linear regression model is built. We then derive the statistical properties of the OLS estimator. In particular, we point out that the coefficient of determination, R^2, is not a suitable criterion for model selection, because it is always nondecreasing with the number of regressors. Suitable model selection criteria, such as AIC, BIC and adjusted R^2, are introduced. We show that conditional on the regressor matrix $\mathbf{X}$, the OLS estimator $\hat{\theta}$ is BLUE. Under an additional conditional normality assumption, we can derive the normal distribution for $\hat{\theta}$, the χ^2_{n-p} distribution for $(n-p)s^2/\sigma^2$, as well as the conditional independence between $\hat{\theta}$ and s^2, in finite samples. These statistical properties form the basis to construct the popular t-test and F-test for hypotheses of linear restrictions on model parameters.

When there exist conditional heteroskedasticity and/or autocorrelation, the OLS estimator is still unbiased but is no longer BLUE, and $\hat{\theta}$ and s^2 are not independent. Under the assumption of a known variance-covariance matrix up to an unknown scale parameter, one can transform the linear regression model by correcting conditional heteroskedasticity and eliminating autocorrelation, so that the transformed model has conditionally homoskedastic and uncorrelated errors. The OLS estimator of this transformed model is called the GLS estimator, and it is BLUE. The t-test and F-test based on the GLS estimation are applicable.

EXERCISE 10

10.1. Suppose $\mathbf{Y} = \mathbf{X}\theta_0 + \varepsilon$, $\mathbf{X}'\mathbf{X}$ is nonsingular. Let $\hat{\theta} = (\mathbf{X}'\mathbf{X})^{-1}\mathbf{X}'\mathbf{Y}$ be the OLS estimator and $e = \mathbf{Y} - \mathbf{X}\hat{\theta}$ be the $n \times 1$ estimated residual vector. Define a $n \times n$ projection matrix $\mathbf{P} = \mathbf{X}(\mathbf{X}'\mathbf{X})^{-1}\mathbf{X}'$ and $\mathbf{M} = \mathbf{I} - \mathbf{P}$, where $\mathbf{I}$ is a $n \times n$ identity matrix. Show:

(1) $\mathbf{X}'e = 0$;

(2) $\hat{\theta} - \theta_0 = (\mathbf{X}'\mathbf{X})^{-1}\mathbf{X}'\varepsilon$;

(3) $\mathbf{P}$ *and* $\mathbf{M}$ *are symmetric and idempotent* (*i.e.,* $\mathbf{P}^2 = \mathbf{P}, \mathbf{M}^2 = \mathbf{M}$), $\mathbf{PX} = \mathbf{X}$, and $\mathbf{MX} = 0$;

(4) $SSR(\hat{\theta}) \equiv e'e = \mathbf{Y}'\mathbf{M}\mathbf{Y} = \varepsilon'\mathbf{M}\varepsilon$.

10.2. Consider a bivariate linear regression model

$$Y_i = \alpha_0 + \beta_0 X_{1i} + \varepsilon_i$$
$$= X_i'\theta_0 + \varepsilon_i, \qquad i = 1, \ldots, n,$$

where $X_i = (1, X_{1i})'$, $\theta_0 = (\alpha_0, \beta_0)'$, and ε_i is an error term.

(1) Let $\hat{\theta} = (\hat{\alpha}, \hat{\beta})'$ be the OLS estimator. Show that $\hat{\alpha} = \bar{Y} - \hat{\beta}\bar{X}_1$, and

$$\hat{\beta} = \frac{\sum_{i=1}^n (X_{1i} - \bar{X}_1)(Y_i - \bar{Y})}{\sum_{i=1}^n (X_{1i} - \bar{X}_1)^2}$$
$$= \frac{\sum_{i=1}^n (X_{1i} - \bar{X}_1)Y_i}{\sum_{i=1}^n (X_{1i} - \bar{X}_1)^2}$$
$$= \sum_{i=1}^n C_i Y_i,$$

where $\bar{X}_1 = n^{-1} \sum_{i=1}^n X_{1i}$, and $C_i = (X_{1i} - \bar{X}_1)/\sum_{i=1}^n (X_{1i} - \bar{X}_1)^2$.

(2) Suppose $\mathbf{X}$ and ε are independent. Show $\text{var}(\hat{\beta}|\mathbf{X}) = \sigma^2/[(n-1)S_{X_1}^2]$, where $S_{X_1}^2$ is the sample variance of $\{X_{1i}\}_{i=1}^n$ and $\sigma^2 = \text{var}(\varepsilon_i)$. Thus, the more variations in $\{X_{1i}\}$, the more accurate estimation for β_0.

(3) Let $\hat{\rho}$ denote the sample correlation between Y_i and X_{1i}; namely,

$$\hat{\rho} = \frac{\sum_{i=1}^n (X_{1i} - \bar{X}_1)(Y_i - \bar{Y})}{\sqrt{\sum_{i=1}^n (X_{1i} - \bar{X}_1)^2 \sum_{i=1}^n (Y_i - \bar{Y})^2}}.$$

Show $R^2 = \hat{\rho}^2$. Thus, the squared sample correlation between $\{Y_i\}_{i=1}^n$ and $\{X_{1t}\}_{i=1}^n$ is the fraction of the sample variation in Y_i that can be explained by the predictor $\hat{Y}_i$. This result also indicates that R^2 is a measure of the strength of sample linear association between $\{Y_i\}_{i=1}^n$ and $\{X_{1i}\}_{i=1}^n$.

10.3. For the OLS estimation of the linear regression model $Y_i = X_i'\theta_0 + \varepsilon_i$, where X_i is a $p \times 1$ regressor vector, show $R^2 = \hat{\rho}_{Y\hat{Y}}^2$, the squared sample correlation between $\{Y_i\}_{i=1}^n$ and $\{\hat{Y}_i\}_{i=1}^n$.

10.4. For the OLS estimation of a linear regression model $Y_i = X_i'\theta_0 + \varepsilon_i$, the adjusted R^2, denoted as $\bar{R}^2$, is defined as

$$\bar{R}^2 = 1 - \frac{e'e/(n-K)}{(\mathbf{Y} - \bar{Y}\mathbf{i})'(\mathbf{Y} - \bar{Y}\mathbf{i})/(n-1)},$$

where $\mathbf{i} = (1, \ldots, 1)'$ is a $n \times 1$ vector with each element equal to unity. Show

$$\bar{R}^2 = 1 - \frac{n-1}{n-p}(1 - R^2).$$

10.5. Does a high value of R^2 imply a precise OLS estimation for the true parameter θ_0 in a linear regression model $Y_i = X_i'\theta_0 + \varepsilon_i$? Explain.

10.6. [Effect of Multicollinearity] Consider a regression model

$$Y_i = \alpha_0 + \beta_{10}X_{1i} + \beta_{20}X_{2i} + \varepsilon_i, \qquad i = 1, \ldots, n.$$

Suppose $\mathbf{X}'\mathbf{X}$ is nonsingular, $E(\varepsilon|\mathbf{X}) = 0$ and $E(\varepsilon\varepsilon'|\mathbf{X}) = \sigma^2\mathbf{I}$. Let $\hat{\theta} = (\hat{\alpha}, \hat{\beta}_1, \hat{\beta}_2)'$ be the OLS estimator. Show:

$$\text{var}(\hat{\beta}_1|\mathbf{X}) = \frac{\sigma^2}{(1 - \hat{\rho}^2)\sum_{i=1}^n (X_{1i} - \bar{X}_1)^2},$$

$$\text{var}(\hat{\beta}_2|\mathbf{X}) = \frac{\sigma^2}{(1 - \hat{\rho}^2)\sum_{i=1}^n (X_{2i} - \bar{X}_2)^2},$$

where $\bar{X}_1 = n^{-1}\sum_{i=1}^n X_{1i}, \bar{X}_2 = n^{-1}\sum_{i=1}^n X_{2i}$, and

$$\hat{\rho}^2 = \frac{\left[\sum_{i=1}^n (X_{1i} - \bar{X}_1)(X_{2i} - \bar{X}_2)\right]^2}{\sum_{i=1}^n (X_{1i} - \bar{X}_1)^2 \sum_{i=1}^n (X_{2i} - \bar{X}_2)^2}.$$

Note that high sample correlation between $\{X_{1i}\}$ and $\{X_{2i}\}$ will lead to imprecise estimation of $\hat{\beta}_1$ and $\hat{\beta}_2$.

10.7. Consider the linear regression model

$$Y_i = X_i'\theta_0 + \varepsilon_i, \qquad i = 1, \ldots, n,$$

where $X_i = (1, X_{1i}, \ldots, X_{ki})'$. Suppose $\mathbf{X}'\mathbf{X}$ is nonsingular, $E(\varepsilon|\mathbf{X}) = 0$ and $E(\varepsilon\varepsilon'|\mathbf{X}) = \sigma^2\mathbf{I}$. Let R_j^2 is the coefficient of determination of regressing the explanatory variable X_{ji} on all other explanatory variables $\{X_{li}, 0 \leq l \leq k, l \neq j\}$. Show

$$\text{var}(\hat{\beta}_j|\mathbf{X}) = \frac{\sigma^2}{(1 - R_j^2)\sum_{i=1}^n (X_{ji} - \bar{X}_j)^2},$$

where $\bar{X}_j = n^{-1}\sum_{i=1}^n X_{ji}$ is the sample mean of $\{X_{ji}\}_{i=1}^n$. The factor $1/(1 - R_j^2)$ is called the Variance Inflation Factor (VIF). It measures the degree of multicollinearity among explanatory variables in X_i.

10.8. Consider the following linear regression model

$$Y_i = X_i'\theta_0 + u_i, \qquad i = 1, \ldots, n,$$

where $u_i = \sigma(X_i)\varepsilon_i$, X_i is nonstochastic, and $\sigma(X_i)$ is a positive function of X_i such that

$$\Omega = \begin{bmatrix} \sigma^2(X_1) & 0 & 0 & \ldots & 0 \\ 0 & \sigma^2(X_2) & 0 & \ldots & 0 \\ 0 & 0 & \sigma^2(X_3) & \ldots & 0 \\ \ldots & \ldots & \ldots & \ldots & \ldots \\ 0 & 0 & 0 & \ldots & \sigma^2(X_n) \end{bmatrix} = \Omega^{\frac{1}{2}}\Omega^{\frac{1}{2}},$$

with

$$\Omega^{\frac{1}{2}} = \begin{bmatrix} \sigma(X_1) & 0 & 0 & \ldots & 0 \\ 0 & \sigma(X_2) & 0 & \ldots & 0 \\ 0 & 0 & \sigma(X_3) & \ldots & 0 \\ \ldots & \ldots & \ldots & \ldots & \ldots \\ 0 & 0 & 0 & \ldots & \sigma(X_n) \end{bmatrix}.$$

Assume that $\{\varepsilon_i\}$ is IID $N(0,1)$. Then $\{u_i\}$ is independently $N(0,\sigma^2(X_i))$. This differs from the classical assumption that $\varepsilon|\mathbf{X} \sim N(0,\sigma^2\mathbf{I})$, because $\{u_i\}$ exhibits conditional heteroskedasticity.

Let $\hat{\theta}$ denote the OLS estimator for θ_0.

(1) Is $\hat{\theta}$ unbiased for θ_0?

(2) Show $\mathrm{var}(\hat{\theta}) = (\mathbf{X}'\mathbf{X})^{-1}\mathbf{X}'\Omega\mathbf{X}(\mathbf{X}'\mathbf{X})^{-1}$.

Now consider an alternative estimator

$$\tilde{\theta} = (\mathbf{X}'\Omega^{-1}\mathbf{X})^{-1}\mathbf{X}'\Omega^{-1}Y$$

$$= \left[\sum_{i=1}^n \sigma^{-2}(X_i)X_iX_i'\right]^{-1}\sum_{i=1}^n \sigma^{-2}(X_i)X_iY_i.$$

(3) Is $\tilde{\theta}$ unbiased for θ_0?

(4) Show $\mathrm{var}(\tilde{\theta}) = (\mathbf{X}'\Omega^{-1}\mathbf{X})^{-1}$.

(5) Is $\mathrm{var}(\hat{\theta}) - \mathrm{var}(\tilde{\theta})$ positive semi-definite (PSD)? Which estimator, $\hat{\theta}$ or $\tilde{\theta}$, is more efficient?

(6) Is $\tilde{\theta}$ a best linear unbiased estimator (BLUE) for θ_0?

[Hint: There are several approaches to answering this question. A simple one is to consider the transformed model

$$Y_i^* = X_i^{*\prime}\theta_0 + \varepsilon_i, \qquad i = 1, \ldots, n,$$

where $Y_i^* = Y_i/\sigma(X_i), X_i^* = X_i/\sigma(X_i)$. This model is obtained from model $Y_i = X_i'\theta_0 + u_i$ by dividing by $\sigma(X_i)$. In matrix notation, the transformed model can be written as

$$\mathbf{Y}^* = \mathbf{X}^*\theta_0 + \varepsilon,$$

where the $n \times 1$ vector $\mathbf{Y}^* = \Omega^{-\frac{1}{2}}\mathbf{Y}$ and the $n \times p$ matrix $\mathbf{X}^* = \Omega^{-\frac{1}{2}}\mathbf{X}$.]

(7) Construct two test statistics for the null hypothesis $\mathbb{H}_0 : R\theta_0 = r$, where R is a $1 \times p$ nonstochastic vector and r is a known scalar. One test is based on $\hat{\theta}$, and the other is based on $\tilde{\theta}$. What are the finite sample distributions of your test statistics under $\mathbb{H}_0$? Can you tell which test is more powerful?

(8) Construct two test statistics for the null hypothesis $\mathbb{H}_0 : R\theta_0 = r$, where R is a $J \times p$ matrix with $J > 1$. One test is based on $\hat{\theta}$, and the other is based on $\tilde{\theta}$. What are the finite sample distributions of your test statistics under $\mathbb{H}_0$?

10.9. Consider a linear regression model

$$Y_i = X_i'\theta_0 + \varepsilon_i, \qquad i = 1, \ldots, n.$$

Suppose we are interested in testing the null hypothesis $\mathbb{H}_0 : R\theta_0 = r$, where R is a nonstochastic $J \times p$ matrix, and r is a $J \times 1$ vector. The F-test statistic is defined as

$$F = \frac{(R\hat{\theta} - r)'[R(\mathbf{X}'\mathbf{X})^{-1}R']^{-1}(R\hat{\theta} - r)/J}{s^2}.$$

Show

$$F = \frac{(\tilde{e}'\tilde{e} - e'e)/J}{e'e/(n-p)},$$

where $e'e$ is the sum of squared residuals from the unrestricted model, and $\tilde{e}'\tilde{e}$ is the sum of squared residuals from the restricted regression model subject to the constraint of $R\theta = r$.

10.10. Show that the F test statistic is equivalent to a quadratic form in $\tilde{\lambda}$, where $\tilde{\lambda}$ is the Lagrange multiplier in the constrained OLS estimation for the linear regression $\mathbf{Y} = \mathbf{X}\theta_0 + \varepsilon$. This result implies that the F-test is equivalent to a Lagrange multiplier test.

10.11. The F-test statistic is defined as

$$F = \frac{(R\hat{\theta} - r)'[R(\mathbf{X}'\mathbf{X})^{-1}R']^{-1}(R\hat{\theta} - r)/J}{s^2}.$$

Show

$$F = \frac{\sum_{i=1}^{n}(\hat{Y}_i - \tilde{Y}_i)^2/J}{s^2} = \frac{(\hat{\theta} - \tilde{\theta})'\mathbf{X}'\mathbf{X}(\hat{\theta} - \tilde{\theta})/J}{s^2},$$

where $\hat{Y}_i = X_i'\hat{\theta}, \tilde{Y}_i = X_i'\tilde{\theta}$, and $\hat{\theta}$ and $\tilde{\theta}$ are the unrestricted and restricted OLS estimators respectively. This indicates that the F-test is proportional to the sum of squared deviations between the fitted values of the unrestricted and restricted models.

10.12. Consider a classical linear regression model

$$Y_i = X_i'\theta_0 + \varepsilon_i$$

$$= \alpha_0 + \sum_{j=1}^{k}\beta_{j0}X_{ji} + \varepsilon_i, \qquad i = 1,\ldots,n,$$

where $\theta_0 = (\alpha_0, \beta_{10}, \ldots, \beta_{k0})'$, and $\mathbf{X}'\mathbf{X}$ is nonsingular. Suppose we are interested in testing the null hypothesis that all slope coefficients are jointly zero:

$$\mathbb{H}_0 : \beta_{10} = \beta_{20} = \cdots = \beta_{k0} = 0.$$

Then the F-statistic can be written as

$$F = \frac{(\tilde{e}'\tilde{e} - e'e)/k}{e'e/(n - k - 1)},$$

where $e'e$ is the sum of squared residuals from the above unrestricted model, and $\tilde{e}'\tilde{e}$ is the sum of squared residuals from the following restricted model

$$Y_i = \alpha_0 + \varepsilon_i.$$

(1) Show

$$F = \frac{R^2/k}{(1 - R^2)/(n - k - 1)},$$

where R^2 is the coefficient of determination of the unrestricted model.

(2) Suppose in addition $\varepsilon|\mathbf{X} \sim N(0, \sigma^2\mathbf{I})$. Show that under $\mathbb{H}_0$, as $n \to \infty$,

$$(n - k - 1)R^2 \xrightarrow{d} \chi_k^2.$$

10.13. *[Structural Changes]* Suppose $\mathbf{X}'\mathbf{X}$ is nonsingular. Consider the following linear regression model with the whole sample:

$$Y_i = X_i'\theta_0 + (D_i X_i)'\alpha^o + \varepsilon_i, \qquad i = 1, \ldots, n,$$

where the time dummy variable $D_i = 0$ if $i \leq n_1$ and $D_i = 1$ if $i > n_1$, and $1 < n_1 < n$. This model can be written as two separate models:

$$Y_i = X_i'\theta_0 + \varepsilon_i, \qquad i = 1, \ldots, n_1$$

and

$$Y_i = X_i'(\theta_0 + \alpha^o) + \varepsilon_i, \qquad i = n_1 + 1, \ldots, n.$$

Let SSR_u, SSR_1, SSR_2 denote the sums of squared residuals of the above three regression models. Show

$$SSR_u = SSR_1 + SSR_2.$$

This identity implies that estimating the first regression model with a time dummy D_i over the whole sample via OLS is equivalent to estimating two separate regression models over two subsample periods respectively.

10.14. A quadratic polynomial regression model

$$Y_i = \alpha_0 + \beta_{10} X_i + \beta_{20} X_i^2 + \varepsilon_i, \qquad i = 1, \ldots, n,$$

is fitted to a data. Suppose the P-value for the OLS estimate was 0.67 for β_1 and 0.84 for β_2. Can one accept the hypothesis that β_1 and β_2 are jointly zero at the 5% significance level? Explain.

10.15. Suppose $Y = \mathbf{X}\theta_0 + \varepsilon$, where $\varepsilon|\mathbf{X} \sim N(0, \sigma^2 \mathbf{V})$, $\mathbf{V} = V(\mathbf{X})$ is a known $n \times n$ nonsingular matrix, and $0 < \sigma^2 < \infty$ is unknown. It can be shown that the variances of the OLS $\hat{\theta}$ and GLS $\hat{\theta}^*$ are respectively:

$$\text{var}(\hat{\theta}|\mathbf{X}) = \sigma^2 (\mathbf{X}'\mathbf{X})^{-1} \mathbf{X}'\mathbf{V}\mathbf{X}(\mathbf{X}'\mathbf{X})^{-1},$$

$$\text{var}(\hat{\theta}^*|\mathbf{X}) = \sigma^2 (\mathbf{X}'\mathbf{V}^{-1}\mathbf{X})^{-1}.$$

Show $\text{var}(\hat{\theta}|\mathbf{X}) - \text{var}(\hat{\theta}^*|\mathbf{X})$ is positive semi-definite (PSD).

10.16. Suppose a data generating process is given by

$$Y_i = \beta_{10} X_{1i} + \beta_{20} X_{2i} + \varepsilon_i$$

$$= X_i'\theta_0 + \varepsilon_i, \qquad i = 1, \ldots, n,$$

where $\theta_0 = (\beta_{10}, \beta_{20})'$, $X_i = (X_{1i}, X_{2i})'$, $E(X_i X_i')$ is nonsingular, and $E(\varepsilon_i|X_i) = 0$. For simplicity, we further assume that $E(X_{2i}) = 0$ and $E(X_{1i} X_{2i}) \neq 0$.

Now we specify a bivariate linear regression model

$$Y_i = \beta_{10} X_{1i} + u_i.$$

(1) Show that if $\beta_{20} \neq 0$, then $E(Y_i|X_i) = X_i'\theta_0 \neq E(Y_{1i}|X_{1i})$. That is, there exists an omitted variable X_{2i} in the bivariate regression model.

(2) Show that $E(Y_i|X_{1i}) \neq \beta_1 X_{1i}$ for all values of β_1. That is, the bivariate linear regression model is misspecified for $E(Y_i|X_{1i})$.

10.17. Suppose a data generating process is given by

$$Y_i = \beta_{10}X_{1i} + \beta_{20}X_{2i} + \varepsilon_i = X_i'\theta_0 + \varepsilon_i, \qquad i = 1, \ldots, n,$$

where $\theta_0 = (\beta_{10}, \beta_{20})', X_i = (X_{1i}, X_{2i})'$, $\mathbf{X}'\mathbf{X}$ is nonsingular and $\varepsilon|\mathbf{X} \sim N(0, \sigma^2 \mathbf{I})$. Denote the OLS estimator by $\hat{\theta} = (\hat{\beta}_1, \hat{\beta}_2)'$.

If $\beta_{20} = 0$ and we know it, then we can consider a simpler regression model

$$Y_i = \beta_{10}X_{1i} + \varepsilon_i.$$

Denote the OLS of this simpler regression as $\tilde{\beta}_1$.

Compare the relative efficiency between $\hat{\beta}_1$ and $\tilde{\beta}_1$. Which estimator has a smaller MSE? Give your reasoning.

10.18. Consider a linear regression model $\mathbf{Y} = \mathbf{X}\theta_0 + \varepsilon$, where $\varepsilon|\mathbf{X} \sim N(0, \sigma^2 \mathbf{V})$, where $\mathbf{V} = V(\mathbf{X})$ is a known $n \times n$ nonsingular matrix, and $0 < \sigma^2 < \infty$ is unknown. Is the OLS estimator $\hat{\theta}$ BLUE? Explain.

10.19. Consider a linear regression model $\mathbf{Y} = \mathbf{X}\theta_0 + \varepsilon$, where $\varepsilon|\mathbf{X} \sim N(0, \sigma^2 \mathbf{V})$, $\mathbf{V} = V(\mathbf{X})$ is a known $n \times n$ nonsingular matrix, and $0 < \sigma^2 < \infty$ is unknown. The GLS estimator $\hat{\theta}^*$ is defined as the OLS estimator of the transformed model

$$\mathbf{Y}^* = \mathbf{X}^*\theta_0 + \varepsilon^*,$$

where $\mathbf{Y}^* = \mathbf{CY}, \mathbf{X}^* = \mathbf{CX}, \varepsilon^* = \mathbf{C}\varepsilon$, and $\mathbf{C}$ is a $n \times n$ nonsingular matrix from the factorization $\mathbf{V}^{-1} = \mathbf{CC}'$. Is the coefficient of determination R^2 for the transformed model always positive? Explain.

10.20. Suppose $\mathbf{Y} = \mathbf{X}\theta_0 + \varepsilon$, where $\mathbf{X}'\mathbf{X}$ is nonsingular, $\varepsilon|\mathbf{X} \sim N(0, \mathbf{V})$, and $\mathbf{V} = V(\mathbf{X})$ is a known $n \times n$ finite and positive definite matrix. Define the GLS estimator $\hat{\theta}^* = (\mathbf{X}'\mathbf{V}^{-1}\mathbf{X})^{-1}\mathbf{X}'\mathbf{V}^{-1}\mathbf{Y}$.

(1) Is $\hat{\theta}^*$ BLUE?

(2) Put $\mathbf{X}^* = \mathbf{CX}$ and $s^{*2} = e^{*\prime}e^*/(n-p)$, where $e^* = \mathbf{Y} - \mathbf{X}^*\hat{\theta}^*$, $\mathbf{C}'\mathbf{C} = \mathbf{V}^{-1}$. Do the usual t-test and F-test defined as

$$T^* = \frac{R\hat{\theta}^* - r}{\sqrt{s^{*2}R(\mathbf{X}^{*\prime}\mathbf{X}^*)^{-1}R'}}, \quad \text{for } J = 1,$$

$$F^* = \frac{(R\hat{\theta}^* - r)'[R(\mathbf{X}^{*\prime}\mathbf{X}^*)^{-1}R']^{-1}(R\hat{\theta}^* - r)/J}{s^{*2}}$$

follow the t_{n-K} and $F_{J,n-K}$ distributions respectively under the null hypothesis $\mathbb{H}_0 : R\theta_0 = r$? Explain.

(3) Construct two alternative test statistics:

$$\tilde{T}^* = \frac{R\hat{\theta}^* - r}{\sqrt{R(\mathbf{X}^{*\prime}\mathbf{X}^*)^{-1}R'}}, \text{ for } J = 1,$$

$$\tilde{Q}^* = (R\hat{\theta}^* - r)'[R(\mathbf{X}^{*\prime}\mathbf{X}^*)^{-1}R']^{-1}(R\hat{\theta}^* - r).$$

What distributions will these test statistics follow under the null hypothesis $\mathbb{H}_0 : R\theta_0 = r$? Explain.

(4) Which set of tests, (T^*, F^*) or $(\tilde{T}^*, \tilde{Q}^*)$, is more powerful at the same significance level? For simplicity, compare the tests T^* and $\tilde{T}^*$ when $J = 1$. Explain. [Hint: The t-distribution has a heavier tail than $N(0,1)$ and so has a larger critical value at a given significance level.]

10.21. Consider a linear regression model

$$Y_i = X_i'\theta_0 + \varepsilon_i, \qquad i = 1, \ldots, n,$$

where $\varepsilon_i = \sigma(X_i)v_i$, X_i is a $p \times 1$ nonstochastic vector, and $\sigma(X_i)$ is a positive function of X_i, and $\{v_i\}$ is IID $N(0,1)$.

Let $\hat{\theta} = (\mathbf{X}'\mathbf{X})^{-1}\mathbf{X}'\mathbf{Y}$ denote the OLS estimator for θ_0.

(1) Is $\hat{\theta}$ unbiased for θ_0?

(2) Find $\text{var}(\hat{\theta}) = E[(\hat{\theta}-E\hat{\theta})(\hat{\theta}-E\hat{\theta})']$. [Hint: You may find the following notations useful: $\Omega = \text{diag}\{\sigma^2(X_1), \sigma^2(X_2), \ldots, \sigma^2(X_n)\}$, i.e., Ω is a $n \times n$ diagonal matrix with the i-th diagonal component equal to $\sigma^2(X_i)$ and all off-diagonal components equal to zero.]

Now consider the transformed regression model

$$\frac{1}{\sigma(X_i)}Y_i = \frac{1}{\sigma(X_i)}X_i'\beta^0 + v_i, \qquad i = 1, \ldots, n,$$

or equivalently

$$Y_i^* = X_i^{*\prime}\theta_0 + v_i, \qquad i = 1, \ldots, n,$$

where $Y_i^* = \sigma^{-1}(X_i)Y_i$ and $X_i^* = \sigma^{-1}(X_i)X_i$.

Denote the OLS estimator of this transformed model as $\tilde{\theta}$.

(3) Show $\tilde{\theta} = (\mathbf{X}'\Omega^{-1}\mathbf{X})^{-1}\mathbf{X}'\Omega^{-1}\mathbf{Y}$.

(4) Is $\tilde{\theta}$ unbiased for θ_0?

(5) Find $\text{var}(\tilde{\theta})$.

(6) Which estimator, $\hat{\theta}$ or $\tilde{\theta}$, is more efficient in terms of MSE? Give your reasoning.

(7) Use the difference $R\tilde{\theta} - r$ to construct a test statistic for the null hypothesis $\mathbb{H}_0 : R\theta_0 = r$, where R is a $J \times p$ nonstochastic matrix, r is $J \times 1$, and $J > 1$. What is the finite sample distribution of your test statistic under $\mathbb{H}_0$? Give your reasoning.

10.22. Consider a linear regression model

$$Y_i = X_i'\theta_0 + \varepsilon_i, \qquad i = 1, \ldots, n,$$

where X_i is a $p \times 1$ nonstochastic regressor vector, θ_0 is a $p \times 1$ unknown vector, and $\{\varepsilon_i\}$ follows an AR(q) process, namely,

$$\varepsilon_i = \sum_{j=1}^{q} \alpha_j \varepsilon_{i-j} + v_i, \quad \{v_i\} \sim \text{IID} \ (0, \sigma_v^2).$$

We assume that the autoregressive coefficients $\{\alpha_j\}_{j=1}^{q}$ are known but σ_v^2 is unknown.

(1) Find a BLUE estimator for θ_0; Explain.

(2) Construct a test statistic for the null hypothesis $\mathbb{H}_0 : R\theta_0 = r$ and derive its sampling distribution under $\mathbb{H}_0$, where R is a known $J \times p$ nonstochastic matrix and r is a known $J \times 1$ nonstochastic vector. Discuss the cases of $J = 1$ and $J > 1$ respectively.

Chapter 11

Conclusion

Abstract: In this chapter, we first summarize the main contents what we have covered and then point out directions for further studies in modern statistics and econometrics.
Key words: Cross-sectional econometrics, Modern econometrics, Modern statistics, Panel data analysis, Time series analysis.

What have we learnt in this book so far? We have covered the following topics:

- introduction to statistics;
- foundation of probability theory;
- random variables and univariate probability distributions;
- important probability distributions;
- random vectors and multivariate probability distributions;
- introduction to sampling theory;
- convergence concepts and limit theorems;
- parameter estimation and evaluation;
- hypothesis testing;
- classical linear regression analysis.

Chapter 1 is a general introduction to statistics and econometrics. We present two fundamental axioms on which econometrics is built, namely, any economy can be viewed as a stochastic system governed by some probability law, and economic and financial data observed in practice are realizations of this stochastic system. The purpose of econometric analysis is to make inference of the probability law of an economic system based on the observed data, and use the obtained knowledge of the probability law to test economic theories and economic hypotheses, and predict the future behavior of the economy. It is important to bear in mind that some assumptions, such as homogeneity and stationarity, have to be made in order to implement statistical analysis for economic systems. It is rather difficult, although perhaps not impossible, to check these maintained assumptions.

The book consists of two parts: Chapters 2 to 5 present probability theory, and Chapters 6 to 10 present statistical theory. Chapter 2 lays down the foundation of probability theory. We introduce basic concepts in probability theory. In particular, we define the probability space for a random experiment. A probability space consists of the sample space, the sigma field, and the probability function; it completely characterizes the probability law of a random experiment. Chapter 2 is very general but abstract and it introduces the basic concepts which are the building blocks for probability theory and will be very helpful in understanding some concepts developed in later chapters.

Chapter 3 introduces random variables and univariate probability distributions. We use mathematical tools to formulate the concepts introduced in Chapter 2. In particular, we introduce the concept of a random variable, which is a mapping from the original space to a new sample space. The probability function of a random variable is an induced probability measure based on the probability function defined on the original sample space. A more convenient characterization is the cumulative distribution function of a random variable. We study two types of random variables — discrete random variables and continuous random variables. For discrete random variables, the probability mass function can be used to describe the probability law. Both the probability mass function and the cumulative distribution function are equivalent: they can be recovered from each other. For a continuous random variable, the probability distribution can be characterized by a probability density function. Given a probability density function, one can obtain the cumulative distribution function by integration. And given a cumulative distribution function, one can obtain a probability density function by differentiation. Given a continuous cumulative distribution function, there may exist an infinite number of probability density functions. However, these probability density functions yield the same probability statements, and one can always choose a most convenient probability density function (e.g., the one that is the smoothest). Unlike a probability mass function, a probability density function is not a probability measure, but it is proportional to the probability that the random variable takes values in a small interval centered at a point of interest.

We can obtain important characteristics from a probability distribution function, such as various moments (e.g., mean, variance, skewness and kurtosis) and quantiles (e.g., median), which are widely used in economics and finance. For a probability distribution whose moment generating function exists, we can obtain various moments by differentiating the moment generating function at the origin. The moment generating function can also be used to uniquely characterize the probability distribution. The moment generating function may not exist for some probability distributions, but the characteristic function always exists for any distribution, and it plays a similar role to the moment generating function.

In Chapter 4, we introduce a class of commonly used parametric discrete distributions and continuous distributions, discussing their properties and pointing out where they can be applied in economics and finance. Examples of discrete distributions include Bernoulli, Binomial, Negative binomial, Geometric and Poisson distributions, and the examples of continuous distributions include Uniform, Beta, Normal, Cauchy, Lognormal, Gamma, Chi-squares, Exponential and Double Exponential distributions.

In Chapter 5, we introduce random vectors and multivariate probability distributions. A multivariate probability distribution not only contains information about the univariate distributions — the so-called marginal distributions, but also contains the relationship between the components of a random vector. Like in the univariate case, we introduce the joint cumulative distribution function for a random vector, a joint probability mass function for a discrete random vector, and a joint probability density function for a continuous random vector. Also, we introduce the conditional probability mass function and the conditional probability density function.

With these tools to characterize a joint distribution, we introduce the concepts of covariance and correlation to measure the linear association between two random variables, and discuss their connection with and difference from the concept of independence. We define a joint moment generating function which can be differentiated at the origin to obtain the covariance and be used to characterize a joint probability distribution. We also use the conditional probability distributions to define conditional expectations, such as conditional moments and conditional quantiles. In particular, we discuss the conditional mean and conditional variance and their economic applications. We also introduce the multivariate normal distribution and discuss its properties.

The basic idea of statistical analysis is to use a subset or sample information to infer knowledge of the data generating process. Chapter 6 is an introduction to sampling theory. We introduce the concepts of populations, random samples, statistics, and sampling distributions of statistics. We use the sample mean and sample variance as the leading examples to illustrate the relevant statistical concepts. This allows us to introduce important distributions such as the Student t distribution and F distribution, both of which are closely associated with a random sample from a normal distribution population. Finally, we introduce the sufficiency principle and sufficient statistics, which are very useful techniques of data reduction and are very important to understand the properties of many statistics.

The sampling distributions of statistics play a very important role in statistical inference. However, the sampling distribution of a statistic is usually rather difficult to obtain, unless strong assumptions are made on the population and the statistic. A convenient approach to tackling this difficulty is use of asymptotic theory, which permits us to obtain the approximate distributions of a statistic when the sample size is large. In Chapter 7, we introduce basic asymptotic analytic tools. In particular, we introduce a variety of convergence concepts, laws of large numbers and central limit theorems. We also introduce the Delta method, which is useful to derive the asymptotic distribution of a nonlinear statistic.

Parameter estimation and hypothesis testing are two most important tasks in statistical inference. In Chapter 8, we first discuss methods to estimate population parameters. We introduce two important estimation methods — the maximum likelihood estimation and the (generalized) method of moments estimation. Their asymptotic distributions are obtained, which can be used for statistical inference, such as confidence interval estimation and hypothesis testing. We also discuss the methods to evaluate different estimators using the mean squared error criterion. In particular, we discuss the best linear unbiased estimator which does not require the knowledge of the population distribution model, and the Cramer-Rao lower bound method which requires the knowledge of the population distribution model.

In Chapter 9, we discuss basic concepts of hypothesis testing and present the well-known Neyman-Pearson lemma which states that the likelihood ratio test is uniformly most powerful for simple hypotheses. Three popular testing principles — the Wald test, the Lagrange Multiplier test (also called the efficient score test), and the likelihood ratio test are introduced. A link between the likelihood ratio test and a sufficient statistic is established. Using the asymptotic tools introduced in Chapter 7, we investigate the asymptotic properties of these three test statistics. They are all asymptotically Chi-square distributed and are asymptotically equivalent to each other under the null hypothesis.

Chapter 10 presents the statistical theory for classical linear regression models. We discuss the key assumptions for the classical linear regression models, and investigate the statistical properties of the OLS estimator, which is a best linear unbiased estimator. We develop the Student t-test and F-test for hypotheses on model parameters and discuss various applications and examples in economics and econometrics. We also discuss the properties of the GLS estimator when there exist conditional heteroskedasticity and autocorrelation of a known form (up to an unknown scalar factor).

For important probability and statistical concepts, methods and theory, we have offered intuitions, explanations and applications from an economic perspective. In particular, we have provided many examples in economics and finance to illustrate how probability and statistics are useful in economic analysis. Emphasis of economic intuition, explanation and application for the introduced probability and statistical concepts and tools is a most important feature that distinguishes this book from other probability and statistics textbooks.

In statistical theory, we have focused on basic statistical inference problems based on an IID random sample. One extension is to relax the identical distribution assumption to allow for certain degrees of heterogeneity across observations, which is common in microeconometrics.

Another extension is to relax the independence assumption to allow for temporal dependence across the observations, as often encountered in macroeconomics and finance. This usually occurs in a time series context, and therefore is called time series analysis.

Our statistical analysis is based on the assumption that the population distribution coincides with a parametric distribution model whose functional form is known up to some unknown parameters. This assumption may be too strong in many economic applications. One solution is to use nonparametric approaches, which do not require any parametric functional form and simply let data tell what is the appropriate model. Obviously, this approach requires a large data set because it has to check various possible functional forms. This is called nonparametric analysis.

There have been increasing data sets which combine both cross-sectional units and time periods together. These are called panel data. Associated analysis is called the econometrics of panel data. See Hsiao (2016) for an excellent introduction to panel data analysis. Indeed, in the current era of Big Data, various forms of data have emerged, including the so-called unstructured data, such as those in form of text, image, voice and video. These new types of data will call for new theory and methods in statistics and econometrics.

All these are future subjects in modern statistics and econometrics.

EXERCISE 11

11.1. Summarize the main concepts, tools, methods and theories discussed in this book.

11.2. Why is it important to use statistical analysis in economics and social statistics?

11.3. Why is it important to have "stochastic thinking" or "statistical thinking" in economic analysis? By stochastic or statistical thinking, it is meant to view the economic system as a stochastic process or a random experiment.

11.4. Why is it important to offer tuition and explanation for probability and statistical concepts from an economic perspective?

11.5. In this book, a maintained assumption in most of the discussion is the IID assumption. Discuss the merits and undesired features of this assumption from both statistical and economic perspectives.

Bibliography

Akaike, H., (1973). Information Theory and an Extension of the Maximum Likelihood Principle, in *Second International Symposium on Information Theory*, Ed. B. Petrov and F. Csake, Budapest: Akademiai Kiado.

Bartle, R.G., (1966). *The Elements of Integration*, New York: John Wiley and Sons.

Bartle, R.G., (1976). *The Elements of Real Analysis*, 2nd Edition, New York: John Wiley and Sons.

Bekaert, G., C. Erb, C. Harvey and T. Viskanta, (1998). Distributional Characteristics of Emerging Market returns and Asset Allocation, *Journal of Portfolio Management*, 24, 102-116.

Behboodian, J., (1990). Examples of Uncorrelated Dependent Random Variables Using a Bivariate Mixture, *American Statistician*, 44, 218.

Billingsley, P., (1995). *Probability and Measure*, 3rd Edition, New York: John Wiley and Sons.

Black, F. and M. Scholes, (1973). The Pricing of Options and Corporate Liabilities, *Journal of Political Economy*, 81, 637-654.

Bollerslev, T., (1986). Generalized Autoregressive Conditional Heteroskedasticity, *Journal of Econometrics*, 31, 307-327.

Bollerslev, T., (1987). A Conditionally Heteroskedastic Time Series Model for Speculative Prices and Rates of Return, *Review of Economics and Statistics*, 69, 542-547.

Bond, S.A., (2000). Asymmetry and Downside Risk in Foreign Exchange Markets, Working Paper, Department of Land Economy, University of Cambridge, Cambridge, U.K.

Bortkiewicz, L. von, (1898). *Das Gesetz der Kleinen Zahlen*, Leipzig: Teubner.

Brennan, M., (1993). Agency and Asset Pricing, Working Paper, UCLA and London Business School.

Breusch, T.S. and A.R. Pagan, (1980). The Lagrange Multiplier Test and its Applications to Model Specification in Econometrics, *Review of Economic Studies*, 47, 239-253.

Campbell, J.Y. and M. Yogo, (2006). Efficient Tests of Stock Return Predictability, *Journal of Finance Economics*, 81, 27-60.

Casella, G. and R.L. Berger, (2002). *Statistical Inference*, 2nd Edition, Pacific Grove, California: Duxbury.

Chan, K.C., G.A. Karolyi, F.A. Longstaff and A.B. Sanders, (1992). An Empirical Comparison of Alternative Models of the Short-term Interest Rate, *Journal of Finance*, 47, 1209-1227.

Chang, R., L. Kaltani and N. Loayza, (2005). Openness can Be Good for Growth: The Role of Policy Complementarities, NBER Working Paper no. 11787.

Cherubini, U., E. Luciano and W. Vecchiato, (2004). *Copula Methods in Finance*, Wiley Finance Series, Hoboken, NJ: John Wiley and Sons.

Cox, D.R., (1972). Regression Models and Life-Tables, *Journal of Royal Statistical Society. Series B (Methodological)*, 34, 187-220.

Cox, J.C., J.E. Ingersol and S.A. Ross, (1985). An Intertemporal General Equilibrium Model of Asset Prices, *Econometrica*, 53, 363-384.

Das, S.R., (2002). The Surprise Element: Jumps in Interest Rate, *Journal of Econometrics*, 106, 27-65.

Devroye, L., (1985). The Expected Length of the Longest Probe Sequence for Bucket Searching When the Distribution Is not Uniform, *Journal of Algorithms*, 6, 1-9.

Ding, Z., C.W.J. Granger and R.F. Engle, (1993). A Long Memory Property of Stock Market Returns and a New Model, *Journal of Empirical Finance*, 1, 83-106.

Douglas, J.B., (1980). *Analysis with Standard Contagious Distributions*, Burtonsville, Maryland: International Cooperative Publishing House.

Duffie, D., J. Pan, and K. Singleton, (2000). Transform Analysis and Asset Pricing for Affine Jump-Diffusions, *Econometrica*, 68, 1343-1376.

Dykstra, R.L. and J.E. Hewett, (1972). Examples of Decompositions of Chi-Squared Variables, *American Statistician*, 26, 42-43.

Engle, R.F., (1982). Autoregressive Conditional Heteroscedasticity with Estimates of the Variance of United Kingdom Inflation, *Econometrica*, 50, 987-1007.

Engle, R.F., (1984). Wald, Likelihood Ratio and Lagrange Multiplier Tests in Econometrics, Chapter 13 in *Handbook of Econometrics*, Vol. 2, Ed. Z. Griliches and M.D. Intriligator, Amsterdam: North-Holland, 775-826.

Engle, R.F. and J.R. Russell, (1998). Autoregressive Conditional Duration: A New Model for Irregularly Spaced Transaction Data, *Econometrica*, 66, 1127-1162.

Fama, E.F., (1965). The Behavior of Stock Market Prices, *Journal of Business*, 38, 34-105.

Fama, E.F., (1970). Efficient Capital Market: A Review of Theory and Empirical Work, *Journal of Finance*, 25, 383-417.

Feller, W., (1971). *An Introduction to Probability Theory and its Applications*, 2nd Edition, New York: John Wiley and Sons.

Gallant, A.R., (1997). *An Introduction to Econometric Theory*, Princeton, NJ: Princeton University Press.

Gardner, P.S., (1961). Adenovirus and Intussusception, *British Medical Journal*, 2, 495-496.

Gibrat, R., (1930). Une Loi Des Repartions Econmiques: L'effet Proportionelle, Bulletin de Statistique General, France, 19, 469.

Gibrat, R., (1931). Les Inegalites Economiques Paris: Libraire du Recueil Sirey.

Granger, C.W.J., (1969). Investigating Causal Relations by Econometric Models and Cross-spectral Methods, *Econometrica*, 37, 428-438.

Granger, C.W.J., (1980). Long-memory Relationships and the Aggregation of Dynamic Models, *Journal of Econometrics*, 14, 227-238.

Granger, C.W.J., (1999). *Empirical Modeling in Economics: Specification and Evaluation*, London: Cambridge University Press.

Hall, P., (1992). *The Bootstrap and Edgeworth Expansion*, New York: Springer.

Hamilton, J.D., (1994). *Time Series Analysis*, Princeton N.J.: Princeton University Press.

Harrison, A., (1996). Openness and Growth: A Time-series, Cross-country Analysis for Developing Countries, *Journal of Development Economics*, 48, 419-447.

Harvey, C.R. and A. Siddique, (2000). Conditional Skewness in Asset Pricing Tests, *Journal of Finance*, 55, 1263-1295.

Hansen, B.E., (1994). Autoregressive Conditional Density Estimation, *International Economic Review*, 35, 705-730.

Hausman, J.A., B.H. Hall, and Z. Griliches, (1984). Econometric Models for Count Data with an Application to the Parents R and D Relationship, *Econometrica*, 52, 909-938.

Hong, Y., (1999). Hypothesis Testing in Time Series via the Empirical Characteristic Function: A Generalized Spectral Density Approach, *Journal of the American Statistical Association*, 94, 1201-1220.

Hong, Y., (2017). Manuscript, *Advanced Econometrics*, Department of Economics and Department of Statistical Sciences, Cornell University.

Hong, Y. and H. Li, (2005). Nonparametric Specification Testing for Continuous-Time Models with Applications to Term Structure of Interest Rates, *Review of Financial Studies*, 18, 37-84.

Hong, Y. and H. White, (1995). Consistent Specification Testing Via Nonparametric Series Regression, *Econometrica*, 63, 1133-1159.

Hong, Y. and H. White, (2005). Asymptotic Distribution Theory for Nonparametric Entropy Measures of Serial Dependence, *Econometrica*, 73, 837-901.

Hsiao, C., (2016). *Panel Data Analysis*, 2nd Edition, Cambridge University Press: New York.

Judge, G.G., R.C. Hill, W.E. Griffiths, H. Lutkepohl and T.-C. Lee, (1988). *Introduction to the Theory and Practice of Econometrics*, 2nd Edition, New York: Wiley.

Kendall, M.G. and A. Stuart, (1961). *The Advanced Theory of Statistics, Vol. 2*, London: Griffin.

Kiefer, N.M., (1988). Economic Duration Data and Hazard Functions, *Journal of Economic Literature*, 26, 646-679.

Kingman, J.F.C. and S.J. Taylor, (1966). *Introduction to Measure and Probability*, Cambridge: Cambridge University Press.

Kraus, A. and R.H. Litzenberger, (1976). Skewness Preference and the Valuation of Risk Assets, *Journal of Finance*, 31, 1085-1100.

Lancaster, T., (1990). *The Econometric Analysis of Transition Data*, Cambridge: Cambridge University Press.

Lee, T.H., H. White and C.W.J. Granger, (1993). Testing for Neglected Nonlinearity in Time Series Models: A Comparison of Neural Network Methods and Alternative Tests, *Journal of Econometrics*, 56, 269-290.

Liapounov, A.M., (1901). Nouvelle Forme du Theoreme Sur la Limite des Probabilites. Mem. Acad. Sci. St. Petersburg, 8, 1-24.

Loayza, N., P. Fajnzylber and C. Calderon, (2005). *Economic Growth in Latin America and the Caribbean: Stylized Facts, Explanations, and Forecasts*, Washington D.C.: The World Bank.

Lukacs, E., (1970). *Characteristic Functions*, 2nd Edition, London: Griffin.

Mandelbrot, B., (1963). The Variation of Certain Speculative Prices, *Journal of Business*, 36, 394-419.

Markowitz, H.M., (1991). Foundations of Portfolio Theory, *Journal of Finance*, 46, 469-477.

McClean, S.I., (1976). A Continuous-time Population Model with Poisson Recruitment, *Journal of Applied Probability*, 13, 348-354.

Merton, R.C., (1976). Option Pricing When Underlying Stock Returns are Discontinuous, *Journal of Financial Economics*, 3, 125-144.

Neftci, S.N., (1984). Are Economic Time Series Asymmetric over the Business Cycles?, *Journal of Policy Economy*, 92, 307-328.

Nelson, D.R., (1991). Conditional Heteroskedasticity in Asset Returns: A New Approach, *Econometrica*, 59, 347-370.

Newey, W.K. and D. McFadden, (1994). Large Sample Estimation and Hypothesis Testing, *Handbook of Econometrics*, 4, 2111-2245.

O'Neill, B. and W.T. Wells, (1972). Some Recent Results in Lognormal Parameter Estimation Using Grouped and Ungrouped Data, *Journal of American Statistical Association*, 67, 76-80.

Parzen, E., (1960). *Modern Probability Theory and its Applications*, New York: John Wiley and Sons.

Poisson, S.D., (1837). Recherches Sur la Probabilite des Jugements en Matiere Criminelle et en Matiere Civile, Precedees des Regles Generales du Calcul des Probabilities, Bachelier, Paris.

Rao, C.R., (1959). Some Problem Involving Linear Hypotheses in Multivariate Analysis, *Biometrika*, 46, 49-58.

Rao, N.M., K.S. Shurpalekar, E.E. Sundarvalli and T.R. Doraiswamy, (1973). Flatus Production in Children Fed Legume Diets, *PAG (Protein Advisory Group) Bull*, 3, 53.

Resnik, S.I., (1999). *A Probability Path*, Boston: Birkhauser.

Ripley, B., (1987). *Stochastic Simulation*, New York: John Wiley and Sons.

Robinson, P.M., (1991). Consistent Nonparametric Entropy-based Testing, *Review of Economic Studies*, 58, 437-453.

Robinson, P.M., (1994). Semiparametric Analysis of Long-memory Time Series, *Annals of Statistics*, 22, 515-539.

Rodriguez, F. and D. Rodrik, (2000). Trade Policy and Economic Growth: A Skeptic's Guide to the Cross-national Evidence, *NBER Macroeconomics Annual 2000*, Cambridge, MA: MIT Press, 261-324.

Singleton, J.C. and J. Wingender, (1986). Skewness Persistence in Common Stock Returns, *Journal of Financial and Quantitative Analysis*, 21, 335-341.

Schwarz, G., (1978). Estimating the Dimension of a Model, *Annals of Statistics*, 6, 461-464.

Vasicek, O., (1977). An Equilibrium Characterization of the Term Structure, *Journal of Finance Economics*, 5, 177-188.

Venn, J., (1881). *Symbolic Logic*, London: The MacMillan Company.

White, H., (1982). Maximum Likelihood Estimation of Misspecified Models, *Econometrica*, 50, 1-25.

White, H., (1984). *Asymptotic Theory for Econometricians*, New York: Academic Press.

White, H., (1994). *Estimation, Inference and Specification Analysis*, New York: Cambridge University Press.

White, H., (2001). *Asymptotic Theory for Econometricians*, Revised Edition, New York: Academic Press.

White, H. and M. Stinchcombe, (1991). Adaptive Efficient Weighted Least Squares with Dependent Observations, In W. Stahel *et al.* (Eds.), *Directions in Robust Statistics and Diagnostics*, New York: Springer-Verlag, 337-363.

White, H. and J. Wooldridge, (1990). Some Results on Sieve Estimation with Dependent Observations, In Barnett W.A., J. Powell, G. Tauchen (Eds.), *Non-parametric and Semi-parametric Methods in Econometrics and Statistics*, Cambridge: Cambridge University Press, 459-493.

Young, A., (1971). Demographic and Ecological Models for Manpower Planning in Aspects of Manpower Planning, Ed. D.J. Bartholomew and B.R. Morris, London: English Universities Press.

Index

Printed in the United States
By Bookmasters